THE STRUCTURE OF
SOCIOLOGICAL THEORY

FIFTH EDITION

THE STRUCTURE OF SOCIOLOGICAL THEORY

Jonathan H. Turner

University of California, Riverside

With contributions by
Alexandra Maryanski and Stephan Fuchs

WADSWORTH PUBLISHING COMPANY

Belmont, California
A Division of Wadsworth, Inc.

Sociology Editor: Serina Beauparlant
Editorial Assistant: Marla Nowick
Production Editor: Sara Hunsaker   *Ex Libris*
Print Buyer: Martha Branch
Copy Editor: Micky Lawler
Compositor: Impressions, Inc.
Cover Design: Anne Kellejian

Printed in the United States of America

3 4 5 6 7 8 9 10—95 94 93

Library of Congress Cataloging in Publication Data
Turner, Jonathan H.
    The structure of sociological theory /
Jonathan H. Turner.—5th ed.
      p.   cm.
    ISBN 0-534-13842-X
    1. Sociology.   I. Title.
  HM24.T84 1990
  301—dc20   90-12027
                    CIP

ISBN 0-534-13842-X

TO
Clara Dean
MY TYPIST AND FRIEND FOR 20 YEARS

ABOUT THE AUTHOR

Jonathan H. Turner is professor of Sociology at the University of California at Riverside. He received his Bachelor of Arts degree from the University of California at Santa Barbara in 1965 and his Ph.D. from Cornell University in 1968. He is the author of sixteen books on a variety of topics, including sociological theory, American society, social organization, social stratification, and ethnic relations. He is currently writing three books on theoretical sociology, one dealing with the bio-cultural basis of human evolution, another on macrostructural processes in societies, and a third book on a general theory of social organization.

PREFACE

This has become a huge book, but, in writing this fifth edition, my goal has been to increase comprehensiveness without sacrificing depth of coverage. Hence the book is bigger, and in this case I hope that bigger does indeed mean better. Of course, I do not imagine that a student can read every chapter during a one-quarter or semester course, but at least the material is there for reference, and it is there in depth.

Depth of coverage is what has always distinguished this book. There are many fine texts on theory—perhaps more than in any other area of sociology—but most of them provide a review of the contours of theories, sacrificing considerable depth for breadth. My strategy has always been to analyze *the structure of* a theory in detail. Previously I sacrificed comprehensiveness to do so, but this time around I made few compromises. No one book can, however, cover every theorist and theoretical perspective in the world of sociology, but I have tried to give a detailed example of every major theoretical approach—at least the general ones—in sociology.

There are many specific theories on particular topics—deviance, family, urbanization, world systems, organizations, societal evolution, group dynamics, and the like. These are not covered here; rather, five broad perspectives or traditions are the rubrics for grouping those theories that seek to explain human interaction and organization in general.

The five general perspectives, or rubrics, organizing the chapters of this book are functional, conflict, exchange, interactionist, and structural theorizing. Within each of these labels exists a diversity of theories and theorists. I have sought to examine the most prominent and relevant.

What have I changed in this new edition? I have updated every chapter and, in several cases, significantly rewritten portions of existing chapters. But the major change is the addition of 11 new chapters. I have added chapters on general systems theory, human ecology, sociobiology, rational choice theory, dramaturgy, network analysis, and structuralism. These chapters add breadth to the earlier edition, correcting for some obvious omissions. In three of these new chapters I have taken on a coauthor, since I was extending myself beyond my knowledge base. Thus the chapters on sociobiology and network analysis were coauthored with Alexandra Maryanski, and the chapter on structuralism was coauthored with Stephan Fuchs.

In addition to these new substantive chapters, I have provided something that was missing from the last edition: a conclusion and synthesis. In the four closing chapters I present some of my ideas on how to go about developing sociological theory; in so doing, I hope to break down some of the barriers that partition and divide theorists in sociology. I try to indicate how we can develop micro, macro, and meso theories, using the theories examined in five parts of the book as a basis for synthesis. I do not provide a startling new breakthrough in these chapters; rather, I suggest lines and avenues that offer, in my view, the best prospects for theoretical synthesis.

In sum, then, I think that this is a much more complete book than its predecessors. It is now 20 years—almost to the day (as I write this preface)—since I began work on the first edition of *The Structure of Sociological Theory*. Theory has changed a great deal over those 20 years, and I trust that each edition, and especially this new one, provides the reader with a real sense for *the structure of sociological theory* in the world today.

Jonathan H. Turner
Riverside, California

CONTENTS

ix

THE STRUCTURE OF
SOCIOLOGICAL THEORY

CHAPTER 1

Sociological Theory

Diversity and Disagreement

SOCIOLOGY AND SCIENCE

Theory is a "story" about how and why events in the universe occur. Sociological theory thus seeks to explain how and why humans behave, interact, and organize themselves in certain ways. When stated in this way, few would disagree; but, as soon as we question *what kind* of story is to be developed by sociology, controversy and acrimony immediately surface. Sociologists in general, and social theorists in particular, do not agree on such basic issues as what kind of knowledge about human interaction and organization can be developed, what procedures can and should be used in developing explanations, what ends or goals are to be served by sociological knowledge, or even what phenomena should be the topics of our explanations.

The center of this storm is *science*, which over the last few hundred years has gained ascendance in many disciplines as the most useful way to understand the universe. From its early beginnings as a coherent field, sociology attached itself to the rising star of science. Indeed, the titular founder of sociology, Auguste Comte, recognized in the early 1800s that the status of a "science of society" was precarious.[1] To defend sociology from its many critics and to legitimate his claims that the emergence of sociology as a science was now possible, he posited a "law of the three stages." In the religious stage, interpretations of events are initially provided by religious beliefs or by reference to the activities of sacred and supernatural forces. Out of religion comes a metaphysical stage in which logic, mathematics, and other formal systems of reason come to dominate how events are interpreted. And out of these gains

---

[1] Auguste Comte, *System of Positive Philosophy*, vol. 1 (Paris: Bachelier, 1930). Subsequent portions were published between 1831 and 1842. For a more detailed analysis of Comte's thought, see Jonathan H. Turner, Leonard Beeghley, and Charles Powers, *The Emergence of Sociological Theory* (Belmont, CA: Wadsworth, 1989).

FIGURE 1–1 Types of Knowledge

		Is knowledge to be empirical?	
		Yes	No
Is knowledge to be evaluative?	Yes	*Ideologies;* or beliefs that state the way the world should be	*Religions;* or beliefs that state the dictates of supernatural forces
	No	*Science;* or the belief that all knowledge is to reflect the actual operation of the empirical world	*Logics;* or the various systems of reasoning that employ rules of calculation

in formal reasoning in the metaphysical stage emerges the possibility for "positivism" or a scientific stage, where formal statements are critically examined against carefully collected facts. Comte argued that the accumulation of knowledge about each domain of the universe—the physical, the chemical, the biological, and, finally, the social—passed successively through these three stages. Patterns of human organization were to be the last such domain to move into the positive stage, and in 1830 he trumpeted the call for the use of science to develop knowledge about human affairs.

I mention Comte's advocacy because the issues that he raised still haunt sociology today. More than 150 years after Comte pronounced sociology to be in the positive phase, he is denounced in many quarters as naive and just plain wrong. True, his point of view has many supporters, including me, but his faith that there could be a "natural science of society"[2] is hardly shared by all. Hence the question of whether or not sociology can be a science is *the* basic issue over which sociologists find themselves in disagreement.

One way to gain some perspective on the question of sociology as a science is to examine Figure 1-1, in which I summarize the kinds of general belief systems that have been used to interpret events and to generate knowledge about human affairs.[3] The typology asks two basic questions: (1) is the search for knowledge to be evaluative or neutral? and (2) is the knowledge developed to pertain to actual empirical events and processes, or is it to be about non-empirical realities? In other words, should knowledge tell us what *should be* or *what is*? And should it make reference to the observable world or to other, less observable, realms? If knowledge is to tell us what should exist (and, by

---

[2]I have taken this phrase from A. R. Radcliffe-Brown's *A Natural Science of Society* (Glencoe, IL: Free Press, 1948).

[3]I am borrowing the general idea from Talcott Parsons' *The Social System* (New York: Free Press, 1951). For another version of this typology, see my *The Science of Human Organization* (Chicago: Nelson-Hall, 1985), Chapter 2.

implication, what should not occur) in the empirical world, then it is ideological knowledge. If it informs us about what should be but does not pertain to observable events, then the knowledge is religious, or about forces and beings in another realm of existence. If knowledge is neither empirical nor evaluative, then it is a formal system of logic, such as mathematics. And if it is about empirical events and is nonevaluative, then it is science.

I concede that this typology is crude, but it makes the essential point: there are different ways to look at, interpret, and develop knowledge about the world. Science is only one way. Science is based upon the presumption that knowledge can be value free, that it can explain the actual workings of the empirical world, and that it can be revised on the basis of careful observations of empirical events. These characteristics distinguish science from other beliefs about how we should generate understanding and insight.[4] However, even this portrayal of science is questioned by many who would regard it as rather idealized. For these critics, values always figure into what we study. The empirical world is not "just there"; rather, it is filtered through concepts and presuppositions. And rarely are facts dispassionately collected to test theories; indeed, there are always organizational politics, revolving around vested interests and resources, that influence what "facts" we collect and how we interpret them.[5]

Perhaps even more fundamental than criticisms of science's assertion of "neutrality" and "objectivity" is the belief among many sociologists that *the very nature of the social universe* precludes science as a useful mode of inquiry. Humans are creative and can change the nature of their world; as a consequence,

---

[4]Among several fine introductory works on the nature of scientific theory, the discussion in this chapter draws heavily upon Paul Davidson Reynolds's excellent *A Primer in Theory Construction* (Indianapolis: Bobbs-Merrill, 1971). For excellent introductory works, see Arthur L. Stinchcombe, *Constructing Social Theories* (New York: Harcourt, Brace & World, 1968), pp. 3–56; Karl R. Popper, *The Logic of Scientific Discovery* (New York: Harper & Row, 1959); David Willer and Murray Webster, Jr., "Theoretical Concepts and Observables," *American Sociological Review* 35 (August 1970), pp. 748–57; Hans Zetterberg, *On Theory and Verification in Sociology*, 3rd ed. (Totowa, NJ: Bedminster Press, 1965); Jerald Hage, *Techniques and Problems of Theory Construction in Sociology* (New York: John Wiley, 1972); Walter L. Wallace, *The Logic of Science in Sociology* (Chicago: Aldine, 1971); Robert Dubin, *Theory Building* (New York: Free Press, 1969); Jack Gibbs, *Sociological Theory Construction* (Hinsdale, IL: Dryden Press, 1972); Herbert M. Blalock, Jr., *Theory Construction: From Verbal to Mathematical Formulations* (Englewood Cliffs, NJ: Prentice-Hall, 1969); Nicholas C. Mullins, *The Art of Theory: Construction and Use* (New York: Harper & Row, 1971); Bernard P. Cohen, *Developing Sociological Knowledge: Theory and Method* (Chicago: Nelson-Hall, 1989).

[5]For example, there is a growing conviction among some sociologists that science is much like any other thought system in that it is devoted to sustaining a particular vision, among a community of individuals called scientists, of what is "really real." Science simply provides one interesting way of constructing and maintaining a vision of reality, but there are other, equally valid, views among different communities of individuals. Obviously, I do not accept this argument, but I will explore it in more detail in various chapters. For some interesting explorations of the issues, see Edward A. Tiryakian, "Existential Phenomenology and the Sociological Tradition," *American Sociological Review* 30 (October 1965), pp. 674–88; J. C. McKinney, "Typification, Typologies, and Sociological Theory," *Social Forces* 48 (September 1969), pp. 1–11; Alfred Schutz, "Concept and Theory Formation in the Social Sciences," *Journal of Philosophy* 51 (April 1954), pp. 257–73; Harold Garfinkel, *Studies in Ethnomethodology* (Englewood Cliffs, NJ: Prentice-Hall, 1967); George Psathas, "Ethnomethods and Phenomenology," *Social Research* 35 (September 1968), pp. 500–520.

they can obviate and make obsolete or irrelevant theories that purport to explain human interaction and organization. There are no timeless, universal properties of human organization, and hence there can be no laws like those in the natural sciences. Humans constantly remake their universe in ways that render scientific theories inappropriate.

These and many other arguments against science will appear in the chapters to follow. My point here is to sound a warning that the science and anti-science factions are engaged in a long, and often long-winded, philosophical debate. More practically, the fact that there is disagreement over sociology's status as a science makes a review of sociological theory highly problematic, for the vehicle for developing scientific knowledge *is* theory. Scientific theory provides an interpretation of events, but this interpretation must be constantly checked and rechecked against the empirical facts. Yet, if the whole enterprise of science is questioned, then sociological theories that tell us how and why events occur will be very diverse. Depending upon what kind of science (if any) sociology is considered to be, our "theories" will vary. Some theories will look like those that Auguste Comte envisioned—that is, very much like those in the natural sciences. Other theoretical approaches will be very different because they are formulated by those who have serious reservations about a social science that fits neatly into the lower-left box in Figure 1-1.

These considerations present me with very real problems in defining theory in sociology; but, undaunted, let me try to present a minimal portrayal of what I think all sociological theories have in common. Then we can return to the controversial issues that divide theorists and that make for different kinds of theoretical activity in sociology.[6]

THE ELEMENTS OF THEORY

Theory is a mental activity. As I have already indicated, it is a process of developing ideas that can allow us to explain how and why events occur. Theory is constructed with several basic elements or building blocks: (1) concepts, (2) variables, (3) statements, and (4) formats. Although there are many divergent claims about what theory is or should be, these four elements are common to all of them. Let me examine each of these elements in more detail.

Concepts: The Basic Building Blocks of Theory

Theories are built from concepts. Most generally, concepts denote phenomena; in so doing, they isolate features of the world that are considered, for the moment at hand, important. For example, notions of atoms, protons, neutrons, and the like are concepts pointing to and isolating phenomena for certain analytical purposes. Familiar sociological concepts would include group, formal

---

[6]For my views on these controversial issues, see: Jonathan H. Turner, "In Defense of Positivism," *Sociological Theory* 3 (Fall 1985), pp. 24–30; and Stephan Fuchs and Jonathan H. Turner, "What Makes a Science Mature?" *Sociological Theory* 4 (Fall 1986), pp. 143–50.

organization, power, stratification, interaction, norm, role, status, and social-ization. Each term is a concept that embraces aspects of the social world that are considered essential for a particular purpose.

Concepts are constructed from definitions.[7] A definition is a system of terms, such as the sentences of a language, the symbols of logic, or the notation of mathematics, that inform investigators as to the phenomenon denoted by a concept. For example, the concept *conflict* has meaning only when it is defined. One possible definition might be: *Conflict is interaction among social units in which one unit seeks to prevent another from realizing its goals.* Such a def-inition allows us to visualize the phenomenon that is denoted by the concept. It enables all investigators to "see the same thing" and to understand what it is that is being studied.

Thus, concepts that are useful in building theory have a special charac-teristic: they strive to communicate a uniform meaning to all those who use them. However, since concepts are frequently expressed with the words of everyday language, it is difficult to avoid words that connote varied meanings—and hence point to different phenomena—for varying groups of scientists. It is for this reason that many concepts in science are expressed in technical or more "neutral" languages, such as the symbols of mathematics. In sociology, expression of concepts in such special languages is sometimes not only im-possible but also undesirable. Hence the verbal symbols used to develop a concept must be defined as precisely as possible in order that they point to the same phenomenon for all investigators. Although perfect consensus may never be attained with conventional language, a body of theory rests on the premise that scholars will do their best to define concepts unambiguously.

I should stress that the concepts of theory reveal a special characteristic: *abstractness.*[8] Some concepts pertain to concrete phenomena at specific times and locations. Other, more abstract, concepts point to phenomena that are not related to concrete times or locations. For example, in the context of small-group research, *concrete concepts* would refer to the persistent interactions of particular individuals, whereas an *abstract* conceptualization of such phenom-ena would refer to those general properties of face-to-face groups that are not tied to particular individuals interacting at a specified time and location. Whereas abstract concepts are not tied to a specific context, concrete concepts are. In building theory, abstract concepts are crucial, although we will see shortly that theorists disagree considerably on this issue.

Abstractness poses a problem: how do we attach abstract concepts to the ongoing, everyday world of events? Although it is essential that some of the concepts of theory transcend specific times and places, it is equally critical that there be procedures for making these abstract concepts relevant to observable

---

[7]For more detailed work on concept formation, see Carl G. Hempel, *Fundamentals of Concept Formation in Empirical Science* (Chicago: University of Chicago Press, 1952).

[8]For useful and insightful critique of sociology's ability to generate abstract concepts and theory, see David and Judith Willer, *Systematic Empiricism: Critique of Pseudoscience* (Englewood Cliffs, NJ: Prentice-Hall, 1973).

situations and occurrences. After all, the utility of an abstract concept can be demonstrated only when the concept is brought to bear on some specific empirical problem encountered by investigators; otherwise, concepts remain detached from the very processes they are supposed to help investigators understand. Just how to attach concepts to empirical processes, or the workings of the real world, is an area of great controversy in sociology. Some argue for very formal procedures for attaching concepts to empirical events. Those of this persuasion contend that abstract concepts should be accompanied by a series of statements known as *operational definitions,* which are sets of procedural instructions telling investigators how to go about discerning phenomena in the real world that are denoted by an abstract concept. Others argue that the nature of our concepts in sociology precludes such formalistic exercises. At best, concepts can be only sensitizing devices that must change with alterations of social reality, and so we can only intuitively and provisionally apply abstract concepts to the actual flow of events. To emulate the natural sciences in an effort to develop formal operations for attaching concepts to reality is to ignore the fact that social reality is changeable; it does not reveal invariant properties like the other domains of the universe.[9] Thus, to think that abstract concepts denote enduring and invariant properties of the social universe and to presume, therefore, that the concept itself will never need to be changed is, at best, naive.[10]

And so the debate rages, taking many different turns. I need not go into detail here, since these issues will be brought out again and again as I move into the substance of sociological theory in subsequent chapters. For the present, I want only to draw the approximate lines of battle.

Variables as an Important Type of Concept

When used to build theory, two general types of concepts can be distinguished: (1) those that simply label phenomena and (2) those that refer to phenomena that differ in degree.[11] Concepts that merely label phenomena would include such commonly employed abstractions as *dog, cat, group, social class,* and *star.* When stated in this way, none of these concepts reveals the ways in which the phenomena they denote vary in terms of such properties as size, weight, density,

---

[9]For examples of this line of argument, see Herbert Blumer, *Symbolic Interaction: Perspective and Method* (Englewood Cliffs, NJ: Prentice-Hall, 1969); or Anthony Giddens, *New Rules of Sociological Method* (New York: Basic Books, 1977).

[10]For the counterargument, see Jonathan H. Turner, "Toward a Social Physics: Reducing Sociology's Theoretical Inhibitions," *Humboldt Journal of Social Relations* 7 (Fall/Winter 1979-80), pp. 140-55; "Returning to Social Physics," *Perspectives in Social Theory,* vol. 2 (1981); "Some Problematic Trends in Sociological Theorizing," *The Wisconsin Sociologist* 15 (Spring/Summer 1978), pp. 80-88; and *Societal Stratification: A Theoretical Analysis* (New York: Columbia University Press, 1984).

[11]Reynolds, *Primer in Theory Construction,* p. 57; see also Stinchcombe, *Constructing Social Theories,* pp. 38-47, for a discussion of how concepts point not only to variable properties of phenomena but also to the interaction effects of interrelated phenomena. For an interesting discussion of the importance of variable concepts and for guidelines on how to use them, see J. Hage, *Techniques and Problems of Theory Construction.*

velocity, cohesiveness, or any of the many criteria used to inform investigators about differences in degree among phenomena.

Those who believe, as I do, that sociology can be like other sciences prefer concepts that are translated into variables—that is, into states that vary. We want to know the variable properties—size, degree, intensity, amount, and so forth—of events denoted by a concept. For example, to note that an aggregate of people is a group does not indicate what type of group it is or how it compares with other groups in terms of such criteria as size, differentiation, and cohesiveness. And so, some the concepts of scientific theory should denote the *variable* features of the world. To understand events requires that we visualize how variation in one phenomenon is related to variation in another. Others, who are less enamored by efforts to make sociology a natural science, are less compulsive about translating concepts into variables. They are far more interested in whether or not concepts sensitize and alert investigators to important processes than they are in converting each concept into a metric that varies in some measurable way. They are not, of course, against the conversion of ideas into variables, but they are cautious about efforts to translate each and every concept into a metric.

Theoretical Statements and Formats

To be useful, the concepts of theory must be connected to one another. Such connections among concepts constitute theoretical statements. These statements specify the way in which events denoted by concepts are interrelated, and at the same time they provide an interpretation of how and why events should be connected. When these theoretical statements are grouped together, they constitute what I term a *theoretical format*. I use the word *format* because it is general and can describe many different ways to organize theoretical statements. Indeed, in sociological theory there is relatively little consensus over just how to organize theoretical statements into a format. And, in fact, much of the theoretical controversy in sociology revolves around differences over the best way to develop theoretical statements and to group them together into a format. Depending on one's views about what kind of science, if any, sociology can be, the structure of theoretical statements and their organization into formats differ dramatically. Let me review the range of opinion on the matter.

I think that there are four basic approaches in sociological theory for generating theoretical statements and formats: (1) meta-theoretical schemes, (2) analytical schemes, (3) propositional schemes, and (4) modeling schemes. Figure 1-2 summarizes the relations among these schemes and the basic elements of theory. Concepts are constructed from definitions; theoretical statements link concepts together; and statements are organized into four basic types of formats. However, these four formats can be executed in a variety of ways, and so in reality there are more than just four strategies for developing theoretical statements and formats. Moreover, these various strategies are not always mutually exclusive, for, in executing one of them, we are often led to another as a kind of "next step" in building theory. Yet—and this is a point that I think

FIGURE 1-2 The Elements of Theory in Sociology

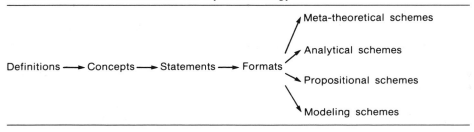

is crucial—these various approaches are often viewed as antagonistic, and the proponents of each strategy have spilled a great deal of ink sustaining the antagonism. Moreover, even within a particular type of format there is constant battle over the best way to develop theory. For me, this acrimony represents a great tragedy because in a mature science—which, sad to say, sociology is not—these approaches are viewed as highly compatible. Before pursuing this point further, I will delineate in more detail each of these approaches.

Meta-theoretical schemes This kind of theoretical activity is more comprehensive than ordinary theory. Meta-theoretical schemes are not, by themselves, theories that explain specific classes of events; rather, they explicate the basic issues that a theory must address. In many sociological circles, meta-theory is considered an essential prerequisite to adequate theory building,[12] even though the dictionary definition of *meta* emphasizes "occurring later" and "in succession" to previous activities.[13] Furthermore, in most other sciences, meta-theoretical reflection has occurred *after* a body of formal theoretical statements has been developed. It is typically after a science has used a number of theoretical statements and formats successfully that scholars begin to ask: What are the underlying assumptions about the universe contained in these statements? What strategies are demanded by, or precluded from, these statements and their organization into formats? What kind of knowledge is generated by these statements and formats, and, conversely, what is ignored? In sociological theory, however, advocates of meta-theory usually emphasize that we cannot develop theory until we have resolved these more fundamental epistemological and metaphysical questions.

My opinion is that such meta-theorizing has put the cart before the horse, but I emphasize that my opinion is probably in the minority among social theorists. For those who emphasize meta-theory, several preliminary issues must be resolved. These include: (1) What is the basic nature of human activity about which we must develop theory? For example, what is the basic nature of human beings? What is the fundamental nature of society? What is the

---

[12]For a forceful and scholarly advocacy of this point, see Jeffrey C. Alexander, *Theoretical Logic in Sociology*, 4 vols. (Berkeley: University of California Press, 1982–84).

[13]Webster's *New Collegiate Dictionary* (Springfield, MA: G. & C. Merriam, 1986).

fundamental nature of the bonds that connect people to one another and to society? (2) What is the appropriate way to develop theory, and what kind of theory is possible? For instance, can we build highly formal systems of abstract laws, as in physics, or must we be content with general concepts that simply sensitize and orient us to important processes? Can we rigorously test theories with precise measurement procedures, or must we use theories as interpretative frameworks that cannot be tested by the same procedures as in the natural sciences? (3) What is the critical problem on which social theory should concentrate? For instance, should we examine the processes of social integration, or must we concentrate on social conflict? Should we focus on the nature of social action among individuals, or is the major question one of structures of organization? Should we stress the power of ideas, like values and beliefs, or must we focus on the material conditions of people's existence?

A great deal of what is defined as sociological theory involves trying to answer these questions. The old philosophical debates—idealism versus materialism, induction versus deduction, causation versus association, subjectivism versus objectivism, and so on—are reevoked and analyzed with respect to social reality. At times, meta-theorizing has been true to the meaning of *meta* and has involved a reanalysis of previous scholars' ideas in light of these philosophical issues.[14] The favorite targets of such analyses are Karl Marx, Max Weber, Émile Durkheim, and, in recent decades, Talcott Parsons (the subject of Chapter 3). The idea behind reanalysis is to summarize the metaphysical and epistemological as assumptions of the scholars' work and to show where the schemes went wrong and where they still have utility. Furthermore, on the basis of this assessment, there are some recommendations in reanalyses as to how we should go about building theory and what this theory should be.

In my view, such reanalysis can be useful when it stimulates actual theory—that is, efforts to explain social events. However, meta-theorizing often gets bogged down in weighty philosophical matters and immobilizes theory building. The enduring philosophical questions persist, I would imagine, because they are not resolvable. One must just take a stand on the issues and see what kinds of insights can be generated. But meta-theory often stymies as much as stimulates theoretical activity because it embroils theorists in inherently unresolvable and always debatable controversies. Of course, my opinion is not shared by some. For our present purposes, the more important conclusion is that a great deal of sociological theory is, in fact, meta-theoretical activity.

Analytical schemes   Much theoretical activity in sociology consists of concepts organized into a classification scheme that denotes the key properties, and interrelations among these properties, in the social universe. There are many different varieties of analytical schemes, but they share an emphasis on classifying basic properties of the social world. The concepts of the scheme

---

[14]Alexander's work is more in this tradition. See also Richard Münch, *Theory of Action: Reconstructing the Contributions of Talcott Parsons, Émile Durkheim, and Max Weber* (Frankfurt: Suhrkamp, 1982).

FIGURE 1–3 Types of Analytical Schemes

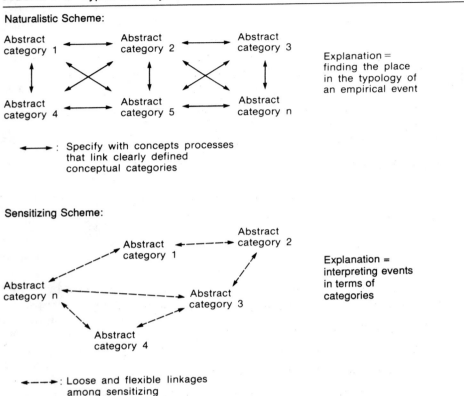

Naturalistic Scheme:

Explanation =
finding the place
in the typology of
an empirical event

◀——▶ : Specify with concepts processes
that link clearly defined
conceptual categories

Sensitizing Scheme:

Explanation =
interpreting events
in terms of
categories

◀——▶: Loose and flexible linkages
among sensitizing
conceptual categories

chop up the universe; then, the ordering of the concepts gives the social world a sense of order. Explanation of an empirical event comes whenever a place in the classificatory scheme can be found for the empirical event.

There are, however, wide variations in the nature of the typologies in analytical schemes. I think that there are two basic types: *naturalistic schemes*, which try to develop a tightly woven system of categories that is presumed to capture the way in which the invariant properties of the universe are ordered,[15] and *sensitizing schemes*, which are more loosely assembled congeries of concepts intended only to sensitize and orient researchers and theorists to certain critical processes. Figure 1–3 summarizes these two types of analytical approaches. Naturalistic/positivistic schemes assume that there are timeless and universal processes in the social universe, as much as there are in the physical and biological realms. The goal is to create an abstract conceptual typology

---

[15]Talcott Parsons' work is of this nature, as we will see in the next chapter. See also Münch.

that is isomorphic with these timeless processes. In contrast, sensitizing schemes are sometimes more skeptical about the timeless quality of social affairs: concepts and their linkages must always be provisional and sensitizing because the nature of human activity is to change those very arrangements denoted by the organization of concepts into theoretical statements.[16] Hence, except for certain very general conceptual categories, the scheme must be flexible and capable of being revised as circumstances in the empirical world change. t best, then, explanation is simply an interpretation of events by seeing them an instance or example of the provisional and sensitizing concepts in the ?me.

Often it is argued that analytical schemes are a necessary prerequisite for ping other forms of theory. Until one has a scheme that organizes the ties of the universe, it is difficult to develop propositions and models ecific events. For without the general analytical framework—it is asked nents of this view—how can a theorist or researcher know what to As I will argue at the end of this book, there is some merit to this lthough we must be careful not to begin with such an elaborate cheme that it is impossible to use it for developing other forms of I will argue that sensitizing analytical schemes can represent a begin theorizing (and, by implication, naturalistic schemes are laborate to stimulate theorizing outside the parameters imposed itself).[17]

al schemes   A proposition is a theoretical statement that ction between two or more variables. It tells us how variation accounted for by variation in another. For example, the ent "group solidarity is a positive function of external oups" says that, as group conflict increases, so does the arity among members of the respective groups involved o properties of the social universe denoted by variable ity" and "conflict," are connected by the proposition value, so does the other.

vary perhaps the most of all theoretical approaches. two dimensions: (1) the level of abstraction and (2) anized into formats. Some are highly abstract and denote any particular case but all cases of a type and conflict are abstract because no particular and solidarity is addressed). In contrast, other o empirical facts and simply summarize rela- lar case (for example, as World War II pro- increased). Propositional schemes vary not

ternative. See his *The Constitution of Society* (Berke-

mes, see Jonathan H. Turner, *A Theory of Social* Press, 1988).

only in terms of abstractness but also by virtue of how propositions are laced together into a format. Some are woven together by very explicit rules; others are merely loose bunches or congeries of propositions.

I think that, by using these two dimensions, several different types of propositional schemes can be isolated: (a) axiomatic formats, (b) formal formats, and (c) various empirical formats. The first two (axiomatic and formal formats) are clearly theoretical, whereas various empirical formats are simply research findings that test theories. But I emphasize that these more empirical types of propositional schemes are often considered theory by practicing sociologists, and so I have included them for our discussion here.

(a) An axiomatic organization of theoretical statements involves the following elements. First, it contains a set of concepts. Some of the concepts are highly abstract; others, more concrete. Second, there is always a set of existence statements that describe those types and classes of situations in which the concepts and the propositions that incorporate them apply. These existence statements make up what are usually called the *scope conditions* of the theory. Third—and most nearly unique to the axiomatic format—propositional statements are stated in a hierarchical order. At the top of the hierarchy are *axioms* or highly abstract statements, from which *all* other theoretical statements are logically derived. These latter statements are usually called *theorems* and are logically derived in accordance with varying rules from the more abstract axioms. The selection of axioms is, in reality, a somewhat arbitrary matter, but usually they are selected with several criteria in mind. The axioms should be consistent with one another, although they do not have to be logically interrelated. The axioms should be highly abstract; they should state relationships among abstract concepts. These relationships should be lawlike in that the more concrete theorems derived from them have not been disproved by empirical investigation. And the axioms should have an intuitive plausibility so that their truth appears to be self-evident.

The end result of tight conformity to axiomatic principles is an inventory or set of interrelated propositions, each derivable from at least one axiom and usually more abstract theorems. There are several advantages to this kind of theory construction. First, highly abstract concepts, encompassing a broad range of related phenomena, can be employed. These abstract concepts do not have to be directly measurable since they are logically tied to more specific and measurable propositions that, when empirically tested, can indirectly subject the more abstract propositions and the axioms to empirical tests. By virtue of this logical interrelatedness of the propositions and axioms, theory can be more efficient since the failure to refute a particular proposition lends credence to other propositions and to the axioms. Second, the use of a logical system to derive propositions from abstract axioms can also generate new propositions that point to previously unknown or unanticipated relationships among social phenomena.

But I emphasize that there are some fatal limitations on the use of axiomatic theory in sociology. In terms of strict adherence to the rules of axiomatic theory (the details of which are not critical for my purposes here), most concepts and propositions in sociology cannot be legitimately employed because the concepts are not stated with sufficient precision and because they cannot be incorporated into propositions that state unambiguously the relationship between concepts. Axiomatic theory also requires controls on all potential extraneous variables so that the tight logical system of deduction from axiom to empirical reality is not contaminated by extraneous factors. Sociologists can rarely create such controls.[18] Thus, axiomatic theory can be used only when very precise definitions of concepts exist, when concepts are organized into propositions using a precise calculus that specifies relations unambiguously, and when the contaminating effects of extraneous variables are eliminated.

These limitations are often ignored in propositional theory building, and the language of axiomatic theory is employed (axioms, theorems, corollaries, and the like); but these efforts are, at best, pseudoaxiomatic schemes.[19] In fact, I think it is best to call them formal propositional schemes[20]—the second type of proposition strategy listed earlier.

(*b*) Formal theories are, in essence, watered-down versions of axiomatic schemes. The idea is to develop highly abstract propositions that are used to explain some empirical event. The propositions are usually grouped together and seen as higher-order laws, and the goal of explanation is to see empirical events as an instance or example of this "covering law." Deductions from the law are made, but they are much looser, rarely conforming to the strict rules of axiomatic theory. Moreover, there is a recognition that extraneous variables cannot always be excluded, and so the propositions usually have the disclaimer "other things being equal." That is, if other forces do not impinge, then the relationship among concepts in the proposition should hold true. For example, my earlier example of the relationship between conflict and solidarity might be one abstract proposition in a formal system. Thus a formal scheme might say "Other things being equal, group solidarity is a positive function of conflict." Then we would use this law to explain some empirical event—say, for instance, World War II (the conflict variable) and nationalism in America (the solidarity variable). And we might find an exception to our rule or law, such as America's involvement in the Vietnam War, that contradicts the principle, forcing its revision or the recognition that "all things were not equal." In this case we might revise the principle by stating a condition under which it holds true: when parties to a conflict perceive the conflict as a threat to their welfare, then the level of solidarity of groups is a positive function of their degree of conflict. Thus the Vietnam War did not produce internal solidarity in America because it was not defined as a threat to the general welfare (whereas, for the North Vietnamese, it was a threat and produced solidarity).

---

[18]For more details of this argument, see Lee Freese, "Formal Theorizing," *Annual Review of Sociology* 6 (1980), pp. 187–212; and Herbert L. Costner and Robert K. Leik, "Deductions from Axiomatic Theory," *American Sociological Review* 29 (December 1964), pp. 19–35.

[19]See, for example, Peter Blau's excellent *Inequality and Heterogeneity: A Primitive Theory of Social Structure* (New York: Free Press, 1977).

[20]See Freese.

The essential idea here is that, in formal theory, an effort is made to create abstract principles. These principles are often clustered together to form a group of laws from which we make rather loose deductions to explain empirical events. Much like axiomatic systems, formal systems are hierarchical, but the restrictions of axiomatic theory are relaxed considerably. Most propositional schemes in sociological theorizing are, therefore, of this formal type.

(c) Yet, much of what is defined as theory in sociology is more empirical. These empirical formats consist of generalizations from specific events in particular empirical contexts. For example, Golden's Law states that "as industrialization increases, the level of literacy in the population increases." Such a proposition is not very abstract; it is filled with empirical content—industrialization and literacy. Moreover, it is not about a timeless process, since industrialization is only a few hundred years old and literacy emerged, at best, only 6,000 years ago. There are many such generalizations in sociology that are considered theoretical. They represent statements of empirical regularities that scholars think are important to understand. Indeed, most substantive areas and subfields of sociology are filled with these kinds of propositions.

Strictly speaking, however, these are not theoretical. They are too tied to empirical contexts. In fact, they are generalizations that are in need of a theory to explain them. Yet I caution that my opinion about such empirical generalizations is not shared by all. And if we ask scholars working in substantive areas if their generalizations are theory, a good many would answer affirmatively.

There are other kinds of empirical generalizations, however, that raise fewer suspicions about their theoretical merits. These are often termed *middle-range theories,* as we will see in Chapter 4, because they are more abstract than a research finding and because their empirical content pertains to variables that are also found in other domains of social reality.[21] For example, a series of middle-range propositions from the complex organization's literature might be stated: "Increasing size of a bureaucratic organization is positively related to (a) increases in the complexity (differentiation) of its structure, (b) increases in the reliance on formal rules and regulations, (c) increases in the decentralization of authority, and (d) increases in span of control for each center of authority."[22] These principles (the truth of which is not at issue here) are more abstract than Golden's Law because they denote a whole class of phenomena—organizations. They also deal with more generic variables—size, differentiation, centralization of power, spans of control, rules, and regulations—that have existed in all times and all places. Moreover, these variables could be stated more abstractly to apply to all organized social systems, not just bureaucratic organizations. For instance, I can visualize a more abstract law that states: "In-

---

[21]See Chapter 4 on Robert K. Merton's work. In particular, consult his *Social Theory and Social Structure* (New York: Free Press, 1975).

[22]I have borrowed this example from Peter M. Blau's "Applications of a Macrosociological Theory" in *Mathematizche Analyse von Organisationsstrukktaren und Prozessen* (Internationale Wissenschaftliche Fachkonferenz, vol. 5, March 1981).

creasing size of a social system is positively related to (*a*) increases in levels of system differentiation, (*b*) increases in the codification of norms, (*c*) increases in the decentralization of power, and (*d*) increases in the spans of control for each center of power." Now, I am not asserting that these propositions are true, but I see them as interesting laws that can be tested out in many diverse empirical contexts, not just bureaucratic organizations. The central point here is that some empirical generalizations have more theoretical potential than others. If their variables are relatively abstract and if they pertain to basic and fundamental properties of the social universe that exist in other substantive areas of inquiry, then it is more reasonable, I believe, to consider them theoretical.

In sum, then, there are three basic kinds of propositional schemes: axiomatic, formal, and various types of empirical generalizations. These propositional schemes are summarized in Figure 1–4. Although axiomatic formats are elegant and powerful, sociological variables and research typically cannot conform to their restrictions. Instead, we must rely upon formal formats that generate propositions stating abstract relations among variables and then make loosely structured "deductions" to specific empirical cases. Finally, there are empirical formats that consist of generalizations from particular substantive areas, and these are often considered theories of that area. Some of these theories are little more than summaries of research findings that require a theory to explain them. Others are more middle range and have more potential as theory because they are more abstract and pertain to more generic classes of variables.

Modeling schemes At times it is useful to draw a picture of social events. Some models are drawn with neutral languages such as mathematics, in which the equation is presumed to map and represent empirical processes.[23] I prefer to visualize such equations as propositions (formal statements of relations among variables) *unless* they can be used to generate a picture or some form of graphic representation of processes. There is no clear consensus on what a model is, but in sociological theory there is a range of activity that involves representing concepts and their relations as a "picture" that models what are considered the important elements of a social process. Perhaps I am being too restrictive here, but my purpose is to distinguish an important type of theoretical activity.

A model, then, is a diagrammatic representation of social events. The diagrammatic elements of any model include: (1) concepts that denote and highlight certain features of the universe; (2) the arrangement of these concepts in visual space so as to reflect the ordering of events in the universe; and (3)

---

[23]Actually, these are typically "regression equations" and would not constitute modeling as I think it should be defined. A series of differential equations, especially as they are simulated or otherwise graphically represented, would constitute a model. Computer simulations represent, I think, an excellent approach to modeling. See, for example, Robert A. Hanneman, *Computer-Assisted Theory Building: Modeling Dynamic Social Systems* (Newbury Park, CA: Sage, 1988).

FIGURE 1–4   Types of Propositional Schemes

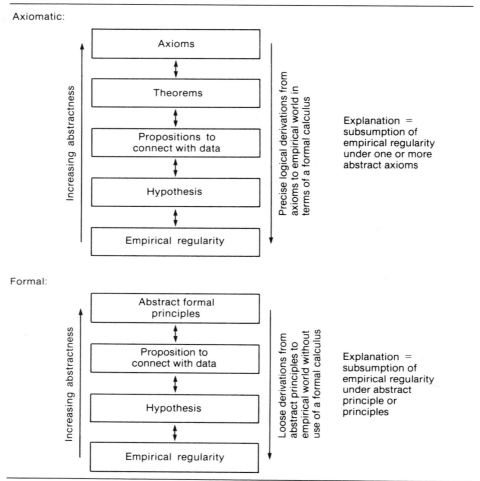

symbols that mark the connections among concepts, such as lines, arrows, vectors, and so on. The elements of a model may be weighted in some way, or they may be sequentially organized to express events over time, or they may represent complex patterns of relations, such as lag effects, threshold effects, feedback loops, mutual interactions, cycles, and other potential ways in which properties of the universe affect one another.[24]

---

[24]Good examples of such models are in my *Societal Stratification* and *A Theory of Social Interaction*. For examples of more empirical, yet still analytical, models, see Gerhard and Jean Lenski, *Human Societies* (New York: McGraw-Hill, 1982). See also the numerous analytical models in Randall Collins, *Theoretical Sociology* (San Diego: Harcourt Brace Jovanovich, 1988).

FIGURE 1–4 *(continued)*

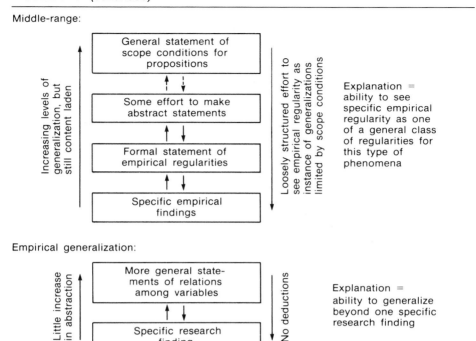

Middle-range:

Empirical generalization:

In sociology, I think that most diagrammatic models are constructed to emphasize the causal connections among properties of the universe. That is, they are designed to show how changes in the values of one set of variables are related to changes in the values of other variables. Models are typically constructed when there are numerous variables whose causal interrelations an investigator wants to highlight.

With the issue far from clear, my sense is that sociologists generally construct two different types of models, which I will term *analytical* models and *causal* models. This distinction is, I admit, somewhat arbitrary, but it is a necessary one if we are to appreciate the kinds of models that are constructed in sociology. The basis for making this distinction is twofold: First, some models are more abstract than others in that the concepts in them are not tied to any particular case, whereas other models reveal concepts that simply summarize statistically relations among variables in a particular data set. Second, more abstract models almost always reveal more complexity in their representation of causal connections among variables. That is, one will find feedback loops, cycles, mutual effects, and other connective representations that complicate the causal connections among the variables in the model and make them difficult to summarize with simple statistics. In contrast, the less abstract models typically depict a clear causal sequence among empirical variables.[25] They typically reveal independent variables that effect variation in some dependent variable; and, if the model is more complex, it might also highlight intervening variables and perhaps even some interaction effects among the variables.

Thus, analytical models are more abstract, they highlight more generic properties of the universe, and they portray a complex set of connections among variables. In contrast, causal models are more empirically grounded; they are more likely to devote particular properties of a specific empirical case; and they are likely to present a simple lineal view of causality. These modeling strategies are summarized in Figure 1-5.

Causal models are typically drawn in order to provide a more detailed interpretation of an empirical generalization. They are designed to sort out the respective influences of variables, usually in some temporal sequence, as they operate on some dependent variable of interest. At times a causal model becomes a way of representing the elements of a middle-range theory so as to connect these elements to the particulars of a specific empirical context. For example, if we wanted to know why the size of a bureaucratic organization is related to its complexity of structure in a particular empirical case of a growing organization, we might translate the more abstract variables of size and complexity into specific empirical indicators and perhaps try to introduce other variables that also influence the relationship between size and complexity in this empirical case. The causal model thus becomes a way to represent with more clarity the empirical association between size and complexity in a specific context.[26]

Analytical models are usually drawn to specify the relations among more abstract and generic processes. Often they are used to delineate the processes that operate to connect the concepts of an axiomatic or, more likely, a formal theory.[27] For example, we might construct a model that tells us more about the processes that operate to generate the relationship between conflict and solidarity or between size and differentiation in social systems. Additional concepts would be introduced, and their weighted, direct, indirect, feedback, cyclical, lagged, and other patterns of effect on one another would be diagramed. In this way, the analytical model tells us more about how and why properties of the universe are connected. In addition to specifying processes among formal propositions, analytical models can be used to describe processes that connect variables in the propositions of a middle-range theory. For example, we might use a model to map out how organization size and complexity are connected by virtue of other processes operating in an organization.

Of course, we can construct analytical models or causal models for their own sake, without reference to an empirical generalization, a middle-range

---

[25]The "path analysis" that was so popular in American sociology in the 1970s is a good example of such modeling techniques.

[26]For an example of a model for these variables, see Peter M. Blau's "A Formal Theory of Differentiation in Organizations," *American Sociological Review* 35 (April 1970), pp. 201–18. See also Chapter 12.

[27]Ibid. is a good example.

FIGURE 1–5 Types of Modeling Schemes

Analytical models:

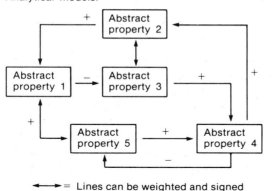

Explanation =
ability to map crucial
connections (perhaps
weighted) among basic
properties of a
specific class of
process or phenomenon

←——▶ = Lines can be weighted and signed

Causal models:

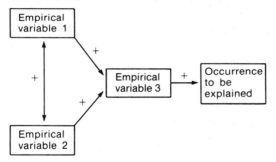

Explanation =
tracing of causal connections
among measured variables
accounting for variation in
the occurrence of interest

←——▶ = Lines usually state a statistical association among variables

theory, or a formal/axiomatic theory. We may simply prefer modeling to propositional formats. One of the great advantages of modeling is that it allows representation of complex relations among many variables in a reasonably parsimonious fashion. To say the same thing as a model, a propositional format might have to write complex equations or use many words. Thus, by itself, modeling represents a tool that many theorists find preferable to alternative theoretical schemes.

In sum, then, I have isolated four general and a number of more specific ways in which sociologists organize concepts and theoretical statements into formats. I believe that this description summarizes the range of approaches, although we could, no doubt, quibble over the various distinctions that I have drawn. But my purpose is not to draw finely tuned distinctions nor to enter a dialogue with philosophers of science. Rather, I want only to provide a general perspective for understanding the diversity of activities that sociologists call "theory." We cannot understand the structure of sociological theory without some sense for the underlying, and typically unarticulated, theoretical schemes employed by individual theorists.

As I have mentioned frequently in this description of the elements of theory, these various schemes are often interrelated. Moreover, I have expressed, at least implicitly, personal biases in favor of some schemes. Before closing this opening chapter and moving on to the actual substance of sociological theory, therefore, I would like to explore the relative merits of these theoretical approaches.

AN ASSESSMENT OF THEORETICAL APPROACHES

Abstractness and Scope of Various Theoretical Schemes

In Figure 1-6, I have arrayed the major types of approaches to building theory along two dimensions: (1) their level of abstraction and (2) their degree of scope. That is, how abstract are the concepts and statements, and how broad a range of phenomena do they encompass? The more the concepts of a theoretical strategy are free of reference to any specific empirical case, the higher is its level of abstraction; and, the greater the range of substantive phenomena encompassed by the concepts and statements of an approach, the greater is its scope. Naturally, these are not mutually exclusive, since the process of abstraction by its very nature leads to the inclusion of more cases and, hence, increases the scope of phenomena denoted by concepts. I can certainly be criticized for plotting along nonexclusive dimensions, but my point in Figure 1-6 is not to create an elegant typology. Rather, I simply want to communicate that abstract schemes encompass more substantive phenomena in the social universe. But I also want to emphasize that abstraction per se does not always lead to a breadth of coverage. A theoretical format can be highly abstract in that it contains no direct empirical referents, but it can denote only a delimited range of phenomena. Thus I think that axiomatic and formal propositions tend to be highly abstract, but they typically pertain to a limited range of phenomena. In contrast, analytical schemes and meta-theory are no more abstract than formal or axiomatic theories, but they usually contain more concepts and statements denoting a broader range of social phenomena. And so, although I admit to an overlap in the dimensions of Figure 1-6, the figure communicates my essential point: theoretical schemes vary primarily in terms of two interrelated dimensions, abstraction and scope. Some address a wide range of phenomena with abstract concepts, others only a limited range with equally abstract concepts. Still others are not abstract at all and are thereby limited in scope. And some are more abstract and cover varying ranges of events.

My sense is that, in terms of abstraction alone, the various approaches cluster at four diverse levels. Empirical generalizations and causal models are tied to specific empirical contexts. A generalization summarizes a particular set of research findings, whereas a simple causal model typically involves correlational statements linking empirically measured variables. Middle-range

FIGURE 1–6 Variations in Types of Theoretical Formats

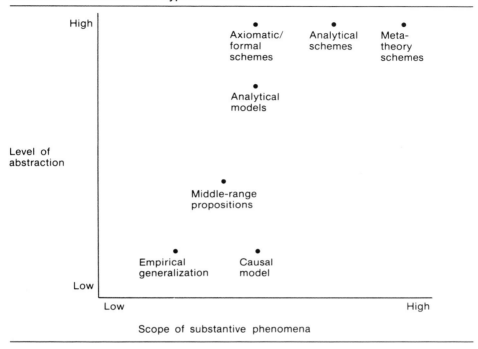

propositions are more abstract because they seek to explain events for a whole class of phenomena. Yet the concepts of middle-range theory reveal empirical content that limits their abstractness. An analytical model is generally more abstract than a middle-range theory because it introduces generic properties of the social universe to explain some general class of empirical events. Axiomatic/formal propositional schemes, analytical schemes, and meta-theory are even more abstract in that they reveal no empirical content about specific times, places, or contexts. They are usually about the basic and universal properties of social organization, human action, or social interaction, without reference to any particular pattern of organization, form of action, or context of interaction.

Of course, this is only a rough grouping. I am well aware that there is variation, but I would submit that the relative position of these three groupings in regard to their respective levels of abstraction is roughly captured in Figure 1-6. For example, an axiomatic scheme may be somewhat less abstract than a meta-theory, but it is not likely to become less abstract than an analytical model or a middle-range theory. What distinguishes theoretical approaches at any given level of abstraction is their scope, or the range of phenomena covered by theoretical statements. The scope of empirical generalizations tends to be limited since they emanate from research findings that, by their nature, focus on a limited range of phenomena (since, after all, there are only so many things that can be studied in even the largest research project). Causal models often add extra variables in an effort to explain more variance, and so they frequently expand the scope somewhat of an empirical generalization. Middle-range theories try to explain a whole class of phenomena—for example, delinquency, revolutions, ethnic antagonism, and urbanization. They are, therefore, broader in scope than empirical generalizations and causal models, but the very goal of middle-range theory is to limit scope by trying to explain only one class of events. Analytical models are usually broader than middle-range theories, but they tend to be limited because they too are about a specific range of phenomena—industrialization, ethnic relations, political centralization, differentiation in organizations, and similar topics. They are somewhat broader in scope because it is easier to include more variables in a model than in a series of middle-range propositions. Yet analytical models can become highly abstract and pertain to only more general processes—thereby placing them on the same level of abstraction as axiomatic, analytical, and meta-theoretical schemes.

Axiomatic/formal propositional schemes tend to be limited in range because they focus on certain generic processes and ignore others. For example, a formal theory might be about differentiation, consensus formation, conflict, exchange, behavior, action, interaction, and similar basic processes in the social universe. Although the proponents of these schemes often claim that their propositions explain much more of reality, it is more reasonable to see these abstract propositions as pertaining to only a limited range of phenomena. Analytical schemes in sociological theory tend to be all-encompassing, seeking to explain human action, interaction, and organization in one grand scheme. They are rarely immodest efforts; and, although they differ rather dramatically in their content, they are all similar in their presumption that they encompass all of the social universe that needs to be explained. Meta-theories are very broad but diffuse and imprecise. They explicate all that needs to be explained and present a philosophical justification for developing a particular form of explanation. But they tend to be rather grandiose, and this fact makes them so broad as to be somewhat vacuous.

Relative Merits of Diverse Theoretical Approaches

My belief is that theory should be abstract. That is, the less substantive the content in the concepts, the better they are. For, if theories are filled with empirical referents, they are tied to specific contexts and hence are not so useful as those that view specific empirical contexts as instances or examples of a more basic underlying process. Many theorists in sociology, however, would disagree with me on this score, and I will return to this point of contention shortly.

I also believe that theory should be such that it can be proven wrong by empirical tests. As a general platitude, few would disagree with this statement. But as a more practical matter of how we should construct theories to be proven wrong, there is enormous disagreement. My sense is that theories must be sufficiently precise in the definitions of concepts and in the organization of concepts into statements that they can be, in principle, measured and tested. It is only through the generation of precise theoretical statements and efforts at their refutation that scientific knowledge can be generated. For me, then, what distinguishes good theoretical statements from bad ones is that they are *created to be proven wrong*. A theory that, in principle, cannot be proven wrong is not very useful. It becomes a self-sustaining dogma that is accepted on faith. A theory must allow for an understanding of events, and hence it must be tested against the facts of the world. If a theoretical statement is proven wrong by empirical tests, science has advanced. When a theory is rejected, then one less possible line of inquiry will be required in search of an answer to the question "why?" By successively eliminating incorrect statements, those that survive attempts at refutation offer, for the present at least, the most accurate picture of the real world. Although having one's theory refuted may cause professional stigma, refutations are crucial to theory building. It is somewhat disheartening, therefore, that some scientists appear to live in fear of such refutation. For in the ideal scientific process, just the opposite should be the case, as Karl Popper has emphasized:

> Refutations have often been regarded as establishing the failure of a scientist, or at least of his theory. It should be stressed that this is an inductive error. Every refutation should be regarded as a great success; not merely as a success of the scientist who refuted the theory, but also of the scientist who created the refuted theory and who thus in the first instance suggested, if only indirectly, the refuting experiment.[28]

Even statements that survive refutation and hence bring professional prestige to their framers are never fully proven. It is always possible that the next empirical test could disprove them. Yet, if statements consistently survive empirical tests, they have high credibility and are likely to be at the core of a theoretical body of knowledge. But, as I have now phrased the issue, many sociological theorists would disagree. Moreover, most philosophers of science would argue that this process of refutation is idealized and, in fact, rarely occurs in the actual operation of science.

Despite these reservations, however, my view is this: we should proceed *as if* we can develop theoretical statements that are highly abstract and, at the same time, sufficiently precise so as to be testable. Again, as I will document in the chapters to come, a majority of social theorists disagree with me on this score. I have injected my personal views because it is important to understand the biases with which I approach the review and analyses of social theory. Moreover, these biases are the central issue around which the debate over the best approach to developing theory and knowledge rages. Since I have presented my biases, let me elaborate on them by assessing the merits of various approaches in terms of my prejudices.

From my point of view, empirical generalizations and causal models of empirically operationalized variables are not theory at all. They are useful

---

[28]Karl R. Popper, *Conjectures and Refutations* (New York: Basic Books, 1962), p. 243.

summaries of data that need a theory to explain them. Some would argue that theory can be built *from* such summaries of empirical regularities. That is, we can induce from the facts the more general properties that these facts illustrate. I doubt this, but many disagree. I grant that familiarity with empirical regularities is crucial to developing more abstract and comprehensive theoretical statements, but I doubt if this process of mechanically raising the level of abstraction from empirical findings will produce interesting theory. A much more creative leap of insight is necessary, and so I would not suggest that theory building begin with a total immersion in the empirical facts. I suspect that, once buried in the facts, one rarely rises above them.

At the other extreme, meta-theory is like empirical facts: interesting but counterproductive. Many scholars feel just the opposite, however. They argue that we must have meta-theory before we try to develop more precise theoretical statements. My opinion is that this has rarely been the case in science, primarily because meta-theory gets bogged down in enduring philosophical issues. The result is that proponents of various philosophical camps become so preoccupied with their debate (which by its nature can never end) that they never get around to developing theory.

Analytical schemes often suffer from the same problems as meta-theory. Naturalistic schemes have, I am convinced, a tendency to become overly concerned with their architectural majesty. In an effort to construct an orderly scheme that mirrors at an abstract level the empirical world in all its dimensions, naturalistic schemes get ever more complex; as new elements are added to the scheme, efforts to reconcile new portions with the old take precedence over making the scheme testable. Moreover, the scheme as a whole is impossible to test because relations among its elements cover such a broad range of phenomena and are rarely stated with great precision. And when imprecision is compounded by the abstractness, then empirical tests are infrequent because it is not clear how to test any portion of the scheme. Yet, despite these problems, creators of analytical schemes view them as a necessary prerequisite for developing testable theoretical statements. In their view, one needs to know what is important to test, and only through an ordered scheme is this possible.

In contrast, sensitizing schemes are typically constructed as a loose framework of concepts to interpret events and to see if they yield greater understanding of how and why these events occur. Obviously, I do not agree with those who see such schemes as an alternative to science, but I have found many of these sensitizing schemes very insightful. Yet, much like naturalistic approaches, sensitizing schemes also become self-reinforcing because they are so loosely structured and so often vague (albeit suggestive and insightful) that the facts can always be bent to fit the scheme. Hence the scheme can never be refuted or, I suspect, revised on the basis of actual empirical events. Thus I see sensitizing schemes as most useful when they are used to orient us to important phenomena and then are elaborated upon with propositions and analytical models.

Let me now turn to axiomatic/formal propositional formats, analytical models, and middle-range propositions. As I have already indicated, axiomatic theorizing is, for the most part, impractical in sociology. In my view, formal theorizing is the most useful approach because it contains abstract concepts that are linked with sufficient precision so as to be testable. Analytical models can be highly insightful, but they are hard to test as a whole. They contain too many concepts, and their linkages are too diverse to be directly tested. So it is reasonable to ask: in what sense can they be useful for sociological theorizing? My view is that an analytical model can best be used to specify the processes by which concepts in a formal proposition are connected.[29] For example, if a proposition states that the "degree of differentiation" is a function of "system size," the model can tell us why and how size and differentiation are connected. That is, we can get a better sense for the underlying processes by which size increases differentiation (and perhaps vice versa). Alternatively, analytical models can also be a starting point for formal theorizing. By isolating basic processes and mapping their interconnections, we can get a sense for the important social processes about which we need to develop formal propositions. And although the model as a whole cannot be easily tested (because it is too complex to be subject to a definitive test), we can decompose it into abstract propositional statements that are amenable to definitive tests.

Middle-range propositions are, I feel, less useful as places to begin theory building. They tend to be filled with empirical content, much of which does not pertain to the more basic, enduring, and generic features of the social universe.[30] For example, "theory of ethnic antagonism" is often difficult to translate into a more general proposition or model on conflict. Moveover, scholars working at this middle range tend to become increasingly empirical as they seek to devise ways to test their theories. Their propositions become, I have found, ever more like empirical generalizations as more and more research content is added. There is no logical reason why substantive and empirical referents cannot be taken out of middle-range theories and the level of abstraction raised, but such has infrequently occurred. The reason for this is that, because the generalizations are content filled and stated propositionally, they are readily tested; thus there is a bias to test the propositions rather than to extract their more generic content and create more abstract propositions.

In Figure 1–7 I have summarized these conclusions in the right column. Meta-theory and naturalistic analytical schemes are interesting philosophy but poor theory. Sensitizing analytical schemes, formal propositional statements, and analytical models offer the best place to begin theorizing, especially if interplay among them is possible. Middle-range theories have rarely realized their theoretical potential, tending to move toward empirical generalizations as opposed to formal propositions. Causal models and empirical generalizations are useful in that they give theorists some sense of empirical regularities, but

---

[29]I have tried to illustrate this strategy in my *Societal Statification* and *A Theory of Social Interaction*.

[30]I doubt if this was Merton's intent when he formulated this idea, but my sense is that his advocacy became a legitimation for asserting that empirical generalizations were "theory."

FIGURE 1-7   Relations among Theoretical Approaches and Potential for Building Theory

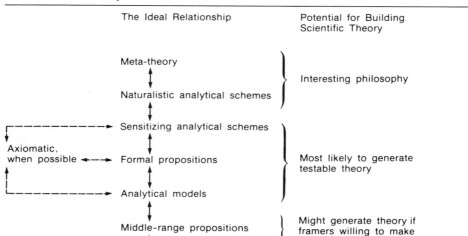

by themselves and without creative leaps in scope and abstraction they are not theoretical. They are usually data in need of a theory.

In the left column I have presented my idealized view of the proper place of each theoretical approach for generating knowledge.[31] If we begin to accumulate bodies of formal laws (perhaps on the basis of leads provided by a sensitizing scheme), then it is desirable to extract out the key concepts and look at these as the basic sensitizing and orienting concepts of sociology (much as magnetism, gravity, relativity, and the like were for early-20th-century physics). We may even want to construct a formal analytical scheme and ponder on its meta-theoretical implications. In turn, such pondering can help reformulate or clarify analytical schemes, which can perhaps help construct new, or reverse old, formal propositions. But without a body of formal laws to pull meta-theory and analytical schemes back into the domain of the testable, they become hopelessly self-sustaining and detached from the very reality they are supposed to help clarify.

For building theory, the most crucial interchange is, I believe, between formal propositions and analytical models. There is a creative synergy between translating propositions into models and translating models into propositions.

---

[31]I should emphasize that this is not how things actually work; the diagram represents my *wish* for how sociological theory should be developed.

This is the level at which most theory should be couched. But, as I will document in the following chapters, a good many theorists clearly disagree.

Middle-range propositions can inspire analytical models that, in turn, encourage the development of more abstract propositions. Such will be the case, however, only if there already exists a body of useful abstract propositions that can serve as an inducement to middle-range theorists to raise the level of abstraction of their empirically laden statements. Conversely, middle-range theories could potentially be one vehicle by which more abstract propositions—such as on conflict processes—are attached to specific empirical problems—such as ethnic conflict. Thus, middle-range theories become part of the deductive calculus of a formal theoretical system. And, finally, the techniques of research generalizations and causal modeling of empirical variables can help test the implication of formal theories. They become the lowest-order generalizations in the deductive system of formal theory. They might also stimulate inductive efforts to develop more abstract propositions of greater scope, but, as I indicated earlier, I have my doubts that this will occur.

Thus, in my view, there is no necessary conflict among these various approaches to accumulating knowledge. But in actual fact there is enormous conflict among those (including me and almost all other theorists) who claim that one of these activities is the most essential for developing social theory. Much of the theoretical literature in sociology is consumed with the debates among scholars working with one or the other of these strategies. Indeed, I would imagine that my comments in this section are just more fuel for a fire that shows little potential for extinguishing itself.

THE STATE OF SOCIOLOGICAL THEORY

Prevailing Theoretical Schemes in Sociology

The preceding comments express my opinions, but they have also summarized the state of sociological theorizing. Most theory in sociology is of two basic types: analytical schemes and loose systems of formal propositions. There is some analytical modeling at a general theoretical level, but it is usually part of a system of propositions or general analytical scheme. In specific substantive subfields there is considerable modeling and middle-range theorizing. Yet I cannot cover in this volume all of the theories in various subfields. Moreover, much of this work is insufficiently abstract to be true theory; rather, most of it represents a series of summaries of the empirical generalizations that need theoretical explanation. At the other extreme, I will not analyze any of the meta-theoretical schemes that can be found in the theoretical literature. I find these too philosophical to constitute interesting theory. However, if meta-theoretical considerations are part of a general analytical scheme or system of formal propositions, then I will explore these considerations in light of their implications for the theoretical scheme.

Thus, in the pages to follow, the core structure of sociological theory revolves around analytical schemes and formal propositional systems. At times, analytical models and meta-theoretical issues are part of these analytical and propositional schemes, but the essence of sociological theory today is a series of naturalistic analytical schemes, sensitizing analytical schemes, and variously structured systems of abstract formal propositions.

Enduring Controversies in Sociological Theorizing

In addition to this diversity of schemes employed by various theorists, there is enormous controversy over a number of substantive and strategic issues. Some of this controversy reflects the implicit assumptions of the schemes per se, but even more of it is the result of different preferences among individual theorists. Let me review some of these controversies, because they will surface again and again in subsequent chapters.

Can sociology be a science?   This issue will always confront us in a review of sociological theories. The basic line of argument is: Can sociology be like the other natural sciences, or is there something unique about human affairs?[32] If the social universe is considered to be like other realms, such as that analyzed by physics or biology, then theory tends toward propositional and naturalistic analytical schemes. If human action, interaction, and organization are seen as fundamentally different and as not amenable to analysis with the procedures of natural science, then sensitizing analytical schemes are most likely to be used.

This issue is probably the most divisive in sociological theorizing. At present I would guess that sociological theorists are about equally split, although I suspect that the natural-science advocates have lost ground in recent years to those who do not feel that the theoretical procedures of the other sciences are relevant to human beings. I consider this trend unfortunate, but obviously my feelings are not shared by a growing number of theorists.

Should sociological theory be micro or macro?   This question addresses the issue of whether sociological theory should be about the micro actions and interactions of individuals or the macro social structures that such actions and interactions create.[33] Should we study and theorize about people in concrete settings, or should we stand back and look at the institutional complexes of society? Micro theorists typically accuse macro theorists of reifying social structure—that is, of making a reality out of a concept. They see social structure as nothing more than the micro processes of action and interaction among individuals. In their defense, macro theorists see micro the-

---

[32]See notes 9 and 10 for examples of the two sides on this issue.

[33]This has reemerged as a major issue in sociological theory. For some recent commentaries on it, see K. Knorr-Cetina and A. V. Cicourel, eds., *Advances in Social Theory and Methodology: Toward an Integration of Micro- and Macro-sociologies* (Boston: Routledge & Kegan Paul, 1981); Jonathan H. Turner, "Theoretical Strategies for Linking Micro and Macro Processes: An Evaluation of Seven Approaches," *Western Sociological Review* 14, no. 1 (1983), pp. 4–15.

orists as never seeing the forest through the trees. Micro theorists, they argue, are so busy studying interaction that they fail to see that such interaction is constrained by the structures of society.

There have been many conceptual efforts to resolve this debate by developing approaches that address both micro and macro processes.[34] But these efforts typically generate further controversy, as the synthesizing scheme is seen by its critics as having a micro or macro bias. Proponents of micro, macro, or synthetic efforts tend to employ all the approaches outlined earlier. I sense that those with a micro bias tend to use sensitizing analytical schemes, whereas those theorists who focus on macro structural processes tend to employ formal propositional schemes and naturalistic analytical schemes. Yet there are so many exceptions to this generalization that I hesitate to make it. But there is at least a slight tendency along these lines.

Is sociology a paradigmatic science?　　This issue is less burning than the other two. But it is an issue when organizing a discussion of various theoretical approaches. Some argue that sociological theory can be divided into several paradigms that hold different views of reality and advocate varying theoretical as well as research strategies.[35]

My feeling is that the concept of paradigm is too strong. As employed by its most visible exponent, Thomas Kuhn,[36] I suspect that no theoretical approach is sufficiently coherent, precise, and established as to constitute a paradigm of abstract concepts and laws as well as verified research findings. If anything, sociology is pre-paradigmatic.

Moreover, the concept of paradigm has been so overused that it has lost any meaning, at least for me. I think that, at best, sociology has a series of perspectives or orientations that guide theoretical activity, but these perspectives do not constitute paradigms in the same sense as this term has been employed in the natural sciences.

THEORETICAL PERSPECTIVES IN SOCIOLOGY

Much of what is labeled sociological theory is, in reality, only a loose clustering of implicit assumptions, some basic concepts, and various kinds of theoretical statements and formats. But none of these is dominant or sufficiently precise to constitute a paradigm. Sometimes assumptions are stated explicitly and serve to inspire abstract theoretical statements containing well-defined concepts, but most sociological theory constitutes a verbal "image of society" rather than a

---

[34]For examples, see Giddens, *The Constitution of Society,* and Randall Collins, *Conflict Sociology* (New York: Academic Press, 1975). As we will see shortly, many specific perspectives besides these two examples try to provide a reconciliation of micro and macro processes.

[35]The best of these arguments is George Ritzer's *Sociology: A Multiple Paradigm Science* (Boston: Allyn & Bacon, 1975). For my views on meta-theory, see "The Misuse and Use of Metatheory," *Sociological Forum*, 1990, in press. This issue of *Sociological Forum* is devoted to meta-theory.

[36]Thomas Kuhn, *The Structure of Scientific Revolutions* (Chicago: University of Chicago Press, 1962, 1970).

rigorously constructed set of theoretical statements organized into a logically coherent format. Thus a great deal of "theory" is really a general perspective or orientation for looking at various features of the social world.

The fact that there are many such perspectives in sociology presents me with a problem of exposition. This is compounded when we recognize that the perspectives blend into one another, sometimes rendering it difficult to analyze them separately. My solution to this dilemma is to limit arbitrarily the number of perspectives covered and, at the same time, to act as if they were separable. Accordingly, in the sections to follow I will analyze five general sociological perspectives or orientations: (1) functional theorizing, (2) conflict theorizing, (3) exchange theorizing, (4) interactionist theorizing, and (5) structuralist theorizing. These general perspectives do not constitute paradigms because the various theorists working within these traditions often disagree over the best strategy, over whether sociology can be a science, and over whether or not the micro, macro, or some combination of the two should be emphasized. Indeed, the debates *within* orientations are frequently far more acrimonious than those between orientations.

I will focus on these perspectives for a number of reasons: (1) They are the most general perspectives in sociology and underlie most specific perspectives in the field. (2) These perspectives are also the most widespread and influential—the subjects of much analytical elaboration and, of course, criticism. (3) Each of these perspectives, at various times, has been proclaimed by its more exuberant proponents as the only one that could take sociology out of its theoretical difficulties. Therefore, each must be considered in a book attempting to assess the structure of sociological theorizing.

For each of these perspectives, I will examine its emergence and then its dominant contemporary practitioners. In so doing, I will emphasize a number of key topics: (1) the assumptions about the nature of the social world that a perspective and its advocates hold; (2) the theory-building strategy typically advocated by those working with a perspective; (3) the image of social processes that a perspective reveals; (4) the key concepts and propositions developed within the perspective; and (5) the types of theoretical statements generated within a perspective.

In sum, I think it is safe to say that sociological theorizing is in its intellectual infancy. Yet my review of its major orientations will demonstrate that theory in sociology has great potential for developing useful knowledge about the social universe.

PART 1

◆

Functional Theorizing

CHAPTER 2

◆

The Emergence
of Functionalism

Even today, classical economic ideas dominate sociological theory and social thought in general.[1] From this economic perspective, humans are seen as rational beings who try to maximize their gains and minimize their losses. And I suspect that many still view social life as a kind of marketplace where people buy and sell their qualities in hopes of making a psychic profit. Indeed, social life is a competitive game of people rationally pursuing their interests, with social order somehow emerging out of these clashes of self-interest.

This view of humans is frequently termed *utilitarianism* because there is the assumption that actors are rational and that they try to maximize their "utilities" or rewards and gratifications. Adam Smith is most commonly associated with this perspective because he was the first to conceptualize analytically the dynamics of competitive markets and because he postulated an "invisible hand of order" as emerging from open competition in free markets.[2] Although utilitarianism pervades much of our thinking today, it was even more dominant in the last century. And just as sociology today must overcome the limitations of this narrow view of humans and social organization, so it had to confront utilitarianism in the last century. In fact, I do not think it an exaggeration to say that sociology's first theoretical perspective—functionalism—emerged as a reaction against utilitarianism. In questioning utilitarianism, sociology pursued an alternative: organicism.

FUNCTIONALISM AND THE ORGANISMIC ANALOGY

In early-19th-century sociology, humans no longer were viewed as rational and calculating entrepreneurs in a free, open, unregulated, and competitive mar-

---

[1]As I will outline later, exchange theory typically begins with these assumptions. But it also penetrates other perspectives, as we will come to see.

[2]Adam Smith, *An Inquiry into the Nature and Causes of the Wealth of Nations* (London: Davis, 1805; originally published in 1776). Only Volume 1 contains these extreme statements. Subsequent portions of the book are quite sociological in nature.

ketplace. Nor was the doctrine of the "invisible hand of order" considered a very adequate explanation of how social organization could emerge out of free and unbridled competition among individuals. Although utilitarianism remained a prominent social doctrine for the entire 19th century, the first generation of French sociologists had ceased to accept the assumption that social order would automatically be forthcoming if only free competition among individuals was left intact.

The disenchantment with utilitarianism was aided in France, and to a lesser extent in all of continental Europe, by the disruptive social changes wrought by industrialization and urbanization. Coupled with the political instability of the late 18th century, as revealed most dramatically by the violent French Revolution, early-19th-century social thinkers in France displayed a profound concern with the problems of maintaining the social order. Although each phrased the question somewhat differently, I think it fair to conclude that all social thinkers asked similar questions: Why and how is society possible? What holds society together? What makes societies change?[3]

Whether in France or elsewhere in Europe, the answers to these fundamental questions were shaped by events occurring in the biological sciences. It was in the 19th century that biological discoveries were to alter significantly the social and intellectual climate of the times. For example, as many of the mysteries of the human body were being discovered, the last vestiges of mysticism surrounding the body's functioning were being laid to rest. The diversity of the animal species was finally being systematically recorded under the long-standing classification procedures outlined by the Swedish biologist Carolus Linnaeus. And most importantly, conceptions of evolution, culminating in the theories of Wallace and Darwin, were stimulating great intellectual and social controversy. Since it was in this social and intellectual milieu that sociology as a self-conscious discipline was born, it is not surprising that conceptions of social order were influenced by a preoccupation with biology.

The Organicism of Auguste Comte

Auguste Comte (1798–1857) is usually credited as being the founder of sociology. Philosophizing about humans and society had, of course, long been a preoccupation of lay people and scholars alike, but it was Comte who advocated a "science of society" and coined the term *sociology*. And although Comte's work was soon to fall into neglect and obscurity and he was to live out his later years in frustration and bitterness, his work profoundly influenced social thought. I think it regrettable that few recognize this influence, even today. Yet, despite Comte's current obscurity, I mark the emergence of the functionalist perspective with his work.[4]

---

[3]In fact, Adam Smith had originally posed the question that all French thinkers were to ask: as societies become more complex and differentiated, and as actors live in different worlds, what "force" can hold the social fabric together?

[4]Auguste Comte, *The Course of Positive Philosophy* (1830–1842). References are to the more commonly used edition that Harriet Martineau condensed and translated, *The Positive Philosophy of Auguste Comte*, vols. 1, 2, and 3 (London: Bell & Sons, 1898; originally published in 1854).

Like most French thinkers of his time, Comte was preoccupied with propagating order and harmony out of the chaos created by the French Revolution. He attacked the individualism of utilitarian doctrines so prominent in England and carried forward Rousseau's and Saint-Simon's desire to develop a "collective philosophy"—one that would provide the principles for creating social consensus. In so doing, however, he articulated the principles of science as they should be applied to society.

Comte felt that human evolution in the 19th century had reached the "positive stage" in which empirical knowledge could be used to understand the social world and to create a better society. Comte thus became an advocate of the application of the scientific method to the study of society—a strategy that, in deference to Comte, is still termed *positivism* in the social sciences. This application of the scientific method was to give birth to a new science, sociology.

Comte's entire intellectual life represented an attempt to legitimate sociology. His efforts on this score went so far as to construct a "hierarchy of the sciences," with sociology as the "queen" of the sciences. Although this hierarchy allowed Comte to assert the importance of sociology and thereby separate it from social philosophy, his most important tactic for legitimating sociology was to borrow terms and concepts from the highly respected biological sciences. Sociology was thus initiated and justified by appeals to the biological sciences— a fact that will help explain why functionalism was sociology's first and, until the 1970s, most dominant theoretical orientation.

Seeing the affinity between sociology and biology as residing in their common concern with organic bodies, Comte divided sociology into social "statics," or morphology, and "dynamics," or social growth and progress. But Comte was convinced that, although "Biology has hitherto been the guide and preparation for Sociology ... Sociology will in the future ... (provide) the ultimate systematization of Biology."

Comte visualized that, with initial borrowing of concepts from biology, and later with the development of positivism in the social sciences, the principles of sociology would inform biology. Thus, sociology must first recognize the correspondence between the individual organism in biology and the social organism in sociology:

> We have thus established a true correspondence between the Statical Analysis of the Social Organism in Sociology, and that of the Individual Organism in Biology. ... If we take the best ascertained points in Biology, we may decompose structure anatomically into *elements, tissues,* and *organs.* We have the same things in the Social Organism; and may even use the same names.[5]

Comte then began to make clear analogies between specific types of social structures and the biological concepts:

---

[5]Auguste Comte, *System of Positive Polity or Treatise on Sociology* (London: Burt Franklin, 1875; originally published in 1851), pp. 239–40.

I shall treat the Social Organism as definitely composed of the Families which are the true elements or cells, next the Classes or Castes which are its proper tissues, and lastly, of the cities and Communes which are in real organs.[6]

Most of Comte's organismic analogies came in his later works, which were highly flawed.[7] And so I would be remiss if I left Comte at this point, for he was much more than a simple-minded organicist. My main concern is, of course, with functionalism, but let me pause for a closer look at Comte's positivism, which contains the basic tenets of the abstract propositional approach that I outlined in the last chapter as well as a critique of both causal and functional analysis. Comte preferred the name *social physics* over the current name of our discipline—sociology.[8] He wanted to mold sociology after Newtonian physics in which abstract theoretical principles are used to interpret empirical events. Contrary to many contemporary critics of positivism, who use the term *positivism* as an epithet for *raw empiricism,* Comte argued:

The next great hindrance to the use of observation is empiricism which is introduced into it by those who, in the name of impartiality, would interdict the use of any theory whatever. . . . No real observation of any kind of phenomena is possible, except as far as it is first directed, and finally interpreted, by some theory.[9]

What was such theory to look like? His answer is, I think, very important because he warned against the very functional analysis that he helped initiate in his later works, while he cautioned against excessive concern with the causal modeling so prominent in contemporary sociology. As he stressed:

The first characteristic of Positive Philosophy is that it regards all phenomena as subject to invariable natural *laws.* Our business is,—seeing how vain any research into what are called *causes,* whether first or final,—to pursue an accurate discovery of these laws, with a view to reducing them to the smallest possible number.[10]

In this passage, "final causes" is Comte's term for functions. We should not, he argued, analyze processes in terms of their consequences; nor should we search for their origins as "first causes." Rather:

Our real business is to analyze accurately the circumstances of phenomena, and to connect them by the natural relations of succession and resemblance. The best illustration of this is the case of the doctrine of Gravitation.[11]

Thus I think it fair to say that Comte's view of positivism rejected the functionalism and the more extreme organicism that were to emerge in the

---

[6]Ibid., pp. 241–42.

[7]By the time he wrote *Positive Polity* in 1851, Comte was, in a very real sense, a broken intellectual. Far more important, I feel, is the first work on *Positive Philosophy.*

[8]He was forced to abandon his preferred name because the Belgian statistician Adolphe Quetelet had already usurped the label for his statistical analyses.

[9]*Positive Philosophy,* vol. 1, p. 2.

[10]Ibid., p. 5.

[11]Ibid., p. 6.

later decades of the 19th century. Indeed, Émile Durkheim was to "turn Comte on his head" and stress that the very essence of adequate scientific explanation was causal and functional analysis.[12] I can only imagine how Comte would turn over in his grave if he had known what Durkheim did to his views. Yet, despite his eloquent advocacy for a social physics modeled after the natural sciences, Comte did reintroduce organic analogies into sociological inquiry, and he did see sociology as closely allied with the biological sciences.[13] And so we must conclude that it was with Comte that functional theory begins.

In the period between Comte's decline and Durkheim's ascendance as the most forceful advocate of functionalism, Herbert Spencer was to codify functional analysis into a more explicit theoretical strategy. Let me now turn to a brief review of Spencer's contribution.

The Analytical Functionalism of Herbert Spencer

Herbert Spencer (1820-1903) is a stigmatized figure in contemporary sociology. I think that this is a great tragedy because much contemporary theorizing owes an unacknowledged debt to Spencer.[14] Moreover, analyses of Spencer's work tend to focus on the weakest portion of his sociology, its functionalism. But unlike Durkheim two decades later, the anthropological functionalists of the early decades of this century, or the modern functionalists of the contemporary era, Spencer's functional analysis was comparatively recessive. Yet, in what may seem like a contradiction, Spencer anticipated in those few pages devoted to functional analysis all of the main features of modern functionalism.

Like many contemporary functionalists, Spencer saw the universe as divided into realms or domains. For Spencer, these basic realms are the inorganic (physical, chemical), the organic (biological, psychological), and the superorganic (sociological).[15] His great philosophical project was to generate a series of abstract principles or laws—what he termed the *first* or *cardinal principles*—that could explain all of these realms.[16] Needless to say, he was a bit over-ambitious on this score, but the general idea was to explain social processes with abstract laws or principles. The content of these principles borrows from the physics of his time, not the biology. Yet it is his biological analogizing for which we most remember Spencer; and, when making these analogies, he in-

---

[12]Émile Durkheim, *The Rules of the Sociological Method* (New York: Free Press, 1938; originally published in 1895).

[13]Organicism was not original with Comte, of course. Plato and other more distant thinkers had also made organismic analogies; Comte simply reintroduced this idea.

[14]See my *Herbert Spencer: Toward a Renewed Appreciation* (Beverly Hills, CA: Sage, 1985) for a more detailed presentation of this line of argument.

[15]Herbert Spencer, *The Principles of Sociology* (1874-1896). This work has been reissued in varying volume numbers. References in this chapter are to the three-volume edition (the third edition) issued by D. Appleton and Company, New York, in 1898. In reading this long work, it is much more critical to note the parts (numbered I through VII) than the volumes, since pagination can vary with editions.

[16]These are contained in his *First Principles* (New York: A. C. Burt, 1880; originally published in 1862).

troduced analytical functionalism. But I emphasize again that Spencer saw both the organic and superorganic realms as obeying the same abstract laws, or first principles. Indeed, his organicism is secondary to his effort at making deductions from his abstract first principles.[17]

Spencer published his monumental *Principles of Biology* before his first sociological works.[18] As a consequence, I think, he wanted to demonstrate that both organic and superorganic bodies reveal "parallels in principles of organization" that could be deduced from the first principles. And so it is not surprising that he compared societies and organisms in terms of their similarity and dissimilarity. Among the points of similarity he emphasized:[19]

1. As organic and superorganic bodies increase in size, they increase in structure. That is, they become more complex and differentiated.
2. Such differentiation of structures is accompanied by differentiation of functions. Each differentiated structure comes to serve distinctive functions for sustaining the "life" of the systemic whole.
3. Differentiated structures and functions require in both organic and superorganic bodies integration through mutual dependence. Each structure can be sustained only through its dependence upon others for vital substances.
4. Each differentiated structure in both organic and superorganic bodies is, to a degree, a systemic whole by itself (i.e., organs are composed of cells and societies of groupings of individuals); thus the larger whole is always influenced by the systemic processes of its constituent parts.
5. The structures of organic and superorganic bodies can "live on" for a while after the destruction of the systemic whole.

These points of similarity between organism and society, Spencer argued, must be qualified for their points of "extreme unlikeness":[20]

1. There are great differences in the degree of connectedness of the parts, or structures, in organic and social wholes. In superorganic wholes, there is less direct and continuous physical contact and more dispersion of parts than in organic bodies.
2. There are differences in the modes of contact between organic and superorganic systems. In the superorganic there is much more reliance upon symbols than in the organic.[21]

---

[17]For a more detailed analysis of these, see Jonathan H. Turner, Leonard Beeghley, and Charles Powers, *The Emergence of Sociological Theory* (Belmont, CA: Wadsworth, 1989).

[18]Herbert Spencer, *The Principles of Biology* (New York: D. Appleton, 1864–1867).

[19]See my *Herbert Spencer: Toward a Renewed Appreciation*, Chapter 4. See also his *Principles of Sociology*, vol. I, part II, pp 449–57. The bulk of Spencer's organicism is on these few pages of a work that spans more than 2000 pages, and yet this is what we most remember about Spencer.

[20]*Principles of Sociology*, part II, pp. 451–62.

[21]Spencer's theory of symbolism is commonly ignored by contemporary sociologists who simply accept Durkheim's critique. Part I of *Principles* contains a very sophisticated analysis of symbols.

3. There are differences in the levels of consciousness and voluntarism of parts in organic and superorganic bodies. All units in society are conscious, goal seeking, and reflective, whereas only one unit can potentially be so in organic bodies.

As Spencer continued to analogize the points of similarity between organicism and societies, he began to develop what I call *requisite functionalism.* That is, organic and superorganic bodies reveal certain universal requisites that must be fulfilled in order for them to adapt to an environment. Moreover, these same requisites exist for all organic and superorganic systems. Let me quote Spencer on this point:

> Close study of the facts shows us another striking parallelism. Organs in animals and organs in societies have internal arrangements framed on the same principle.
>
> Differing from one another as the viscera of a living creature do in many respects, they have several traits in common. Each viscus contains appliances for conveying nutriment to its parts, for bringing it materials on which to operate, for carrying away the product, for draining off waste matters; as also for regulating its activity.[22]

It is not hard to see the seeds of an argument for universal functional requisites in this passage. Indeed, on the next page from this quote, Spencer argued that "it is the same for society" and proceeded to list the basic functional requisites of societies. For example, each superorganic body

> has a set of agencies which bring the raw material . . . ; it has an apparatus of major and minor channels through which the necessities of life are drafted out of the general stocks circulating through the kingdom . . . ; it has appliances . . . for bringing those impulses by which the industry of the place is excited or checked; it has local controlling powers, political and ecclesiastical, by which order is maintained and healthful action furthered.

Even though these universal requisites are not so clearly separated as they were to become in modern functional approaches, the logic of the analysis is clear. First, there are certain universal needs or requisites that structures function to meet. These revolve around (*a*) securing and circulating resources, (*b*) producing usable substances, and (*c*) regulating and integrating internal activities through power and symbols. Second, each system level—group, community, region, or whole society—reveals a similar set of needs. Third, the important dynamics of any empirical system revolve around processes that function to meet these universal requisites. Fourth, the level of adaptation of a social unit to its environment is determined by the extent to which it meets these functional requisites.

Thus, by recognizing that certain basic or universal needs must be met, analysis of organic and superorganic systems is simplified. One examines processes with respect to needs for integrating differentiated parts, needs for sus-

---

[22]*Principles of Sociology,* part II, p. 477.

taining the parts of the system, needs for producing and distributing information and substances, and needs for political regulation and control. In simple systems these needs are met by each element of the system; but, when structures begin to grow and to become more complex, they are met by distinctive types of structures that specialize in meeting one of these general classes of functions. And as societies become highly complex, structures become even more specialized and meet only specific subclasses of these general functional needs.

The logic behind this form of requisite functionalism guided much of Spencer's substantive analysis. And it is the essence of functional analysis today. The list of basic needs to be met varies among theorists, but the mode of the analysis remains the same: examine specific types of social processes and structures in terms of the needs or requisites that they meet.

FUNCTIONALISM AND ÉMILE DURKHEIM

We should not be surprised that, as the inheritor of a long French tradition of social thought, especially Comte's organicism, Émile Durkheim's (1858–1917) early works were heavily infused with organismic terminology. Although his major work, *The Division of Labor in Society*, was sharply critical of Herbert Spencer, many of Durkheim's formulations were clearly influenced by the 19th-century intellectual preoccupation with biology.[23] Aside from the extensive use of biologically inspired terms, Durkheim's basic assumptions reflected those of the organicists: (1) Society was to be viewed as an entity in itself that could be distinguished from and was not reducible to its constituent parts. In conceiving of society as a reality, *sui generis*, Durkheim in effect gave analytical priority to the social whole. (2) Although such an emphasis by itself did not necessarily reflect organismic inclinations, Durkheim, in giving causal priority to the whole, viewed system parts as fulfilling basic functions, needs, or requisites of that whole. (3) The frequent use of the notion "functional needs" is buttressed by Durkheim's conceptualization of social systems in terms of "normal" and "pathological" states. Such formulations, at the very least, connote the view that social systems have needs that must be fulfilled if "abnormal" states are to be avoided. (4) In viewing systems as normal and pathological, as well as in terms of functions, there is the additional implication that systems have equilibrium points around which normal functioning occurs.

Durkheim recognized all of these dangers and explicitly tried to deal with several of them. First, he was clearly aware of the dangers of teleological analysis—of implying that some future consequence of an event causes that very

---

[23]Émile Durkheim, *The Division of Labor in Society* (New York: Macmillan, 1933; originally published in 1893). Durkheim tended to ignore the fact that Spencer wore several intellectual hats. He reacted to Spencer's advocacy of utilitarianism, seemingly ignoring the similarity between Spencer's organismic analogy and his own organic formulations as well as the close correspondence between their theories of symbols. For more details on this line of argument, see: Jonathan H. Turner, "Émile Durkheim's Theory of Social Organization," *Social Forces*, 68 (3, 1990) pp. 1–15, and "Spencer's and Durkheim's Principles of Social Organization," *Sociological Perspectives* 27 (January 1984), pp. 21–32.

event to occur. Thus he warned that the causes of a phenomenon must be distinguished from the ends it serves:

> When, then, the explanation of a social phenomenon is undertaken, we must seek separately the efficient cause which produces it and the function it fulfills. We use the word "function" in preference to "end" or "purpose," precisely because social phenomena do not generally exist for the useful results they produce.[24]

Thus, despite giving analytical priority to the whole and viewing parts as having consequences for certain normal states and hence meeting system requisites, Durkheim remained aware of the dangers of asserting that all systems have "purpose" and that the need to maintain the whole causes the existence of its constituent parts. Yet Durkheim's insistence that the function of a part for the social whole always be examined sometimes led him, and certainly many of his followers, into questionable teleological reasoning. For example, even when distinguishing *cause* and *function* in his major methodological statement, he leaves room for an illegitimate teleological interpretation: "Consequently, to explain a social fact it is not enough to show the cause on which it depends; we must also, at least in most cases, show its function in the establishment of social order."[25] In this summary phrase I think that the words "in the establishment of" could connote that the existence of system parts can be explained only by the whole, or social order, that they function to maintain. From this view it is only a short step to outright teleology: the social fact in question is caused by the needs of the social order that the fact fulfills. Such theoretical statements do not necessarily have to be illegitimate, for it is conceivable that a social system could be programmed to meet certain needs or designated ends and thereby have the capacity to cause variations in cultural items or "social facts" in order to meet these needs or ends. But if such a system is being described by an analyst, it is necessary to document how the system is programmed and how it operates to cause variations in social facts to meet needs or ends. As the above quotation illustrates, Durkheim did not have this kind of system in mind when he formulated his particular brand of functional analysis; thus he did not wish to state his arguments teleologically.

Despite his warnings to the contrary, Durkheim appears to have taken this short step into teleological reasoning in his substantive works. In his first major work on the division of labor, Durkheim went to great lengths to distinguish between cause (increased population and moral density) and function (integration of society).

However, the causal statements often become fused with functional statements. The argument is, generally, like this: population density increases moral density (rates of contact and interaction); moral density leads to competition, which threatens the social order; in turn, competition for resources results in the specialization of tasks; and specialization creates pressures for mutual in-

---

[24]Émile Durkheim, *The Rules of the Sociological Method,* p. 96.
[25]Ibid., p. 97.

terdependence and increased willingness to accept the morality of mutual obligation. This transition to a new social order is not made consciously, or by "unconscious wisdom"; yet the division of labor is necessary to restore the order that "unbridled competition might otherwise destroy."[26] Hence the impression is left that the threat or the need for social order causes the division of labor. Such reasoning can be construed as an illegitimate teleology, since the consequence or result of the division of labor—social order—is the implied cause of it. At the very least, then, cause and function are not kept as analytically separate as Durkheim so often insisted.

In sum, then, despite Durkheim's warnings about illegitimate teleology, he often appears to waver on the brink of the very traps he wished to avoid. I suspect that the reason for this failing can probably be traced to the organismic assumptions built into this form of sociological analysis. In taking a strong sociologistic position on the question of emergent properties—that is, on the irreducibility of the whole to its individual parts—Durkheim separated sociology from the utilitarianism as well as the naive psychology and anthropology of his day.[27] However, in supplementing this emphasis on the social whole with organismic assumptions of function, requisite, need, and normality/pathology, Durkheim helped weld organismic principles to sociological theory for nearly three-quarters of a century. The brilliance of his analysis of substantive topics, as well as the suggestive features of his analytical work, made a functional mode of analysis highly appealing to subsequent generations of sociologists and anthropologists.

FUNCTIONALISM AND THE ANTHROPOLOGICAL TRADITION

Functionalism might have died with Durkheim except for the fact that anthropologists began to find it an appealing way to analyze simple societies. Indeed, functionalism as a well-articulated conceptual perspective was perpetuated in the first half of the 20th century by the writings of two anthropologists, Bronislaw Malinowski and A. R. Radcliffe-Brown.[28] Each of these thinkers was heavily influenced by the organicism of Durkheim, as well as by their own field studies among primitive societies. Despite the similarities in their intellectual

---

[26]Ibid., p. 35. For a more detailed analysis, see Jonathan H. Turner and Alexandra Maryanski, *Functionalism* (Menlo Park, CA: Benjamin/Cummings, 1979). See also Percy S. Cohen, *Modern Social Theory* (New York: Basic Books, 1968), pp. 35–37.

[27]Robert A. Nisbet, *Émile Durkheim* (Englewood Cliffs, NJ: Prentice-Hall, 1965), pp. 9–102.

[28]For basic references on Malinowski's functionalism, see his "Anthropology," *Encyclopedia Britannica*, supplementary vol. 1 (London and New York, 1936); *A Scientific Theory of Culture* (Chapel Hill: University of North Carolina Press, 1944); and *Magic, Science, and Religion and Other Essays* (Glencoe, IL: Free Press, 1948). For basic references on A. R. Radcliffe-Brown's functionalism, see his "Structure and Function in Primitive Society," *American Anthropologist* 37 (July–September 1935), pp. 58–72; *Structure and Function in Primitive Society* (Glencoe, IL: Free Press, 1952); and *The Andaman Islanders* (Glencoe, IL: Free Press, 1948). See also Turner and Maryanski, *Functionalism*.

backgrounds, however, the conceptual perspectives developed by Malinowski and Radcliffe-Brown reveal a considerable number of dissimilarities.

The Functionalism of A. R. Radcliffe-Brown

Recognizing that "the concept of function applied to human societies is based on an analogy between social life and organic life" and that "the first systematic formulation of the concept as applying to the strictly scientific study of society was performed by Durkheim," Radcliffe-Brown (1881-1955) tried to indicate how some of the problems of organismic analogizing might be overcome.[29] For him, the most serious problem with functionalism was the tendency for analysis to appear teleological. Noting that Durkheim's definition of function pertained to the way in which a part fulfills system needs, Radcliffe-Brown emphasized that, in order to avoid the teleological implications of such analysis, it would be necessary to "substitute for the term 'needs' the term 'necessary condition of existence.' " In doing so, he felt that no universal human or societal needs would be postulated; rather, the question of which conditions were necessary for survival would be an empirical one, an issue that would have to be discovered for each given social system. Furthermore, in recognizing the diversity of conditions necessary for the survival of different systems, analysis would avoid asserting that every item of a culture must have a function and that items in different cultures must have the same function.

Once the dangers of illegitimate teleology were recognized, functional or (to use his term) structural analysis could legitimately proceed from several assumptions: (1) One necessary condition for survival of a society is minimal integration of its parts. (2) The term *function* refers to those processes that maintain this necessary integration or solidarity. (3) Thus, in each society, structural features can be shown to contribute to the maintenance of necessary solidarity. In such an analytical approach, social structure and the conditions necessary for its survival are irreducible. In a vein similar to that of Durkheim, Radcliffe-Brown saw society as a reality in and of itself. For this reason he was usually led to visualize cultural items, such as kinship rules and religious rituals, as explicable in terms of social structure—particularly its need for solidarity and integration. For example, in analyzing a lineage system, Radcliffe-Brown would first assume that some minimal degree of solidarity must exist in the system. Processes associated with lineage systems would then be assessed in terms of their consequences for maintaining this solidarity. The conclusion was that lineage systems provided a systematic way of adjudicating conflict in societies where families owned land, because such a system specified who had the right to land and through which side of the family it would always pass. The integration of the economic system—landed estates owned by families—is thus explained.[30]

---

[29]Radcliffe-Brown, "Structure and Function in Primitive Society," p. 68. This statement is, of course, incorrect, since the organismic analogy was far more developed in Spencer's work.

[30]Radcliffe-Brown, *Structure and Function in Primitive Society,* pp. 31-50. For a secondary analysis of this example, see Arthur L. Stinchcombe, "Specious Generality and Functional Theory," *American Sociological Review* 26 (December 1961), pp. 929-30.

I believe that this form of analysis poses a number of problems that continue to haunt functional theorists. Although Radcliffe-Brown admitted that "functional unity [integration] of a social system is, of course, a hypothesis," he failed to specify the analytical criteria for assessing just how much or how little functional unity is necessary for system survival, to say nothing of specifying the operations necessary for testing this hypothesis. As subsequent commentators were to discover, without some analytical criteria for determining what is and what is not minimal functional integration and societal survival, the hypothesis cannot be tested, even in principle. Thus, what is typically done is to assume that the existing system is minimally integrated and surviving because it exists and persists. Without carefully documenting how various cultural items promote instances of both integration and malintegration of the social whole, such a strategy can reduce the hypothesis of functional unity to a tautology: if one can find a system to study, then it must be minimally integrated; therefore, lineages that are a part of this system must promote its integration. To discover the contrary would be difficult, since the system, by virtue of being a system, is already composed of integrated parts, such as a lineage system. I see a non sequitur in such reasoning, since it is quite possible to view a cultural item like a lineage system as having both integrative and malintegrative (and other) consequences for the social whole. In his actual ethnographic descriptions, Radcliffe-Brown often slips inadvertently into a pattern of circular reasoning: the fact of a system's existence requires that its existing parts, such as a lineage system, be viewed as contributing to the system's existence. Assuming integration and then assessing the contribution of individual parts to the integrated whole lead to an additional analytical problem. Such a mode of analysis implies that the causes of a particular structure—for example, lineages—lie in the system's needs for integration, which is, I think, most likely an illegitimate teleology.

Radcliffe-Brown would, of course, have denied my conclusions. His awareness of the dangers of illegitimate teleology would have seemingly eliminated the implication that the needs of a system cause the emergence of its parts. And his repeated assertions that the notion of function "does not require the dogmatic assertion that everything in the life of every community has a function" should have led to a rejection of tautological reasoning.[31] However, much like Durkheim, what Radcliffe-Brown asserted analytically was frequently not practiced in the concrete empirical analysis of societies. Such lapses were not intended but appeared to be difficult to avoid with functional needs, functional integration, and equilibrium as operating assumptions.[32]

---

[31]See, for example, Radcliffe-Brown, *Structure and Function in Primitive Society.*

[32]A perceptive critic of an earlier edition of this manuscript provided an interesting way to visualize the problems of tautology:

When do you have a surviving social system?
When certain survival requisites are met.
How do you know when certain survival requisites are met?
When you have a surviving social system.

Thus, although Radcliffe-Brown displayed an admirable awareness of the dangers of organicism—especially of the problem of illegitimate teleology and the hypothetical nature of notions of solidarity—he all too often slipped into a pattern of questionable teleological reasoning. Forgetting that integration was only a working hypothesis, he opened his analysis to problems of tautology. Such problems were persistent in Durkheim's analysis; and, despite his attempts to the contrary, their specter haunted even Radcliffe-Brown's insightful essays and ethnographies.

The Functionalism of Bronislaw Malinowski

I think that functionalism would have ended with Radcliffe-Brown because it had very little to offer sociologists attempting to study complex societies. Both Durkheim and Radcliffe-Brown posited one basic societal need—integration—and then analyzed system parts in terms of how they meet this need. For sociologists who are concerned with differentiated societies, this is likely to become a rather mechanical task. Moreover, it does not allow for analysis of those aspects of a system part that are not involved in meeting the need for integration.

It was Bronislaw Malinowski's (1884–1942) functionalism that was to remove these restrictions; by reintroducing Spencer's approach, it offered a way for modern sociologists to employ functional analysis.[33] Malinowski's scheme reintroduced two important ideas from Spencer: (1) the notion of system levels and (2) the concept of different and multiple system needs at each level. In making these two additions, Malinowski made functional analysis more appealing to 20th-century sociological theorists.

In Malinowski's scheme there are three system levels: the biological, the social structural, and the symbolic.[34] At each of these levels one can discern basic needs or survival requisites that must be met if biological health, social-structural integrity, and cultural unity are to exist. Moreover, these system levels constitute a hierarchy, with biological systems at the bottom, social-structural arrangements next, and symbolic systems at the highest level. Malinowski stressed that the way in which needs are met at one system level sets constraints on how they are met at the next level in the hierarchy. Yet he did not advocate a reductionism of any sort; indeed, he thought that each system level reveals its own distinctive requisites and processes meeting these needs. Additionally, he argued that the important system levels for sociological or anthropological analysis are the structural and symbolic. And in this actual discussion it is the social-structural level that receives the most attention. Table 2–1 lists the requisites or needs of the two most sociologically relevant system levels.

---

[33]Don Martindale, *The Nature and Types of Sociological Theory* (Boston: Houghton Mifflin, 1960), p. 459.

[34]Bronislaw Malinowski, *A Scientific Theory of Culture and Other Essays* (London: Oxford University Press, 1964), pp. 71–125; see also Turner and Maryanski, *Functionalism*, pp. 44–57.

TABLE 2–1 Requisites of System Levels

Cultural (Symbolic) System Level

1. Requisites for systems of symbols that provide information necessary to adjust to the environment.
2. Requisites for systems of symbols that provide a sense of control over people's destiny and over chance events.
3. Requisites for systems of symbols that provide members of a society with a sense of a "communal rhythm" in their daily lives and activities.

Structural (Instrumental) System Level

1. The requisite for production and distribution of consumer goods.
2. The requisite for social control of behavior and its regulation.
3. The requisite for education of people in traditions and skills.
4. The requisite for organization and execution of authority relations.

In analyzing the structural system level, Malinowski stressed that institutional analysis is necessary. For Malinowski, institutions are the general and relatively stable ways in which activities are organized to meet critical requisites. All institutions, he felt, have certain universal properties or "elements" that can be listed and then used as dimensions for comparing different institutions. These universal elements are:

1. *Personnel:* Who and how many people will participate in the institution?
2. *Charter:* What is the purpose of the institution? What are its avowed goals?
3. *Norms:* What are the key norms that regulate and organize conduct?
4. *Material apparatus:* What is the nature of the tools and facilities used to organize and regulate conduct in pursuit of goals?
5. *Activity:* How are tasks and activities divided? Who does what?
6. *Function:* What requisite does a pattern of institutional activity meet?

By describing each institution along these six dimensions, Malinowski believed that he had provided a common analytical yardstick for comparing patterns of social organization within and between societies. He even went so far as to construct a list of universal institutions as they resolve not just structural but also biological and symbolic requisites.

In sum, Malinowski's functional approach opened new possibilities for sociologists who had long forgotten Spencer's similar arguments. Malinowski suggested to sociologists that attention to system levels is critical in analyzing requisites; he argued that there are universal requisites for each system level; he forcefully emphasized that the structural level is the essence of sociological analysis; and, much like Spencer before him and Talcott Parsons a decade later (see next chapter), he posited four universal functional needs at this level—economic adaptation, political authority, educational socialization, and social control—that were to be prominent in subsequent functional schemes. Moreover, he provided a clear method for analyzing institutions as they operate to meet functional requisites. It is fair to say, therefore, that Malinowski drew the rough contours for modern sociological functionalism.

FUNCTIONALISM AND THE GHOST OF MAX WEBER

I would be remiss if I did not address the impact of Max Weber (1864–1920) on functional theory. During the latter 19th century and into the early part of this century, Weber developed a particular approach for sociological analysis. His approach in a wide range of substantive areas—economic sociology, stratification, complex organizations, sociology of religion, authority and social change, for example—still guides modern research and theory in these areas. In the development of general theoretical orientations, however, Weber's influence has been less direct. Although his impact on some perspectives is clear, I think that his influence on functionalism is less evident. And yet, because several contemporary functionalists were so important in initially exposing American scholars to Weber's thought, I consider it unlikely that these functionalists' theorizing was not influenced by the power of his approach.

What, then, has been Weber's impact on the emergence of functionalism? Generally, I see two aspects of Weber's work as having had an important influence: (1) his substantive vision of "social action" and (2) his strategy for analyzing social structures. Weber argued that sociology must understand social phenomena on two levels, at the "level of meaning" of the actors themselves and at the level of collective action among groupings of actors. Weber's substantive view of the world and his strategy for analyzing its features were thus influenced by these dual concerns. In many ways, Weber viewed two realities— that of the subjective meanings of actions and that of the emergent regularities of social institutions.[35] Much functionalism similarly addresses this dualism: how do the subjective states of actors influence emergent patterns of social organization, and vice versa?

As I will discuss shortly, Talcott Parsons, in particular, labeled his functionalism *action theory,* and his early theoretical scheme was devoted to analyzing the basic components and processes of the subjective processes of individual actors. But, much like Weber, Parsons and other functionalists were to move to a more macroscopic concern with emergent patterns of collective action.

This shift from the micro to the macro represents only part of the Weberian analytical strategy. One of the most enduring analytical legacies of Weber is his strategy for constructing "ideal types." For Weber, an ideal type represented a category system for "analytically accentuating" the important features of social phenomena. Ideal types are abstractions from empirical reality, and their purpose is to highlight certain common features among similar processes and

---

[35]For basic references on Weber, see his *The Theory of Social and Economic Organization* (New York: Free Press, 1947); "Social Action and Its Types" in *Theories of Society,* ed. Talcott Parsons et al. (New York: Free Press, 1961); Hans Gerth and C. Wright Mills, eds., *From Max Weber: Essays in Sociology* (New York: Oxford University Press, 1958).

structures. Moreover, they can be used to compare and contrast empirical events in different contexts by providing a common analytical yardstick. By noting the respective deviations of two or more concrete, empirical situations from the ideal type, it is possible to compare these two situations and thus better understand them. Thus, for virtually all phenomena studied by Weber—religion, organizations, power, and the like—he constructed an ideal type in order to visualize its structure and functioning. In many ways, I think that the ideal-type strategy corresponds to taxonomic procedures for categorizing species and for describing somatic structures and processes in the biological sciences. I see it as encouraging a concern with conceptual schemes and categories rather than propositions and laws. And so, although Weber's work is devoid of the extensive organismic imagery of Durkheim's or Spencer's, the concern with categorization of different social structures is, I believe, highly compatible with the organismic reasoning of early functionalism. Thus it is not surprising that contemporary functionalists borrowed the substantive vision of the world implied by the concepts of structure and function as well as Weber's use of the taxonomic approach for studying structures and processes.

For functionalists in general and Talcott Parsons in particular, the construction of conceptual schemes remains an important activity. Functionalists elaborately categorize the social world in order to emphasize the importance of some structures and processes for maintaining the social system. For example, much like Weber before him, Parsons first developed a category system for individual social action and then elaborated this initial system of categories into an incredibly complex, analytical edifice of concepts. What is important to recognize is that this strategy of developing first category systems and then propositions about the relationships among categorized phenomena lies at the heart of contemporary functionalism. This emphasis on category systems is, no doubt, one of the subtle ways that Weber's ideal-type strategy continues to influence functional theorizing in sociology.

THE EMERGENCE OF FUNCTIONALISM: AN OVERVIEW

With its roots in the organicism of the early 19th century, functionalism is the oldest and, until recent decades, the dominant conceptual perspective in sociology. The organicism of Comte and later that of Spencer and Durkheim clearly influenced the first functional anthropologists—Malinowski and Radcliffe-Brown—who in turn, with Durkheim's timeless analysis, helped shape the more modern functional perspectives. Coupled with Weber's emphasis on social taxonomies, or ideal types, of both subjective meaning and social structure, a strategy for studying the properties of the "social organism" similarly began to shape contemporary functionalism.

In emphasizing the contribution of sociocultural items to the maintenance of a more inclusive systemic whole, early functional theorists often conceptualized social needs or requisites. The most extensive formulations of this position were those of Malinowski, in which institutional arrangements meet one of various levels of needs or requisites: biological, structural, and symbolic.

For Émile Durkheim and A. R. Radcliffe-Brown, it was important to analyze separately the causes and functions of a sociocultural item, since the causes of an item could be unrelated to its function in the systemic whole. In their analyses of actual phenomena, however, both Durkheim and Radcliffe-Brown lapsed into assertions that the need for integration caused a particular event—for example, the emergence of a particular type of lineage system or the division of labor.

I see their tendency to blur the distinction between cause and function as generating two related problems in the analyses of Durkheim and Radcliffe-Brown: (1) tautology and (2) illegitimate teleology. To say that a structural item (such as the division of labor) emerges because of the need for social integration is a teleological assertion, for an end state (social integration) is presumed to cause the event (the division of labor) that brings about this very end state. Such a statement is not necessarily illegitimate, since, indeed, the social world is rife with systemic wholes that initiate and regulate the very structures and processes maintaining them. However, to assert that the need for integration is the cause of the division of labor is probably an illegitimate teleology: to make the teleology legitimate would require some documentation of the causal chain of events through which needs for integration operate to produce a division of labor. Without such documentation the statement is vague and theoretically vacuous. Assumptions about and taxonomies of system needs and requisites also create problems of tautology. For unless clear-cut and independent criteria can be established to determine when a system requisite is fulfilled or not fulfilled, theoretical statements become circular: a surviving system is meeting its survival needs; the system under study is surviving; a sociocultural item is a part of this system; therefore, it is likely that this item is meeting the system's needs. Such statements are true by definition, since no independent criteria exist for assessing when a requisite is met and whether a given item meets these criteria. To stretch Durkheim's analysis for purposes of illustration, without clear criteria for determining what constitutes integration and what levels of it denote a surviving system, the statement that the division of labor meets an existing system's needs for integration must be true by definition, since the system exists and is therefore surviving and the division of labor is its most conspicuous integrative structure.

In looking back on the theoretical efforts of early functionalists, then, I would see the legacy of their work as follows:

1. The social world was viewed in systemic terms. For the most part such systems were considered to have needs and requisites that had to be met to assure survival.
2. Despite their concern with evolution, thinkers tended to view systems with needs and requisites as having normal and pathological states—thus connoting system equilibrium and homeostasis.
3. When viewed as a system, the social world was seen as composed of mutually interrelated parts; the analysis of these interrelated parts focused on how they fulfilled requisites of systemic wholes and, hence, maintained system normality or equilibrium.

4. By typically viewing interrelated parts in relation to the maintenance of a systemic whole, causal analysis frequently became vague, lapsing into tautologies and illegitimate teleologies.

Much of contemporary functionalism has attempted to incorporate the suggestiveness of early functional analysis—especially the conception of *system* as composed of interrelated parts.[36] At the same time, current forms of functional theorizing have tried to cope with the analytical problem of teleology and tautology, which Durkheim and Radcliffe-Brown so unsuccessfully tried to avoid. In borrowing the 19th-century organicism and in exploiting conceptually the utility of viewing system parts as having implications for the operation of systemic wholes, modern functionalism provided early sociological theorizing with a unified conceptual perspective.

Moreover, in developing a concern with conceptual schemes as opposed to systems of propositions, functional theory often appeared to order the complexities of social structures and processes. Yet the adequacy of this perspective has increasingly been called into question in recent decades. For, as I will emphasize throughout this book, this questioning has often led to excessively polemical and counterproductive debates in sociology. On the positive side, however, the controversy over functional theorizing has also stimulated attempts to expand upon old conceptual perspectives and to develop new perspectives as alternatives to what are perceived to be the inadequacies of functionalism. For the present, I will concentrate in the following chapters on a detailed overview of contemporary functional theorizing.

---

[36]For a more thorough analysis of the historical legacy of functionalism, see Don Martindale's *The Nature and Types of Sociological Theory* and his "Limits of and Alternatives to Functionalism in Sociology," in *Functionalism in the Social Sciences,* American Academy of Political and Social Science Monograph, no. 5 (Philadelphia, 1965), pp. 144–62; see also in this monograph, Ivan Whitaker, "The Nature and Value of Functionalism in Sociology," pp. 127–43.

CHAPTER 3

Analytical Functionalism

Talcott Parsons

Talcott Parsons was probably the most dominant theorist of his time. It is unlikely, I suspect, that any one theoretical approach will so dominate sociological theory again. For in the years between 1950 and the late 1970s, Parsonian functionalism was clearly the focal point around which theoretical controversy raged. Even those who despised Parsons' functional approach could not ignore it; and even now, years after his death and well over two decades since its period of dominance, Parsonian functionalism is still the subject of controversy.[1] To appreciate Parsons' achievement, I think it best to start at the beginning, in 1937, when he published his first major work, *The Structure of Social Action*.[2] Then I will trace the continuity of the scheme as it evolved over the next four decades.[3]

---

[1]Although few appear to agree with all aspects of Parsonian theory, rarely has anyone quarreled with the assertion that he has been the dominant sociological figure of this century. For documentation of Parsons' influence, see Robert W. Friedrichs, *A Sociology of Sociology* (New York: Free Press, 1970); and Alvin W. Gouldner, *The Coming Crisis of Western Sociology* (New York: Basic Books, 1970).

[2]Talcott Parsons, *The Structure of Social Action* (New York: McGraw-Hill, 1937); the most recent paperback edition (New York: Free Press, 1968) will be used in subsequent footnotes.

[3]It has been emphasized again and again that such continuity does not exist in Parsons' work. For the most often quoted source of this position, see Joseph F. Scott, "The Changing Foundations of the Parsonian Action Scheme," *American Sociological Review* 28 (October 1969), pp. 716–35. This position is held to be incorrect in the analysis to follow. In addition to the present discussion, see also Jonathan H. Turner and Leonard Beeghley, "Current Folklore in the Criticisms of Parsonian Action Theory," *Sociological Inquiry* 44 (Winter 1974). See also Parsons' reply and comments on this article, ibid. For more recent comments on this issue, see Dean Robert Gerstein, "A Note on the Continuity of Parsonian Action Theory," *Sociological Inquiry* 46 (Winter 1976); Richard Münch, "Talcott Parsons and the Theory of Action I: The Structure of the Kantian Lore" and "Talcott Parsons and the Theory of Action II: The Continuity of the Development," both in *American Journal of Sociology* 86 (December 1981) and 87 (February 1982), pp. 709–39 and 771–826, respectively.

like a Weber's
ideal types

THE STRUCTURE OF SOCIAL ACTION

In *The Structure of Social Action,* Parsons advocated an "analytical realism" in building sociological theory. Theory in sociology must utilize a limited number of important concepts that "adequately 'grasp' aspects of the external world. . . . These concepts do not correspond to concrete phenomena, but to elements in them which are analytically separable from other elements."[4] Thus, first of all, theory must involve the development of concepts that abstract from empirical reality, in all its diversity and confusion, common analytical elements. In this way, concepts will isolate phenomena from their embeddedness in the complex relations that go to make up social reality.

use of concepts

The unique feature of Parsons' analytical realism is the insistence on how these abstract concepts are to be employed in sociological analysis. Parsons did not advocate the immediate incorporation of these concepts into theoretical statements but rather their use to develop a "generalized system of concepts." This use of abstract concepts would involve their ordering into a coherent whole that would reflect the important features of the "real world." What is sought is an organization of concepts into analytical systems that grasp the salient and systemic features of the universe without being overwhelmed by empirical details. This emphasis upon systems of categories represents Parsons' application of Max Weber's ideal-type strategy for analytically accentuating salient features of the world. Thus, much like in Weber's work, Parsons believed that theory should initially resemble an elaborate classification and categorization of social phenomena that reflects significant features in the organization of these social phenomena. In the terms of my discussion in Chapter 1, Parsons sought to develop a naturalistic/positivistic conceptual scheme. For in Parsons' view the empirical world does reveal fundamental properties that can be isolated and studied by a classificatory conceptual scheme.

concepts
before system of theory building

My sense is that Parsons had more than classification in mind. However, he was advocating the priority of developing systems of concepts over systems of abstract propositions. Concepts in theory should not be incorporated into propositions prematurely. They must first be ordered into analytical systems that are isomorphic with the systemic coherence of reality; then, if one is so inclined, operational definitions can be devised and the concepts can be incorporated into true theoretical statements.

Thus, only after systemic coherence among abstract concepts has been achieved is it fruitful to begin the job of constructing true theory. Parsons' subsequent theoretical and substantive work makes sense, I feel, only after this classificatory strategy is comprehended. For indeed, throughout his intellectual career—from *The Structure of Social Action* to his death—Parsons adhered to this strategy for building sociological theory.[5]

---

[4]Parsons, *Structure of Social Action,* p. 730.

[5]See ibid., especially pp. 3–43, 727–76. For an excellent secondary analysis of Parsons' position and why it does not appeal to the critics, see Enno Schwanenberg, "The Two Problems of Order in Parsons' Theory: An Analysis from Within," *Social Forces* 49 (June 1971), pp. 569–81.

Parsons' strategy for theory building maintains a clear-cut ontological position: the social universe displays systemic features that must be captured by a parallel ordering of abstract concepts. Curiously, the substantive implications of this strategy for viewing the world as composed of systems were recessive in *The Structure of Social Action*. Much more conspicuous were assumptions about the "voluntaristic" nature of the social world.

The "voluntaristic theory of action" represented for Parsons a synthesis of the useful assumptions and concepts of utilitarianism, positivism, and idealism.[6] In reviewing the thought of classical economists, Parsons noted the excessiveness of their utilitarianism: unregulated and atomistic actors in a free and competitive marketplace rationally attempting to choose those behaviors that will maximize their profits in their transactions with others. Such a formulation of the social order presented for Parsons a number of critical problems: Do humans always behave rationally? Are they indeed free and unregulated? How is order possible in an unregulated and competitive system? Yet Parsons saw as fruitful several features of utilitarian thought, especially the concern with actors as seeking goals and the emphasis on the choice-making capacities of human beings who weigh alternative lines of action. Stated in this minimal form, Parsons felt that the utilitarian heritage could indeed continue to inform sociological theorizing. In a similar critical stance, Parsons rejected the extreme formulations of radical positivists, who tended to view the social world in terms of observable cause-and-effect relationships among physical phenomena. In so doing, he felt, they ignored the complex symbolic functionings of the human mind. Furthermore, Parsons saw the emphasis on observable cause-and-effect relationships as too easily encouraging a sequence of infinite reductionism: groups were reduced to the causal relationships of their individual members; individuals were reducible to the cause-and-effect relationships of their physiological processes; these were reducible to physicochemical relationships, and so on, down to the most basic cause-and-effect connections among particles of physical matter. Nevertheless, despite these extremes, radical positivism draws attention to the physical parameters of social life and to the deterministic impact of these parameters on much—but of course not all—social organization. Finally, in assessing idealism, Parsons saw as useful the conceptions of "ideas" as circumscribing both individual and social processes, although all too frequently these ideas are seen as detached from the ongoing social life they were supposed to regulate.

I cannot possibly communicate the depth of scholarship in Parsons' analysis of these traditions. More important than the details of his analysis is the weaving of selected concepts from each of these traditions into a voluntaristic

---

[6]For a more recent analysis of Parsons' work in relation to the issues he raised in *The Structure of Social Action*, see Leon Mayhew, "In Defense of Modernity: Talcott Parsons and the Utilitarian Tradition," *American Journal of Sociology* 89 (May 1984), pp. 1273–1306; and Jeffrey C. Alexander, "Formal and Substantive Voluntarism in the Work of Talcott Parsons: A Theoretical Reinterpretation," *American Sociological Review* 43 (Winter 1978), pp. 177–98.

theory of action.[7] For it is at this starting point that, in accordance with his theory-building strategy, Parsons began to construct a functional theory of social organization. In this initial formulation he conceptualizes voluntarism as the subjective decision-making processes of individual actors, but he views such decisions as the partial outcome of certain kinds of constraints, both normative and situational. Voluntaristic action therefore involves these basic elements: (1) Actors, at this point in Parsons' thinking, are individual persons. (2) Actors are viewed as goal seeking. (3) Actors are also in possession of alternative means to achieve the goals. (4) Actors are confronted with a variety of situational conditions, such as their own biological makeup and heredity as well as various external ecological constraints, that influence the selection of goals and means. (5) Actors are seen to be governed by values, norms, and other ideas in that these ideas influence what is considered a goal and what means are selected to achieve it. (6) Action involves actors making subjective decisions about the means to achieve goals, all of which are constrained by ideas and situational conditions.

Figure 3-1 represents this conceptualization of voluntarism. The processes diagramed are often termed the *unit act,* with social action involving a succession of such unit acts by one or more actors. I believe that Parsons chose to focus on such basic units of action for at least two reasons. First, he felt it necessary to synthesize the historical legacy of social thought about the most basic social process and to dissect it into its most elementary components. Second, given his position on what theory should be, the first analytical task in the development of sociological theory is to isolate conceptually the systemic features of the most basic unit from which more complex processes and structures are built.

Once these basic tasks were completed, I think Parsons began to ask: how are unit acts connected to each other, and how can this connectedness be conceptually represented? Indeed, near the end of *The Structure of Social Action* he recognized that "any atomistic system that deals only with properties identifiable in the unit act . . . will of necessity fail to treat these latter elements adequately and be indeterminate as applied to complex systems."[8] However, only the barest hints of what was to come were evident in those closing pages.

---

[7]At one time there was considerable debate and acrimony over "de-Parsonizing" Weber and Durkheim. The presumption is that Parsons gave a distorted portrayal of these and other figures; therefore it is necessary to reexamine their works in an effort to remove Parsons' interpretation from them. This plea ignores two facts: (1) Parsons never maintained that he was summarizing works; instead, he was using works and selectively borrowing concepts to build a theory of action. (2) All sociologists can read these classic thinkers for themselves and derive their own interpretations; there is no reason that they should be overly influenced by Parsons. For the relevant articles, see Jere Cohen, Lawrence E. Hazelrigg, and Whitney Pope, "De-Parsonizing Weber: A Critique of Parsons' Interpretation of Weber's Sociology," *American Sociological Review* 40 (April 1975), pp. 229–41; and Whitney Pope, Jere Cohen, and Lawrence Hazelrigg, "On the Divergence of Weber and Durkheim: A Critique of Parsons' Convergence Thesis," *American Sociological Review* 40 (August 1975), pp. 417–27.

[8]Parsons, *Structure of Social Action,* pp. 748–49.

FIGURE 3–1   The Units of Voluntaristic Action

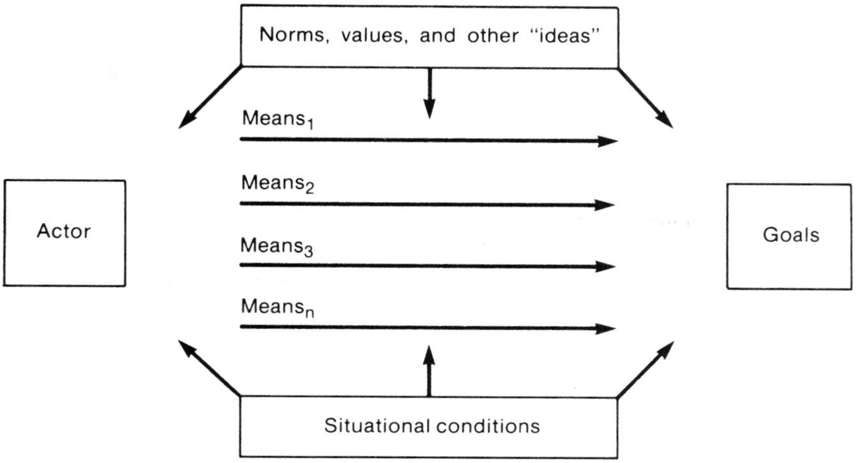

Yet, perhaps only through the wisdom of hindsight, Parsons did offer several clues about the features of these more complex systems. Most notable, near the close of this first work, he emphasized that "the concept of action points again to the *organic* property of action systems" [emphasis added].[9] Buttressed by his strategy for building theory—that is, the development of systems of concepts that mirror reality—it is clear that he intended to develop a conceptual scheme that would capture the systemic essence of social reality.

By 1945, eight years after he published *The Structure of Social Action*, Parsons had become more explicit about the form this analysis should take: "The structure of social systems cannot be derived directly from the actor-situation frame of reference. It requires functional analysis of the complications introduced by the interaction of a plurality of actors."[10] More significantly, this functional analysis should allow notions of needs to enter: "The functional needs of social integration and the conditions necessary for the functioning of a plurality of actors as a 'unit' system sufficiently well integrated to exist as such impose others."[11] Starting from these assumptions, which bear a close resemblance to those of Spencer, Durkheim, Radcliffe-Brown, and Malinowski, Parsons began to develop a complex functional scheme. Let me now attempt to reconstruct this transition from voluntaristic unit acts to a functional scheme emphasizing the systemic properties of action.

[9]Ibid., p. 745.

[10]Talcott Parsons, "The Present Position and Prospect of Systemic Theory in Sociology," *Essays in Sociological Theory* (New York: Free Press, 1949), p. 229.

[11]Ibid.

FIGURE 3–2 Parsons' Conception of Action, Interaction, and Institutionalization

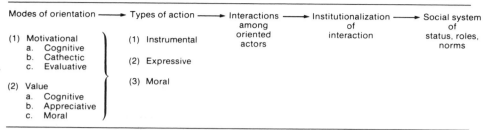

THE SOCIAL SYSTEM

In Figure 3-2 I have summarized the transition from unit acts to social system.[12] This transition occupies the early parts of Parsons' next significant work, *The Social System.*[13] Drawing inspiration from Max Weber's typological approach to this same topic,[14] Parsons views actors as "oriented" to situations in terms of motives (needs and readiness to mobilize energy) and values (conceptions about what is appropriate). There are three types of motives: (1) cognitive (need for information), (2) cathectic (need for emotional attachment), and (3) evaluative (need for assessment). Also, there are three corresponding types of values: (1) cognitive (evaluation in terms of objective standards), (2) appreciative (evaluation in terms of aesthetic standards), and (3) moral (evaluation in terms of absolute rightness and wrongness). Parsons called these *modes of orientation.* Although I often find this discussion vague, the general idea seems to be that the relative salience of these motives and values for any actor creates a composite type of action, which can be one of three types: (1) instrumental (action oriented to realize explicit goals efficiently), (2) expressive (action directed at realizing emotional satisfactions), and (3) moral (action concerned with realizing standards of right and wrong). That is, depending upon which modes of motivational and value orientation are strongest, an actor will act in one of these basic ways. For example, if cognitive motives are strong and cognitive values most salient, their action will be primarily instrumental, although it will also have expressive and moral content. Thus the various combinations and permutations of the modes of orientation—that is, motives and values—produce action geared in one of these general directions.

"Unit acts" therefore involve motivational and value orientations and have a general direction as a consequence of what combination of values and motives prevails for an actor. Thus far Parsons had elaborated only upon his conceptualization of the unit act. Here is the critical next step, which, as I mentioned earlier, was only hinted at in the closing pages of *The Structure of Social Action.*

---

[12]See also my "The Concept of 'Action' in Sociological Analysis," in *Analytical and Sociological Theories of Action,* ed. G. Seeba and Raimo Toumea (Dordrecht, Holland: Reidel, 1985).

[13]Talcott Parsons, *The Social System* (New York: Free Press, 1951).

[14]Max Weber, *Economy and Society,* vol. I (Totowa, NJ: Bedminster Press, 1968), pp. 1–95.

As variously oriented actors (in terms of their configuration of motivational and value orientations) interact, they come to develop agreements and sustain patterns of interaction, which become "institutionalized." Such institutionalized patterns can be, in Parsons' view, conceptualized as a social system. Such a system represents an emergent phenomenon that requires its own conceptual edifice. The normative organization of status-roles becomes Parsons' key to this conceptualization; that is, the subject matter of sociology is the organization of status, roles, and norms. Yet Parsons recognizes that the actors who are incumbent in such status-roles are motivationally and value oriented; thus, as with patterns of interaction, the task now becomes one of conceptualizing these dimensions of action in systemic terms. The result is the conceptualization of action as composed of three "interpenetrating action systems": the cultural, the social, and the personality. That is, the organization of unit acts into social systems requires a parallel conceptualization of motives and values that become, respectively, the personality and cultural systems. The goal of action theory now becomes understanding how institutionalized patterns of interaction (the social system) are circumscribed by complexes of values, beliefs, norms, and other ideas (the cultural system) and by configurations of motives and role-playing skills (the personality system). Later Parsons adds the organismic (subsequently called behavioral) system, but let me not get ahead of the story. For at this stage of conceptualization, analyzing social systems involves developing a system of concepts that, first of all, captures the systemic features of society at all its diverse levels and, second, points to the modes of articulation among personality systems, social systems, and cultural patterns.

To capture conceptually the systemic features of culture, social system, and personality, Parsons wastes little time in introducing notions of functional requisites for each of these basic components of action. Such requisites pertain not only to the internal problems of the action components but also to their articulation with one another. Following both Durkheim's and Radcliffe-Brown's lead, he views integration within and among the action systems as a basic survival requisite. Since the social system is his major topic, Parsons is concerned with the integration within the social system itself and between the social system and the cultural patterns, on the one hand, and between the social system and the personality system, on the other. In order for such integration to occur, at least two functional requisites must be met:

1. A social system must have "a sufficient proportion of its component actors adequately motivated to act in accordance with the requirements of its role system."[15]
2. Social systems must avoid "commitment to cultural patterns which either fail to define a minimum of order or which place impossible demands on people and thereby generate deviance and conflict."[16]

---

[15]Talcott Parsons, *The Social System,* p. 27.
[16]Ibid., pp. 27–28.

Parsons made explicit the incorporation of requisites, which in later works are expanded and made even more prominent. He then attempts to develop a conceptual scheme that reflects the systemic interconnectedness of social systems, although he later returns to the integrative problems posed by the articulation of culture and personality with the social system. Crucial to this conceptualization of the social system is the concept of institutionalization, which refers to relatively stable patterns of interaction among actors in statuses. Such patterns are normatively regulated and infused with cultural patterns. This infusing of values can occur in two ways. First, norms regulating role behaviors can reflect the general values and beliefs of a culture. Second, cultural values and other patterns can become internalized in the personality system and hence affect that system's need structure, which in turn determines an actor's willingness to enact roles in the social system.

Parsons views institutionalization as both a process and a structure. It is significant that he initially discusses the process of institutionalization and only then refers to it as a structure—a fact that is often ignored by critics who contend that action theory is overly structural. Let me emphasize the point again by reference to Figure 3-2. As a process, institutionalization can be portrayed in the following terms: (1) Actors who are variously oriented enter into situations where they must interact. (2) The way actors are oriented is a reflection of their need structure and how this need structure has been altered by the internalization of cultural patterns. (3) Through specific interaction processes—which are not clearly indicated but which by implication include role taking, role bargaining, and exchange—norms emerge while actors adjust their orientations to one another. (4) Such norms emerge as a way of adjusting the orientations of actors to one another, but at the same time they are circumscribed by general cultural patterns. (5) In turn, these norms regulate subsequent interaction, giving it stability. It is through such a process that institutionalized patterns are created, maintained, or altered.

As interactions become institutionalized, a social system can be said to exist, as I have indicated in Figure 3-2. Parsons has typically been concerned with whole societies, but a social system is not necessarily a whole society. Indeed, any organized pattern of interaction, whether a micro or macro form, is termed a social system. When focusing on total societies or large parts of them, Parsons frequently refers to the constituent social systems as subsystems of these larger systemic wholes.

In sum, then, institutionalization is the process through which social structure is built up and maintained. Institutionalized clusters of roles—or, to phrase it differently, stabilized patterns of interaction—constitute a social system. When the given social system is large and is composed of many interrelated institutions, these institutions are typically viewed as subsystems. A total society may be defined as one large system comprising interrelated institutions. At all times, for analytical purposes, it is necessary to remember that a social system is circumscribed by cultural patterns and infused with personality systems.

In his commitment to the development of concepts that reflected the properties of all action systems, Parsons was led to a set of concepts denoting some of the variable properties of these systems. Termed *pattern variables,* they allow for the categorization of the modes of orientation in personality systems, the value patterns of culture, and the normative requirements in social systems. The variables are phrased in terms of polar dichotomies that, depending upon the system under analysis, allow for a rough categorization of decisions by actors, the value orientations of culture, or the normative demands on status roles.

1. *Affectivity/affective neutrality* concerns the amount of emotion or affect that is appropriate in a given interaction situation. Should a great deal of or little affect be expressed?
2. *Diffuseness/specificity* denotes the issue of how far-reaching obligations in an interaction situation are to be. Should the obligations be narrow and specific, or should they be extensive and diffuse?
3. *Universalism/particularism* points to the problem of whether evaluation of others in an interaction situation is to apply to all actors, or should all actors be assessed in terms of the same standards?
4. *Achievement/ascription* deals with the issue of how to assess an actor, whether in terms of performance or on the basis of inborn qualities, such as sex, age, race, and family status. Should an actor treat another on the basis of achievements or ascriptive qualities that are unrelated to performance?
5. *Self-collectivity* denotes the extent to which action is to be oriented to self-interest and individual goals or to group interests and goals. Should actors consider their personal or self-related goals over those of the group or large collectivity in which they are involved?[17]

Some of these concepts, such as self-collectivity, have been dropped from the action scheme, but others, such as universalism-particularism, have assumed greater importance. But I believe that the intent of the pattern variables has remained the same: to categorize dichotomies of decisions, normative demands, and value orientations. However, in *The Social System,* Parsons is inclined to view them as value orientations that circumscribe the norms of the social system and the decisions of the personality system. Thus the structure of the personality and social systems is a reflection of the dominant patterns of value orientations in culture. This implicit emphasis on the impact of cultural patterns on regulating and controlling other systems of action was to become more explicit in his later work, as I will discuss shortly.

For the present, however, it is evident that by 1951 Parsons had already woven a complex conceptual system that emphasizes the process of institu-

---

[17]These pattern variables were developed in collaboration with Edward Shils and were elaborated upon in *Toward a General Theory of Action* (New York: Harper & Row, 1951), pp. 76–98, 203–4, 183–89. Again, Parsons' debt to Max Weber's concern with constructing ideal types can be seen in his presentation of the pattern variables.

FIGURE 3–3 Parsons' Early Conception of Integration among Systems of Action

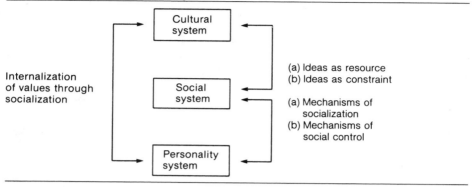

tionalization of interaction into stabilized patterns called social systems, which are penetrated by personality and circumscribed by culture. The profile of institutionalized norms, of decisions by actors in roles, and of cultural value orientations can be typified in terms of concepts—the pattern variables—that capture the variable properties in each of these components of action.

Having built this analytical edifice, Parsons returns to a question, first raised in *The Structure of Social Action*, that guided all his subsequent theoretical formulations: how do social systems survive? More specifically, why do institutionalized patterns of interaction persist? Such questions raise the issue of system imperatives or requisites. For Parsons is asking how systems resolve their integrative problems. The answer to this question is provided by the elaboration of additional concepts that point to the ways that personality systems and culture are integrated into the social system, thereby providing assurance of some degree of normative coherence and a minimal amount of commitment by actors to conform to norms and play roles. In developing concepts of this kind, Parsons begins to stress the equilibrating tendencies of social systems, which I believe was a fatal mistake. For as I will discuss shortly, Parsons left himself open here to severe criticism. But for the moment, let me concentrate on his argument. I have diagrammatically represented his reasoning in Figure 3–3. Now I will fill in the details.

Just how are personality systems integrated into the social system, thereby promoting equilibrium? At the most abstract level, Parsons conceptualizes two mechanisms that integrate the personality into the social system: (1) mechanisms of socialization and (2) mechanisms of social control. It is through the operation of these mechanisms that personality systems become structured so as to be compatible with the structure of social systems. Let me elaborate on each of these mechanisms.

1. In abstract terms, *mechanisms of socialization* are seen by Parsons as the means through which cultural patterns—values, beliefs, language, and other symbols—are internalized into the personality system, thereby

circumscribing its need structure. It is through this process that actors are made willing to deposit motivational energy in roles (thereby willing to conform to norms) and are given the interpersonal and other skills necessary for playing roles. Another function of socialization mechanisms is to provide stable and secure interpersonal ties that alleviate much of the strain, anxiety, and tension associated with acquiring proper motives and skills.

2. *Mechanisms of social control* involve those ways in which status-roles are organized in social systems to reduce strain and deviance. There are numerous specific control mechanisms, including (*a*) institution-alization, which makes role expectations clear and unambiguous while seg-regating in time and space contradictory expectations; (*b*) interpersonal sanctions and gestures, which actors subtly employ to mutually sanction conformity; (*c*) ritual activities, in which actors act out symbolically sources of strain that could prove disruptive while they reinforce dominant cultural patterns; (*d*) safety-valve structures, in which pervasive deviant propens-ities are segregated in time and space from normal institutional patterns; (*e*) reintegration structures, which are specifically charged with bringing back into line deviant tendencies; and, finally, (*f*) the institutionalization into some sectors of a system of the capacity to use force and coercion.

These two mechanisms are thus viewed as resolving one of the most per-sistent integrative problems facing social systems. The other major integrative problem facing social systems concerns how cultural patterns contribute to the maintenance of social order and equilibrium. Again at the most abstract level, Parsons visualizes two ways in which this occurs. (1) Some components of culture, such as language, are basic resources necessary for interaction to occur. Without symbolic resources, communication and hence interaction would not be possible. Thus, by providing common resources for all actors, interaction is made possible by culture. (2) A related but still separable influence of culture on interaction is exerted through the substance of ideas contained in cultural patterns (values, beliefs, ideology, and so forth). These ideas can provide actors with common viewpoints, personal ontologies, or, to borrow from W. I. Thomas, a common "definition of the situation." These common meanings allow inter-action to proceed smoothly with minimal disruption.

Naturally, Parsons acknowledges that the mechanisms of socialization and social control are not always successful, hence allowing deviance and social change to occur. But it is clear that the concepts developed in *The Social System* weight analysis in the direction of looking for processes that maintain the integration and, by implication, the equilibrium of social systems. In Figure 3–4 I have summarized the logic of Parsons' functionalism at this stage in the elaboration of his conceptual scheme.

I see subsequent developments in Parsons' action theory as an attempt to expand upon the basic analytical scheme of *The Social System* that is sum-marized in Figures 3–2, 3–3, and 3–4 while trying to accommodate some of the critics' charges of a static and conservative conceptual bias (see later section).

FIGURE 3–4 Parsons' Mechanism-Equilibrium Functional Analysis

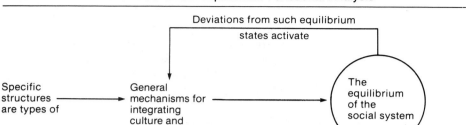

I do not imagine that the many critics of Parsons will ever be silenced, but some interesting elaborations of the scheme occurred in the quarter-century following Parsons' first explicitly functional work. Let me now turn to these.

THE TRANSITION TO FUNCTIONAL IMPERATIVISM

In collaboration with Robert Bales and Edward Shils, Parsons published *Working Papers in the Theory of Action* shortly after *The Social System*. It was in this work that conceptions of functional imperatives came to dominate the general theory of action;[18] and by 1956, with Parsons' and Neil Smelser's publication of *Economy and Society*, the functions of structures for meeting system requisites were well institutionalized into action theory.[19]

During this period, systems of action were conceptualized to have four survival problems, or requisites: adaptation, goal attainment, integration, and latency. *Adaptation* involves the problem of securing from the environment sufficient facilities and then distributing these facilities throughout the system. *Goal attainment* refers to the problem of establishing priorities among system goals and mobilizing system resources for their attainment. *Integration* denotes the problem of coordinating and maintaining viable interrelationships among system units. *Latency* embraces two related problems: pattern maintenance and tension management. Pattern maintenance pertains to the problem of how to ensure that actors in the social system display the appropriate characteristics (motives, needs, role-playing skills, and so forth). Tension management concerns the problem of dealing with the internal tensions and strains of actors in the social system.

---

[18]Talcott Parsons, Robert F. Bales, and Edward A. Shils, *Working Papers in the Theory of Action* (Glencoe, IL: Free Press, 1953).

[19]Talcott Parsons and Neil J. Smelser, *Economy and Society* (New York: Free Press, 1956). These requisites are the same as those enumerated by Malinowski. See previous chapter, Table 2–1.

FIGURE 3–5 Parsons' Functional Imperativism or Requisite Functionalism

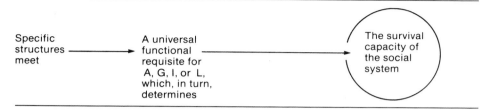

All of these requisites were, I think, implicit in *The Social System*, but they tended to be viewed under the general problem of integration. In Parsons' discussion of integration within and between action systems, problems of securing facilities (adaptation), allocation and goal seeking (goal attainment), and socialization and social control (latency) were conspicuous. Thus, in my opinion, the development of the four functional requisites—abbreviated A, G, I, and L—is not so much a radical departure from earlier works as an elaboration of concepts implicit in *The Social System*.

With the introduction of A, G, I, L, however, I see a subtle shift away from the analysis of structures to the analysis of functions. Structures are now viewed explicitly in terms of their functional consequences for meeting the four requisites. Interrelationships among specific structures are now analyzed in terms of how their interchanges affect the requisites that each must meet. This shift can be seen in Figure 3-5, especially if compared to Figure 3-4.

As Parsons' conceptual scheme becomes increasingly oriented to function, social systems are divided into sectors, each corresponding to a functional requisite—that is, A, G, I, or L. In turn, any subsystem can be divided into these four functional sectors. Then, each of these subsystems can be divided into four functional sectors, and so on. This process of "functional sectorization," if I can invent a word to describe it, is illustrated for the adaptive requisite in Figure 3-6.

Of critical analytical importance in this scheme are the interchanges among systems and subsystems. It is difficult to comprehend the functioning of a designated social system without examining the interchanges among its A, G, I, and L sectors, especially since these interchanges are affected by exchanges among constituent subsystems and other systems in the environment. In turn, the functioning of a designated subsystem cannot be understood without examining internal interchanges among its adaptive, goal attainment, integrative, and latency sectors, especially since these interchanges are influenced by exchanges with other subsystems and the more inclusive system of which it is a subsystem. Thus, at this juncture, as important interchanges among the functional sectors of systems and subsystems are outlined, the Parsonian scheme now begins to resemble an elaborate mapping operation.

THE INFORMATIONAL HIERARCHY OF CONTROL

Toward the end of the 1950s, Parsons turned his attention toward interrelationships *among* (rather than within) what were then four distinct action sys-

FIGURE 3–6 Parsons' Functional Imperativist View of Social Systems

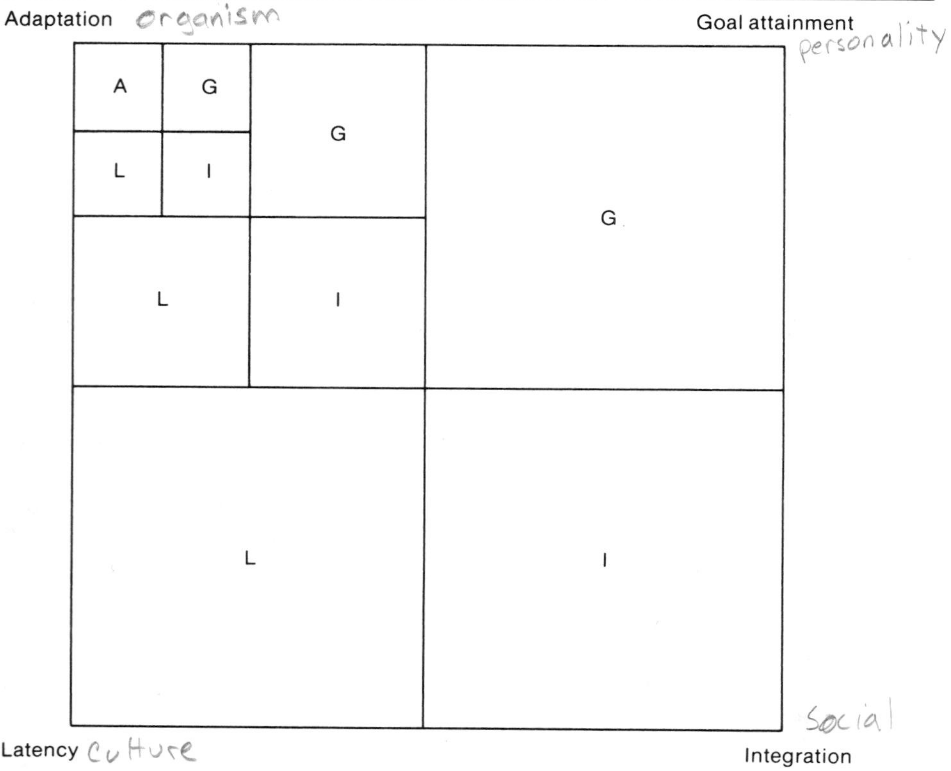

Adaptation *Organism*

Goal attainment *personality*

Latency *Culture*

Integration *Social*

A	G
L	I

G

L | I

G

L | I

L

I

tems: culture, social structure, personality, and organism. In many ways, I think this concern represented an odyssey back to the analysis of the basic components of the unit act outlined in *The Structure of Social Action*. But now each element of the unit act is a full-fledged action system, each confronting four functional problems to resolve: adaptation, goal attainment, integration, and latency. Furthermore, although individual decision making is still a part of action as personalities adjust to the normative demands of status-roles in the social system, the analytical emphasis has shifted to the input/output connections among the four action systems.

It is at this juncture that Parsons begins to visualize an overall action system, with culture, social structure, personality, and organism composing its constituent subsystems.[20] Each of these subsystems is seen as fulfilling one of

[20]Talcott Parsons, "An Approach to Psychological Theory in Terms of the Theory of Action," in *Psychology: A Science*, ed. S. Koch, vol. 3 (New York: McGraw-Hill, 1958), pp. 612–711. By 1961 these ideas were even more clearly formulated; see Talcott Parsons, "An Outline of the Social System," in *Theories of Society*, ed. T. Parsons, E. Shils, K. D. Naegele, and J. R. Pitts (New York: Free Press, 1961), pp. 30–38. See also Jackson Toby, "Parsons' Theory of Social Evolution," *Contemporary Sociology* 1 (September 1972), pp. 395–401.

the four system requisites—A, G, I, L—of the overall action system. The organism is considered to be the subsystem having the most consequences for resolving adaptive problems since it is ultimately through this system that environmental resources are made available to the other action subsystems. As the goal-seeking and decision-making system, personality is considered to have primary consequences for resolving goal-attainment problems. As an organized network of status-norms integrating the patterns of the cultural system and the needs of personality systems, the social system is viewed as the major integrative subsystem of the general action system. As the repository of symbolic content of interaction, the cultural system is considered to have primary consequences for managing tensions of actors and assuring that the proper symbolic resources are available to ensure the maintenance of institutional patterns (latency).

After viewing each action system as a subsystem of a more inclusive, overall one, Parsons begins to explore the interrelations among the four subsystems. What emerges is a hierarchy of informational controls, with culture informationally circumscribing the social system, social structure informationally regulating the personality system, and personality informationally regulating the organismic system. For example, cultural value orientations would be seen as circumscribing or limiting the range of variation in the norms of the social system; in turn, these norms, as translated into expectations for actors playing roles, would be viewed as limiting the kinds of motives and decision-making processes in personality systems; these features of the personality system would then be seen as circumscribing biochemical processes in the organism. Conversely, each system in the hierarchy is also viewed as providing the "energic conditions" necessary for action at the next higher system. That is, the organism provides the energy necessary for the personality system, the personality system provides the energic conditions for the social system, and the organization of personality systems into a social system provides the conditions necessary for a cultural system. Thus the input/output relations among action systems are reciprocal, with systems exchanging information and energy. Systems high in information circumscribe the utilization of energy at the next lower system level, while each lower system provides the conditions and facilities necessary for action in the next higher system. This scheme has been termed a cybernetic hierarchy of control. I have diagrammed it in Figure 3–7.

GENERALIZED MEDIA OF EXCHANGE

Until his death, Parsons maintained his interest in the intra- and intersystemic relationships of the four action systems. Although he was never to develop the concepts fully, he had begun to view these relationships in terms of *generalized symbolic media of exchange*.[21] In any interchange, generalized media are em-

___

[21]Parsons' writings on this topic are incomplete, but see his "On the Concept of Political Power," *Proceedings of the American Philosophical Society* 107 (June 1963), pp. 232–62; Talcott Parsons, "On the Concept of Influence," *Public Opinion Quarterly* 27 (Spring 1963), pp. 37–62; and Talcott

FIGURE 3–7 Parsons' Cybernetic Hierarchy of Control

Overall function	System level	Interrelations among system levels
Latency	Cultural system	Informational controls
Integration	Social system	
Goal attainment	Personality system	
Adaptation	Organismic system	Energic conditions

ployed—for example, money is used in the economy to facilitate the buying and selling of goods. What typifies these generalized media, such as money, is that they are really symbolic modes of communication. The money is not worth much by itself; its value is evident only in terms of what it says symbolically in an exchange relationship.

Thus, what Parsons proposes is that the links among action components are ultimately informational. This means that transactions are mediated by symbols. Parsons' emphasis on information is consistent with the development of the idea of a cybernetic hierarchy of control. Informational exchanges, or cybernetic controls, are seen as operating in at least three ways. First, the interchanges or exchanges *among* the four subsystems of the overall action system are carried out by means of different types of symbolic media; that is, money, power, influence, or commitments. Second, the interchanges *within* any of the four action systems are also carried out by means of distinctive symbolic media. Finally, the system requisites of adaptation (A), goal attainment (G), integration (I), and latency (L) determine the type of generalized symbolic media used in an inter- or intrasystemic exchange.

Within the social system, the adaptive sector utilizes money as the medium of exchange with the other three sectors; the goal-attainment sector employs power—the capacity to induce conformity—as its principal medium of exchange; the integrative sector of a social system relies upon influence—the capacity to persuade; and the latency sector uses commitments—especially the capacity to be loyal. The analysis of interchanges of specific structures within social systems should thus focus on the input/output exchanges utilizing different symbolic media.

Among the subsystems of the overall action system, a similar analysis of the symbolic media used in exchanges should be undertaken, but Parsons never

---

Parsons, "Some Problems of General Theory," in *Theoretical Sociology: Perspectives and Developments,* ed. J. C. McKinney and E. A. Tiryakian (New York: Appleton-Century-Crofts, 1970), pp. 28–68. See also Talcott Parsons and Gerald M. Platt, *The American University* (Cambridge, MA: Harvard University Press, 1975).

clearly described the nature of these media.[22] What he appeared to be approaching was a conceptual scheme for analyzing the basic types of symbolic media, or information, linking systems in the cybernetic hierarchy of control.[23]

PARSONS ON SOCIAL CHANGE

In the last decade of his career, Parsons became increasingly concerned with social change. Built into the cybernetic hierarchy of control is a conceptual scheme for classifying the locus of such social change. What Parsons visualized was that the information/energic interchanges among action systems provide the potential for change within or between the action systems. One source of change may be excesses in either information or energy in the exchange among action systems. In turn, these excesses alter the informational or energic outputs across systems and within any system. For example, excesses of motivation (energy) would have consequences for the enactment of roles and perhaps ultimately for the reorganization of these roles or the normative structure and eventually of cultural value orientations.[24] Another source of change comes from an insufficient supply of either energy or information, again causing external and internal readjustments in the structure of action systems. For example, value (informational) conflict would cause normative conflict (or anomie), which in turn would have consequences for the personality and organismic systems. Thus, inherent in the cybernetic hierarchy of control are concepts that point to the sources of both stasis and change.[25]

To augment this new macro emphasis on change, Parsons utilized the action scheme to analyze social evolution in historical societies. In this context, I think that the first line of *The Structure of Social Action* is of interest: "Who now reads Spencer?" Parsons then answered the question by delineating some of the reasons why Spencer's evolutionary doctrine had been so thoroughly rejected by 1937. Yet, after some 40 years, Parsons chose to reexamine the issue of societal evolution that he had so easily dismissed in the beginning. And in so doing, he reintroduced Spencer's and Durkheim's evolutionary models back into functional theory.

---

[22]For his first attempt at a statement, see Parsons, "Some Problems of General Theory," pp. 61–68.

[23]For a more readable discussion of these generalized media, see T. S. Turner, "Parsons' Concept of Generalized Media of Social Interaction and Its Relevance for Social Anthropology," *Sociological Inquiry* 38 (Spring 1968), pp. 121–34.

[24]There are several bodies of empirical literature that bear on this example. McClelland's work on the achievement motive as initiating economic development in modernizing societies is perhaps the most conspicuous example; see David C. McClelland, *The Achieving Society* (New York: Free Press, 1961).

[25]For a fuller discussion, see Alvin L. Jacobson, "Talcott Parsons: A Theoretical and Empirical Analysis of Social Change and Conflict," in *Institutions and Social Exchange: The Sociologies of Talcott Parsons and George C. Homans,* ed. H. Turk and R. L. Simpson (Indianapolis: Bobbs-Merrill, 1970).

In drawing heavily from Spencer's and Durkheim's insights into societal development,[26] Parsons proposed that the processes of evolution display the following elements:

1. Increasing differentiation of system units into patterns of functional interdependence.
2. Establishment of new principles and mechanisms of integration in differentiating systems.
3. Increasing adaptive capacity of differentiated systems in their environments.

From the perspective of action theory, then, evolution involves: (*a*) increasing differentiation of the personality, social, cultural, and organismic systems from one another; (*b*) increasing differentiation within each of these four action subsystems; (*c*) escalating problems of integration and the emergence of new integrative structures; and (*d*) the upgrading of the survival capacity of each action subsystem, as well as of the overall action system, to its environment.

Parsons then embarked on an ambitious effort in two short volumes to outline the pattern of evolution in historical systems through primitive, intermediate, and modern stages.[27] In contrast with *The Social System*, where he stressed the problem of integration between social systems and personality, Parsons draws attention in his evolutionary model to the inter- and intradifferentiation of the cultural and social systems and to the resulting integrative problems. In fact, each stage of evolution is seen as reflecting a new set of integrative problems between society and culture as each of these systems has become more internally differentiated as well as differentiated from the other. Thus the concern with the issues of integration within and among action systems, so evident in earlier works, was not abandoned but applied to the analysis of specific historical processes.

Even though I think Parsons is vague about the causes of evolutionary change, he sees evolution as guided by the cybernetic hierarchy of controls, especially the informational component. In his documenting of how integrative problems of the differentiating social and cultural systems have been resolved in the evolution of historical systems, the informational hierarchy is regarded as crucial because the regulation of societal processes of differentiation must be accompanied by legitimation from cultural patterns (information). Without such informational control, movement to the next stage of development in an evolutionary sequence will be inhibited.

Thus I see the analysis of social change as an attempt to use the analytical tools of the general theory of action to examine a specific process, the historical

---

[26]See previous chapter for references.

[27]Talcott Parsons, *Societies: Evolutionary and Comparative Perspectives* and *The System of Modern Societies* (Englewood Cliffs, NJ: Prentice-Hall, 1966, 1971, respectively). The general stages of development were first outlined in Talcott Parsons, "Evolutionary Universals in Society," *American Sociological Review* 29 (June 1964), pp. 339–57.

development of human societies. What is of interest in this effort is that Parsons developed many propositions about the sequences of change and the processes that will inhibit or accelerate the unfolding of these evolutionary sequences. It is of more than passing interest that preliminary tests of these propositions indicate that, on the whole, they have a great deal of empirical support.[28] But I must concede that, in many respects, Parsons' discussion does not improve greatly upon Spencer's. In a sense, Parsons simply recapitulates Spencer's analysis.

PARSONS ON "THE HUMAN CONDITION"

Again in a way reminiscent of Spencer's grand theory, Parsons attempted to extend his analytical scheme to all aspects of the universe.[29] In this last conceptual addition, I find it ironic that, as it came to a close, Parsons' work increasingly resembled Spencer's. Except for the opening line in *The Structure of Social Action*—"Who now reads Spencer?"—Parsons ignored him. Indeed, I am not sure that he even realized how closely his analyses of societal evolution and his conceptualization of the "human condition" resembled Spencer's effort of 100 years earlier. At any rate, I see this last effort as more philosophy than sociology. Yet it represents the culmination of Parsons' thought. Parsons began in 1937 with an analysis of the smallest and most elementary social unit, the act. He then developed a requisite functionalism that embraced four action systems: the social, cultural, personality, and what he called the behavioral in later years (he had earlier called this the organismic). Finally, in this desire to understand basic parameters of the human condition, he viewed these four action systems as only one subsystem within the larger system of the universe. This vision is portrayed in Figure 3–8.

As can be seen in Figure 3–8, the universe is divided into four subsystems, each meeting one of the four requisites—that is, A, G, I, or L. The four action systems resolve integrative problems, the organic system handles goal-attainment problems, the physicochemical copes with adaptation problems, and the telic ("ultimate" problems of meaning and cognition) deals with latency problems.

Each of these subsystems employs its own media for intra- and intersubsystem activity. For the action subsystem, the distinctive medium is symbolic meanings; for the telic, it is transcendental ordering; for the organic, it is health; and for the physicochemical, it is empirical ordering (lawlike relations of matter, energy, etc.). There are double interchanges of these media among the four A, G, I, L sectors, with "products" and "factors" being reciprocally exchanged.

---

[28]See Gary L. Buck and Alvin L. Jacobson, "Social Evolution and Structural-Functional Analysis: An Empirical Test," *American Sociological Review* 33 (June 1968), pp. 343–55; A. L. Jacobson, "Talcott Parsons: Theoretical and Empirical Analysis."

[29]Talcott Parsons, *Action Theory and the Human Condition* (New York: Free Press, 1978). See the last chapter and my analysis in "Parsons on the Human Condition," *Contemporary Sociology* 9 (May 1980), pp. 380–83.

FIGURE 3–8   The Subsystems of the Human Condition

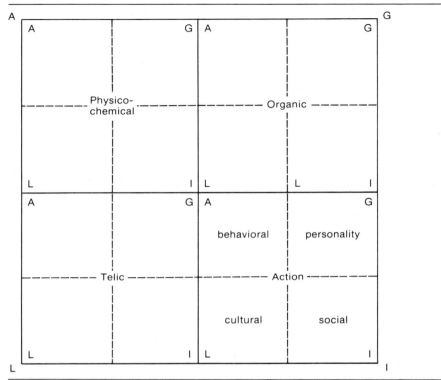

That is, each subsystem of the universe transmits a product to the others, while it also provides a factor necessary for the operation of other subsystems. Let me illustrate with the L (telic) and I (action) interchange. At the product level the telic system provides "definitions of human responsibility" to the action subsystems and receives "sentiments of justification" from the action subsystem. At the factor level the telic provides "categorical imperatives" and receives "acceptance of moral obligations." These double interchanges are, of course, carried out with the distinctive media of the I and A subsystems—that is, transcendental ordering and symbolic meaning, respectively.

The end result of this analysis, is, I feel, a grand metaphysical vision of the universe as it impinges upon human existence. It represents an effort to categorize the universe in terms of systems, subsystems, system requisites, generalized media, and exchanges involving these media. As such it is no longer sociology but philosophy or, at best, a grand meta-theoretical vision. Parsons had indeed come a long way since the humble unit act made its entrance in 1937.

CRITICISMS OF PARSONIAN FUNCTIONALISM

I cannot imagine a social theorist enduring more criticism than Talcott Parsons received. With each step in the scheme's elaboration, criticism mounted, reaching a peak in the late 1960s. And by the early 1970s these critiques had dislodged Parsonian theory from its once-dominant place. Many critiques were unfair, but there was also an element of truth in each. Let me now extract this element with respect to (1) the substantive image of social organization and (2) the logic of his requisite functionalism.

Criticisms of Parsons' Substantive Image of Society

By the early 1960s a number of critics had begun to question whether Parsons' emerging "system of concepts" corresponded to events in the real world. I view such a line of criticism as significant because the Parsonian strategy assumes that it is necessary to elaborate a system of concepts that adequately grasp salient features of the social world. Assertions that the maturing system of concepts inadequately mirrors features of actual social systems represent a fundamental challenge to the naturalistic and positivistic assumptions behind Parsons' conceptual scheme.

Ralf Dahrendorf, who is the subject of Chapter 10, codified this growing body of criticism when he likened functionalism to a utopia.[30] Much like prominent portrayals of social utopias of the past, Dahrendorf asserted, Parsons' concepts point to a world that (a) reveals no developmental history, (b) evidences only consensus over values and norms, (c) displays a high degree of integration among its components, and (d) reveals only mechanisms that preserve the status quo. Such an image of society is utopian because there appears little possibility that ubiquitous phenomena like deviance, conflict, and change could ever occur.

The evidence marshaled by Dahrendorf to support these assertions is, I feel, rather minimal and flimsy. Yet I do not think it difficult to visualize the source of the critics' dismay. With the publication of *The Social System*, the critics charge, Parsons becomes overly concerned with the integration of social systems. In a vein similar to Radcliffe-Brown and Durkheim, Parsons' emphasis on integration involves a disproportionate concern with those processes in social systems that meet this need for integration. In *The Social System*, this concern with integration is evidenced by the tendency to assume, for analytical purposes, a system that is in equilibrium. From this starting point, analysis must then focus on the elaboration of concepts promoting integration and

---

[30]Ralf Dahrendorf, "Out of Utopia: Toward a Reorientation of Sociological Analysis," *American Journal of Sociology* 64 (September 1958), pp. 115–27. This polemic echoed the earlier assessments by others, including David Lockwood, "Some Remarks on 'The Social System,' " *British Journal of Sociology* 7 (June 1950), pp. 134–46; C. Wright Mills, *The Sociological Imagination* (New York: Oxford University Press, 1959), pp. 44–49; and Lewis Coser, *The Functions of Social Conflict* (New York: Free Press, 1956), pp. 1–10.

equilibrium. For example, the extended discussion of institutionalization de-scribes the processes whereby structure is built up, with relatively scant men-tion of concepts denoting the breakdown and change of institutionalized pat-terns. To compound this omission, a discussion of how institutionalized patterns are maintained by the mechanisms of socialization and social control is launched. For the critics, too much emphasis is placed on how socialization assures the internalization of values and the alleviation of strains among actors and how mechanisms of social control reduce the potential for malintegration and deviance. When deviance and change are discussed, the critics contend, they are viewed as residual or, in a way reminiscent of Durkheim, as patho-logical. In fact, deviance, conflict, and change are so alien to the scheme that the social equilibrium is considered to constitute, in Parsons' words, a "first law of social inertia."

The subsequent expansion of concepts denoting four system requisites—adaptation, goal attainment, integration, and latency—has further horrified the critics. For now system processes become almost exclusively viewed in terms of their consequences for meeting an extended list of system needs. In all this concern for the consequences of processes for meeting needs, how is it, the critics ask, that deviance, conflict, and change are to be conceptualized?[31] Are they merely pathological events that occur on those rare occasions when system needs are not met? Or, in reality, are not these phenomena pervasive features of social systems, which are inadequately grasped by the proliferating system of concepts?

The elaboration of the informational hierarchy of control among the overall systems of action and its use to analyze social change did not silence the critics because the only type of change that is conceptualized is evolution, as opposed to revolution and other forms of violent disruption to social systems. Much like Durkheim and Spencer, Parsons' views change as a "progressive" differ-entiation and integration, with the inexorable progress of societal development delayed from time to time by a failure to integrate the differentiating cultural and social systems.[32]

The Logical Criticisms of Requisite Functionalism

The problems of illegitimate teleology and tautology have consumed a consid-erable amount of the literature on functionalism.[33] For the most part, this

---

[31]Leslie Sklair, "The Fate of the Functional Requisites in Parsonian Sociology," *British Journal of Sociology* 21 (March 1970), pp. 30–42.

[32]Perhaps the most critical and scholarly attempt to document the reasons behind these prob-lems in Parsonian action theory is provided by Alvin Gouldner, *The Coming Crisis of Western Sociology* (New York: Basic Books, 1970). However, John K. Rhoads, "On Gouldner's Crisis of Western Sociology," *American Journal of Sociology* 78 (July 1972), pp. 136–54, emphasizes that Gouldner has perceived what he wants to perceive in Parsons' work, ignoring those passages that would connote just the opposite of stasis, control, consensus, and order. See also Rhoads, "Reply to Gouldner," *American Journal of Sociology* 78 (May 1973), pp. 1493-96, which was written in response to Gouldner's defense of his position (Alvin Gouldner, "For Sociology: 'Varieties of Po-litical Expression' Revisited," *American Journal of Sociology* 78 [March 1973], pp. 1063-93, par-ticularly pp. 1083-93).

[33]For analyses of the logic of functionalist inquiry, see R. B. Braithwaite, *Scientific Explanation*

literature holds that, since assumptions of needs and requisites are so prominent in functional theorizing, theoretical statements will too frequently lapse into illegitimate teleologies and tautologies. Typically, conspicuous examples of the functional works of Durkheim, Radcliffe-Brown, and Malinowski are cited to confirm the truth of this assertion, but by implication the efforts of contemporary functionalists are similarly indicted—otherwise, the criticisms would not be worth the considerable efforts devoted to making them. To the extent that this indirect indictment of Parsons' requisite functionalism can be sustained, I see it as a serious criticism. For Parsons' strategy for theory building has revolved around the assumption that his system of concepts can generate testable propositions that account for events in the empirical world. But if such a conceptual system inspires illegitimate teleologies and tautologous propositions, then its utility as a strategy for building sociological theory can be called into question.

The issue of teleology   Parsons always considered action to be goal directed—whether it is a single unit act or the complex informational and energic interchanges among the organismic, personality, social, and cultural systems. Thus, Parsons' conceptualization of goal attainment as a basic system requisite would make inevitable teleological propositions, since for Parsons much social action can be understood only in terms of the ends it is designed to serve. I think that such propositions are often vague, however, because assessing goal-attainment consequences frequently is a way to obscure the specific causal chains whereby goal-attainment sectors in a system activate processes to meet specified end states. Yet a close look at the Parsonian legacy reveals that his many essays are vitally concerned with just how, and through what processes, system processes are activated to meet goal states. For example, Parsons' various works on how political systems strive to legitimate themselves are filled with both analytical and descriptive accounts of how the processes— such as patterns of socialization in educational and kinship institutions—are activated to meet goal-attainment requisites.[34] Although the empirical adequacy

(New York: Harper & Row, 1953), Chaps. 9 and 10; Carl G. Hempel, "The Logic of Functional Analysis," in *Symposium on Sociological Theory*, ed. L. Gross (New York: Harper & Row, 1959), pp. 271–307; Percy S. Cohen, *Modern Social Theory* (New York: Basic Books, 1968), pp. 58–64; Francesca Cancian, "Functional Analysis of Change," *American Sociological Review* 25 (December 1960), pp. 818–27; S. F. Nadel, *Foundations of Social Anthropology* (Glencoe, IL: Free Press, 1951), pp. 373–78; Ernest Nagel, "Teleological Explanation and Teleological Systems," in *Readings in the Philosophy of Science*, ed. H. Feigl and M. Brodbeck (New York: Harper & Row, 1953), pp. 537–58; Phillip Ronald Dore, "Function and Cause," *American Sociological Review* 26 (December 1961), pp. 843–53; Charles J. Erasmus, "Obviating the Functions of Functionalism," *Social Forces* 45 (March 1967), pp. 319–28; Harry C. Bredemeier, "The Methodology of Functionalism," *American Sociological Review* 20 (April 1955), pp. 173–80; Bernard Barber, "Structural-Functional Analysis: Some Problems and Misunderstandings," *American Sociological Review* 21 (April 1956), pp. 129–35; Robert K. Merton, *Social Theory and Social Structure* (Glencoe, IL: Free Press, 1957), pp. 44–61; Arthur L. Stinchcombe, *Constructing Social Theories* (New York: Harcourt, Brace & World, 1968), pp. 80–116; Hans Zetterberg, *On Theory and Verification in Sociology* (Totowa, NJ: Bedminster Press, 1965), pp. 74–79.

[34]For example, see Talcott Parsons, "Authority, Legitimation and Political Action" in *Authority*,

of this discussion can be questioned, Parsons' analysis does not present illegitimate teleologies, for his work reveals a clear concern for documenting the causal chains involved in activating processes designed to meet various end states.

I think that it is the other three requisites—adaptation, integration, and latency—that would seemingly pose a more serious problem of illegitimate teleology. Critics would argue that to analyze structures and processes in terms of their functions for these three system needs compels analysts to state their propositions teleologically, although the processes so described may not be goal directed or teleological. Logically, as several commentators have pointed out, teleological phrasing of propositions in the absence of clear-cut goal-attainment processes does not necessarily make the proposition illegitimate, for at least two reasons.[35]

1. Nagel has argued that phrasing statements in a teleological fashion is merely a shorthand way of stating the same causal relationship nonteleologically.[36] For example, to argue that the relief of anxiety (an end state) is the "latency function" of religion (a present phenomenon) can be rephrased nonteleologically without loss of asserted content: under conditions $C_1, C_2, C_3, \ldots, C_n$, religion (concept x) causes reduction of group anxiety (concept y). Such a form is quite acceptable in that it involves existence and relational statements: under $C_1, C_2, \ldots, C_n$, variations in x cause variations in y. However, other authors have contended that such transposition is not always possible because the existence statements so necessary to such conversion are absent from the statements of functionalists such as Parsons. Without necessary existence statements, the assertion that the function of religion is to reduce group anxiety really means: "The latency needs of the group for low levels of anxiety cause the emergence of religion." I see such statements as illegitimate teleologies since little information is provided about the nature of the "latency purposes" of a given system and the specific causal chains involved in keeping the system in pursuit of its latency goals. Or, if teleology is not intended, then I think that the state-

---

ed. C. J. Friedrich (Cambridge, MA: Harvard University Press, 1958); Parsons, "On the Concept of Power"; and Talcott Parsons, "The Political Aspect of Structure and Process," in *Varieties of Political Theory*, ed. David Easton (Englewood Cliffs, NJ: Prentice-Hall, 1966).

[35]For basic references on the issue, consult G. Bergman, "Purpose, Function and Scientific Explanation," *Acta Sociologica* 5 (1962), pp. 225–28; J. Canfield, "Teleological Explanation in Biology," *The British Journal for the Philosophy of Science* 14 (1964), pp. 285–95; K. Deutsch, "Mechanism, Teleology and Mind," *Philosophy and Phenomenological Research* 12 (1951), pp. 185–223; C. J. Ducasse, "Explanation, Mechanism, and Teleology," in *Readings in Philosophical Analysis*, ed. H. Feigl and W. Sellars (New York: D. Appleton, 1949); D. Emmet, *Function, Purpose and Powers* (London: Routledge-Kegan, 1958); L. S. Fever, "Causality in the Social Sciences," *Journal of Philosophy* 51 (1954), pp. 191–208; W. W. Isajiw, *Causation and Functionalism in Sociology* (New York: Shocken Books, 1968); A. Kaplan, "Noncausal Explanation," in *Cause and Effect*, ed. D. Lerner (New York: Free Press, 1965); I. Scheffler, "Thoughts on Teleology," *The British Journal for the Philosophy of Science* 9 (1958), pp. 265–84; P. Sztompka, "Teleological Language in Sociology," *The Polish Sociological Bulletin* (1969), pp. 56–69, and *System and Function: Toward a Theory of Society* (New York: Academic Press, 1974).

[36]Nagel, "Teleological Explanation and Teleological Systems."

ment is simply vague, offering none of the necessary information that would allow its conversion to a nonteleological form. As Nagel was led to conclude: "To pronounce at once upon the ultimate functions subserved by social fact is to short-circuit explanation and reduce it to generalities which, so prematurely stated, have little significance."[37]

2. Perhaps the most significant defense of Parsons' tendency to phrase propositions teleologically comes from the fact that such propositions point to reverse causal chains that are typical of many social phenomena.[38] By emphasizing that the function served by a structure in maintaining the needs of the whole could cause the emergence of that structure, Parsons' functional imperativism forces analysis to be attuned to those causal processes involved in the initial selection, from the infinite variety of possible social structures, of only certain types of structures. The persistence over time of these selected structures can also be explained by the needs and/or equilibrium states of the whole: those structures having consequences for meeting needs and/or maintaining an equilibrium have a "selective advantage" over those that do not. I do not think that such statements are illegitimate teleologies, for it is quite possible for the systemic whole to exist prior in time to the structures that emerge and persist to maintain that whole. Furthermore, I do not think it necessary to impute purpose to the systemic whole. Just as in the biophysical world, where ecological and population balances are maintained by nonpurposive selective processes (for example, predators increase until they eat themselves out of food and then decrease until the food supply regenerates itself), so social wholes can maintain themselves in a state of equilibrium or meet the imperatives necessary for survival.

This line of argument has led Stinchcombe to summarize:

> Functional explanations are thus complex forms of causal theories. They involve causal connections among . . . variables as with a special causal priority of the consequences of activity in total explanation. There has been a good deal of philosophical confusion about such explanations, mainly due to the theorist's lack of imagination in realizing the variety of reverse causal processes which can select behavior or structures according to their consequences.[39]

I see the above considerations leading to several tentative conclusions about Parsons' scheme and the issue of teleology: (1) The scheme has always been teleological, from the initial conceptualization of unit acts to the four-function paradigm embracing the concept of goal attainment. (2) Contrary to the opinion of his detractors, most of Parsons' theoretical statements can be converted into nonteleological form, such that relevant statements about the conditions under which x varies with y can be discerned. (3) Parsons' work is replete with reverse

---

[37]Nadel, *Foundations of Social Anthropology*, p. 375.
[38]Stinchcombe, *Constructing Social Theories*, pp. 87–93.
[39]Ibid., p. 100.

causal chains in which a systemic whole existing prior in time to the emergence of subsystems causes the perpetuation of a subsystem because of its selective advantages in meeting problems faced by the systemic whole.

Most of the criticisms outlining the dangers of illegitimate teleology in functional theorizing have drawn examples from early functional anthropology, where it is relatively easy to expose questionable teleologies. But I do not see how Parsons' notion of system requisites has led him to this same trap, thus allowing the defenders of the action-theoretic strategy to challenge the critics to find conspicuous instances in Parsons' work where there is an illegitimate teleology.

The issue of tautology   Parsons' conceptualization of four system requisites—adaptation, goal attainment, integration, and latency—is based on the assumption that, if these requisites are not met, the system's survival is threatened. When employing this assumption, however, it is necessary to know what level of failure in meeting each of these requisites is necessary to pronounce a crisis of survival. How does one determine when adaptive needs are not being met? Goal attainment needs? Integrative requisites? And latency needs? Unless there is some way to determine what constitutes the survival and nonsurvival of a system, I believe that propositions documenting the contribution of items for meeting survival requisites become tautologous: the items meet survival needs of the system because it exists and, therefore, must be surviving. Thus, to phrase propositions with regard to system requisites of adaptation, goal attainment, integration, and latency, Parsons must provide either of two types of information: (1) evidence of a nonsurviving system where a particular item did not exist; or (2) specific criteria as to what constitutes survival and nonsurvival in various types and classes of social systems. Without this kind of information, my sense is that propositions employing notions of requisites are likely to be untestable, even in principle. Therefore I would not consider them very useful in building sociological theory. And if Parsons' scheme cannot be readily converted into nontautologous propositions that can be tested, his conceptual approach can be questioned. Parsons argued that systems of concepts must precede systems of propositions, but, if the system of concepts is likely to generate tautologous propositions, then what is its theoretical utility?

The theoretical utility of survival imperatives   Considering the problems of tautology created by using the concept of requisites, I see it as reasonable to ask: what do requisites add to Parsons' theoretical scheme and to the analysis of specific events? For it seems to me that it is possible to document the conditions under which events influence one another in systemic wholes without dragging in notions of survival requisites. In fact, my sense is that Parsons often appears to abandon reference to system requisites when discussing concrete empirical events, causing me to wonder why the requisites are retained in his more formal conceptual edifice.

My answer is that Parsons retained the requisites for strategic reasons: to provide crude and rough criteria for distinguishing important from unimportant social processes. Parsons' entire intellectual career was spent elaborating the complex systems of interrelationships among the basic unit acts he first described in *The Structure of Social Action*. The more the system of concepts was brought to bear on increasingly complex patterns or organization among unit acts, the more Parsons relied upon the requisites to sort out what processes in these complex patterns will help explain the most variance. Thus, Parsons' imperatives constitute not so much a metaphysical entity as a methodological yardstick for distinguishing what is crucial from not crucial among the vast number of potential processes in social systems. Despite the fact that Parsons was unable to specify exact criteria for assessing whether adaptive, goal-attainment, integrative, and latency needs are being met, he used these somewhat vaguely conceptualized requisites to assess the theoretical significance of concrete social phenomena. I am sure that Parsons implicitly employed the requisites this way in his many essays, which even the critics must admit are insightful. And perhaps this fact alone can justify use of the requisites to assess social phenomena.[40]

Furthermore, I can see that the requisites are particularly useful in studying complex empirical systems, because in empirical systems it may be possible to specify more precisely criteria necessary for their survival. With these empirically based requisites as criteria, it may then be possible to distinguish significant from less significant social processes in these systems, thereby assuring more insightful explanations. It appears, then, that, despite some of the logical problems created by their retention, Parsons felt that the strategic value of the requisites in explaining social processes more than compensates for the logical difficulties so frequently stressed by the critics. Other functionalists agree, but I think that we should maintain some skepticism here until functional analyses of empirical events prove superior to less problematic alternatives.

TALCOTT PARSONS: AN OVERVIEW

As it has unfolded over the last decades, the theory of action reveals an enormous amount of continuity—starting with the basic unit act and proliferating into the cybernetic hierarchy of control among the systems of action. Such continuity is the outgrowth of Parsons' particular view of how theory in sociology should be constructed, for he consistently advocated the priority of systems of concepts over systems of propositions. The latter can be useful only when the former task is sufficiently completed. This view, I emphasize, is shared by many—functionalists and nonfunctionalists alike.

Yet the substantive vision of the world connoted by Parsons' concepts and the logical problems imputed to the scheme have stimulated widespread criticism of his functional perspective. In fact, other forms of sociological theorizing

---

[40]For more on my views on this matter, as well as a comment on "neofunctionalism" (e.g., Jeffrey C. Alexander, ed., *Neofunctionalism*, Newbury Park, CA: Sage, 1985), see Jonathan H. Turner and Alexandra Maryanski, "Is Neofunctionalism Functional?" *Sociological Theory* 6 (1, Spring 1988), pp. 110–22.

cannot be understood unless the revulsion to the perspective is appreciated. As I will emphasize in subsequent chapters, many other theoretical perspectives in sociology typically begin with a rejection of Parsonian functionalism and then proceed to build what is considered a more desirable alternative. In fact, Parsons appears to have become the "straw man" of sociological theorizing. No theory in sociology is considered adequate unless it has performed at least some portions of a ritual rejection of Parsons' analytical functionalism.[41]

---

[41]For an interesting review of sociological theory, as a reaction to Parsonian action theory, see Jeffrey C. Alexander, *Twenty Lectures: Sociological Theory Since World War II* (New York: Columbia University Press, 1987).

CHAPTER 4

◆

Empirical Functionalism

Robert K. Merton

THEORIES OF THE MIDDLE RANGE

Just as Talcott Parsons was beginning to embrace a form of requisite functionalism,[1] Robert K. Merton launched a critique of Parsons' functional strategy for building sociological theory.[2] At the heart of this criticism was Merton's contention that Parsons' concern for developing an all-encompassing system of concepts would prove both futile and sterile. To him, Parsons' search for "a total system of sociological theory, in which observations about every aspect of social behavior, organization, and change promptly find their preordained place, has the same exhilarating challenge and the same small promise as those many all-encompassing philosophical systems which have fallen into deserved disuse."[3]

For Merton, such grand theoretical schemes are premature since the theoretical and empirical groundwork necessary for their completion has not been

---

[1]As will be recalled, Parsons in 1945 began to conceptualize unit acts in systemic terms and began to visualize such systems in terms of requisites. See Talcott Parsons, "The Present Position and Prospects of Systemic Theory in Sociology," in *Essays in Sociological Theory* (Glencoe, IL: Free Press, 1949). Moreover, Merton may have introduced Parsons to functional analysis, since Merton conducted a seminar at Harvard, where Parsons was a young instructor, on functional theorists such as Malinowski. In fact, I suspect that Merton was a functional theorist several years before Parsons made the conversion. Yet from the beginning Merton was a critic of the Parsonian approach, although Merton rarely mentions Parsons by name.

[2]Robert K. Merton, "Discussion of Parsons' 'The Position of Sociological Theory,' " *American Sociological Review* 13 (April 1948), pp. 164–68.

[3]Ibid. Most of Merton's significant essays on functionalism have been included, and frequently expanded upon, in Robert K. Merton, *Social Theory and Social Structure* (Glencoe, IL: Free Press, 1949). Quotation taken from page 45 of the 1968 edition of this classic work. Most subsequent references will be made to the articles incorporated into this book. For more recent essays on Merton's work, see Lewis A. Coser, ed., *The Idea of Social Structure* (New York: Free Press, 1975). For excellent overviews of Merton's sociology, see: Piotr Sztompka, *Robert K. Merton: An Intellectual Profile* (New York: St. Martin's, 1986); and Charles Crothers, *Robert K. Merton* (London: Tavistock, 1981).

performed. Just as Einsteinian theory did not emerge without a long cumulative research foundation and theoretical legacy, so sociological theory will have to wait for its Einstein, primarily because "it has not yet found its Kepler—to say nothing of its Newton, Laplace, Gibbs, Maxwell or Planck."[4]

In the absence of this foundation, what passes for sociological theory, in Merton's critical eye, consists of "general orientations toward data, suggesting types of variables which theorists must somehow take into account, rather than clearly formulated, verifiable statements of relationships between specified variables."[5] Strategies advocated by those such as Parsons are not really theory but philosophical systems, with "their varied suggestiveness, their architectonic splendor, and their sterility."[6] However, to pursue the opposite strategy of constructing inventories of low-level empirical propositions will prove equally sterile, thus suggesting to Merton the need for "theories of the middle range" in sociology. As I emphasized in Chapter 1 (see Figure 1-1), such theories are not highly abstract or broad. And I am not as optimistic as Merton that they will yield much theoretical payoff. Yet, despite my reservations, Merton's advocacy has developed a wide following, and so I will try to indicate what made Merton's middle-range strategy so appealing.

In Merton's view, theories of the middle range offer more theoretical promise than Parsons' grand theory. They are couched at a lower level of abstraction and reveal clearly defined and operationalized concepts that are incorporated into statements of covariance for a limited range of phenomena. Although middle-range theories are abstract, they are also connected to the empirical world, thus encouraging the research so necessary for the clarification of concepts and reformulation of theoretical generalizations. Without this interplay between theory and research, Merton contended, theoretical schemes will remain suggestive congeries of concepts, which are incapable of being refuted, while, on the other hand, empirical research will remain unsystematic, disjointed, and of little utility in expanding a body of sociological knowledge. Thus, by following a middle-range strategy, the concepts and propositions of sociological theory will become more tightly organized as theoretically focused empirical research forces clarification, elaboration, and reformulation of the concepts and propositions of each middle-range theory.

From this growing clarity in theories directed at a limited range of phenomena and supported by empirical research can eventually come the more encompassing theoretical schemes. In fact, for Merton, although it is necessary to concentrate energies on the construction of limited theories that inspire research, theorists must also be concerned with "consolidating the special theories into a more general set of concepts and mutually consistent propositions."[7] The special theories of sociology must therefore be formulated with an eye

---

[4]Merton, *Social Theory and Social Structure*, p. 47.

[5]Ibid., p. 42.

[6]Ibid., p. 51.

[7]Merton, *Social Theory and Social Structure* (1957), p. 10.

toward what they can offer more general sociological theorizing. However, just how these middle-range theories should be formulated to facilitate their eventual consolidation into a more general theory poses a difficult analytical problem, for which Merton has a ready solution: a form of functionalism should be utilized in formulating the theories of the middle range. Such functional theorizing is to take the form of a paradigm that would allow for both easy specification and elaboration of relevant concepts, while encouraging systematic revision and reformulation as empirical findings would dictate. Conceived in this way, functionalism became for Merton a method for building not only theories of the middle range but also the grand theoretical schemes that would someday subsume such theories of the middle range. Thus, in a vein similar to Parsons, functionalism for Merton represents a strategy for ordering concepts and for sorting out significant from insignificant social processes. But, unlike Parsons' strategy, Merton's functional strategy requires first the formulation of a body of middle-range theories. Only when this groundwork has been laid should a functional protocol be used to construct more abstract theoretical systems.

Merton's functionalism has never become as dominant as Parsons'. Yet the middle-range strategy is currently the legitimating credo for much theoretical activity. Indeed, as I have argued many times, Merton's middle-range strategy encouraged the proliferation of what I have called "theories of ____" (fill in the blank with any empirical topic).[8] That is, empirical generalizations have been defended as theory, despite the fact that they are not sufficiently abstract. Thus, sociology has spawned a large number of theories about such specific empirical processes as juvenile delinquency, family conflict, race relations, social mobility in America, urbanization, and other empirical events. I am not sure if Merton intended theories of the middle range to be so empirically grounded, but this was the outcome of his advocacy. Virtually none of these theories has been consolidated into more general and abstract propositions because they are too empirical and, therefore, too tied to specific times, places, and contexts. Thus, whereas Merton's functionalism did not catch on, his advocacy for middle-range theory did. I have my doubts as to whether this was for the good of sociological theory, but most sociologists do not share my reservations. Indeed, they see the grounding of theory in research as a great virtue, whereas for me these "theories of" are little more than statements of empirical regularity that need a more abstract set of propositions to explain them. And so the debate over the best strategy for developing theory rages. Since my concern here, however, is with Merton's functionalism, let me now turn to his paradigm for functional analysis.

---

[8]See for examples my *Societal Stratification: A Theoretical Analysis* (New York: Columbia University Press, 1984), Chap. 1; "Returning to Social Physics," *Current Perspectives in Social Theory,* vol. 2 (1981), pp. 187–208; "Toward a Social Physics," *Humboldt Journal of Social Relations* 7 (Spring 1980), pp. 140–55; "Sociology as a Theory Building Enterprise," *Pacific Sociological Review* 22 (October 1979), pp. 427–56; "In Defense of Positivism," *Sociological Theory* 3 (2, 1985).

MERTON'S PARADIGM FOR FUNCTIONAL ANALYSIS

As with most commentators on functional analysis, Merton begins his discussion with a review of the mistakes of early functionalists, particularly the anthropologists Malinowski and Radcliffe-Brown.[9] Generally, Merton saw functional theorizing as potentially embracing three questionable postulates: (1) the functional unity of social systems, (2) the functional universality of social items, and (3) the indispensability of functional items for social systems.[10]

The Functional Unity Postulate

As we can recall from Chapter 2, Radcliffe-Brown, in following Durkheim's lead, frequently transformed the hypothesis that social systems reveal social integration into a necessary requisite or need for social survival. Although it is difficult to argue that human societies do not possess some degree of integration—otherwise they would not be systems—Merton views the degree of integration in a system as an issue to be empirically determined. To assume, however subtly, that a high degree of functional unity must exist in a social system is to define away the important theoretical and empirical questions: What levels of integration exist for different systems? What various types of integration can be discerned? Are varying degrees of integration evident for different segments of a system? And, most importantly, what variety of processes leads to different levels, forms, and types of integration for different spheres of social systems? For Merton, to begin analysis with the postulate of "functional unity" or integration of the social whole can divert attention away not only from these questions but also from the varied and "disparate consequences of a given social or cultural item (usage, belief, behavior pattern, institutions) for diverse social groups and for individual members of these groups."[11]

I am sure that underlying this discussion of the functional unity is an implicit criticism of Parsons' early concern with social integration. As I stressed in Chapter 3, Parsons first postulated only one requisite in his early functional work: the need for integration.[12] Later this postulate was to be expanded into three additional functional requisites for adaptation, goal attainment, and latency. But Parsons' functionalism appears to have begun with the same concerns evident in Durkheim's and Radcliffe-Brown's work, leading Merton to question the "heuristic value" of an assumption that can divert attention away from important theoretical and empirical questions. Thus, instead of the postulate of functional unity, there should be an emphasis on varying types, forms, levels, and spheres of social integration and the varying consequences of the

---

[9]Robert K. Merton, "Manifest and Latent Functions," in *Social Theory and Social Structure* (1968), pp. 74–91.

[10]See Robert K. Merton, *Social Theory and Social Structure,* pp. 45–61.

[11]Merton, *Social Theory and Social Structure* (1968), pp. 81–82.

[12]See Parsons, "Present Position and Prospects of Systemic Theory" and Talcott Parsons, *The Social System* (Glencoe, IL: Free Press, 1951).

existence of items for specified segments of social systems. In this way, Merton begins to direct functional analysis away from concern with total systems and toward an emphasis on how different patterns of social organization within more inclusive social systems are created, maintained, and changed, not only by the requisites of the total system but also by interaction among sociocultural items within systemic wholes.

The Issue of Functional Universality

One result of an emphasis on functional unity was that some early anthropologists assumed that if a social item existed in an ongoing system, it must therefore have positive consequences for the integration of the social system. This assumption tended to result in tautologous statements: a system exists; an item is a part of the system; therefore the item is positively functional for the maintenance of the system.

For Merton, if an examination of empirical systems is undertaken, it is clear that there is a wider range of empirical possibilities. First, items may be not only positively functional for a system or another system item but also dysfunctional for either particular items or the systemic whole. Second, some consequences, whether functional or dysfunctional, are intended and recognized by system incumbents and are thus manifest, whereas other consequences are not intended or recognized and are therefore latent. Thus, in contrast with the assertions of Malinowski and Radcliffe-Brown, Merton proposes the analysis of diverse consequences or functions of sociocultural items—whether positive or negative, manifest or latent—"for individuals, for subgroups, and for the more inclusive social structure and culture."[13] In turn, the analysis of varied consequences requires the calculation of a "net balance of consequences" of items for each other and more inclusive systems. In this way, Merton visualizes contemporary functional thought as compensating for the excesses of earlier forms of analysis by focusing on the crucial types of consequences of sociocultural items for each other and, if the facts dictate, for the social whole.

The Issue of Indispensability

Somewhat out of context and unfairly, I feel, Merton quotes Malinowski's assertion that every cultural item "fulfills some vital function, has some task to accomplish, represents an indispensable part within a working whole"[14] as simply an extreme statement of two interrelated issues in functional analysis: (1) Do social systems have functional requisites or needs that must be fulfilled? (2) Are there certain crucial structures that are indispensable for fulfilling these functions?

---

[13]Merton, *Social Theory and Social Structure* (1968), p. 84.

[14]This quote is from an encyclopedia article in which Malinowski was arguing against ethnocentrism. His more scholarly work (see Chapter 2) is much less extreme. See also Jonathan H. Turner and Alexandra Maryanski, *Functionalism* (Menlo Park, CA: Benjamin/Cummings, 1979).

In response to the first question, Merton provides a tentative yes, but with an important qualification: the functional requisites must be established empirically for specific systems. For actual groups or whole societies it is possible to ascertain the "conditions necessary for their survival," and it is of theoretical importance to determine which structures, through what specific processes, have consequences for these conditions. But to assume a system of universal requisites—as Parsons does—adds little to theoretical analysis, since to stress that certain functions must be met in all systems simply leads observers to describe processes in social systems that meet these requisites. Such descriptions, Merton contends, can be done without the excess baggage of system requisites. It is more desirable to describe cultural patterns and then assess their various consequences in meeting the specific needs of different segments of concrete empirical systems.

Merton's answer to the second question is emphatic: empirical evidence makes the assertion that only certain structures can fulfill system requisites obviously false. Examination of the empirical world reveals quite clearly that alternative structures can exist to fulfill basically the same requisites in both similar and diverse systems. This fact leads Merton to postulate the importance in functional analysis of concern with various types of "functional alternatives," or "functional equivalents," and "functional substitutes" within social systems. In this way, functional analysis would not view as indispensable the social items of a system and thereby would avoid the tautologous trap of assuming that items must exist to assure the continued existence of a system. Furthermore, in looking for functional alternatives, analytical attention would be drawn to questions about the range of items that could serve as functional equivalents. If these questions are to be answered adequately, analysts should then determine why a particular item was selected from a range of possible alternatives, leading to questions about the "structural context" and "structural limits" that might circumscribe the range of alternatives and account for the emergence of one item over another. For Merton, examination of these interrelated questions would thus facilitate the separate analysis of the causes and consequences of structural items. By asking why one particular structure, instead of various alternatives, had emerged, analysts would not forget to document the specific processes leading to an item's emergence as separate from its functional consequences. In this way the danger of assuming that items must exist to fulfill system needs would be avoided.

In looking back at Merton's criticisms of traditional anthropological reasoning and at some contemporary functionalists' implications, I think that much of his assessment of these three functional postulates involves the destruction of "straw men." Yet, in destroying these assumptions, Merton was led to formulate alternative postulates that advocated a concern for the multiple consequences of sociocultural items for one another and for more inclusive social wholes, without a priori assumptions of functional needs or imperatives. Rather, functional analysis must specify (1) the social patterns under consideration, whether a systemic whole or some subpart; (2) the various types of consequences of these patterns for empirically established survival requisites;

and (3) the processes whereby some patterns rather than others come to exist and have the various consequences for one another and for systemic wholes.

With this form of functional analysis, Merton has sought to provide "the minimum set of concepts with which the sociologist must operate in order to carry through an adequate functional analysis."[15] In doing so, Merton hopes that this strategy will allow sociological analysis to avoid some of the mistaken postulates and assumptions of previous attempts using a functional strategy. Although the functional imperativism of Parsons is only briefly assessed in Merton's proposals, it appears that Merton is stressing the need for an alternative form of functional analysis in which there is less concern with total systems and abstract statements of system requisites. Instead, to build theories of the middle range, it is necessary to focus attention on the mutual and varied consequences of specified system parts for one another and for systemic wholes. Although these parts and systemic wholes have conditions necessary for their survival, these conditions must be empirically established. For only through a clear understanding of the actual requisites of a concrete system can the needs of social structures provide a useful set of criteria for assessing the consequences, or functions, of social items. Furthermore, although the analysis of consequences of items is the unique feature of functional analysis, it is also necessary to discover the causal processes that have resulted in a particular item having a specified set of consequences for other items and systemic wholes. To assure adherence to this form of structural analysis, Merton went so far as to outline a set of procedures for executing the general guidelines of his functional orientation.

A PROTOCOL FOR EXECUTING FUNCTIONAL ANALYSIS

To ascertain the causes and consequences of particular structures and processes, Merton insists that functional analysis begin with "sheer description" of individual and group activities. In describing the patterns of interaction and activity among units under investigation, it will be possible to discern clearly the social items to be subjected to functional analysis. Such descriptions can also provide a major clue to the functions performed by such patterned activity. In order for these functions to become more evident, however, additional steps are necessary.

The first of these steps is for investigators to indicate the principal alternatives that are excluded by the dominance of a particular pattern. Such description of the excluded alternatives provides an indication of the structural context from which an observed pattern first emerged and is now maintained—thereby offering further clues about the functions or consequences the item might have for other items and perhaps for the systemic whole. The second analytical step beyond sheer description involves an assessment of the meaning,

---

[15]Merton, *Social Theory and Social Structure* (1968), p. 109.

or mental and emotional significance, of the activity for group members. Description of these meanings may offer some indication of the motives behind the activities of the individuals involved and thereby shed some tentative light on the manifest functions of an activity. These descriptions require a third analytical step of discerning some array of motives for conformity or for deviation among participants, but these motives must not be confused with either the objective description of the pattern or the subsequent assessment of the functions served by the pattern. Yet, by understanding the configuration of motives for conformity and deviation among actors, an assessment of the psychological needs served (or not served) by a pattern can be understood—offering an additional clue to the various functions of the pattern under investigation.

But focusing on the meanings and motives of those involved in an activity can skew analysis away from unintended or latent consequences of the activity. Thus a final analytical step involves the description of how the patterns under investigation reveal regularities not recognized by participants but appearing to have consequences for both the individuals involved and other central patterns or regularities in the system. In this way, analysis will be attuned to the latent functions of an item.

Merton assumes that, by following each of these steps, it will be possible to assess the net balance of consequences of the pattern under investigation, as well as to determine some of the independent causes of the item. These steps assure that a proper functional inquiry will ensue, because postulates of functional unity, assumptions of survival requisites, and convictions about indispensable parts do not precede the analysis of social structures and processes. On the contrary, attention is drawn only to observable patterns of activity, the structural context in which the focal pattern emerged and persists in the face of potential alternatives, the meaning of these patterns for actors involved, the actors' motives for conformity and deviation, and the implications of the particular pattern for unrecognized needs of individuals and other items in the social system. Thus, with this kind of preliminary work, functional analysis will avoid the logical and empirical problems of previous forms of functionalism. And in this way it can provide an understanding of the causes and consequences of system parts for one another and for more inclusive system units.[16] I have diagramed the basic logic of this approach in Figure 4-1.

Figure 4-1 recapitulates the essential elements of Merton's strategy. First of all, only empirical units are to be analyzed, and the part and the social context of the part must be clearly specified. Then the task is to establish the particular survival requisites of the empirical system—that is, what is necessary for this particular empirical system to survive. By assessing the functions or consequences of an item's meeting or not meeting these needs, one can achieve insight into the nature of a part and its social contexts. In addition to this structural analysis must come an analysis of the meaning for participants, particularly since it reveals the psychological needs served or not served by

---

[16]Ibid., p. 136.

FIGURE 4–1 Merton's Net Functional Balance Analysis

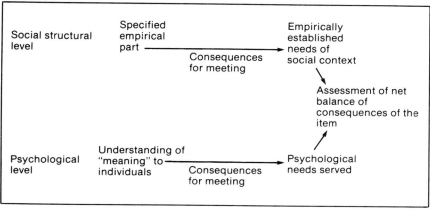

participation in a structure. In this way the net balance of consequences of an item at diverse levels of social organization can be assessed.

ILLUSTRATING MERTON'S FUNCTIONAL STRATEGY

Merton's paradigm and protocol for constructing functional theories of the middle range are markedly free of statements about individual and system needs or requisites. In his protocol statements, Merton approaches the question of the needs and requisites fulfilled by a particular item only after a description of (1) the item in question, (2) the structural context in which the item survives, and (3) its meaning for the individuals involved. With this information it is then possible to establish both the manifest and latent functions of an item, as well as the net balance of functions and dysfunctions of the item for varied segments of a social system. Unfortunately, the implied sequencing of functional analysis is not always performed by Merton, presumably for at least two reasons. First, in selecting an established structure in a system for analysis, the investigator usually assumes that the item persists because it is fulfilling some need. As I will make evident, Merton begins (as opposed to concludes) with this assumption in his analysis of political machines—thus leaving him to conclude that "structure affects function and function affects structure." When description of items begins with an implicit assumption of their functions for fulfilling needs, then I think that the descriptions will be performed in ways assuring confirmation of this implicit assumption. Second, in analyzing the structural context of an item and assessing why it emerges and persists over alternative items, I believe it inevitable that there will be preconceptions of the functions served by an item in order to know why it fulfills a set of needs better than would various alternatives. Otherwise, I do not see how it is possible to determine what potential alternatives could exist to substitute for the present item.

For at least these two reasons, then, execution of Merton's strategy is difficult. Let me indicate how this is so by examining Merton's illustration of it. This illustration involves an analysis of American political machines. Much like that of his anthropological straw men, such as Radcliffe-Brown and Malinowski, Merton's need to analyze separately the causes and functions of structural items is not as evident in his actual account of empirical events.

Merton begins his analysis of American political machines with the simple question: "How do they manage to continue in operation?"[17] Following this interesting question is an assumption reminiscent of Malinowski's functional analysis:

> Preceding from the functional view, therefore, that we should *ordinarily* (not invariably) expect persistent social patterns and social structures to perform positive functions *which are at the time not fulfilled by other patterns and structures*, the thought occurs that perhaps this publicly maligned organization is, *under present conditions*, satisfying basic latent functions. [Merton's emphasis.][18]

The fact that the word *ordinarily* is qualified by the parenthetical phrase *not invariably* is perhaps enough for Merton to escape the charge of tautology: if an item persists in a surviving system, it must therefore have positive functions. Yet my sense is that Merton implicitly argues that, if an enduring item does not fulfill manifest functions, then it fulfills latent functions. This leads me to recall Merton's portrayal of Malinowski's dictum that "every custom, material object, idea and belief fulfills some vital function, has some task to accomplish." For Merton this assumption becomes translated into the dictum that social items that do not fulfill manifest functions must therefore fulfill latent ones; and, as is added in a footnote, if the item has dysfunctions for some segments of the population, its persistence implies that it ordinarily must have positive functions for meeting the needs of other segments.

In fairness to Merton's suggestive analysis of political machines, I emphasize that he offered it only as an illustration of the usefulness of the distinction between manifest and latent functions. It was not intended as a full explication of his functional paradigm or protocol, but only as an example of how attention to latent functions can provide new insights into the operation of political machines. However, it appears that Merton's commitment to a clear protocol is not enough to preclude an inadvertent lapse—so typical of earlier functionalists—into the postulates of "universality of functions" and "functional indispensability." Thus Merton appears to begin his concrete analysis with a set of postulates that he earlier had gone to great lengths to discredit, resulting in this central assumption:

> The key structural function of the Boss is to organize, centralize and maintain in good working condition "the scattered fragments of power" which are at

---

[17]Ibid., p. 125.
[18]Ibid., pp. 125–26.

present dispersed through our political organization. By this centralization of political power, *the boss can satisfy the needs of diverse subgroups* in the larger community which are not adequately satisfied by legally devised and culturally approved social structures.[19]

For Merton, political machines emerge in the structural context of a system in which power is decentralized to the extent that it cannot be mobilized to meet the needs of significant population segments. The causal processes by which machines arise in this power vacuum to pick up the "scattered fragments of power" involve a sequence of events in which political machines are seen as able to satisfy the needs of diverse groups more effectively than "legally devised and culturally approved social structures." Logically, this form of analysis is not necessarily tautologous or an illegitimate teleology—as some critics might charge—because Merton appears to be asserting that political machines at one time had a selective advantage over alternative structures in meeting needs of certain segments in a system. This kind of *reverse causal chain,* to use Stinchcombe's term,[20] is a legitimate form of causal analysis, since system needs are seen as existing prior in time to the events they cause—in this instance, the emergence of political machines in the American social structure. Furthermore, I do not think it necessary to impute purposes—although at times purpose is certainly involved—to the segments of the system affected by the machines. A political machine may be seen as a chance event that had a selective advantage over alternatives in a spiraling process similar to expansions and contradictions of predator populations that grow rapidly until they eat themselves out of prey. Clearly, the emergence of a political machine is both a purposive and nonpurposive process in which the machine meets the prior needs of a population, which has signaled to the machine leaders or bosses the efficacy of their expanded pursuits (purpose) in meeting its needs. Eventually, in a spiraling process of this nature, the original needs of the population, which caused the emergence and expansion of political machines and big-city bosses, may recede in causal significance, whereas the needs of the well-established political machine cause certain activities that have little consequence (or perhaps a dysfunctional consequence) for the needs of the population.

This kind of causal argument appears to be Merton's intent. Unfortunately, I think that his overriding concern is with discerning the functions of the political machines; consequently, he obscures the causal analysis, for, as he is prone to remark, "Whatever its specific historical origins, the political machine persists as an apparatus for satisfying unfulfilled needs of diverse groups in the population."[21] By bypassing these specific causal chains involved in the emergence of American political machines, Merton is left with the relatively simple and ad hoc task of cross-tabulating the needs of a population and the activities of the political machine that fulfill them.

---

[19]Ibid., p. 126.

[20]Arthur L. Stinchcombe, *Constructing Social Theories* (New York: Harcourt, Brace & World, 1968), p. 100.

[21]Merton, *Social Theory and Social Structure* (1968), p. 127.

For example, the political machine fulfills the needs of deprived classes by providing vital services through the local neighborhood ward heeler, including "food baskets and jobs, legal and extralegal advice, setting to rights minor scrapes with the law, helping the bright poor boy to a political scholarship in a local college, looking after the bereaved," and so on. The political machine, according to Merton, can provide these services more effectively than can various alternatives, such as welfare agencies, settlement houses, and legal-aid clinics, because it offers these services in a personal way through the neighborhood ward heeler with a minimum of questions, red tape, and abuse to people's self-respect. For other populations, such as the business community, the political machines provide another set of needed services—namely, political regulation and control of unbridled competition among corporations and businesses without undue governmental interference in the specific operations of economic enterprises. By virtue of controlling various public agencies and bureaus, the big-city boss can rationalize and organize relations among economic organizations while preventing too much governmental scrutiny into their various illegal activities. The political machine can perform this function more effectively than can legal governmental alternatives because it recognizes the need of economic organizations for both regulation and noninterference in certain activities. In contrast, legally constituted government agencies would recognize only the former need—thus giving the political machine a selective advantage over legally constituted government. Similarly, the machine can organize and rationalize illegal economic enterprises concerned with providing illicit services, including gambling, drugs, and prostitution, whereas legally constituted governmental agencies cannot condone (to say nothing of organizing) this kind of prevalent activity. Thus, for both legal and illegal businesses, the political machine provides protection by assuring a stable marketplace, high profits, and selective governmental regulation. Finally, for another population—notably, the deprived—the political machine provides opportunities for social mobility in a society where monetary success is a strong cultural value but where actual opportunities for such success are closed to many deprived groups. Thus, by opening the doors to social mobility for members of deprived groups who do not have "legitimate" opportunities, the political machine meets the needs of the deprived while assuring itself of loyal, committed, and grateful personnel.

I think that Merton's functional explanation of the persistence of political machines has considerable plausibility. For, indeed, the existence of political machines in America was correlated with a relatively ineffective federal establishment, deprived urban masses, high demand for illegal services, and high degrees of economic competition. But I see most of Merton's account as statements of correlation, dressed in functional assumptions about how the needs of diverse subgroups led to the emergence and persistence of political machines. Statements of correlation are obviously not causal statements. To the extent that it simply notes the correlation between social needs and political machines, Merton's analysis will be of little utility in building theoretical statements of the form: under $C_1, C_2, C_3, \ldots, C_n$, x causes variation in y. There are many *implied* causal chains in Merton's analysis, but his failure to make them explicit detracts from his analysis. As was emphasized in Chapter 2, Durkheim's concrete analysis of the division of labor lapsed into statements implying, at the very least, that the "need for social order" caused the division of labor.[22] Without explicit statements or analytical models on how the need for order caused the division of labor, the analysis constituted an illegitimate teleology.

The difficulties that the founders of functionalism had in separating cause and function are clearly recognized by Merton and, presumably, provided the impetus to his explicating a paradigm and protocol for functional analysis. Yet, much like his predecessors, I see Merton as abandoning the very protocol that would keep cause and function separated. Merton indicates that the emergence and persistence of political machines occur in response to needs, without documenting very precisely the causal chains through which needs cause the emergence and persistence of an event.

I think that Merton also falls into the problem of tautology that he imputed—incorrectly, I emphasize—to Malinowski's functional analysis. By assuming that "ordinarily" persistent structures serve positive functions for meeting the needs of some population segment, Merton indicates that, if an item persists in an existing system, then it is functional (perhaps only latently) for some groups. I find it surprising that Merton falls back onto this postulate, since he went to such great lengths to sound the warning against it. Yet Merton's analysis of political machines does not start with a description of the phenomenon; nor does he initially address the structural context in which it exists. Rather, Merton begins with the assumption that political machines exist to fulfill a function—if not a manifest function, then a latent one.

This criticism of Merton's analysis of a concrete phenomenon does not mean that, with more specification of causal processes, charges of tautology and illegitimate teleology could be avoided. Indeed, with more specification of the historical origins of the political machines and of the feedback processes between machines and the population segments they serve, Merton's account could be rephrased in less suspicious causal terms. This fact leads to an important question: why did Merton fail to specify the causal chains that would make his propositions less suspicious? One answer is that Merton offered this account of political machines only as an illustration of the utility in the concept of *latent* functions. As an illustration, the account would naturally be brief and would not involve a thorough explication of the emergence of political machines in America. Merton's awareness of the problems inherent in previous functional analysis would lend credence to this argument, for how could he fall into the very traps that he sought to avoid?

However, Durkheim's and Radcliffe-Brown's similar failure to avoid completely the logical problems they clearly understood raises the more fundamental question: is there something about functional analysis that encourages theorists to short-circuit causal explanation? I think that there is something

---

[22]Émile Durkheim, *The Rules of Sociological Method* (Glencoe, IL: Free Press, 1938), p. 96.

seductive about functional analysis that leads us consistently into these problems. Thus I would not recommend it as a mode of theoretical explanation. Even as Merton tried to make functionalism more empirical, he fell into the logical problems of illegitimate teleology and tautology. Yet, even with these problems, there has been a revival of functional analysis in recent years; so, despite my misgivings, many theorists continue to use a functional approach.[23] Let me now turn to the most prominent of these "neofunctionalisms."

---

[23]For a defense of neofunctionalism, see: Jeffrey C. Alexander, ed., *Neofunctionalism* (Newbury Park, CA: Sage, 1985); and Jeffrey C. Alexander and Paul Colomy, "Toward Neo-functionalism," *Sociological Theory* 3 (2, Fall 1985), pp. 11–23.

CHAPTER 5

Neofunctionalism

Niklas Luhmann

Jeffrey C. Alexander and others have heralded the arrival of "neofunctional-ism."[1] This vision is a revised and revitalized functional approach that, ac-cording to Alexander, is "nothing so precise as a set of concepts, a method, a model, or an ideology"[2] but rather, a concern with theory that emphasizes (1) analytical levels, especially cultural, structural, and individual; (2) systems and subsystems; (3) normative processes; (4) social differentiation; and (5) inter-relations among "institutional spheres." What is noticeably absent in this list of guidelines is the notion of functional needs and requisites—what I see as *the* defining feature of functional theorizing.[3] Alexander and others have, much like Kingsley Davis[4] before them, downplayed the analysis of phenomena in terms of their consequences for system needs—the most problematic charac-teristic of functional analysis. But notions of system needs and requisites are what make functional analysis unique and distinctive; if one simply removes them from analysis, then the approach is no longer functional. Thus, to some extent, neofunctionalism is nonfunctionalism.

---

[1]Jeffrey C. Alexander, ed., *Neofunctionalism* (Newbury Park, CA: Sage, 1985); and Jeffrey C. Alexander and Paul Colomy, "Toward Neo-functionalism," *Sociological Theory* 3 (2, Fall 1985), pp. 11–23, "Neofunctionalism Today: Restructuring a Theoretical Tradition," in *Frontiers of Social Theory,* edited by G. Ritzer (New York: Columbia University Press, 1990), and *Differentiation Theory and Social Change: Comparative and Historical Perspectives,* edited by J. C. Alexander and P. Colomy (New York: Columbia University Press, 1990). See also: Paul Colomy, ed., *Functionalist Sociology: Classic Statements* (London: Edward Elgar, 1990), and *Neofunctionalist Sociology: Contemporary Statements* (London: Edward Elgar, 1990).

[2]Alexander, *Neofunctionalism,* p. 9.

[3]For my critique of Alexander's argument, see: Jonathan H. Turner and Alexandra Maryanski, "Is Neofunctionalism Functional?" *Sociological Theory* 6 (1, Spring-Summer 1988), pp. 110–21. See also our *Functionalism* (Menlo Park, CA: Benjamin/Cummings, 1978).

[4]Kingsley Davis, "The Myth of Functional Analysis," *American Sociological Review* 25 (1959), pp. 757–72.

When we actually look at the works of those seen as part of the neofunctionalist camp,[5] however, it is evident that needs and requisites are often part of their work. Thus, in selecting a theorist to discuss as a "neofunctionalist," I picked one who meets the list of features cited above and who, at the same time, downplays the notion of functional requisites, leaving them implicit. By Alexander's criteria and by my assertion that functional requisites must be part of the approach, even if only implicitly, the obvious choice of a neofunctionalist is the German sociological theorist Niklas Luhmann.

Since he was a student of Talcott Parsons, we would expect Luhmann's scheme to be highly analytical and abstract. Yet Luhmann is very critical of Parsonian action theory because it increasingly became "overly concerned with its own architecture." That is, Parsons kept elaborating categories and thereby created a complex analytical edifice that became ever more divorced from empirical reality. But, in contrast to Robert Merton, Luhmann does not retreat into a wholly empirical functionalism in which each system's varying needs must be empirically established in all their particulars. Rather, his alternative is to construct an abstract conceptual scheme that is "relatively simple" but that, at the same time, can be used in "highly complex research programs" and guide the analysis of a wide diversity of empirical events.

As I will discuss, Luhmann has used his conceptual scheme to analyze a variety of empirical processes, and although its architecture is not as complex as Parsonian action theory, it nonetheless tends to bend empirical reality to its purposes. Even so, I find it an intriguing approach that documents the staying power of functional analysis in sociological theory. In this chapter, therefore, my goal is to summarize the basic elements of Luhmann's functionalism and to ascertain the extent to which it offers a viable alternative to earlier forms of functional analysis.

LUHMANN'S "GENERAL SYSTEMS" APPROACH[6]

System and Environment

Luhmann employs what he terms a *general systems* approach. Such an approach stresses the fact that human action becomes organized and structured into systems. When the actions of several people become interrelated, a social system can be said to exist. The basic mechanism by which actions become interrelated so as to create social systems is communication via symbolic codes, such as words and, as I will describe shortly, other media. All social systems exist in multidimensional environments, which pose potentially endless com-

---

[5]For example, scholars such as Neil J. Smelser, S. I. Eisenstadt, Bernard Barber, Adrian Hayes, Dean Gerstein, Richard Münch, Victor Lidz, Paul Colomy, and others.

[6]Luhmann has published extensively, but most of his work is in German. The best sample of his work in English is his *The Differentiation of Society*, trans. S. Holmes and C. Larmore (New York: Columbia University Press, 1982).

plexity with which a system must deal. To exist in a complex environment, therefore, a social system must develop mechanisms for reducing complexity, lest the system simply merge with its environment. These mechanisms involve selecting ways and means for reducing complexity. Such selection creates a boundary between a system and its environment, thereby allowing it to sustain patterns of interrelated actions. Selection involves a process of choosing how to reduce the complexity of the environment.

The basic functional requisite in Luhmann's analysis is thus "the need to reduce the complexity of the environment in relation to a system of interrelated actions." I suspect that Luhmann would not accept this translation of his ideas into the metaphor of traditional functionalism, but, in fact, this requisite pervades his analysis. All social processes are analyzed with respect to their functions for reducing complexity vis-à-vis an environment. Processes that function in this way are typically defined as *mechanisms* in a manner reminiscent of Talcott Parsons' early discussion in *The Social System*[7] (see Chapter 3). Indeed, as I will document, the bulk of Luhmann's sociology revolves around discussions of such mechanisms—differentiation, ideology, law, symbolic media, and other critical elements of his scheme.

Dimensions of the Environment

There are three basic dimensions along which the complexity of the environment is reduced by these mechanisms: (1) a temporal dimension, (2) a material dimension, and (3) a symbolic dimension. More than most social theorists, Luhmann is concerned with time as a dimension of the social universe. Time always presents a system with complexity because it reaches into the past, because it embodies complex configurations of acts in the present, and because it involves the vast horizons of the future. Thus a social system must develop mechanisms for reducing the complexity of time. It must find a way to order this dimension by developing procedures to orient actions to the past, present, and future.[8]

Luhmann is also concerned with the material dimension of the environment—that is, with all of the possible relations among actions in potentially limitless physical space. Luhmann always asks: What mechanisms are developed to order interrelated actions in physical space? What is the structure and form of such ordering of relations?

Luhmann visualizes the third dimension of human systems as the symbolic. Of all the complex symbols and their combinations that humans can conceivably generate, what mechanisms operate to select some symbols over others and to organize them in some ways as opposed to the vast number of potential alternatives? What kinds of symbolic media are selected and used by a social system to organize social actions?

---

[7]Talcott Parsons, *The Social System* (New York: Free Press, 1951).

[8]Luhmann, *The Differentiation,* Chap. 12.

Thus the mechanisms of a social system that operate to reduce complexity and thereby maintain a boundary between the system and the environment function along three dimensions, the temporal, material, and symbolic. The nature of a social system—its size, form, and differentiation—will be reflected in the mechanisms that it uses to reduce complexity along these dimensions.

Types of Social Systems

A social system exists any time the actions of individuals are "meaningfully interrelated and interconnected," thereby setting them off from the temporal, material, and symbolic environment by virtue of the selection of functional mechanisms. Out of such processes come three basic types of social systems: (1) interaction systems, (2) organization systems, and (3) societal systems.[9]

Interaction systems   An interaction system emerges when individuals are co-present and perceive each other. The very act of perception is a selection mechanism that sorts from a much more complex environment, creating a boundary and setting people off as a system. Such systems are elaborated by the use of language in face-to-face communication, thereby reducing complexity even further along the temporal, material, and symbolic dimensions. For example, Luhmann would ask: How does the language and its organization into codes shape people's perceptions of time? Who is included in the conversation? And what codes and agreements guide conversation and other actions?

Interaction systems reveal certain inherent limitations and vulnerabilities, however. First, only one topic can be discussed at a time, lest the system collapse as everyone tries to talk at once (which of course frequently occurs). Second, the varying conversational resources of participants often lead to competition over who is to talk, creating inequalities and tensions that can potentially lead to conflict and system disintegration. Third, talk and conversation are time-consuming because they are sequential; as a result, an interaction system can never be very complex.

Thus, interaction systems are simple because they involve only those who can be co-present, perceived, and talked to; they are vulnerable to conflict and tension; and they consume a great deal of time. In order for a social system to be larger and more complex, additional organizing principles beyond perceptions of co-presence and sequential talk are essential.

Organization systems   These systems coordinate the actions of individuals with respect to specific conditions, such as work on a specific task in exchange for a specific amount of money. They typically have entry and exit rules (for example, come to work for this period of time and leave with this much money), and their main function is to "stabilize highly 'artificial' modes of behavior over a long stretch of time." They resolve the basic problem of

---

[9]Ibid., pp. 71–89.

reconciling the motivations and dispositions of individuals and the need to get certain tasks done. An organization is not dependent upon the moral commitment of individuals; nor does it require normative consensus. Rather, the entrance/exit rules specify tasks in ways that allow individuals to do what is required without wholly identifying with the organization.

Organization systems are thus essential to a complex social order. They reduce environmental complexity by organizing people (1) *in time* by generating entrance and exit rules and by ordering activities in the present and future; (2) *in space* by creating a division of labor, which authority coordinates; and (3) *in symbolic terms* by indicating what is appropriate, what rules apply, and what media, such as money or pay, are to guide action. In his delineation of organization systems, Luhmann stresses that complex social orders do not require consensus over values, beliefs, or norms to be sustained; they can operate quite effectively without motivational commitments of actors. In fact, their very strength—flexibility and adaptability to changing environmental conditions—depends upon delimited and situational commitments of actors, along with neutral media of communication, such as money.[10]

Societal systems   These systems cut across interaction and organization systems. A societal system is a "comprehensive system of all reciprocally accessible communication actions."[11] Historically, societal systems have been limited by geopolitical considerations, but today Luhmann sees a trend toward one world society. I find Luhmann's discussion on the societal system rather vague, but the general idea can be inferred from his analysis of more specific topics. Let me summarize: Societal systems use highly generalized communication codes, such as money and power, to reduce the complexity of the environment. In so doing, they set broad limits on how and where actions are to be interrelated into interaction and organization systems. They also organize the way time is perceived and how actions are oriented to the past, present, and future.

System Differentiation, Integration, and Conflict

These three systems—interaction, organization, and societal—cannot be totally separated, since "all social action obviously takes place in society and is ultimately possible only in the form of interaction."[12] Indeed, in very simple societies they are fused together; but, as societies become larger and more complex, these systems become clearly differentiated from and irreducible to one another. Organizations become distinctive with respect to (1) their functional domains (government, law, education, economy, religion, science), (2) their entrance/ exit rules, and (3) their reliance upon distinctive media of communication

---

[10]In making this assertion, Luhmann directly attacks Parsons. See Luhmann, *The Differentiation*, Chap. 3.

[11]Ibid., p. 73.

[12]Ibid., p. 79.

(money, truth, power, love, etc.). As a consequence, they cannot be reduced to a societal system. Interaction systems follow their own laws, for rarely do people in their conversations strictly follow the guidelines of organizations and society.

In fact, the differentiation of these systems poses a number of problems for the more inclusive system. First, there is the problem of what Luhmann calls "bottlenecks." Interaction systems are slow, sequentially organized patterns of talk, and they follow their own dynamics as people use their resources in conversations. As a result, they often prevent organizations from operating at high levels of efficiency. As people interact, they develop informal agreements and take their time, with the specific tasks of the organization going un- or underperformed. Similarly, as organization systems develop their own structure and programs, their interests often collide, and they become "bottlenecks" to action requirements at the societal level. Second, there is the problem of conflict in differentiated systems. Interactants may disagree on topics; they may become jealous or envious of those with conversational resources. And since interaction systems are small, they cannot become sufficiently complex "to consign marginals to their borders or to otherwise segregate them." At the organizational level, diverse organizations can pursue their interests in ways that are disruptive to both the organization and the more inclusive societal system.

Yet, countervailing these disruptive tendencies are processes that function to maintain social integration. One critical set of processes is the "nesting" of system levels inside each other. Actions within an interactive system are often nested within an organization system, and organizational actions are conducted within a societal system. Hence the broader, more inclusive, system can operate to promote integration in two ways: (1) it provides the temporal, material, and social premises for the selection of actions; and (2) it imposes an order or structure on the proximate environment around any of its subsystems. For example, an organizational system distributes people in space and in an authority hierarchy; it orients them to time; it specifies the relevant communication codes; and it orders the proximate environment (other people, groupings, offices, etc.) of any interaction system. Similarly, the functional division of a society into politics, education, law, economy, family, religion, and science determines the substance of an organization's action, while it orders the proximate environment of any particular organization. For example, societal differentiation of a distinctive economy delimits what any economic organization can do. Thus a corporation in a capitalist economy will use money as its distinctive communications media; it will articulate with other organizations in terms of market relations; it will organize its workers into bureaucratic organizations with distinctive entrance and exit rules ("work for money"); and it will be oriented to the future, with the past as only a collapsed framework to guide present activity in the pursuit of future outcomes (such as profits and promotions).

In addition to these nesting processes, integration is promoted by the deflection of people's activities across different organizations in diverse functional domains. When there are many organizations in a society, none consumes an individual's sense of identity and self since people's energies are dispersed across several organization systems. As a consequence of their piecemeal involvement, members are unlikely to be emotionally drawn into conflict among organization systems; and, when individual members cannot be pulled emotionally into a conflict, its intensity and potential for social disruption are lessened. Moreover, because interaction systems are distinct from the more inclusive organization, any conflict between organizations is often seen by the rank and file as distant and remote to their interests and concerns; it is something "out there" in the environment of their interaction systems, and hence it is not very involving.

Yet another source of conflict mitigation are the entrance/exit rules of an organization. As these become elaborated into hierarchies, offices, established procedures, salary scales, and the like, they reduce the relevance of members' conflicts outside the organization—for example, their race and religion. Such outside conflicts are separated from those within the organization, and as a result their salience in the broader societal system is reduced.

Finally, once differentiation of organizations is an established mechanism in a society, then specific social control organizations—law, police, courts—can be easily created to mitigate and resolve conflicts. That is, the generation of distinct organizations that are functionally specific represents a new "social technology"; once this technology has been used in one context, it can be applied to additional contexts. Thus the integrative problems created by the differentiation and proliferation of organizations create the very conditions that can resolve these problems—the capacity to create organizations to mediate among organizations.

And so, although differentiation of three system levels creates problems of integration and conditions conducive to conflict, it also produces countervailing forces for integration. In making this argument, Luhmann emphasizes that, in complex systems, order is not sustained by consensus on common values, beliefs, and norms. On the contrary, there is likely to be considerable disagreement over these, except perhaps at the most abstract level. I think that this is an important contribution of Luhmann's sociology, for it distinguishes his theoretical approach from that of Talcott Parsons, who, I think, overstressed the need for value consensus in complex social systems. Additionally, Luhmann stresses that individuals' moral and emotional attachment to the social fabric is not essential for social integration. To seek a romantic return to a cohesive community, as Émile Durkheim, Marx, and others have, is impossible in most spheres of a complex society.[13] And, rather than viewing this fact as a pathological state—as concepts like alienation, egoism, and anomie connote—the impersonality and neutrality of many encounters in complex systems can be seen as normal and analyzed less evaluatively. Moreover, people's lack of emotional embeddedness in complex systems gives them more freedom, more options, and more flexibility.[14] It also liberates them from the constraints of

---

[13]See Chapter 2 on Durkheim and Chapter 9 on Marx.

[14]Here Luhmann takes a page from Georg Simmel's *The Philosophy of Money*, trans T. Bottomore and D. Frisby (Boston: Routledge & Kegan Paul, 1978).

tradition, the restrictions of dependency on others, and the indignities of surveillance by the powerful that are so typical of less complex societies.

Communications Media, Reflexivity, and Self-Thematization

Luhmann's system theory stresses the relation of a system to its environment
and the mechanisms used to reduce its complexity. All social systems are
based upon communication among actors as they align their respective modes
of conduct. Because action systems are built from communication, Luhmann
devotes considerable attention to *communications theory,* as he defines it.[15] He
stresses that human communications become reflexive and that this reflexiveness leads to self-thematization. Luhmann thus develops a communications
theory revolving around communication codes and media as well as reflexiveness and self-thematization. I will briefly explore each of these elements in his
theory.

Communication and codes

Luhmann waxes philosophically and met-
aphorically about these concepts, but in the end he concludes that communication occurs in terms of symbols that signal actors' lines of behavior; and
such symbols constitute a code with several properties.[16] First, the organization
of symbols into a code guides the selection of alternatives that reduce the
complexity of the environment. For example, when someone in an interaction
system says that he or she wants to talk about a particular topic, these symbols
operate as a code that reduces the complexity of the system in an environment
(its members will now discuss this topic and not all the potential alternatives).
Second, codes are binary and dialectical in that their symbols imply their
opposite. For example, the linguistic code "be a good boy" also implicitly signals
its opposite—that is, what is not good and what is not male. As Luhmann notes,
"language makes negative copies available" by its very nature. Third, in implying their opposite, codes create the potential for the opposite action—for
instance, "to be a bad boy." In human codes, then, the very process of selecting
lines of action and reducing complexity with a code also expands potential
options (to do just the opposite or some variant of the opposite). This fact
makes the human system highly flexible, because the communications codes
used to organize the system and reduce complexity also contain implicit messages about alternatives.

Communications media

Communication codes function to stabilize
system responses to the environment (while implying alternative responses).
Codes can organize communication into distinctive media that further order
system responses. As a society differentiates into functional domains, distinc-

---

[15]This theory, like his general systems theory, is not very much like what most communications
theorists actually do.

[16]See Luhmann, *The Differentiation,* p. 169.

tive media are used to organize the resources of systems in each domain.[17] For example, the economy uses money as its medium of communication, which guides interactions within and among economic organizations. And so, in an economy, relations among organizations are conducted in terms of money (buying and selling in markets), and intraorganizational relations among workers are guided by entrance/exit rules structured by money (pay for work at specified times and places). Similarly, power is the distinctive communications medium of the political domain; love is the medium of the family; truth, the medium of science; and so on for other functional domains.[18]

There are several critical generalizations implicit in Luhmann's analysis of communications media. First, the differentiation of social systems into functional domains cannot occur without the development of a distinctive medium of communication for that domain. Second, media reduce complexity because they limit the range of action in a system. (For example, love as a medium limits the kinds of relations that are possible in a family system.)[19] Third, even in reducing complexity, media imply their opposite and thus expand potential options, giving systems flexibility (for instance, money for work implies its opposite, work without pay; the use of power implies its opposite, lack of compliance to political decisions).

Reflexivity and self-thematization   The use of media allows for *reflexivity,* or the capacity to examine the process of action as a part of the action itself. With communications media structuring action, it becomes possible to use these media to think about or reflect upon action. Social units can use money to make money; they employ power to decide how power is to be exercised; they can analyze love to decide what is true love; they can use truth to specify the procedures to get at truth; and so on. Luhmann sees this reflexivity as a mechanism that facilitates adaptation of a system to its environment. It does so by ordering responses and reducing complexity, on the one hand, while it provides a system with the capacity to think about new options for action, on the other. For example, it becomes possible to mobilize power to think about new and more adaptive ways to exercise power in political decisions, as is the case when a society's political elite create a constitutional system based on a separation of powers.

As communications media are used reflexively, they allow for what Luhmann terms *self-thematization.* Using media, a system can come to conceptualize itself and relations with the environment in terms of a "perspective" or "theme." Such self-thematization reduces complexity by providing guidelines about how to deal with the temporal, material, and symbolic dimensions of the environment. It becomes possible to have a guiding perspective on how to orient to time, to organize people in space, and to order symbols into codes.

---

[17]Obviously Luhmann is borrowing Parsons' idea about generalized media. See Chapter 3.

[18]Much like Parsons, this analysis of communications media is never fully explicated or systematically discussed for all functional domains.

[19]We will see shortly, however, that money is the sole exception here.

For example, money and its reflexive use for self-thematization in a capitalist economy create a concern with the future, an emphasis on rational organization of people, and a set of codes emphasizing impersonal exchanges of services and commodities. The consequence of these self-thematizations is for economic organizations to reduce the complexity of their environments and, thereby, to coordinate social action more effectively.

Luhmann's Basic Approach

In sum, Luhmann's general systems approach revolves around the system/environment distinction. Systems need to reduce the complexity of their environments in terms of their perceptions about time, their organization of actors in space, and their use of symbols. Processes that reduce complexity are conceptualized as functional mechanisms. There are three types of systems: interaction, organization, and societal. All system processes occur through communications that can develop into distinctive media and allow for reflexivity and self-thematization in a system.

This is, I feel, a fair summary of Luhmann's general systems approach. He uses these concepts as a kind of metaphor for the analysis of specific empirical topics. Yet, as a metaphor, I find his analysis rather vague and imprecise. Seemingly, in reacting against the restrictive architecture of Parsonian action theory, Luhmann has created an imprecise set of concepts with which to analyze social phenomena. Much of this vagueness remains even as Luhmann addresses particular empirical phenomena. But still he is able to develop stimulating insights.

I will now turn to Luhmann's actual analysis of social processes in order to illustrate how he employs this general systems scheme. Virtually all of his work with this scheme has been on the related processes of social evolution and societal differentiation, especially of organization systems in the political, legal, and economic domains of a society.

LUHMANN'S CONCEPTION OF SOCIAL EVOLUTION

Since Luhmann's substantive discussions are cast into an evolutionary framework, I think it wise to begin by extracting from his diverse writings the key elements of this evolutionary approach. Like other evolutionary theorists, Luhmann views evolution as the process of increasing differentiation of a system in relation to its environment.[20] Such increased differentiation allows a system to develop more flexible relations to its environment and, as a result, to increase its level of adaptation. As systems differentiate, however, there is the problem of integrating diverse subsystems; as a consequence, new kinds of mechanisms emerge to sustain the integration of the overall system. But, unlike

---

[20]This is essentially Parsons' definition (see Chapter 3). It was Spencer's and Durkheim's as well (see Chapter 2).

most evolutionary theorists, Luhmann uses this general image of evolution in a way that adds several new twists to previous evolutionary approaches.

The Underlying Mechanisms of Evolution

Luhmann is highly critical of the way traditional theory has analyzed the process of social differentiation.[21] First, traditional theories—from Marx and Durkheim to Parsons—all imply that there are limits as to how divided a system can be, and so they all postulate an end to the process, which, in Luhmann's view, is little more than an evaluative utopia. Second, traditional theories over-stress the importance of value consensus as an integrating mechanism in differentiated systems. Third, these theories see many processes, such as crime, conflict, dissensus over values, and impersonality, as deviant or pathological; however, they are, in fact, inevitable in differentiated systems. Fourth, previous theories have great difficulty handling the persistence of social stratification, viewing it as a source of evil or as a perpetual conflict-producing mechanism.

Luhmann's alternative to these evolutionary models is to use his systems theory to redirect the analysis of social differentiation. Like most functionalists, he analogizes to biology, but not to the physiology of an organism; rather, his analogies are to the processes delineated in the theory of evolution. Thus he argues for an emphasis of those processes that produce (1) variation, (2) selection, and (3) stabilization of traits in societal systems.[22] The reasoning here is that sociocultural evolution is like other forms of biological evolution. Social systems have mechanisms that are the functional equivalents of those in biological evolution. These mechanisms generate variation in the structure of social systems, select those variations that facilitate adaptation of a system, and stabilize these adaptive structures.[23]

Luhmann argues that the "mechanism for variation" inheres in the process of communication and in the formation of codes and media. Since all symbols imply their opposite, there is always the opportunity to act in new ways (a kind of "symbolic mutation"). The very nature of communication permits alternatives, and at times people act in terms of these alternatives, thereby producing new variations. Indeed, compared to the process of biological mutation, the capacity of human systems for variation is much greater than in biological systems.

The "mechanism for selection" can be found in what Luhmann vaguely terms *communicative success*. I find his discussion here rather imprecise, to say the least, but the general idea is that certain new forms of communication facilitate increased adjustment to an environment by reducing its complexity

---

[21]Luhmann, *The Differentiation*, pp. 256–57.

[22]Luhmann's interpretation of the synthetic theory of evolution in biology is, at best, loose and inexact. Again, it is more metaphorical than precise.

[23]Luhmann, *The Differentiation*, p. 265. Luhmann seems completely unaware that Herbert Spencer in his *The Principles of Sociology* (New York: D. Appleton, 1885; originally published in 1874) performed a similar, and more detailed, analysis 100 years ago.

while allowing for more flexible responses to the environment. For example, the creation of money as a medium greatly facilitated adaptation of systems and subsystems to the environment, as did the development of centralized power to coordinate activity in systems. And, because they facilitated survival and adaptation, they were retained in the structure of the social organism.

The "stabilization mechanism" resides in the very process of system formation. That is, new communication codes and media are used to order social actions among subsystems, and, in so doing, they create structures, such as political systems and economic orders, that regularize for a time the use of the new communications media. For example, once money is used, it creates an economic order revolving around markets and exchange that, in turn, feeds back and encourages the extension of money as a medium of communication. It is out of this reciprocity that some degree of continuity and stability in the economic system ensues.

Evolution and Social Differentiation

For Luhmann, sociocultural evolution involves differentiation in seven senses.

1. Evolution is the increasing differentiation of interaction, organization, and societal systems from one another. That is, interaction systems increasingly become distinct from organization systems, which in turn are more clearly separated from societal systems. Although these system levels are nested in each other, they also operate in terms of their unique dynamics.

2. Evolution involves the internal differentiation of these three types of systems. Diverse interaction systems multiply and become different from one another (for example, compare conversations at work, at a party, at home, and at a funeral). Organization systems increase in number and specialize in different activities (compare economic with political organizations, or contrast different types of economic organizations, such as manufacturing and retail organizations). And the societal system becomes differentiated from the organization and interaction systems that it comprises. Moreover, there is an evolutionary trend, Luhmann claims, toward only a one world society.

3. Evolution involves the increasing differentiation of societal systems into functional domains, such as economy, polity, law, religion, family, science, and education. Organization subsystems within these domains are specialized to deal with a limited range of environmental contingencies, and, in being specialized, they can better deal with them. The overall result for a societal system is increased adaptability and flexibility in its environment.

4. Functional differentiation is accompanied by (and is the result of) the increasing use of distinctive media of communication. For example, organization systems in the economy employ money, those in the polity or government exercise power, those in science depend upon truth, and those in the family domain use love.

5. There is a clear differentiation during evolution among the persons, roles, programs, and values. Individuals are entities separated from the roles and organizations in which they participate. One plays many roles, and each involves only a segment or part of a person's personality and sense of self; in fact, many roles are played with little or no investment of oneself in them. Moreover, most roles persist whether or not any one individual plays them, thereby emphasizing their separation from the person. Such roles are increasingly grouped together into an ever-increasing diversity of what Luhmann calls programs (work, family, play, politics, consumption, etc.) that typically exist inside different kinds of organization systems operating in a distinctive functional domain. Additionally, these roles can be shuffled around into new programs, emphasizing the separation of roles and programs. Finally, societal values become increasingly abstract and general, with the result that they do not pertain to any one functional domain, program, role, or individual.[24] They exist as very general criteria that can be selectively invoked to help organize roles into programs or to mobilize individuals to play roles; however, their application to roles and programs is made possible by additional mechanisms such as ideologies, laws, technologies, and norms. For, by themselves, societal values are too general and abstract for individuals to use in concrete situations. Indeed, one of the most conspicuous features of highly differentiated systems is the evolution of mechanisms to attach abstract values to concrete roles and programs.

6. Evolution involves the movement through three distinctive forms of differentiation: (a) segmentation, (b) stratification, and (c) functional differentiation.[25] That is, the five processes outlined above have operated historically to create, Luhmann believes, only three distinctive forms of differentiation. When the simplest societies initially differentiate, they do so segmentally in that they create like and equal subsystems that operate very much like the ones from which they emerged. For example, as it initially differentiates, a traditional society will create new lineages, or new villages, that duplicate previous lineages and villages. But segmentation limits a society's complexity and, hence, its capacity to adapt to its environment. And so, alternative forms of differentiation are selected during sociocultural evolution. Further differentiation creates stratified systems in which subsystems vary in terms of their power, wealth, and other resources. These subsystems are ordered hierarchically, and this new form of structure allows for more complex relations with an environment but imposes limitations on how complex the system can become. For as long as the hierarchical order must be maintained, the options of any subsystem

---

[24]Luhmann is borrowing here from Émile Durkheim's analysis in *The Division of Labor in Society* (New York: Free Press, 1949; originally published in 1893) as well as from Talcott Parsons' discussion of "value generalization." See Chapter 3.

[25]Luhmann, *The Differentiation*, pp. 229–54.

are limited by its place in the hierarchy.[26] Thus pressures build for a third form of differentiation, the functional. Here communication processes are organized around the specific function to be performed for the societal system. Such a system creates inequalities because some functions have more priority for the system (e.g., economics over religion). This inequality is, however, fundamentally different from that in hierarchically ordered or stratified systems. In a functionally differentiated society, the other sub-systems are part of the environment of any given subsystem—for example, organizations in the polity, law, education, religions, science, and family domains are part of the environment of the economy. And although the economy may have functional priority in the society, it treats and responds to the other subsystems in its environment as equals. Thus inequality in functionally differentiated societies does not create a rigid hierarchy of subsystems; as a consequence, it allows for more autonomy of each sub-system, which in turn gives them more flexibility in dealing with their respective environments.[27] The overall consequence of such subsystem au-tonomy is increased flexibility of the societal system to adjust and adapt to its environment.

7. Evolutionary differentiation increases the complexity of a system and its relationship with the environment. In so doing, it escalates the risks, as Luhmann terms the matter, of making incorrect and maladaptive decisions about how to relate to an environment. For with increased com-plexity there is an expanded set of options for a system, but there is a corresponding chance that the selection of options will be dysfunctional for a system's relationship to an environment. For example, any organi-zation in the economy must make decisions about its actions, but there are increased alternatives and escalated unknowns, resulting in expanded risks. In Luhmann's view, the ever-increasing risk level that accompanies evolutionary differentiation must be accompanied by mechanisms to reduce risk, or at least by the perception or sense that risk has been reduced. Thus evolution always involves an increase in the number and complexity of risk-reducing mechanisms. Such mechanisms also decrease the complexity of a system's environment because they select some options over others. For example, a conservative political ideology is a risk-reducing mechanism because it selects some options from more general values and ignores others. In essence, an ideology assures decision makers that the risks are reduced by accepting the goals of the ideology.[28]

Before proceeding further, let me now summarize these elements of Luhmann's view of evolution in terms of how they change a society's and its constituent subsystem's relation to the temporal, material, and symbolic di-mensions of the environment. Temporally, Luhmann argues that social evo-

---

[26]Ibid., p. 235.

[27]I am not asserting, of course, that this assertion by Luhmann is true.

[28]Luhmann, *The Differentiation*, p. 151.

lution and differentiation lead to efforts at developing a chronological metric, or a standardized way to measure time (clocks, for example). Equally fundamental is a shift in people's perspective from the past to the future. The past becomes highly generalized and lacks specific dictates of what should be done in the present and the future. For, as systems become more complex, the past cannot serve as a guide to the present or future because there are too many potential new contingencies and options. The present sees time as ever more scarce and in short supply; thus people become more oriented to the future and the consequences of their present actions. Materially, social differentiation involves: (1) the increasing separation of interaction, organization, and societal systems; (2) the compartmentalization of organization systems into functional domains; (3) the growing separation of person, role, program, and values; and (4) the movement toward functional differentiation and away from segmentation and stratification. And, symbolically, communication codes become more complex and organized as distinctive media for a particular functional domain. Moreover, they increasingly function as risk-reducing mechanisms for a universe filled with contingency and uncertainty.

It is with this overall view of sociocultural evolution and differentiation that Luhmann has approached the study of specific organizational systems. As he has consistently argued, an analytical framework is only as good as the insights into empirical processes that it can generate. I find Luhmann's framework much more complex than he contends such a framework should be, and I see it as more metaphorical than analytical.[29] However, it has allowed him to analyze political, legal, and economic processes in functionally differentiated societies in a very intriguing way.

THE FUNCTIONAL DIFFERENTIATION OF SOCIETY

Politics as a Social System

As societies grow more complex, new structures for reducing complexity emerge. Old processes, such as appeals to traditional truths, mutual sympathy, exchange, and barter, become ever more inadequate. A system that reaches this point of differentiation, Luhmann argues, must develop the "capacity to make binding decisions." Such capacity is generated out of the problems of increased complexity, but it also becomes an important condition for further differentiation.

In order to make binding decisions, the system must use a distinctive medium of communication: power.[30] Power is defined by Luhmann as "the possibility of having one's own decisions select alternatives or reduce complexity for others." Thus, whenever one social unit selects alternatives of action for other units, power is being employed as the medium of communication.

---

[29]Indeed, my discussion has simplified it considerably.
[30]Luhmann, *The Differentiation*, p. 151.

The use of power to make binding decisions functions to resolve conflicts, to mitigate tensions, and to coordinate activities in complex systems. Societies that can develop political systems capable of performing these functions are better able to deal with their environments. Several conditions, Luhmann believes, facilitate the development of this functional capacity. First, there must be time to make decisions; the less time an emerging political system is allowed, the more difficulty it will have in becoming autonomous. Second, the emerging political system must not confront a single power block in its environment, such as a powerful church. Rather, it requires an environment of multiple subsystems whose power is more equally balanced. So, the more the power in the political subsystem's environment is concentrated, the more difficult is its emergence as an autonomous subsystem. Third, the political system must stabilize its relations with other subsystems in the environment in two distinctive ways: (1) at the level of diffuse legitimacy, so that its decisions are accepted as its proper function; and (2) at the level of daily transactions among individuals and subsystems.[31] That is, the greater the problems of a political system in gaining diffuse support for its right to make decisions for other subsystems and the less salient the decisions of the political system for the day-to-day activities, transactions, and routines of system units, then the greater will be its problems in developing into an autonomous subsystem.

Thus, to the extent that a political system has time to develop procedures for making decisions, confront multiple sources of mitigated power, and achieve diffuse legitimacy as well as relevance for specific transactions, then the more it can develop into an autonomous system and the greater will be a society's capacity to adjust to its environment. In so developing, the political system must achieve what Luhmann calls *structural abstraction,* or the capacity to (1) absorb multiple problems, dilemmas, and issues from a wide range of system units and (2) make binding decisions for each of these. Luhmann sees the political system as "absorbing" the problems of its environment and making them internal to the political system. Several variables, he argues, determine the extent to which the political system can perform this function: (1) the degree to which conflicts are defined as political (instead of moral, personal, etc.) and therefore in need of a binding decision, (2) the degree of administrative capacity of the political system to coordinate activities of system units, and (3) the degree of structural differentiation within the political system itself.

This last variable is the most crucial in Luhmann's view. In response to environmental complexity and the need to absorb and deal with problems in this environment, the political system must differentiate along three lines: (1) the creation of a stable bureaucratic administration that executes decisions, (2) the evolution of a separate arena for politics and the emergence of political parties, and (3) the designation of the public as a relevant concern in making binding decisions. Such internal differentiation increases the capacity of the political system to absorb and deal with a wide variety of problems; as a consequence, it allows for greater complexity in the societal system.

---

[31]Ibid., pp. 143–44.

This increased complexity of the political and societal systems also increases the risks of making binding decisions that are maladaptive. For, as complexity increases, there are always unknown contingencies. Therefore, political systems not only develop mechanisms such as internal differentiation for dealing with complexity, but they also develop mechanisms for reducing risk or the perception of risk. One mechanism is the growing reflexiveness of the political process—that is, its increased reflection upon itself. Such reflection is built into the nature of party politics where the manner and substance of political decisions are analyzed and debated. Another mechanism is what Luhmann calls the *positivation of law,* or the creation of a separate legal system that makes "laws about how to make laws" (more on this shortly in the next section). Yet another mechanism is ideology or symbolic codes that select which values are relevant for a particular set of decisions. A related mechanism is the development of a political code that typifies and categorizes political decisions into a simple typology.[32] For example, the distinction between progressive and conservative politics is, Luhmann argues, an important political code in differentiated societies. Such a code is obviously very general, but this is its virtue because it allows very diverse political acts and decisions to be categorized and interpreted with a simple dichotomy, thereby giving political action a sense of order and reducing perceptions of risk. Luhmann goes so far as to indicate that it is a system's capacity to develop a political code, more than consensus of values, that leads to social order. For in interpreting actions in terms of the code, a common perspective is maintained, but it is a perspective based on differences—progressive versus conservative—rather than on commonality and consensus. Thus, complex social orders are sustained by their very capacity to create generalized and binary categories for interpreting events rather than by value consensus.

Still another mechanism for reducing risks is arbitrary decision making by elites. However, although such a solution achieves order, it undermines the legitimacy of the political system in the long run because system units come to resent and to resist arbitrary decision making. And a final mechanism is invocation of a traditional moral code (for example, fundamentalistic religious values) that, in Luhmann's terms, "remoralizes" the political process. But when such remoralization occurs, the political system must de-differentiate because strict adherence to a simple moral code precludes the capacity to deal with complexity (an example of this process would be Iran's return to a theocracy from its previously more complex political system).

In sum, then, I think it fair to say that Luhmann uses his conceptual metaphor to analyze insightfully specific institutional processes, such as government. Yet he does not use his scheme in a rigorous deductive sense; much like Parsons before him, he employs the framework as a means for denoting and highlighting particular social phenomena. Although I think that much of his analysis of political system differentiation is "old wine in new bottles,"

---

[32]Ibid., pp. 168–89.

there is a shift in emphasis and, as a result, some intriguing but imprecise insights. In a similar vein, Luhmann analyzes the differentiation of the legal system and the economy.

The Autonomy of the Legal System

As I discussed earlier, Luhmann visualizes social evolution as involving a separation of persons, roles, programs, and values. For him, differentiation of structure occurs at the level of roles and programs. Consequently, there is the problem of how to integrate values and persons into roles organized into programs within organization systems. The functional mechanism for mobilizing and coordinating individuals to play roles is law, whereas the mechanism for making values relevant to programs is ideology.[33] Thus, because law regulates and coordinates people's participation in roles and programs and because social differentiation must always occur at the roles level, it becomes a critical subsystem if a society is to differentiate and evolve. That is, a society cannot become complex without the emergence of an autonomous legal system to specify rights, duties, and obligations of people playing roles.[34]

A certain degree of political differentiation must precede legal differentiation, since there must be a set of structures to make decisions and enforce them. But political processes often impede legal autonomy, as is the case when political elites have used the law for their own narrow purposes. For legal autonomy to emerge, therefore, political development is not enough. Two additional conditions are necessary: (1) "the invocation of sovereignty," or references by system units to legal codes that justify their communications and actions; and (2) "lawmaking sovereignty," or the capacity of organizations in the legal system to decide just what the law will be.

If these two conditions are met, then the legal system can become increasingly reflexive. It can become a topic unto itself, creating bodies of procedural and administrative law to regulate the enactment and enforcement of law. In turn, such procedural laws can themselves be the subject of scrutiny. Without this reflexive quality the legal system cannot be sufficiently flexible to change in accordance with shifting events in its environment. Such flexibility is essential because only through the law can people's actions be tied to the roles that are being differentiated. For example, without what Luhmann calls the "positivization of law," or its capacity to change itself in response to altered circumstances, new laws and agencies (e.g., workers' compensation, binding arbitration of labor/management disputes, minimum wages, health and safety) could not be created to regulate people's involvement in roles (in this case, work roles in a differentiating economy).

---

[33]Ibid., pp. 90–137.

[34]This is essentially the same conclusion Parsons reached in his description of evolution in *Societies: Evolutionary and Comparative Perspectives* and *The System of Modern Societies* (Englewood Cliffs, NJ: Prentice-Hall, 1966 and 1971, respectively).

Thus, positivization of the law is a critical condition for societal differentiation. It reduces complexity by specifying relations of actors to roles and relegating cooperation among social system units. But it reduces complexity in a manner that presents options for change under new circumstances; thus it becomes a condition for the further differentiation of other functional domains, such as the economic.

The Economy as a Social System

I think that Luhmann defines the economy in a needlessly elliptical way. For Luhmann, the economy has the function of "deferring a decision about the satisfaction of needs while providing a guarantee that they will be satisfied and so utilizing the time thus acquired."[35] The general idea seems to be that economic activity—production and distribution of goods and services—functions to satisfy basic or primary needs for food, clothing, and shelter as well as derived or secondary needs for less basic goods and services. But it does so in a way not fully appreciated in economic analysis: it restructures humans' orientation to time because economic action is oriented to the satisfaction of future needs. Present economic activity is typically directed at future consumption; so, when a person works and a corporation acts in a market, they are doing so to guarantee that their future will be unproblematic.

I think that Luhmann's definition of the economic subsystem is less critical than his analysis of the processes leading to the creation of an autonomous economic system in society. In traditional and undifferentiated societies, Luhmann argues, only small-scale solutions are possible with respect to doing something in the present to satisfy future needs. One solution is stockpiling of goods, with provisions for the redistribution of stocks to societal members or trade with other societies.[36] Another solution is mutual assistance agreements among individuals, kin groups, or villages. But such patterns of economic organization are very limited because they merge familial, political, religious, and community activity. Only with the differentiation of distinctly economic roles can more complexity and flexibility be structured into economic action. The first key differentiation along these lines is the development of markets with distinctive roles for buyers and sellers.

A market performs several crucial functions. First, it sets equivalences or the respective values of goods and services. Second, it neutralizes the relevance of other roles—for instance, the familial, religious, and political roles of parties in an exchange. Value is established in terms of the qualities of respective goods, not the positions or characteristics of the buyers and sellers.[37] Third, markets inevitably generate pressures for a new medium of communication that is not tied to other functional subsystems. This medium is money, and it

---

[35]Luhmann, *The Differentiation*, p. 194.

[36]Ibid., p. 197.

[37]Luhmann fails to cite the earlier work of Georg Simmel on these matters. See Simmel's *The Philosophy of Money*.

allows for quick assessments of equivalences and value in terms of an agreed-upon metric. In sum, then, markets create the conditions for the differentiation of distinctly economic roles, for their separation and insulation from other societal roles, and for the creation of a uniquely economic medium of communication.

Money is a very unusual medium, Luhmann believes, because it "transfers complexity." Unlike other media, money is distinctive because it does not reduce complexity in the environment. For example, the medium of power is used to make decisions that direct activity, thereby reducing the complexity of the environment. The medium of truth in science is designed to simplify the understanding of a complex universe. And the medium of love in the family circumscribes the actions and types of relations among kindred and, in so doing, reduces complexity. In contrast, money is a neutral vehicle that can always be used to buy and sell many different things. It does not limit; in fact, it opens options and creates new opportunities. For example, to accept money for a good or for one's work does not reduce the seller's or worker's options. The money can be used in many different ways, thereby preserving and even increasing the complexity of the environment. Money thus sets the stage for—indeed it encourages—further internal differentiation in the economic subsystem of a society.

In addition to transferring complexity, Luhmann sees money as dramatically altering the time dimension of the environment. Money is a liquid resource that is always "usable in the future." When we have money, it can be used at some future date—whether the next minute or the following year. Money thus collapses time, since it is to be used in the future, hence making the past irrelevant; and the present is defined in terms of what will be done with money in the future. However, this collapsing of time can come about only if (1) money does not inflate over time and (2) it is universally used as the medium of exchange (that is, barter, mutual assistance, and other traditional forms of exchange do not still prevail).[38]

Like all media of communication, money is reflexive. It becomes a goal of reflection, debate, and action itself. We can buy and sell money in markets; we can invest money to make more money; we can condemn money as the root of evil or praise it as a goal that is worth pursuing; we can hoard it in banks or spread it around in consumptive activity. This reflexive quality of money, coupled with its capacity to transfer complexity and reorient actors to time, is what allows money to become an ever more dominant medium of communication in complex societies. Indeed, the economy becomes the primary subsystem of complex societies because its medium encourages constant increases in complexity and growth in the economic system. As a consequence, the economy becomes a prominent subsystem in the environment of other functional subsystems—that is, science, polity, family, religion, and education. In fact, it becomes something that must always be dealt with by these other subsystems.

---

[38]Luhmann, *The Differentiation*, p. 207.

This growing complexity of the economic subsystem increases the risks in human conduct. The potential for making a mistake in providing for a person's future needs or a corporation's profits increases because the number of unknown contingencies dramatically multiplies. Such escalated risks generate pressures, Luhmann argues, for their reduction through the emergence of specific mechanisms. The most important of these mechanisms is the tripart internal differentiation of the economy around (1) households, (2) firms, and (3) markets.[39] There is a "structural selection" for this division, Luhmann believes, because these are structurally and functionally different. Households are segmental systems (structurally the same) and are the primary consumption units. Firms are structurally diverse and the primary productive units. And markets are not so much a unit as a set of processes for distributing goods and services. Luhmann is a bit vague on this point, but it seems that there is strength in this correspondence of basically different structures with major economic functions. Households are segmented structurally and are functionally oriented to consumption; firms are highly differentiated structurally and are functionally geared to production; and markets are processually differentiated in terms of their function to distribute different types of goods and services. Such differentiation reduces complexity, but at the same time it allows for flexibility: households can change consumption patterns, firms can alter production, and markets can expand or contract. And, since they are separated from one another, each has the capacity to change and redirect its actions, independent of the others. It is this flexibility that allows the economic system to become so prominent in modern industrial societies.

Yet, Luhmann warns, the very complexity of the economy and its importance for other subsystems create pressures for other risk-reducing mechanisms. One of these is intervention by government so that power is used to make binding decisions on production, consumption, and distribution as well as on the availability of money as a medium of communication. The extensive use of this mechanism, Luhmann believes, reduces risk and complexity in the economy at the expense of its capacity to meet needs in the future and to make flexible adjustments to the environment.

AN ASSESSMENT OF LUHMANN'S NEOFUNCTIONALISM

I think the critical questions in assessing Luhmann's functionalism are: Does it avoid the problems of functional analysis that I have outlined in previous chapters? And does it add any new insights into the dynamics of social systems? Let me begin with the first question.

It seems to me that Luhmann simply ignores the problems of functional analysis. He assumes functional needs and imperatives for reducing complexity and for adaptation to the environment. Then he simply asserts that mecha-

---

[39]Ibid., p. 216.

nisms emerge to meet these needs. I feel there is much here that is illegitimately teleological: system needs miraculously produce the structures to meet these needs; or, in other words, effects produce causes. Luhmann does invoke a "social selection" argument in that pressures for adaptation and for reducing complexity create "selection pressures" for some of those system mechanisms that will reduce complexity and increase adaptation. But for any given mechanism these selection processes are rarely articulated. For example, if "positive law" has selective advantages, then what exactly were the selection pressures for its emergence? What happened historically to societies that did and did not develop positive law? And what are the reasons why they did or did not? Without answers to such questions, the theoretical explanation of an event is an illegitimate teleology. Effects produce the things that cause these effects.

More problematic, I think, is the view of explanation in Luhmann's approach: develop a conceptual scheme of categories and use these categories to describe social events. For me, Luhmann ends up simply describing with new words what has occurred historically without really explaining how and why these events occurred (except to state that they meet system needs or requisites). For example, to assert that ideologies meet the need to attach values to roles and programs in complex societies simply says that ideologies become prominent in complex societies. But it does not explain how and why this occurs.

Additionally, in an effort to avoid the complex architecture of Parsons' analytical scheme, Luhmann constructs a very loose, metaphorical, and frequently vague alternative. Even more than is the case with Parsons, I suspect, one must internalize the scheme and use it as a general and sensitizing metaphor for interpreting events. As a result, one does not explain by deduction, but by intuition. One senses that this or that process is a "mechanism" that "serves this or that function." Explanation becomes a matter of negotiating the right intuition among communities of Luhmann scholars who have internalized and who accept his concepts, categories, and metaphors.

Moreover, how would one ever refute Luhmann's scheme? When categories are vague and imprecise, when they are often metaphorical, and when explanation is by intuitive use of the scheme, could one ever construct a "test" of the scheme that might throw its utility into doubt? I sense that any such effort would be rejected by adherents to Luhmann's approach as "an incorrect interpretation of what he really meant." Thus the scheme will always remain intact, and all assaults on its virtue will be defined away.

In light of these problems, what does the scheme offer to sociological theorizing? I think that as an explanation of events it has little to offer cumulative theory, but, much like Parsons' scheme, it is often used to produce rich and robust substantive insights. Probably the most useful portions of the scheme are (1) the implicit attack on the evaluative substance of much evolutionary thinking, (2) the recognition that complex social systems are integrated in ways other than value consensus, and (3) the emphasis that social processes operate along temporal, material, and symbolic dimensions. In these areas there are many interesting generalizations to be developed. None of them needs the systems and functional trappings of Luhmann's scheme. One can, for example, analyze entrance/exit rules independently of viewing them as mechanisms for reducing complexity in organization systems. Or one can examine the effects of money on cognitive orientations to time without all the jargon about system adaptation.

Thus the scheme does produce insights that alternative theoretical approaches have not generated. So, if one finds Luhmann's functional metaphors useful, I see little wrong in using them to isolate interesting social processes. But once this is done, the system's jargon and its functional trappings *should be abandoned*. Then the process can be examined in ways that can yield testable propositions. Unfortunately, once one is committed to a system's functional scheme, it is typically difficult to abandon it. As a result, the scheme becomes a grand, self-reinforcing world view, and explanation becomes the intuitive use of the scheme to interpret and describe events.

Luhmann's systems functionalism is very much like other functionalisms. It is intriguing and seductive. But it can only explain by categorizing events as an instance of this or that element in the scheme. However, the scheme itself can never be tested against the facts of the empirical world. And so it is interesting metaphysics but poor theory.

CHAPTER 6

◆

General Systems Functionalism

EARLY SYSTEMS APPROACHES

One of the more cosmic goals of intellectual activity is to interpret all dimensions of the universe in terms of a common core of concepts. This has always been one of the appeals of religion, as well as of various philosophical systems. It is not surprising, therefore, that there would be efforts within science—the third of Auguste Comte's three modes of thought—to forge an intellectual scheme or "general system" of thought that could cut across the natural and social sciences. In a sense, Comte's famous "hierarchy of the sciences" represented an effort to articulate the relations of the sciences,[1] but the first sociological general systems theorist was Herbert Spencer.

As I noted in Chapter 2, Spencer saw himself as a philosopher who desired to demonstrate the unity of all realms of the universe with a common set of "first principles."[2] As with general systems theorists today, Spencer first turned to the physics of his time and culled from his understanding of this science a set of highly abstract laws, or principles, that could explain the physical, psychological,[3] organic,[4] superorganic (societal),[5] and even moral[6] dimensions of the universe. He conceptualized these laws under the rubrics of evolution and dissolution, as I emphasize in the next chapter.[7] In more modern terms, these

---

[1]For example, in Auguste Comte, *Positive Philosophy,* translation by H. Martineau (London: Bell & Sons, 1896; originally published between 1830 and 1842). Comte saw mathematics as the language of science, and he believed that the cumulation of knowledge in each science law in the hierarchy set the stage for the development of the next science up in the hierarchy. Hence sociology was seen to be possible only *after* the development of biology.

[2]Herbert Spencer, *First Principles* (New York: A. C. Burt, 1880; originally published in 1862).

[3]Herbert Spencer, *The Principles of Psychology* (New York: D. Appleton, 1880; originally published in 1855).

[4]Herbert Spencer, *The Principles of Biology* (New York: D. Appleton, 1897; originally published in 1866).

[5]Herbert Spencer, *The Principles of Sociology* (New York: D. Appleton, 1898; originally published in serial form from 1874 to 1896).

[6]Herbert Spencer, *The Principles of Ethics* (New York: D. Appleton, 1892-98).

[7]See pages 132 to 136.

processes are viewed in terms of *entropy* and *negative entropy*—that is, de-structuring to randomness (entropy) and structuring to complexity (negative entropy).

Spencer saw the matter of the universe as always in a process of (1) aggregating to form more complex structures, (2) fluctuating between phases of structuring and destructuring in a precarious equilibrium, and (3) disintegrating as elements begin to destructure and lose their coherence. Thus the systems of the universe were viewed by Spencer as composed of varying types of matter—physical, organic, psychical, moral—moving in space with "force" or energy and building up structure to the point where energy is dissipated and dissolution sets in. In the process of structuring, equilibrium is often achieved as forces of energy hold each other in balance, but eventually the equilibrium is disrupted, either by its own internal dissipation or by the introduction of new inputs of matter and energy from the environment.

I need not dwell on the details of Spencer's analysis,[8] because many of the parallels that he drew among systems of the universe are dated. Yet, more than any other thinker of his time, he anticipated the intent of modern systems theory, as we will come to appreciate in this chapter. Although some versions of general systems theory are decidedly antifunctionalist,[9] many are clearly within the functionalist orientation. For as points of unity between organismic and societal systems have been sought, the isomorphisms between these realms turned out to be their structuring in terms of common functions. That is, what common activities must organic and societal systems perform in order to survive in an environment? Thus, modern systems theory created a new kind of organismic analogy, and, as was the case for Spencer, this analogy usually brought along its inevitable sidekick: functionalism.

GENERAL SYSTEMS THEORY

The development of general systems theory occurred as one manifestation of fundamental changes in the nature of scientific analysis. Rather than viewing the universe in simple cause-and-effect terms, theorists increasingly recognized that any cause/effect relationship occurs within a more complex system of relationships. At all levels of reality, from subatomic physics to the analysis of world systems on earth or the formation of galaxies in the universe, phenomena are analyzed not in isolation but in terms of their connections to a larger system. Moreover, such systems of interconnected events reveal emergent properties that make the whole greater than the sum of its parts. In the modern era this philosophical position was given its most forceful expression by Alfred North Whitehead[10] in the 1920s and culminated in sociology with functionalism.

---

[8]See Jonathan H. Turner, *Herbert Spencer: A Renewed Appreciation* (Newbury Park, CA: Sage, 1985), for a more detailed argument.

[9]For example, Walter Buckley, *Sociology and Modern Systems Theory* (Englewood Cliffs, NJ: Prentice-Hall, 1967).

[10]Alfred North Whitehead, *Science and the Modern World,* Lowell Lectures 1925 (New York: Macmillan, 1953).

Among modern scientists, Ludwig von Bertalanffy was probably the first to argue explicitly for a general systems approach.[11] But in the post–World War II period this movement gained considerable impetus with a number of important insights by Norbert Wiener[12] in cybernetics, Shannon and Weaver[13] in information theory, and Neumann and Morgenstern[14] in game theory. Coupled with the rediscovery of Walter Cannon's[15] concept of homeostasis and a rapid development of computer sciences and their engineering applications, the base was established for the development of general systems theory.[16]

Of course, Comte's use of the organismic analogy, Spencer's analogies to the principles of physics[17] as well as his analysis of the social and biological spheres,[18] Pareto's equilibrium analysis,[19] and Durkheim's forceful cry that society is a reality, *sui generis,*[20] all emphasized the importance of viewing sociocultural phenomena in systemic terms. But in the early 1950s there was a genuine belief that systems *in general,* at all levels of reality, might reveal certain common properties and processes. If this were true, then the partitions among the sciences could be broken down.

Thus, in 1954, the Society for General Systems Research was organized to "(1) investigate the isomorphy of concepts, laws, and models in various fields, and to help in useful transfers from one field to another; (2) encourage the development of adequate theoretical models in fields which lack them; (3) minimize the duplication of theoretical effort in different fields; (4) promote the unity of science through improving communication among specialists."[21] The impact of the Society and its advocacy have been far-reaching.[22] Indeed, Parsonian action theory[23] incorporated the metaphor of information theory and cybernetics in its "informal hierarchy of control," but such adoptions of systems

---

[11]See his discussion in Ludwig von Bertalanffy, *General Systems Theory* (New York: Braziller, 1968), pp. 28–29. Also, see early references to his work in the bibliography, pp. 256–57.

[12]Norbert Wiener, *Human Use of Human Beings: Cybernetics and Society* (Garden City, NY: Doubleday, 1954; originally published in 1949).

[13]Claude Shannon and Warren Weaver, *The Mathematical Theory of Communication* (Urbana: University of Illinois Press, 1949).

[14]John von Neumann and O. Morgenstern, *Theory of Games and Economic Behavior* (Princeton, NJ: Princeton University Press, 1947).

[15]Walter B. Cannon, *The Wisdom of the Body* (New York: W. W. Norton, 1932) and "Organization for Physiological Homeostasis," *Physiological Review* 9 (1929), pp. 397–402.

[16]A term that was first used by Bertalanffy, *General Systems Theory.*

[17]See Herbert Spencer, *First Principles.*

[18]See Herbert Spencer, *Principles of Biology* and *Principles of Sociology.*

[19]Vilfredo Pareto, *Manual on Political Economy* (New York: A. M. Kelly, 1971; originally published in 1906 and 1909); *Treatise on General Sociology* (New York: Harcourt, Brace, 1935; originally published in 1916); *The Rise and Fall of Elites* (Totowa, NJ: Bedminster Press, 1968; originally published in 1901).

[20]See Émile Durkheim, *The Rules of the Sociological Method* (New York: Free Press, 1938; originally published in 1895).

[21]Bertalanffy, *General Systems Theory,* p. 15.

[22]Its official publication is the annual *General Systems: Yearbook of the Society for General Systems* (more typically listed under *General Systems Yearbook*).

[23]See Chapter 3.

theoretic concepts tend to be metaphorical. Yet there are a few who have sought to demonstrate in more precise terms the utility of systems concepts for understanding sociocultural phenomena.[24]

BASIC CONCEPTS OF SYSTEMS THEORY

Much as Herbert Spencer recognized in the 19th century, modern systems theorists stress that all systems reveal certain common properties. We can best visualize these common properties under the following topic headings: (1) energy/matter and information, (2) entropy and negative entropy, (3) organization and system, (4) open and closed systems, (5) cybernetic systems, and (6) system levels.

Energy, Matter, and Information

Systems theory begins with the concepts of energy/matter and information. Matter is anything that occupies physical space; energy and matter are, of course, equivalent in the sense of Einstein's famous equation $E = mc^2$, but, more practically, energy is the capacity to generate movement of matter and information. Information is oftentimes defined narrowly in terms of the number and pattern of binary bits involved in sending a signal, but at other times the meaning, or substantive content and significance, of signals is also seen as involved in a definition of information.

The great insight over the last 50 years is that information is a key element in most systems. Whether contained in the RNA and DNA molecules of a living cell, in the learned memory cells of a living organism, or in the cultural traditions of a society, the release of energy and the organization of matter are guided by informational controls. The analysis of information can become highly technical, but most systems theorists begin with the simple insight that much of the universe involves the organization of matter and energy by information.

Entropy and Negative Entropy

The Second Law of Thermodynamics of classical physics states that matter and energy tend toward their least organized state. Thus organization of a solar

---

[24]In particular, see: Alfred Kuhn, *The Logic of Social Systems* (San Francisco: Jossey-Bass, 1974); Walter Buckley, *Sociology and Modern Systems Theory* and his edited volume, *Modern Systems Research for the Behavioral Scientist* (Chicago: Aldine, 1968); James Grier Miller, *Living Systems* (New York: McGraw-Hill, 1978); Richard N. Adams, *The Eighth Day: Social Evolution and the Self-Organization of Energy* (Austin: University of Texas Press, 1988); Kenneth Boulding, *Ecodynamics: A New Theory of Societal Evolution* (Beverly Hills, CA: Sage, 1978); C. West Churchman, *The Systems Approach* (New York: Dell, 1968); Erich Jantsch, ed., *The Evolutionary Vision: Toward a Unifying Paradigm of Physical, Biological, and Sociological Evolution* (Boulder, CO: Westview Press, 1981); Kenneth E. Bailey, *Social Entropy Theory* (Albany, NY: SUNY Press, 1989). See the bibliographies of these works for an overview of the diversity of work on general systems theory. As is also evident, the general systems and human ecology perspectives overlap, at least to some degree.

system, an organic body, or a society is viewed as a temporary state of affairs, since in the long run these entities will lose their organization and degenerate. This process of "de-organization" is termed *entropy;* systems theory studies "negative entropy" because it is concerned with how energy, matter, and information become organized into coherent systems.

Organization and System

Organization refers to the ways in which identifiable units of energy, matter, and information are interrelated. When such interrelations reveal some degree of coherence, they constitute a system. The main focus of all systems theory thus becomes one of isolating different types of units, or subsystems, of a larger system and then specifying the nature of their relations to one another. For example, this analysis might examine the properties of a particular subsystem, such as its organization of matter and energy by information. Then it might focus on the inputs and outputs of energy/matter and information to and from this subsystem. Finally, it might explore the pattern of organization among this and various other subsystems whose inputs and outputs influence one another.

Open and Closed Systems

Closed systems do not engage in exchanges of energy, matter, or information with their surrounding environment. Their "boundary" is thus relatively impermeable. In contrast, open systems engage in exchanges across their boundaries with the surrounding environment. Open systems are most likely to be negatively entropic, since they take in energy, matter, and information; then they use them to build greater levels of organization. The study of open systems will therefore involve an analysis of the internal properties of the system (its subsystems and interrelations), the properties and processes of its boundary, the nature of the environment, and the exchanges of energy, matter, and information between the system and its environment.

Cybernetic Systems

Systems that act upon and receive inputs from the environment are often cybernetic. The term *cybernetic* denotes the process whereby a system is self-regulating in an environment. Originally the concept of cybernetics referred to systems that *(a)* reveal "normal" internal states that they seek to maintain, *(b)* exist in environments that exert pressures to alter these normal states, *(c)* have internal mechanisms or processes for correcting deviations from these normal states, and *(d)* evidence procedures for receiving information or feedback from the environment in order to check the state of the systems against the pressures of the environment.[25] Such feedback was typically viewed as

---

[25]Wiener, *Human Use of Human Beings.*

"negative feedback" because the information is used to record movement of the system beyond its normal limits and to activate corrective processes. Walter Cannon's discussion of homeostasis of temperature levels in the human body represented an early analysis of such a cybernetic system (although he did not have available to him the term *cybernetic*).[26] A thermostatic system for controlling room temperature is another example of a mechanical cybernetic system. And Talcott Parsons' concern in *The Social System* with "mechanisms of social control" is another application of cybernetic ideas to sociocultural systems.[27]

Later the concept of cybernetic systems was extended to include "positive feedback" or "deviation amplifying" feedback.[28] Not all systems are programmed to maintain themselves within certain limits or parameters. Rather, they have goals that involve growth, expansion, and increased adaptation. Or they may not have any goals except perhaps undefined states of survival. Such systems receive feedback on their current actions in an environment and on the extent to which these actions help realize those goals that facilitate adaptation and survival. When this feedback is positive and indicates that current actions are facilitating survival or the realization of goals, then they will be continued and escalated. And if these renewed actions also have positive results, they will be further escalated. In this escalating and cyclical process, transformation in, rather than self-regulation of, systems is likely. In biological systems the process of speciation by natural selection is one example of how positive feedback processes in ecological systems operate to generate a new species. Random mutations in organisms create attributes that facilitate survival in a given environment; subsequent mutations of this type further facilitate survival; and so on, until a new species can be distinguished from its ancestors. In contrast, an imperialistic society that successfully conquers its neighbors can use this positive feedback to seek further conquests of more distant neighbors, thereby transforming itself as it conquers more and more societies.

Most organic and social systems are cybernetic in that they engage, often simultaneously, in processes of self-regulation and deviation amplification. They use negative feedback to engage in self-regulation and positive feedback to increase their level of organization or their capacity to realize goals and adapt to a particular environment.

System Levels

Although general systems theory seeks to develop a common set of concepts and principles for understanding all systemic phenomena, there is a clear recognition that there are different types of systems. Such types are typically viewed as emergent levels of organization, with each emergent level revealing

---

[26]Cannon, *Wisdom of the Body.*

[27]See Chapter 3.

[28]Magoroh Maruyama, "The Second Cybernetics: Deviation Amplifying Mutual Causal Processes," *American Scientist* 51 (1963), pp. 164–79.

at least some distinctive properties that mark it off from lower- and higher-level systems. For example, organic systems are built from inorganic systems and are not understandable in terms of the properties of these inorganic systems. Similarly, among organic systems there are distinctive and emergent levels. For instance, James G. Miller argues that there are seven emergent levels among "living systems": the cell, organ, organism, group, organization, society, and supranational.[29] Similarly, Kenneth Boulding isolates the following levels in the "hierarchy of systems": (1) "static structures" such as crystals, molecules, etc.; (2) "clock works" like machines, solar systems, etc.; (3) "control mechanisms" such as thermostats and homeostatic mechanisms in organisms; (4) "open systems" like cells and organisms in general; (5) "lower organisms"; (6) "animals"; (7) "humans" and their capacities for self and thinking; (8) "sociocultural" systems; and (9) "symbolic systems" such as language, logic, mathematics, and so forth.[30] Thus systems theory seeks, on the one hand, to use similar concepts to understand all systemic phenomena, but on the other hand it attempts to isolate generic types of systems and to develop concepts and propositions that apply to distinctive system levels.

These, then, are the guiding ideas behind general systems theories. There are a number of such theories, with the result that it is difficult to select any one for more detailed coverage. In a sense, I have selected the most comprehensive approach—that of James G. Miller[31]—in order to illustrate this perspective; yet I should caution that it is not the most sophisticated or sociological of the current general systems theories.[32] Nonetheless, Miller's approach represents an excellent illustration of both the problems and prospects of general systems ideas as they are applied to sociocultural systems.

LIVING SYSTEMS: JAMES G. MILLER'S ANALYSIS

The most ambitious general systems approach since Herbert Spencer's is that developed by James Grier Miller. His efforts have culminated in *Living Systems,* an encyclopedic work of over 1000 pages. This effort is reminiscent of Spencer's Synthetic Philosophy, which sought to isolate common concepts and propositions for various system levels. But, in contrast to Spencer, who was also willing to address inorganic systems, Miller confines his analyses to what he perceives as the seven generic levels of "living systems." To qualify as a living system, phenomena must evidence the following characteristics:[33]

1. They must be open systems.

---

[29]James G. Miller, *Living Systems.*

[30]Kenneth Boulding, *The Image* (Ann Arbor: University of Michigan Press, 1956).

[31]Miller, *Living Systems.*

[32]For example, I see Kenneth Boulding's *Ecodynamics,* Richard N. Adams' *The Eighth Day,* and Kenneth Bailey's *Social Entropy Theory* as better theories.

[33]Miller, *Living Systems.* I have collapsed some of Miller's categories, but the criteria are the same.

2. They must maintain a state of negative entropy in their environment.
3. They must reveal a minimum level of differentiation into integrated subparts.
4. They must evidence informational control through genetic material composed of DNA or reveal a charter.
5. They must reveal the chemical compounds of all organic life.
6. They must evidence a decider subsystem which regulates and controls other subsystems and relations with the environment.

As noted above, there are seven distinctive system levels of *life* that reveal these characteristics: (1) supranationals or systems of societies, (2) societies, (3) organizations, (4) groups, (5) organisms, (6) organs, and (7) cells. These seven levels constitute a hierarchy in that each is composed of all those below it in the hierarchy. For example, organs are composed of cells; organisms are constructed from cells and organs; organizations are composed of cells, organs, organisms, and groups; and so on to the most inclusive system level, a system of societies. Moreover, each system level will evidence greater numbers of subsystems and more complex arrangements of these subsystems. Hence each new level will reveal its own unique and more complex qualities. But all system levels evidence certain common subsystems. Miller outlines 19 such subsystems, which are defined in Table 6-1.[34] As is evident, these subsystems are concerned with how either energy/matter or information is processed. The basic idea is that subsystems perform unique "functions" in regulating such critical matters as inputs, internal coordination, outputs, boundaries, and reproduction of new units. Thus Miller's goal is to create a common category system for analyzing critical processes in the seven levels of living systems.[35] Near the end of his book this intent is made explicit in a large table in which examples of the 19 subsystems are offered for each of the seven system levels. This table is reproduced in Table 6-2.[36]

Miller views Table 6-1 as the equivalent of the periodic table in chemistry in that it orders the crucial "elements" of living systems, primarily in terms of the functions of each subsystem for the large systemic whole. The scholarship and breadth of Miller's analysis of each cell in Table 6-2 are difficult to communicate in a short summary of his work. But more significant than these classificatory efforts, however, are his attempts at developing some propositions about critical processes occurring at various sytem levels.[37] These propositions are phrased as "hypotheses," primarily because they are inducted from systematic reviews of the research literature for each system level. As a result, the

---

[34]Ibid., p. 3.

[35]In many ways Miller and Talcott Parsons both use a biological model, or classificatory view of theory. Miller's system has more categories, but the intent is the same as in Parsons' scheme.

[36]Miller, *Living Systems,* pp. 1028-29.

[37]Ibid., pp. 89-119.

TABLE 6–1 Functional Subsystems of All Living Systems

Subsystems Which Process Both Energy/Matter and Information

1. *Reproducer:* the subsystem which is capable of giving rise to other systems similar to the one it is in.

2. *Boundary:* the subsystem at the perimeter of a system that holds together the components which make up the system, protects them from environmental stresses, and excludes or permits entry to various sorts of energy/matter and information.

Subsystems Which Process Energy/Matter	*Subsystems Which Process Information*
3. *Ingestor:* the subsystem which brings energy/matter across the system boundary from the environment.	11. *Input transducer:* the sensory subsystem which brings markers bearing information into the system, changing them to other energy/matter forms suitable for transmission within it.
	12. *Internal transducer:* the sensory subsystem which receives, from subsystems or components within the system, markers bearing information about significant alterations in those subsystems or components, changing them to other energy/matter forms of a sort which can be transmitted within it.
4. *Distributor:* the subsystem which carries inputs from outside the system or outputs from its subsystems around the system to each component.	13. *Channel and net:* the subsystem composed of a single route in physical space, or multiple interconnected routes, by which markers bearing information are transmitted to all parts of the system.
5. *Converter:* the subsystem which changes certain inputs to the system into forms more useful for the special processes of that particular system.	14. *Decoder:* the subsystem which alters the code of information input to it through the input transducer or internal transducer into a *private* code that can be used internally by the system.
6. *Producer:* the materials being synthesized for growth, damage repair, replacement of components of the system, the energy for outputs of products or information.	15. *Associator:* the subsystem which carries out the first stage of the learning process, forming enduring associations among items of information in the system.
7. *Energy/matter storage:* the subsystem which retains in the system, for different periods of time, deposits of various sorts of energy/matter.	16. *Memory:* the subsystem which carries out the second stage of the learning process, storing various sorts of information in the system for different periods of time.
	17. *Decider:* the executive subsystem which receives information inputs from all other subsystems and transmits to them information outputs that control the entire system.
	18. *Encoder:* the subsystem which alters the code of information input to it from other information processing subsystems, from a *private* code used internally by the system into a *public* code which can be interpreted by other systems in its environment.

TABLE 6–1 *(continued)*

Subsystems Which Process Energy/Matter	*Subsystems Which Process Information*
8. *Extruder:* the subsystem which transmits energy/matter out of the system in the forms of products or wastes.	19. *Output transducer:* the subsystem which puts out markers bearing information from the system, changing markers within the system into other energy/ matter forms which can be transmitted over channels in the system's environment.
9. *Motor:* the subsystem which moves the system or parts of it in relation to part or all of its environment or moves components of its environment in relation to each other.	
10. *Supporter:* the subsystem which maintains the proper spatial relationship among components of the system, so that they can interact without weighting each other down or crowding each other.	

propositions are somewhat disjointed. To communicate Miller's intent, I have rephrased a number of the most central hypotheses and presented them in Table 6–3.

The propositions in Table 6–3 represent hypotheses that seem to apply to several types of living systems. One can see how they might be relevant to various types of social forms, although it is not clear that they improve upon existing sociological formulations. For example, proposition 1 restates Spencer's and Durkheim's old "size, differentiation, and integration" propositions (see, for example, Chapter 7, pp. 132–134), as these have been extended in several literatures, particularly organization theory[38] and human ecology.[39] Moreover, in looking at other propositions, it is not clear that translating sociological terms into "energy" and "matter" adds very much to analysis; indeed, in an effort to achieve isomorphism with other living systems, these translations make sociology even more vague than it already is.

Yet there is still something fascinating in the ability to view all living systems in terms of certain common functions. And of course this has always

---

[38]See, for example, "A Formal Theory of Differentiation in Organizations," *American Sociological Review* 35 (1970), pp. 201–18; Jerald Hage, Michael Aiken, and Cora Bagley Marrett, "Organizational Structure and Communications," *American Sociological Review* 36 (1971), pp. 860–71; Gerry E. Hendershot and Thomas F. James, "Size and Growth as Determinants of Administrative-Production Ratios in Organizations," *American Sociological Review* 37 (1972), pp. 149–53; Thomas F. James, "System Size and Structural Differentiation in Formal Organizations," *Sociological Quarterly* 16 (1975), pp. 124–30; Theodore R. Anderson and Seymour Warkov, "Organizational Size and Functional Complexity," *American Sociological Review* 26 (1961), pp. 23–28; Marshall M. Meyer, "Size and the Structure of Organizations," *American Sociological Review* 37 (1972), pp. 434–41.

[39]For example, see: Amos H. Hawley, *Human Ecology* (Chicago: University of Chicago Press, 1986); Brian J. Berry and John D. Kasarda, *Contemporary Urban Ecology* (New York: Macmillan, 1977); Amos H. Hawley, ed., *Societal Growth: Processes and Implications* (New York: Free Press, 1979); John D. Kasarda, "The Structural Implications of Social System Size," *American Sociological Review* 39 (1974), pp. 19–28.

TABLE 6–2 Examples of Functional Subsystems in Living Systems

Level / Subsystem	Cell	Organ	Organism	Group	Organization	Society	Supranational System
Reproducer	Chromosome	*None:* downwardly dispersed to cell level	Genitalia	Mating dyad	Group that produces a charter for an organization	Constitutional convention	Supranational system which creates another supranational system
Boundary	Cell membrane	Capsule of viscus	Skin	Sergeant at arms	Guard of an organization's property	Organization of border guards	Supranational organization of border guards
Ingestor	Gap in cell membrane	Input artery of organ	Mouth	Refreshment chairman	Receiving department	Import company	Supranational system officials who operate international ports
Distributor	Endoplasmic reticulum	Blood vessels of organ	Vascular system	Mother who passes out food to family	Driver	Transportation company	United Nations Childrens Fund (UNICEF), which distributes food to needy children
Converter	Enzyme in mitochondrion	Parenchymal cell	Upper gastrointestinal tract	Butcher	Oil refinery operating group	Oil refinery	European Atomic Energy Community (Euratom), concerned with conversion of atomic energy
Producer	Enzyme in mitochondrion	Parenchymal cell	*Unknown*	Cook	Factory production unit	Factory	World Health Organization (WHO)

Energy/matter storage	Adenosine triphosphate (ATP)	Intercellular fluid	Fatty tissues	Family member who stores food	Stock-room operating group	Warehouse company	International Red Cross, which stores materials for disaster relief
Extruder	Gap in cell membrane	Output vein of organ	Urethra	Cleaning woman	Delivery department	Export company	Component of the International Atomic Energy Agency (IAEA) concerned with waste extrusion
Motor	Microtubule	Muscle tissue of organ	Muscle of legs	*None:* Laterally dispersed to all members of group who move jointly	Crew of machine that moves organization personnel	Trucking company	Transport component of the North Atlantic Treaty Organization (NATO)
Supporter	Microtubule	Stroma	Skeleton	Person who physically supports others in group	Group that operates organization's building	National officials who operate public buildings and land	Supranational officials who operate United Nations buildings and land
Input transducer	Specialized receptor site of cell membrane	Receptor cell of sense organ	Exteroceptive sense organ	Lookout	Telephone operator group	Foreign news service	News service that brings information into supranational system
Internal transducer	Repressor molecule	Specialized cell of sinoatrial node of heart	Receptor cell that responds to changes in blood states	Group member who reports group states to decider	Inspection unit	Public opinion polling agency	Supranational inspection organization
Channel and net	Cell membrane	Nerve net of organ	Components of neural network	Group member who communicates by signals through the air to other members	Private telephone exchange	National telephone network	Universal Postal Union (UPU)

TABLE 6–2 Examples of Functional Subsystems in Living Systems *(continued)*

Level / Subsystem	Cell	Organ	Organism	Group	Organization	Society	Supranational System
Decoder	Molecular binding site	Receptor or second-echelon cell of sense organ	Cells in sensory nuclei	Interpreter	Foreign-language translation group	Language-translation unit	Supranational language-translation unit
Associator	*Unknown*	*Unknown*	*Unknown*	*None:* laterally dispersed to members who associate for group	*None:* downwardly dispersed to individual persons, organism level	Teaching institution	Supranational university
Memory	*Unknown*	*Unknown*	*Unknown*	Adult in a family	Filing departments	Library	United Nations library
Decider	Regulator gene	Sympathetic fiber of sinoatrial node of heart	Part of cerebral cortex	Head of a family	Executive office	Government	Council of Ministers of the European Communities
Encoder	Component producing hormone	Presynaptic region of output neuron of organ	Temporoparietal area of dominant hemisphere of human brain	Person who composes a group statement	Speech-writing department	Press secretary	United Nations Office of Public Information
Output transducer	Presynaptic membrane	Presynaptic region of output neuron of organ	Larynx	Spokesman	Public relations department	Office of national spokesman	Offical spokesman of the Warsaw Treaty Organization

TABLE 6–3 Some Basic Laws of Living Systems

I. The larger a *living system*, the more likely are (pp. 108–9):*
  A. patterns of structural differentiation among components.
  B. decentralized centers of decision making.
  C. interdependence of subsystems.
  D. elaborate adjustment processes.
  E. differences in input/output sensitivity of components.
  F. elaborate and varied outputs.

II. The greater the level of structural differentiation of components in a living system, the greater is:
  A. the number of echelons or ranks among subsystems (p. 92).
  B. the segregation of functions (p. 109).
  C. the ratio of information transmitted within rather than across boundaries (pp. 93, 103).

III. The greater the ratio of information to energy/matter processed across system boundaries as negative and positive feedback, the more likely a living system is to survive in its environment (p. 94).

IV. The more hierarchically differentiated a living system into echelons, the more likely is:
  A. the presence of discordant information among differentiated subsystems (p. 109).
  B. the utilization by *decider* subsystems (high-ranking) of information from memory banks than from lower-ranking echelons (pp. 99–100).

V. The greater the number of channels for processing information in a structurally differentiated living system, the less likely are:
  A. errors in transmission and reception of information (p. 96).
  B. strains and tensions among subsystems (pp. 97, 107).

VI. The greater the level of stress experienced by a living system from its environment, the greater is (pp. 106–7):
  A. the number of components devoted to its alleviation.
  B. the less manifest previous tensions among internal system components.
  C. the deviation of processes within the system, and each of its subsystems, from previously normal states.
  D. the difficulty of returning the system and subsystem processes to previously normal states after alleviation of the strain.

VII. The greater the level of segregation of subsystems in a living system, the greater is their level of conflict and the greater is the total information and/or energy/matter mobilized in each subsystem for resolving the conflict and the less information or energy/matter available for achieving overall system goals (p. 107).

*Page numbers in parentheses refer to pages in *Living Systems* where hypotheses are initially presented. *Note:* The above propositions have been rephrased and regrouped for more efficient presentation.

been the appeal of functionalism: to see events as embedded in, and part of, systemic wholes that must meet certain requisites in order to survive in an environment. What, then, can we conclude about general systems theory?

CONCLUSION

Some general systems approaches, as I mentioned earlier, are not functional. But most of these are not very *systemic,* either. What they usually involve is a modeling of variables with causal arrows and feedback loops in order to show the configurations of effects among various forces. For example, many of the models that I have drawn thus far as well as those to be drawn in subsequent chapters would be a general systems approach in these loose terms. But general systems theory is more than just abstract modeling of complex causal effects— direct, indirect, and feedback—among variables. It is also a search for iso- morphisms across different units and their ordering into systemic wholes. Miller and others[40] who take this next step usually begin to talk in terms of energy, matter, information, and related ideas as ways of bridging diverse, and perhaps very different, phenomena. My sense is that translating sociological concepts into another language in order to create conceptual unity across the sciences makes these concepts less precise. For example, notions of "infor- mation" do not communicate the properties denoted by more traditional con- cepts like norms, values, beliefs, ideology, etc.

This general systems strategy may also make the mistake of earlier thinkers, such as Herbert Spencer, in assuming more unity in the universe than actually exists. Moreover, even if there are isomorphisms, these similarities are phrased so generally that, in generating explanations of particular social events, they will yield only vague insights. That is, such general points of isomorphism across system levels will not help very much in explaining the particular phe- nomena of most interest to sociologists (or, for that matter, to biologists, psy- chologists, or economists).[41]

Although there have been several efforts to apply general systems concepts to sociological phenomena, these all suffer from the problems of importing precise mathematical formulations that are *not* isomorphic with social events. For example, efforts to adopt decision theory, game theory, information theory, analysis of servomechanisms, and similar bodies of precise concepts have never proven very useful. And they probably never will, since they do not *fit* well with the way complex sociocultural systems actually operate.

As a consequence of these shortcomings, the initial euphoria in the founding of the Society for General Systems Research has not been sustained. Indeed, to the extent that a cross-disciplinary perspective has been maintained in so- ciology, it has been allied with narrower and less cosmic approaches, such as network analysis (see Chapter 27) and human ecology (see Chapter 7), or it is attached to older theoretical perspectives, such as exchange theory, that seek only to cut across the social sciences.

---

[40]For excellent examples, see: Tom R. Burns and Helena Flam, *The Shaping of Social Orga- nization* (Newbury Park, CA: Sage, 1987); Kenneth D. Bailey, "Equilibrium, Entropy and Ho- meostasis: A Multidisciplinary Legacy," *Systems Research* 1 (1, 1984), pp. 25–43 and *Social En- tropy Theory* (Albany, NY: SUNY Press, 1989); Kenneth F. Berrier, *General and Social Systems* (New Brunswick, NJ: Rutgers University Press, 1968); Roger E. Cavallo, "Systems Research Move- ment," *General Systems Bulletin*, Special Issue, 9 (3, 1979); Carl Slawski, "Evaluating Theories Comparatively," *Zeitschrift fur Soziologie* 3 (4, 1974), pp. 397–408.

[41]Alfred Kuhn, "Differences vs. Similarities in Living Systems," *Contemporary Sociology* 8 (September 1979), pp. 691–96.

CHAPTER 7

Ecological Functionalism

Amos H. Hawley

ORIGINS OF THE ECOLOGICAL PERSPECTIVE

Ecological analysis in sociology borrows a critical insight from biology:[1] social life revolves around the adaptation of a population to its environment. Such adaptation occurs through the organization of a population into a system of interrelations. This organization, in turn, circumscribes the flow of resources within the system as well as between the system and its environment. For human ecology, then, the most relevant unit of analysis is a complex whole—an ecosystem—consisting of an organized population of actors adjusting to its physical, biological, and social environment.

Although modern ecological theory self-consciously borrows and extends ideas from 20th-century bioecology, I believe that the basic thrust of this orientation has existed since the origins of sociology. For example, Charles Montesquieu[2] argued in 1721 that the structure of a society, especially its governmental forms and value system, is the result of "physical" and "moral" causes. Among physical causes are climate, soil fertility, and population size; moral causes include external relations with other societies and internal cultural forces, such as manners, religion, commerce, and money. Thus Montesquieu viewed society as embedded in, and constrained by, demographic and environmental processes; although the term *ecology* had yet to be coined, his perspective could certainly be viewed as ecological.

---

[1]See, for general overviews of society and ecosystem processes: Otis Dudley Duncan and Leo F. Schnore, "The Ecosystem," in *Handbook of Sociology*, ed. C.A. Faris (Chicago: Rand McNally, 1964); Jonathan H. Turner, "The Eco-system: The Interrelationship of Society and Nature," in *Understanding Social Problems*, eds. D.H. Zimmerman, D.C. Wieder, and S. Zimmerman (New York: Praeger, 1976), pp. 292–321.

[2]Charles Montesquieu, *The Spirit of Laws* (London: Colonial Press, 1900; originally published in 1748).

Borrowing from Montesquieu and a long line of French thinkers, such as Jean Jacques Rousseau, Auguste Comte, and Alexis de Tocqueville, Émile Durkheim[3] carried forward this ecological line of thought and mixed it with organismic thinking.[4] For Durkheim the "body social" could be viewed as differentiating in response to those forces that increased its material and moral density. Population growth, spatial concentration, and competition were viewed by Durkheim as the primary causes of increases in material density, whereas improvements in communication and transportation technologies are seen as the causes of moral density (rates of contact and interaction). Thus, in Durkheim's eye, the macro dynamics of a society, particularly its differentiation of parts and functions, are to be understood in terms of population, physical setting, competition, transportation systems, and communication technologies.[5] Most of these variables were to become an integral part of modern human ecology.

I believe that many of Durkheim's ideas on ecology were borrowed from Herbert Spencer, who had initiated his sociological work 20 years before Durkheim. Although one can find hints of Durkheim's ecological argument in the French lineage, especially Montesquieu, it was Spencer who developed the most systematic theory. Yet, because Durkheim spent so much time criticizing Spencer's utilitarianism, it is easy to overlook the fact that his functional and ecological arguments are the same as Spencer's.[6] So, if we search for the first sociological theorist who developed a sophisticated ecological approach, the credit must go to Spencer.[7] What, then, was Spencer's perspective? As I noted in Chapter 2, Spencer transformed Auguste Comte's vague organismic analogy into an explicit analogy, and out of this approach he formulated a functional mode of analysis:[8] human organization is a system of parts; this system must meet certain requisites—regulation, production and operation, and distribution—to survive in an environment; hence, to understand any part of the system, it is necessary to know how it meets at least one of these system requisites. This functional approach, however, was blended with several other lines of theorizing.

One of these is ecological. For Spencer, societal systems exist in a social and physical environment, and in his vast empirical project—*Descriptive*

---

[3]Émile Durkheim, *The Division of Labor in Society* (New York: Free Press, 1947; originally published in 1893).

[4]See Jonathan H. Turner, Leonard Beeghley, and Charles Powers, *The Emergence of Sociological Theory* (Belmont, CA: Wadsworth, 1989), pp. 287–304.

[5]See ibid., p. 319, for a model of these processes. Also see Jonathan H. Turner, "Émile Durkheim's Theory of Social Organization," *Social Forces*, (to be published 1990), for an even more detailed model of Durkheim's ecological thinking.

[6]See, for more details, Jonathan H. Turner, "Durkheim's and Spencer's Principles of Social Organization: A Theoretical Note," *Sociological Perspectives* 27 (January 1984), pp. 21–32.

[7]See Jonathan H. Turner, *Herbert Spencer: A Renewed Appreciation* (Newbury Park, CA: Sage, 1985).

[8]See Herbert Spencer, *The Principles of Sociology* (New York: Appleton-Century-Crofts, 1885; originally initiated in 1874), part 2 of vol. 1.

Sociology[9]—the first entries for any society are always its physical, biological, and social (other societies) environment. The organization of a society will reflect, Spencer argued, the way in which it has adapted to these three dimensions of its environment; his detailed descriptions of how such adaptation occurs are particularly insightful, especially when we recognize that they were initially made in the 1870s. Thus the functions of any structure in a society revolve around the way it facilitates the processes of regulation, operation, production, and distribution (communication and transportation) in a given physical, biological, and social environment. Indeed, Spencer often shifted the unit of analysis from society to population and visualized a society as a "social species" adapting to an environment. His famous phrase "survival of the fittest"[10] denotes his view that adaptation involves a competitive struggle not just among individuals within society (as Durkheim was wont to overemphasize) but also among populations organized into societies (as competing "species"). This is the reason that the analysis of war is so prominent in Spencer's sociology, since he viewed war as one manifestation of efforts by populations to adjust to their social environment and, through the expropriation of the economic surplus of those who were conquered, to their physical and biological environment as well. Whereas Spencer abhorred war on moral grounds, he saw it as one of the ecological forces that had dramatically changed the structure of societies during the course of human evolution.[11]

Another line of ecological thought is found in Spencer's famous "law of evolution," which was formulated over 20 years *before* he initiated his sociological analysis.[12] What often goes unrecognized is that this law represents an analogy to the physics of his time, a fact that accounts for the rather strange vocabulary. Thus, evolution is a process of "... integration of matter and concomitant dissipation of motion, during which the matter passes from an indefinite incoherent homogeneity to a definite coherent heterogeneity; and during which the retained motion undergoes a parallel transformation."[13] I should emphasize that Spencer also conceptualized "dissolution," which is the reverse of these processes, or what today we would conceptualize as entropy (i.e., the breakdown of entities from differentiated to homogeneous states).[14] Spencer saw this "law" as applying to all systems in the universe—astronomical, biological, psychological, sociological, and even ethical. Our concern, of course,

---

[9]There were numerous volumes of this project, initiated in 1873 and completed long after Spencer's death in 1934. For an overview, see Jonathan H. Turner and Alexandra Maryanski, "Sociology's Lost Human Relations Area Files," *Sociological Perspectives* 31 (January 1988), pp. 19–34.

[10]Herbert Spencer, *Social Statics* (New York: Appleton-Century-Crofts, 1888; originally published in 1851).

[11]Spencer, *The Principles*.

[12]Herbert Spencer, *First Principles* (New York: A. C. Burt, 1880; originally published in 1862).

[13]Ibid., p. 343.

[14]Ibid., pp. 445–49. See last chapter for more details on how this concept is used in the sociological analysis of general systems.

is how this law and its various corollaries anticipate the ecological approach within sociology. Let me elaborate.[15]

For Spencer, human evolution involved an "aggregation of matter"—in the case of humans, populations of actors. The "force" behind this aggregation—war and conquest, migrations, political treaties, natural population increases—is "retained" and escalates competition, which in turn causes social selection and social differentiation. Differentiation is thus the result of the competition among actors in the system, each seeking a niche where it can better survive (the actors in this process can be either individuals or collective units like organizations and communities). Differentiation creates integrative problems, however. And so there are selection pressures for the "integration of matter" and "dissipation of motion" that caused competition and differentiation in the first place. Such integration occurs through centralization of power to "regulate" the internal processes of the system and through "mutual interdependence," especially of exchange relations in markets (where actors need one another's resources). Such regulation through centralization of power and mutual interdependence creates a more "coherent" and "differentiated" system.

These ideas were to become a part of ecological analysis—that is, population growth, competition, differentiation, and integration through interdependence and power—and they come to us from Spencer. Moreover, Spencer was to add more specific ideas about the evolution of populations in an environment.[16] For example, Spencer argued that differentiation of actors occurs with respect to certain basic functions—regulation, operation (production of goods and services as well as reproduction of system units), and distribution (of information and materials). Another example can be seen in Spencer's notion of system equilibrium, an idea that is prominent in much contemporary ecology. For Spencer, motion dissipates as it is integrated, and so a population will reach an equilibrium unless there is an external "force"—war, migration, population growth, changes in the physical environment, new products from other societies—to disturb this equilibrium and set the process of evolution (or dissolution) in motion. Yet another example of specific ideas in Spencer that are a part of ecological analysis is the notion of system phases and dialectics. For example, Spencer saw societies as cycling between relatively centralized (concentrated power) and decentralized (increased autonomy of units) states.[17] This cycling was dialectical: centralized systems stagnate and create resentments that generate pressures for lessened control and decentralization, whereas decentralized systems diversify and differentiate to the point of causing severe integrative problems of coordination and control, a situation that creates pressures for centralization of power and reregulation. Still another specific idea in Spencer is his notion that, when systems grow in size and initially differentiate regulatory and operative processes, their problems of transportation, movement,

---

[15]See Turner, *Herbert Spencer*, for a more detailed analysis.

[16]See ibid. for a brief overview.

[17]This was the message behind his famous distinction between "militaristic" and "industrial" societies.

and communication increase, causing selection pressures for new kinds of distributive units.

All of the more cosmic as well as these specific ideas in Spencer's work reappear in contemporary ecological theory. Yet, if we look for the more immediate sources of inspiration for human ecology, it is a small group of urban sociologists, mostly in Chicago, in the 1920s and 1930s. As Hawley has noted:[18]

> In their search for order in the turbulent urban centers of America ... sociologists were stimulated by work then being done by bioecologists. ... Those researchers showed that plant species adapt to their environment by distributing themselves over a localized area in a pattern which enables them to engage in contemporary uses of habitat and resources. That idea opened a vista to an understanding of what was occurring in the burgeoning industrial city. For then it was apparent that various subpopulations were jostling for spatial positions from which they could perform their diverse functions in an unfolding division of labor.

What is missing from Hawley's assessment is the fact that, before World War I, Spencer was the most cited figure in American sociology.[19] It is inconceivable, then, that his ideas did not exert considerable influence, not even in conjunction with those of theorists in bioecology (who, I should emphasize, may have taken them from Spencer's widely read *The Principles of Biology*). At any rate, whatever the direct line of influence, Chicago sociology appears to have "downsized" Spencer's theory and applied it to problems of competition, differentiation, succession, and integration of ethnic groups and other subcultures in the bustling city of Chicago.[20] Thus, Robert Park, Ernest Burgess, and Roderick McKenzie began to apply Spencer-like ideas to specific empirical cases. Later, others like William F. Ogburn, Howard Odum, Otis D. Duncan, and Amos Hawley would carry their ideas further. Somewhere in this lineage Spencer was lost; as a result, the "upsizing" of human ecology from urban sociology into a more general theoretical approach appears to have occurred without great inspiration from Spencer.

Nonetheless, whatever its precise origins, modern human and social ecology is now a fertile area for theorizing; and it assures that much of Spencer's legacy will be retained, if only in spirit. Ecological analysis has typically concentrated on spatial and community processes[21] or organizational dynamics within and

---

[18]Amos H. Hawley, "Human Ecology: Persistence and Change," *American Behavioral Science* 24 (January 1981), p. 423.

[19]Stephen P. Turner and Jonathan H. Turner, *The Impossible Science: An Institutional Analysis of American Sociology*, (Newbury Park, CA: Sage, 1990).

[20]Robert E. Park, "The City," *American Journal of Sociology* 20 (1916), pp. 577–612; "Human Ecology," *American Journal of Sociology* 42 (1936) pp. 1–15; *The City* (Chicago: University of Chicago Press, 1925). Robert E. Park and Ernest W. Burgess, *Introduction to the Science of Sociology* (Chicago: University of Chicago Press, 1921). Roderick McKenzie, *The Metropolitan Community* (New York: McGraw-Hill, 1933).

[21]For a recent review, see: W. Parker Frisbie and John D. Kasarda, "Spatial Processes," in *Handbook of Sociology*, ed. Neil J. Smelser (Newbury Park, CA: Sage, 1988); and John D. Kasarda, *Contemporary Urban Ecology* (New York: Macmillan, 1977).

between bureaucratic/corporate units.[22] Yet, alongside of this has been a more general theory of human ecology that transcends these more specific areas of inquiry. Indeed, Hawley notes that the perspective of bioecology forces a more general orientation, emphasizing ". . . the structure of relationships within a population and between the population and its environment."[23] Although numerous scholars have worked on developing this more general ecological theory, Amos Hawley has clearly been most influential, and so it is his mature theory that will occupy my attention.

AMOS H. HAWLEY'S HUMAN ECOLOGY PERSPECTIVE

For almost 50 years, Amos Hawley[24] has sought to articulate an ecological perspective for sociological analysis. His approach has stimulated an enormous amount of conceptually informed empirical research, particularly in the areas of organizational and urban analysis. Yet, as noted above, Hawley has recognized that there is a more general paradigm undergirding the analysis of communities and organizations; more than any other scholar, he has articulated the theoretical base of this paradigm. Hawley's ideas can be found in many places, although I think that a recent theoretical "essay,"[25] coupled with qualifications from earlier work, can provide us with the materials for an analysis of the ecological paradigm.

Production, Transportation, and Communication

The continuity of his theory with early work within the functional and evolutionary theory is immediately evident in several key assumptions, which are stated by Hawley as follows:[26]

---

[22]For examples, see: Charles E. Bidwell and John D. Kasarda, *The Organization and Its Ecosystem* (Greenwich, CT: Jai Press, 1985); James R. Lincoln, "Organizational Differentiation in Urban Communities," *Social Forces* 57 (1970), pp. 15–30; Michael T. Hannan and John H. Freeman, "The Population Ecology of Organizations," *American Journal of Sociology* 82 (1977), pp. 929–64.

[23]Hawley, "Human Ecology: Persistence and Change," p. 423.

[24]See, for example, Amos H. Hawley, "Ecology and Human Ecology," *Social Forces* 27 (1944), pp. 398–405; *Human Ecology: A Theory of Community Structure* (New York: Ronald, 1950); "Human Ecology," in *International Encyclopedia of the Social Sciences*, ed. D. C. Sills (New York: Crowell, Collier and Macmillan, 1968); *Urban Society: An Ecological Approach* (New York: Ronald, 1971 and 1981); "Human Ecology: Persistence and Change," *The American Behavioral Scientist* 24 (3, January 1981), pp. 423–44; "Human Ecological and Marxian Theories," *American Journal of Sociology* 89 (1984), pp. 904–17; "Ecology and Population," *Science* 179 (March 1973), pp. 1196–1201; "Cumulative Change in Theory and History," *American Sociological Review* 43 (1978), pp. 787–97; "Spatial Aspects of Populations: An Overview," in *Social Demography*, eds. K. W. Taueber, L. L. Bumpass, and J. A. Sweet (New York: Academic Press, 1978); "Sociological Human Ecology: Past, Present and Future," in *Sociological Human Ecology*, eds. M. Micklin and H. M. Choldin (Boulder, CO: Westview Press, 1980); and, most significantly, *Human Ecology: A Theoretical Essay* (Chicago: University of Chicago Press, 1986).

[25]Hawley, *Human Ecology*.

[26]Hawley, "Human Ecological and Marxian Theories," p. 905.

(1) Adaptation to environment proceeds through the formation of a system of interdependencies among the members of a population;

(2) system development continues, ceteris paribus, to the maximum complexity afforded by the existing facilities for transportation and communication; and

(3) system development is resumed with the introduction of new information which increases the capacity for movement of materials, people, and messages and continues until that capacity is fully utilized.

Hawley terms these assumptions, respectively, the *adaptive, growth*, and *evolution* "propositions." These assumptions resurrect in altered form ideas developed by Herbert Spencer and Émile Durkheim.[27] In order to survive and adapt in an environment, human populations become differentiated and integrated by a system of mutual interdependencies. The size of a population and the complexity of social organization for that population are limited by its knowledge base, particularly with respect to transportation and communication technologies. Populations cannot increase in size, nor elaborate the complexity of their patterns of organization, without expansion of knowledge about (1) communication and (2) movement of people and materials. Hawley comes to conceptualize the combined effects of transportation and communication technologies as *mobility costs*.

Linked to transportation and communication technologies is another variable, *productivity*. Curiously, in his most recent theoretical essay[28] this variable is somewhat subordinate, whereas in earlier statements it is highlighted. I see no great contradiction or dramatic change in conceptualization in this more recent statement, and so I will reintroduce the productivity variable in more explicit terms. Basically, there is a reciprocal set of relations between production of materials, information, and services, on the one side, and the capacity of a system to move these products to other system units, on the other. The development of new transportation and communication technologies encourages expanded production, whereas the expansion of production burdens existing capacities for mobility and thereby stimulates a search for new technologies. There is also a more indirect linkage among productivity, growth, and evolution, because productivity ". . . constitutes the principal limiting condition on the extent to which a system can be elaborated, on the size of the population that can be sustained in the system, and on the area or space that the system can occupy."[29] Thus, to support a larger, more differentiated, population in a more extended territory requires the capacity to (1) produce more goods and services and (2) distribute these goods and services through transportation and communication technologies. If productivity cannot be increased and/or if the mobility costs of transportation and communication cannot be reduced, then there is an upper limit on the size, scale, and complexity of the system.

---

[27]Spencer, *The Principles*; Durkheim, *The Division of Labor*.
[28]Hawley, *Human Ecology*.
[29]Hawley, "Human Ecology," p. 332.

Thus I am simply reinserting the productivity variable, because it appears to get somewhat "lost" in more recent work. For, as Hawley consistently emphasized in earlier essays,[30] "cumulative change, or growth of the system, presupposes an increase in productivity. . . ." Or, as he stated not long ago,[31] "the adaptation proposition of ecology puts the production of sustenance in a central position in an ecosystem or social system." With this point taken, let me now return to the more general conceptual framework proposed by Hawley.

The Environment

An ecosystem is "an arrangement of mutual dependencies in a population by which the whole operates as a unit and maintains a viable environmental relationship."[32] The environment is the source of energy and materials for productivity, but the environment reveals more than a biophysical dimension. There is also an "ecumenic" dimension composed of the "ecosystems or cultures possessed by peoples in adjacent areas and beyond."[33] The conceptualization of the environment, however, presents a number of analytical problems. For example, statements like the following are somewhat confusing:[34]

> It should be noted that the term "environment" has no fixed denotation. It is a generic concept for whatever is external to and potentially influential upon a unit under study. . . . Thus the act of defining refers one back to the thing environed. That thing, from the standpoint of ecology, is a population which is organized or in the process of organization. The clarity of the environmental definition can be no greater than that of the environed unit.

Thus, in order for the environment to be conceptualized, we need to know the boundaries and parameters of the organized unit or units under investigation. This is not a problem when doing empirical research, because one can trace the effects of the environment on a group, community, organization, nation/state, or some other unit by examining those physical, biological, and sociocultural forces that impinge upon, and require adjustments by, the unit under investigation. However, at a more general theoretical level, how do we distinguish environment and unit of organization? The reason that this question becomes problematic is that the interchanges between environment and organized population become crucial to understanding ecosystem dynamics. But, if we cannot specify conceptually what the unit and environment are, then the theoretical argument can become somewhat muddled. For example, take a statement like the following:[35]

---

[30]Ibid., p. 335.

[31]Hawley, "Human Ecological and Marxian Theories," p. 907.

[32]Hawley, *Human Ecology*, p. 26.

[33]Ibid., p. 13.

[34]Hawley, "Ecology and Population," pp. 1197–8.

[35]Hawley, "Human Ecological and Marxian Theories," p. 912.

Ecology posits an external origin of change, for a thing cannot cause itself. Change is induced when an environmental input, that is, new information, impinges on and is synthesized with existing information. The dialectic of concern for ecology is the system-environment interaction.

Hawley seems to preclude self-generating changes by such a statement, but is this really so? If productive units in a system generate new technologies, does the information come from other system units "outside" these productive units or from the more remote environment? Or, if productive units discover new productive technology in dealing with the environment, is this environmentally induced or internally induced change? Or, if centers of power in a system force changes in productive activity, is this environmental change, or is it internally induced change when the entire population as it is organized into a system whole is considered the relevant unit? Or, if production changes the biophysical environment in ways that force alterations in productive technologies, is this environmentally or internally induced change?

These kinds of questions are not easily answered because "the environment" changes depending on what unit of analysis is selected. Although Hawley is correct that Marxian theory may rely too much on internally generated sources of innovation and change,[36] ecological theory invokes an equally mysterious source of change: the environment. What needs to be recognized, I think, is that there is a complex interplay between system and environment and that sometimes changes come from predominantly internal processes within the system and, at other times, from the exogenous environment. This line of argument is consistent with the biological origins of human ecology, since internally generated dialectical change can be seen as the social system equivalent of "mutations" that can increase or decrease the capacity for adaptation in a given environment. And, such a position does not obviate the presumption that environmental changes that influence the inputs of energy, matter, and information can also be the source of change. At any rate, whatever the merits of this digression, the critical point is that "ecology posits an external origin of change," and so, if one is examining a population as a whole, such as one organized into a nation/state, then change would presumably have to come from the physical, social, or biological environment *of this systemic whole.* Empirically, I do not think that this position would hold up. Indeed, contrary to Hawley, I think that "a thing can cause itself," or at least a transformation in itself, but this point of emphasis is not a central part of ecological theory. Perhaps, as a compromise, it would be wiser to see the emphasis on exogenous sources of change as not a dogmatic assertion but simply a viewpoint or conceptual tactic—which may be what Hawley intended to connote all along. In fact, I suspect that the point that Hawley wishes to emphasize is this: informational inputs from the environment will join and interact with existing information, creating a new synthesis that will have consequences for reorganizing the system under investigation.

---

[36]Ibid.

The Functionalism in Human Ecology

In Hawley's view, the arrangement of mutual dependencies of a population in an environment is conceptualized as classes or types of *units* that form *relations* with one another with respect to *functions*.[37] Functions are defined as "repetitive activity that is reciprocated by another or other repetitive activities." Of particular importance are *key functions*, which are repetitive activities "directly engaged with the environment," and as such they transmit environmental inputs (materials and information) to other "contingent functions" (or repetitive activities joining units in a relation).[38] Hawley visualizes that a relatively small number of key functions exists, and "to the extent that the principle of key functions does not obtain, the system will be tenuous and incoherent."[39] A system is thus composed of functional units, a few of which have direct relations with the environment and perform key functions. Most other units, therefore, must "secure access to the environment indirectly through the agency of the key function."[40]

For example, production is a key function, and in earlier essays Hawley seemed to see productivity as *the* primary key function. Yet there are obviously other key functions—political, military, and perhaps ideological—that also influence the flow of resources to and from the environment. And, as a result, other functional units gain access to the environment only through their interconnections with those units engaged in these various key functions. Thus the relations of units that form the structure of a population are conceptualized in terms of functions and key functions, or classes of reciprocated repetitive activity that join units together. This is, of course, another way of denoting specialization and differentiation of various types in clusters of activity. Just why this particular terminology is proposed cannot be answered, but it begins to transform the ecology perspective into an explicitly functional analysis.

How is this done, especially since the term *function* appears to denote only recurring and reciprocated activities? What begins to slip into Hawley's analysis is a stronger version of the term *function*. In part, this stronger version of function is implied by the concept of "key function." A key function regulates inputs of energy, materials, and information into the system, and it is not hard to see how the next step on the road to functionalism is made: certain key functions are necessary for adaptation and survival. Hawley himself makes this slip when he notes:[41]

> We might suppose, for purposes of illustration, that every instance of collective life is sustained by a mix of activities that produce sustenance and related materials, distribute the products among the participants, maintain the number of units required to produce and distribute the products, and exercise the

---

[37]Hawley, "Human Ecology" and *Human Ecology*, p. 32.

[38]Hawley, *Human Ecology*, p. 34.

[39]Hawley, "Human Ecology," p. 332.

[40]Ibid.

[41]Hawley, *Human Ecology*, p. 32.

TABLE 7-1 General Propositions on Functions in Ecosystems

I. The more a function (recurrent and reciprocated activity) mediates critical environmental relationships (key function), the more it determines the conditions under which all other functions are performed.

II. The more proximate is a function to a key function, the more the latter constrains the other, and vice versa.

III. The more a function is a key function, the greater is the power of those actors and units involved in this function, and vice versa.

IV. The more differentiated are functions, the greater is the proportion of all functions indirectly related to the environment.

V. The greater the number of units using the products of a function and the less the costs of the skills used in the function, the greater is the number of units in the population engaged in this function.

VI. The greater the mobility costs (for communication and transportation) associated with a function, the more stable are the number of, and the interrelations among, units implicated in this function.

VII. The more stable the number of, and interrelations among, units implicated in functions, the more a normative order corresponds to the functional order.

controls needed to assure an uninterrupted performance of all tasks with a minimum of friction.

It is easy to read notions of functional requisites into this and many similar passages in Hawley's work. Indeed, the requisites look very much like those proposed by Spencer—production, regulation, distribution, and sustenance. At any rate, I do not see this functionalism as a fatal flaw—as some might—but rather as a characteristic of Hawley's theory that we should recognize. As we will see, however, Hawley obviates most of the problems that functionalism causes (see pp. 72–77) when he translates his ideas into a series of "hypotheses." Table 7-1 restates in somewhat modified form some of the most critical of these propositions that can be pulled from his analysis thus far.[42] These propositions represent my best effort to extract the more generic statements from Hawley's propositions. I see the propositions in Table 7-1 and in subsequent tables as the more abstract "laws" from which Hawley's many hypotheses could be derived.

The basic ideas in the propositions of Table 7-1 are these: key functions, or those that mediate exchanges with the environment, disproportionately influence other functions and hold power over these other functions (for example, units involved in key economic or political functions in a society usually hold more power and influence than others, because they are engaged in interchanges, respectively, with the physical and social environment); the more proximate is a function to a key function, the greater is this influence; conversely, the more remote from a key function, the less is this influence on a function. Differentiation of key functions decreases other functions' direct access to the environment, since such access is now mediated by units involved in key functions (for example, most people do not grow their own food or provide

---

[42]Ibid., pp. 43–44.

their own military defense in highly differentiated societies). As mobility costs for personnel and materials needed for functions increase, their number and relations stabilize; under these conditions, a normative order can develop to regulate the internal relations of functions, as well as their interrelations.

Equilibrium and Change

An ecosystem is thus a population organized in ways to adapt to an environment, with change in this system being defined by Hawley as "a shift in the number and kinds of functions or as a rearrangement of functions in different combinations."[43] In contrast to change, growth is "the maturation of a system through the maximization of the potential for complexity and integration implicit in the technology for movement and communication possessed at a given point in time,"[44] whereas evolution is "the occurrence of new structural elements from environmental inputs that lead to synthesis of new with old information and a consequent increase in the scope of the accessible environment."[45] This series of definitions presents a picture of ecosystem dynamics as revolving around (1) the internal rearrangement of functions, (2) the increase of complexity to the maximum allowed by an existent level of communication and transportation technologies, and (3) the receipt of environmental inputs, especially new information that increases transportation and communication (or capacity for mobility) and that, as a consequence, increases the scale and complexity of the ecosystem (to the limits imposed by the new technologies). There is an image, then, of a system in equilibrium that is then placed into disequilibrium by new knowledge about production as it influences mobility (of people, materials, and information). Somewhat less clear is where this new information comes from. Must it be totally exogenous (from other societies, migrants, changes in biophysical forces that generate new knowledge)? Or can the system itself generate the new information through a particular array of functions? As I emphasized above, it would seem that both can be the source of change, and yet the imagery of Hawley's model[46] connotes a system that must be disrupted from the outside if it is to evolve and develop new levels of structural complexity. Internal dialectical processes, or self-transforming processes that increase technology, seem to be underemphasized as crucial ecosystem dynamics. Table 7–2 summarizes the more abstract "laws" that can be culled from Hawley's hypotheses on these dynamics.[47]

These propositions reinforce the emphasis in human ecology that the source of change is exogenous, residing particularly in the "ecumenic" environment. It is from this environment that new knowledge will come and then become

---

[43]Ibid., p. 46.

[44]Ibid., p. 52.

[45]Ibid.

[46]Ibid., p. 59. Hawley "boxes" lists of variables and then draws arrows among the boxes, but not among the variables within each box. Hence, detailed causal arguments need to be inferred.

[47]Ibid., pp. 85–87.

TABLE 7–2 General Propositions, Change, Growth, and Evolution in Ecosystems

I. The greater the exposure of an ecosystem to the ecumenic environment (other societies or cultures of other societies), the greater is the probability of new information and knowledge penetrating the system and, hence, the greater is the probability of change, growth, and evolution.
II. The more new information increases the mobility of people, materials, and information, as well as production, as it influences mobility, the more likely that change will be cumulative, or evolutionary, up to the limits of complexity allowed by the new information as it is translated into technologies for production, transportation, and communication.
III. The more new information improves various mobility and productive processes at differential rates, the more the slower rate of technology will impose limits on the faster-changing technology.
IV. The more a system approaches the scale and complexity allowed by technologies, the slower is the rate of change, growth, and evolution and the more likely is the system to achieve a state of closure (equilibrium) in its ecumenic environment.

"synthesized" with the existing knowledge base. Such synthesized knowledge will then change production, transportation, and communication in ways that allow the system to increase its complexity, size, and territory. Yet new knowledge can introduce change only to a point. If some technologies lag behind others, the rate of change will be pulled down by the lower technology. And eventually the maximal size, scale, and complexity of the system will be reached, unless new technologies are inserted into the system from the environment. Thus systems that have grown to the maximum size, scope, and complexity allowed by production, transportation, and communication technologies will achieve an *equilibrium*. New knowledge from the environment can disrupt the equilibrium when it is used to achieve increases in productivity and mobility. But each technology has limits on how much growth and evolution it can facilitate; when this limit is reached, the system will tend to reequilibrate.

The concept of *equilibrium* is most problematic, although I think that Hawley employs it only as a heuristic device. The notion of equilibrium is, for Hawley, meant to connote "the balance of nature, denoting a tendency toward stabilization of the relative numbers of diverse organisms within the web of life and their several claims on the environment. . . ."[48] Yet Hawley recognizes that "equilibrium . . . is a logical construct"[49] and it connotes that ecological systems tend toward stability—a line of argument that, as we can recall, got Talcott Parsons' functionalism into hot water. But Hawley tends to mitigate against the concept when he talks about "partial equilibriums" in the same way that Parsons retreated to notions of "moving equilibriums." Equilibrium becomes a more viable assertion when the environment is the only source of change, since only "external disturbances" can disrupt the system. And if the environment is constant, or relatively so, then the system can easily be seen as reaching an equilibrium point. But if the very nature of organization within

[48]Hawley, "Human Ecology," p. 329.
[49]Ibid., p. 334.

FIGURE 7–1 Elaboration of Hawley's Model

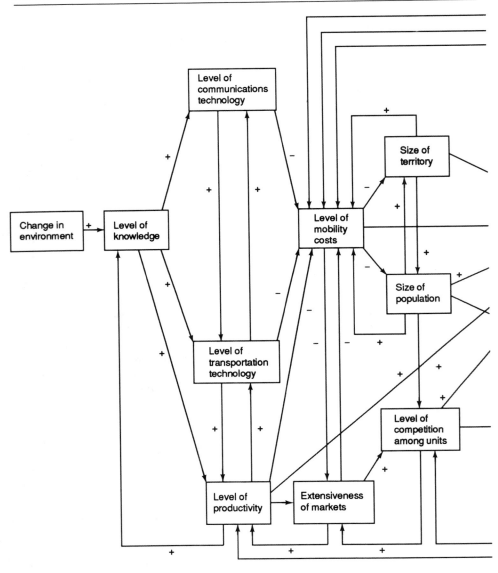

and among functional units contains the seeds of change, then the notion of equilibrium is less viable, especially as systems become complex and reveal many points of disarticulation, tension, conflict, and contradiction.

FIGURE 7–1   *(continued)*

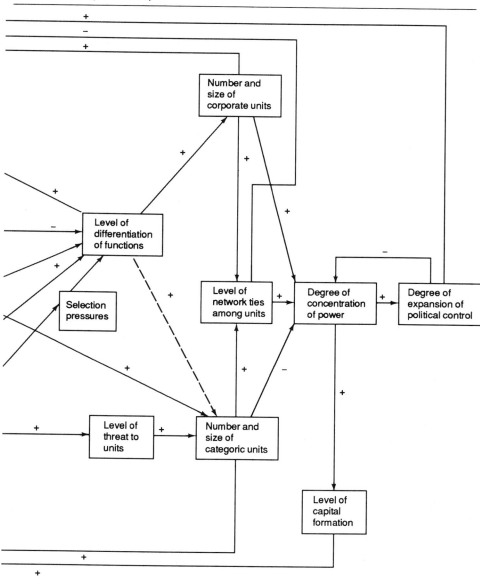

Growth and Evolution

The most interesting portions of ecological theory are those dealing with growth and evolution—that is, increasing size, scale, scope, and complexity of the systematic whole in its environment. This analysis builds upon the propositions in Tables 7-1 and 7-2, but it extends them in creative ways and, as a result, goes considerably beyond the early formulations of Spencer and Durkheim.

In Figure 7-1 I have tried to model in a somewhat different way than Hawley the basic dynamics of his ecosystem model during growth and evolution. I am not sure if the model in Figure 7-1 corresponds to Hawley's intent, but it can give us a better sense of the causal interconnections among those processes that are only listed in Hawley's model.

Starting on the far left of the model, growth and evolution are, as outlined above and in Table 7-2, the result of an expanded knowledge base that, in Hawley's view, must come from the ecumenical environment. As the feedback arrows to this variable emphasize, I do not think that this is the only way that the knowledge base is expanded; as Marx and others in the 19th century recognized, many internal system processes are self-transforming and generate the new knowledge that initiates these changes. This is particularly likely in systems that are differentiated, competitive, and market based—variables in the model that I will soon discuss. As the model stresses, increased knowledge causes growth and change when it increases the level of communication and transportation technologies either directly or indirectly, through increasing production (which then causes expansion of these technologies). A very critical variable in Hawley's scheme is mobility costs; for any given technology, there is a cost (time, energy, money, materials) associated with the movement of information, materials, and people. As these costs reach their maximum—that is, all the system can "pay" for them without degenerating—they impose a limit on the scale of the system: the size of its population, the extent of its territory, the level of its productivity, and the level of complexity. Conversely, as the feedback arrows in the model indicate, the size of territory and population will, as they expand and grow, begin to impose higher mobility costs. Eventually these costs will increase to a point at which the population cannot grow or expand its territory—*unless* new communication and transportation technologies that reduce costs are discovered.

Much as Spencer and Durkheim argued, population and territorial size, as these are influenced by mobility costs, cause specialization of functions—what I have termed *differentiation* in the model. As Hawley notes, however, the relationship between population size and differentiation is not unambiguous, but it does create "conditions that foster, if not necessitate, increases in the sizes of subsystems" and, I might add, the number of such subsystems serving various functions. Thus, for Hawley, "the greater the size, the greater the probable support for units with degrees of specialization." And, as he adds: "other pertinent conditions are the rate or volume of intersystem communications, scope of a market, and amount of stability in intersystem relations."[50]

These causal connections are not clearly delineated by Hawley, and so the model represents my best sense of how he argues. The causal paths moving

---

[50]Hawley, *Human Ecology*, pp. 80–81.

TABLE 7-3 A Typology of System Units

Unifying Principle	Relational Structure	
	Corporate	Categoric
Familial	Household-producing and personal service unit	Kin, clan, and tribe
Territorial	Village, city, and ecumene	Polity, neighborhood, ethnic, enclave, ghetto
Associational	Industry, retail, store, school, government	Caste, class, sect, guild, club, union, professional organization

from level of productivity through extensiveness of markets and level of competition to selection pressures and differentiation of functions represent the old Spencerian and Durkheimian argument: expansion of markets increases the level of competitiveness among units and, at the same time, increases the capacity to distribute goods and services as these are constrained by mobility costs (note arrows connecting markets and mobility costs); competition under conditions of increased production and population size allows—indeed encourages—specialization as actors seek their most viable niche. Similarly, the causal arrows moving from communication and transportation technologies through mobility costs, size of territory, and size of population to level of differentiation of functions restate in more sophisticated form Spencer's, but more particularly Durkheim's, argument: changes in communication and transportation technologies reduce mobility costs and allow for population growth and territorial expansion; all of these forces together create selective pressures to adjust and adapt in terms of varying attributes and competencies, especially under conditions of intense competition for resources.

In Hawley's view, differentiation of subunits engaged in various functions occurs along two axes: (1) corporate and (2) categoric.[51] Corporate units are constructed from "symbiotic relations" of mutual dependence among differentiated actors, whereas categoric units are composed of "commensalistic" relations among actors who reveal common interests and who pool their activities in order to adapt more effectively to their environments. Table 7-3 delineates these types of units along an additional dimension—their unifying principle.[52] Thus, as differentiation increases, an ecosystem will represent a complex configuration of corporate and categoric units along various "unifying principles": familial, territorial, associational. It is not clear if this typology is meant to be exhaustive or merely illustrative. Nonetheless, it is provocative.

The dynamics of these two types of units are very different. Corporate units form around functions, or sets of related activities, and are engaged in interchanges with other corporate units. As a consequence of this contact,

---

[51]Ibid., pp. 68–73, and "Human Ecology," pp. 331–32.
[52]Hawley, *Human Ecology*, p. 74.

corporate units tend to resemble one another, especially those that engage in frequent interchanges or are closely linked to corporate units engaged in key functions (interchanges with the environment). Moreover, as they engage in interchanges, corporate units tend toward closure of structure and establishment of clear boundaries. The size and number of such corporate units depend, of course, on the size of the population, the inter- and intraunit mobility costs associated with communication and transportation technologies, the capacity for production and distribution in markets (as constrained by mobility costs), and the level of competition among units.

In contrast, categoric units involve interdependencies that develop "on the basis of similarities among the members of a population";[53] their number and size are related to the size of the population and territory as well as to the level of threat imposed by their environment. As Hawley notes, the nature of the threat can vary—a "task too large for the individual to accomplish in a limited time, such as the harvesting of a crop. . ."; "losing land to an invader"; "possible destruction of a road or other amenity"; "a technological shift that might render an occupation obsolete"; and so on. If a threat is persistent, the actors in a categoric unit will form a more "lasting association," and, if similar units are in competition (say, rival labor unions or ideologically similar political parties), the costs and destructiveness of such competition will eventually lead to their consolidation into a larger categoric unit. Moreover, those categoric units that persist will develop a corporate core to sustain the flow and coordination of resources necessary to deal with the persisting or recurring threat. Categoric units can also get much larger than corporate units because their membership criteria—mere possession of certain characteristics (ethnic, religious, occupational, ideological)—are much more lax than those of corporate units, which recruit members in order to perform certain specialized and interdependent activities or functions. Of course, the size and number of categoric units are still circumscribed by the size and complexity of the ecosystem and, to a lesser extent than with corporate units, by the mobility costs associated with communication and transportation technologies.

As is indicated in the right portion of Figure 7-1, differentiation of categoric and corporate units leads to the consolidation of units into networks, which create larger subsystems. Such regularization of ties among units engaged in similar and symbiotic activities or functions reduces mobility costs, thereby facilitating growth of the ecosystem (note feedback arrow to mobility costs).

Differentiation of units, and their consolidation into larger networks, also has consequences for the concentration of power. Categoric unit formation, and the consolidation of such units into larger networks and subsystems, tends to reduce concentrations of power, since various confederations of categoric units will pose a check on one another. In contrast, corporate units are more likely to cause the concentration of power. This concentration is directly related to the capacity of some units to perform key functions and thereby dictate the conditions under which interrelated functions must operate. This control is

---

[53]Ibid., p. 70.

facilitated by consolidation of networks into subsystems, because such networks connect outlying and remote corporate units, via configurations of successive network ties, to those engaged in key functions. Such connections among corporate units, as they facilitate the concentration of power, enable political control to expand to the far reaches of the ecosystem; in fact, political and territorial boundaries tend to become coterminous in ecosystems. Yet, as Spencer recognized in his dialectical arguments, centralization of power and extension of control can increase mobility costs as rules and regulations associated with efforts at control escalate, setting limits on how complex the ecosystem can become, without a change in communication and transportation technologies (hence the long feedback arrow at the top of Figure 7–1).

Let me add to the model a variable that I see as crucial: capital formation. Hawley does not address this issue extensively, and so I am clearly adding it to his model. Nonetheless, it is important to recognize that concentration of power also consolidates the flow of resources, facilitating capital formation. If not squandered on maintenance of control, defense, or offensive efforts at military expansion, this capital can be used to expand productivity and, indirectly, to change the knowledge and technological base of the system (note long feedback arrows at the bottom of Figure 7–1).

By reconceptualizing Hawley's argument into a more complex model in which causal linkages and feedback processes are delineated in more detail, I show that change does not need to be viewed as only exogenous—that is, coming as an external input from forces in the biophysical or ecumenical environment. In fact, I am not sure if Hawley's emphasis on equilibrium[54] is appropriate for ecosystems involving social actors, since some set of causal paths and feedback loops will inevitably be "out of balance" and will create strains toward growth and evolution (or, perhaps as Spencer emphasized, dissolution). I think that, in particular, the nature and strength of the feedback processes can indicate just how change, growth, and evolution can be self-generating. For example, if concentrations of power reduce mobility costs, then the system can grow, expand, and differentiate further; or, if concentrations of power create capital that changes the nature of production and related technologies, then the system can not only grow but can evolve; or, if new market processes (money, credit, banks, etc) expand so as to stimulate and change the nature of production and relevant technologies, then the system has yet another source of endogenous evolution; or, if efforts to concentrate power and extend control place a drain on resources, raise mobility costs, and deplete capital, then the system will reveal an indigenous source of stagnation or even degeneration. I do not imagine that Hawley would preclude these processes, but there is a certain ambiguity in his view of the source of change as exogenous.

As with Tables 7–1 and 7–2, Table 7–4 extracts from Hawley's many hypotheses those that I see as crucial. As territory and population size increase, as a result of new knowledge about production, transportation, and communication, it is possible and perhaps necessary to differentiate units around

---

[54]Ibid., pp. 22–24.

TABLE 7-4   Basic Propositions of Patterns of Ecosystem Differentiation

---

I. The greater the size of a population and its territory and the greater the selection pressures stemming from competition among members of the population, the greater the differentiation of functions and the number and size of corporate units, up to the maximum allowed by mobility costs.

II. The greater the size of a population and the greater the threats posed by competition and environmental change to actors in similar situations, the greater the number and size of categoric units, up to the maximum allowed by mobility costs.

III. The greater the number and size of categoric and corporate units, the more the potential number of relations increases as a geometric rate and the greater is the amount of time and energy allocated to mobility.

IV. The greater the number of relations among units and the higher the costs of mobility, the more likely are differentiated units to establish networks and combine into more inclusive subsystems, thereby reducing mobility costs.

V. The more differentiated are corporate units, the more centralized is power around those units and subsystems performing key functions, whereas, the more differentiated are categoric units and the greater is their size, the less centralized is power.

VI. The more concentrated is power, and the more prominent are networks and subsystems, the more extensive is political regulation of units in the ecosystem.

---

specific functions. This is particularly true for corporate units, which represent clusters of interdependencies revolving around a particular function. Categoric units form in response to threats, which ultimately stem from the competition that results from increases in population size, productivity, and markets. As the number of differentiated units in a system increases, the number of relations increases at an exponential rate, increasing mobility costs. Corporate and categoric units both tend to consolidate into larger networks, forming subsystems and reducing mobility costs. But the effects of corporate and categoric unit differentiation and consolidation on the concentration of power vary. Corporate units consolidate, centralize, and extend power and regulation, whereas categoric units form power blocks that diffuse power in a system of checks and balances.

As with the other propositions, the many specific hypotheses in Hawley's scheme can, I sense, be deduced from these and from the scenario delineated above.[55] Thus the propositions in Tables 7-1, 7-2, and 7-4 do not do full justice to the depth and extent of Hawley's scheme. Rather, they are intended as more abstract statements about the underlying dynamics of the many hypotheses (some 63 are listed in his latest work, plus numerous assumptions—many of which could also be converted to testable hypotheses). Nonetheless, I think that we now have at least a general sense for ecological theory and that we are now in a position for a few concluding observations.

CONCLUSION

For me, ecological theory retains some of the important ideas of early sociology. One of these ideas is the obvious but often ignored fact that "society" represents

---

[55]Ibid., pp. 106–8, 123–4.

an adaptation of the human species to its environment. Another related idea is that it is not possible to understand human social organization without reference to the interchanges between environment and internal social structure. Yet another crucial idea is that the basic dynamics of a society revolve around (1) aggregation of actors in physical space, competition, and differentiation; (2) integration through subsystem formation and centralization of power. Still another useful point is the emphasis on population size, territory, productivity, communication and transportation technologies, and competition as important causes of those macro structural processes—differentiation, conflict, class formation, consolidation of power, and the like—that have long interested sociologists. Finally, a significant, though problematic, idea is that the altered flow of resources—energy, information, materials—into the system is the ultimate source of social system growth and evolution.

In a field consumed with microreductionism and assertions about the indeterminant nature of humans and their social constructions, it is always a pleasure to see a theory focus on invariant (always there) forces. And the fact that ecological theorizing avoids the long-winded philosophical embellishments of much current sociological theory is an added bonus. Yet, as I have mentioned, there are several problems with ecological theory.

One of these is the assumption—even if it is acknowledged as only a point of emphasis—of exogenous change. It is an unnecessary assumption, because ecological theory is quite capable of conceptualizing both internal dialectical processes as sources of change, especially as these processes create new kinds of environmental interchanges. Another problematic idea is the notion of equilibrium, which is also an unnecessary assumption, even if viewed as only a provisional heuristic. Systems reveal relative degrees of stasis or change in structure, and this is all that we need to say. The attempt to analogize to the "balance of nature" metaphor in bioecology is probably not warranted in sociology (and, I suspect, it is also unwarranted in biology). Finally, the notions of function and key function are problematic but perhaps necessary. What probably needs to be done is to make the analysis even more functional and specify in the abstract the key functions of all social systems adapting to an environment. Then each of these can be seen as a variable; that is, varying numbers and types of units, to varying degrees, can be analyzed in terms of how they are organized to deal with key functions and how this organization of key functions influences the operation of other units. This is, of course, what human ecology does right now, but it often leaves the "functional requisites" of the system as implicit. Since these requisites are part of the analysis, they might as well be made explicit.

In sum, I think human ecology is perhaps the most viable of the functional perspectives. Unlike most functionalisms, it does not construct majestic and cumbersome conceptual edifices. Instead, it operates with a relatively small set of concepts and views them as variables. The result is that human ecology develops testable propositions and, in the process, stifles present mandates in social theory to digress into philosophical discourse. Human ecology, then, is a theoretical perspective worthy of further development.

CHAPTER 8

◆

Biological Functionalism

Pierre van den Berghe*

THE SOCIOBIOLOGICAL CHALLENGE

Functional theorizing began with an analogy to organisms: a pattern of social organization is like a biological organism in that its constituent parts function to maintain the organic whole in its environment. Auguste Comte, who, as we saw in Chapter 2, coined the term *sociology* in the early 19th century, was the first to link the new science of sociology to the rising star of biology.[1] He saw that, by viewing sociology as arising from biology, it would be possible to legitimate this new science in a hostile intellectual world. For, clearly, the 19th century belonged to biology as the mechanisms of biological speciation and evolution were increasingly worked out. Comte recognized that, if sociology could be seen as having certain affinities to these discoveries in biology, it could take its place at the table of science. Indeed, as was quoted in Chapter 2, Comte went so far as to proclaim that, while "Biology has hitherto been the guide and preparation for Sociology . . . Sociology will in the future . . . (provide) the ultimate systematization of Biology."[2]

There is a dramatic irony here because in the last two decades some biologists, and even a number of social philosophers, are now claiming just the opposite: biology will provide the ultimate systematization of sociology! For example, in popularizing the term *sociobiology*, E. O. Wilson declared this new science to involve "the systematic study of the biological basis of all social behavior." And he added that "It may not be too much to say that sociology and the other social sciences, as well as the humanities, are the last branches

---

*This chapter is coauthored with Alexandra Maryanski.

[1]Auguste Comte, *The Course of Positive Philosophy* (London: Bell & Sons, 1896; originally translated in 1854 by Harriet Martineau, but written in French between 1830 and 1842).

[2]Ibid.

of biology waiting to be included in the Modern Synthesis."[3] Although many sociologists have, in essence, scoffed at such pretensions, sociobiology and its various branches have become an important theoretical perspective in biology and, increasingly, in the social sciences. How, then, could biologists "turn Comte on his head" and proclaim sociology as their field?

An answer to this question is complex, for several reasons. First, sociobiology is now only a name for various research activities that seek to understand behavior and social organization in terms of biological processes. Second, many scholars working within this tradition reject the term *sociobiology*, preferring such labels as "evolutionary biology," "behavioral biology," "evolutionary ecology," or "behavioral ecology." These labels provide a sense for this perspective in that they focus our attention on evolutionary processes, especially population genetics, and how these operate to "explain" behavior and organization of animal species, including humans. Third, as this concern with population genetics might imply, the models developed by sociobiologists are highly formal and complex, making them difficult to understand and communicate. Yet, after all the mathematical dust settles, they reintroduce some old intellectual friends: *utilitarianism*, where emphasis is on the efforts of rational actors to maximize utilities, and *functionalism*, where concern is with how processes operate to promote survival of a system in its environment. As can be recalled from Chapter 2, functionalism emerged partly as a reaction *against* utilitarianism, which was seen as overemphasizing rationality, efforts at maximization, competition in markets, and invisible "hands of order" arising from the competitive struggles of maximizing and rational actors. Sociobiology performs, as we will come to appreciate, an interesting marriage between an implicit utilitarianism and functionalism; and although it does so through extremely sophisticated models, often drawn from various game theories, it is nonetheless an effort at intellectual alchemy.

Thus, describing this emerging biological challenge to sociological theory is not easy, and so we will require some preliminary work on its historical origins. We must therefore go back to the 19th century to see from where the concepts of this new challenge come.

THE ORIGINS OF SOCIOBIOLOGY

Herbert Spencer and Biological Sociology

Some modern-day biologists are not aware that Herbert Spencer wrote works on biology before he turned to sociology. Moreover, they are often not cognizant of the fact that Spencer anticipated their efforts to marry functionalism to utilitarianism, although R. A. Fisher, who set the stage for sociobiology, drew considerable inspiration from Spencer's ideas on social and biological evolu-

---

[3]Edward O. Wilson, *Sociobiology: The New Synthesis* (Cambridge, MA: Harvard University Press, 1975); see also his *On Human Nature* (Cambridge, MA: Harvard University Press, 1978), p. 4.

tion.[4] For, as some sociobiologists have recognized, Spencer analogized to biology on two levels: (1) the population-ecology level and (2) the organismic level. Each of these analogies is summarized below.

(1) Spencer was enormously influenced by Thomas Malthus' essay on the principle of population emphasizing that any "population, when unchecked, increases in a geometrical ratio . . . (while) subsistence increases only in an arithmetical ratio" and leads to conditions favorable to conflict, starvation, pestilence, disease, and death.[5] Such conditions would then "check" or bring back into "balance" the population and enable it to support itself in an environment. Spencer reached a less pessimistic conclusion than Malthus, however, and coined the famous phrase "survival of the fittest" to capture his optimism. For Spencer, conflict and competition among both individuals and societies enable the "most fit" to survive and, hence, promote the elevation of individuals and societies to a higher plane. In this way, Spencer merged what he saw as a biological and moral "law" to utilitarianism—competition among rational actors allows the "fit" to survive and vanquishes the unfit. It is not difficult to see that Spencer comes very close to the notion of natural selection in these ideas, which, for the most part, appear in his work on moral philosophy[6] (they tend to be downplayed in his sociology). Yet the basic metaphor of biological functionalism is implied by Spencer: survival and fitness are the goals of individuals, and to some degree the structure of society reflects the efforts of rational actors to maximize their fitness in competition with others.

(2) When Spencer turned to sociology, he analogized to the organism more than to population dynamics. Society is like an organism in that its constituent "organs" function to meet its needs for survival.[7] Spencer never successfully blended this organismic functionalism to his population-ecology, but in a sense modern sociobiology was to attempt just such a synthesis over 100 years later.

Charles Darwin and Natural Selection

Like Spencer, Darwin was greatly influenced by Malthus' essay on population, especially Malthus' "notion of a natural elimination mechanism." In his early notebooks,[8] for example, Darwin considered the power of the Malthusian force

---

[4]R. A. Fisher, *The Genetical Theory of Natural Selection* (Oxford: Clarendon Press, 1930), pp. 180 ff. See also Herbert Spencer, *The Principles of Sociology* (New York: D. Appleton, 1898; originally published in serial form from 1874 to 1896). See also Jonathan H. Turner and Alexandra Maryanski, *Functionalism* (Menlo Park, CA: Benjamin/Cummings, 1979), pp. 8–14; and Jonathan H. Turner, *Herbert Spencer: A Renewed Appreciation* (Newbury Park, CA: Sage, 1985).

[5]T. R. Malthus, *An Essay on the Principle of Population as It Affects the Future Improvement of Society* (London: Oxford University Press, 1798), pp. 14–15. See also Herbert Spencer's *The Principles of Biology* (New York: D. Appleton, 1899; originally published in serial form between 1864 and 1867).

[6]See especially Herbert Spencer, *Social Status* (New York: D. Appleton, 1888; originally published in 1851).

[7]See Spencer, *Principles of Sociology.*

[8]*Charles Darwin's Notebooks, 1836–1844: Geology, Transmutation of Species, Metaphysical Enquiries.* Notebook E, transcribed and edited by David Kohn (British Museum, New York: Cornell University Press), p. 395.

when he wrote that "no structure will last ... without it(s) ... adaptation to *whole* life ... it will decrease and be driven outwards in the grand crush of population" and "... there is a contest ... and a grain of sand turns the balance." Coupled with his observations as the naturalist on the famous voyages of the *Beagle*, where it became clear to him that somewhat different environments create new variants of species, Darwin posited the concept of *natural selection*. In fact, this was to be the title of his book, but he later settled on the more controversial title *On the Origin of Species*.[9] Darwin held the idea of natural selection for 20 years; only when Alfred Wallace,[10] who had also been inspired by Malthus, came up with a similar idea to explain evolution did Darwin begin to publish his ideas. Darwin's and Wallace's initial papers were read jointly at the Royal Society, and both argued that the natural world reveals a "struggle" among species and that this competition "selects" for survival those organisms that are better equipped to adapt to the conditions imposed by an environment. As a result, such organisms are more likely than less fit organisms to reproduce offspring. The basic assumptions of Darwin's (and Wallace's) model are these:

1. Members of any given species reveal variations in their physical and behavioral traits.
2. Members of any given species tend to produce more offspring than can be supported by an environment.
3. Members of any given species must therefore compete with one another and with other species for resources in an environment.
4. Members of a given species revealing those traits that enable them to compete and secure resources will be more likely to survive and produce offspring, whereas those members evidencing traits that are unsuited to competition and/or ability to secure resources will be less likely to survive and produce offspring.

Thus the environment "selects" those traits of organisms that enable them to compete, secure resources, survive, and reproduce. Evolution is driven by this process of natural selection, as those varying traits of organisms are successively selected for by the environment. In this line of argument there is an important distinction that must be emphasized: selection operates on the *individual organism*, whereas evolution involves the *population of organisms*. That is, it is the individual organism that survives and reproduces, or fails to do so, but it is the population that evolves as a whole. This population consists of all those individual organisms that possess those traits that enabled them to survive and

---

[9]In a letter to Asa Gray on September 5, 1857, Darwin wrote the following: "I think it can be shown that there is such an unerring power at work in *Natural Selection* (the title of my book) which selects exclusively for the good of each organic being" (his emphasis). See George Gaylord Simpson, *The Book of Darwin* (New York: Washington Square Press, 1982), p. 24. Also see Charles Darwin, *On the Origin of Species* (New York: New American Library, Mentor Books, 1958; originally published in 1859).

[10]See Charles Darwin and Alfred Russell Wallace, *Evolution by Natural Selection* (Cambridge, England: Cambridge University Press, 1958).

reproduce. This distinction is critical for understanding not only evolutionary theory but also sociobiology. We must remember, then, that, as different individual organisms, revealing varying traits, are "selected for" by the conditions of the environment, the overall composition of the population of organisms changes, or evolves.

One can see an affinity of these ideas to Spencer's famous phrase "survival of the fittest." But the most direct source of inspiration for sociobiology was the blending of Darwin's discovery of the mechanism of natural selection with Gregor Mendel's overlooked insights into the mechanisms of inheritance. For, despite the power of the concept of natural selection to explain evolution, Darwin's portrayal could not answer some basic questions: What is the source of those variations in organisms that are the objects of natural selection? How are traits passed on from one organism to another? What is the mechanism of inheritance? The answers to these questions sat for 35 years in dusty academic journals in the form of Mendel's short manuscripts on the inheritance of characteristics in garden peas.[11] But it was not until 1900 that the science of genetics was born with the rediscovery of Mendel's work and its independent confirmation by Carl Correns in Germany, Hugo de Vries in Holland, and Erich Tschermak in Austria.

The Genetic Theory of Natural Selection

The word *genetics* was coined to denote the new discoveries about how inheritance operates. Although these discoveries occurred over a number of decades, they are discussed here together. What is critical is the recognition that an increasing understanding of the mechanisms of inheritance would provide the basis for the modern synthesis of evolutionary theory and its recent offspring, sociobiology, or what we will characterize as biological functionalism. What, then, were the new insights from genetics? Let us divide them into two types, (1) those pertaining to individual organisms and (2) those dealing with populations of organisms.

The genetics of the individual   *Genes*, or what Mendel had termed "merkmals," are the basic units of inheritance. It is here that the information regarding the transmission of characteristics is stored. Genes can be dominant or recessive, and the various combinations of these—as worked out by Mendel—determine what traits will be visible for an organism. Yet, even when not visible, the information can still be in the genes since they retain their identity and, as a result, emerge in subsequent generations. Genes are strung along thread-shaped bodies, or chromosomes, within the nuclei of cells. *Alleles* are the alternative or variant forms of a gene that affect the same trait in different ways—

---

[11]Gregor Mendel, "Versuche über pflanzen-hybriden," *Verh. Naturforsch. Ver. in Brünn* 10 (1865), English trans. in *Journ. R. Hort. Soc.* 26 (1901). For a classic historical account of Mendel's discoveries, see W. Bateson, *Mendel's Principles of Heredity* (Cambridge, England: Cambridge University Press, 1909).

for example, potential variations in eye color are on different genes, or alleles (since they affect the variation of the same basic trait). *Genotype* is the sum of all the alleles making up an individual, including those that are visible and those that are stored in the genes but not manifest. *Phenotype* refers to the visible traits of an organism that are regulated by genes.

Thus, at the level of the individual, inheritance is regulated by genes. Those genes of an individual that are visibly expressed are its phenotype, whereas those both expressed and invisible are its genotype, or the complete collection of genes an individual has. Differences in the characteristics of individuals are thus the result of the information on discrete genes that provide the fund of variation; hence, natural selection is the mechanism by which genetic material is preserved or lost. This last point is, as we come to see, critical to understanding biological functionalism.

The genetics of the population   Recall that, whereas selection works on the individual—its phenotype and, as a result, its more inclusive genotype—the breeding population as a whole is what evolves. From a genetic point of view, however, it is not so much a population of individual organisms that evolves but, rather, a cluster of genotypes. Individual organisms come and go, but what is passed on and remains is their genetic information. Those genes that produce traits facilitating individual organisms' survival in an environment will increase the likelihood that such organisms will produce offspring that survive. From a genetic perspective, then, it is the genes that are surviving; the living organisms are simply temporary vessels, carrying a most interesting cargo—genotypes.

As this perspective developed, the concept of *gene pool* was introduced to characterize this shift in emphasis from the individuals who carry the genes to the sum total of their different alleles. A gene pool is thus the pooled sum of all the genotypes in a population of organisms. A species is the most inclusive gene pool; a less inclusive gene pool would be a breeding population in a particular area or region.

What would cause changes in the composition of genes in the pool? One obvious force—which, surprisingly, was not initially recognized—is *natural selection*. As some individual organisms—that is, phenotypes and genotypes—are selected for survival and reproduction, while others are selected against, a shift in the composition of the gene pool can occur. Another force changing the composition of the gene pool is random *mutations*, which add new DNA (the actual codes of information on genes) to the gene pool. Another force is *gene flow*, which results from the movement of individuals to and from different breeding populations (say, for example, intermarriage and offspring among Asians and Caucasians). Yet another force is *genetic drift*, which is the change in the composition of gene frequencies in the pool that results from a decrease in the potential variability of a small population. The overall degree of variation in the gene pool is related to the interactions among these forces. Mutations and gene flow increase variation within a breeding population because they add new genetic material. Natural selection normally reduces variation because it weeds out those phenotypes (and genotypes) less suited to an environment. Similarly, genetic drift reduces variation through the random loss of genetic variance available to a small breeding population.

This shift to pools of genes as opposed to populations of individuals emerged in the 20th century, although the actual term *gene pool* was a midcentury formulation.[12] As early as 1907, for instance, G. H. Hardy and W. Weinberg applied the new discoveries about particulate inheritance to the population level.[13] For them, evolution could be conceptualized as changes in "gene frequencies" (later to be conceptualized within "the gene pool"). In what became known as the Hardy-Weinberg equilibrium model for assessing the degree of evolution, they presented an argument that was to be decisive. If we know the allele frequencies in a population (that is, all the dominant and recessive genes relevant to various traits), we can predict the genotypic frequencies in the next generation, but, to do so, we must assume the following: no natural selection is operating; no mutations are produced; no gene flow resulting from migrations of breeding populations is evident; no genetic drift is to be found; no bias in mating can be observed (it is random); and no limits on population size exist (it is infinite). Obviously such assumptions do not correspond to natural populations, but they allow one to compute the expected distribution of genotypes in the next generation if *none* of these forces of change in the gene pool is operating. Then, by comparing this idealized computation of gene frequencies with the actual frequencies, one gets some indication of how much change or evolution is occurring. If there is not much difference between the idealized prediction and the actual genotypes in the population, then not much evolution is evident; but, if there is a large difference, then the degree of evolution can be measured (by comparing the ideal and actual gene frequencies). What is important about the Hardy-Weinberg Law is that it is based on the fact that evolution (or a change in gene frequencies) can occur only when variations exist in a population upon which selective forces can act. It is these selective differences among genotypes that change the gene frequencies.

---

[12]Although the term *genetics* was coined by Bateson in 1906 as the basic construct to describe individual heredity and variation (see footnote 11), the term *gene pool* was coined by Dobzhansky in 1950 and became the fundamental construct of population genetics. See Theodosius Dobzhansky, "Mendelian Populations and Their Evolution," *American Naturalist* 14 (1950), pp. 401–18. For further readings on the history of genetics, see Theodosius Dobzhansky, *Genetics and the Origin of Species* (New York: Columbia University Press, 3rd rev. ed., 1951), and *Mankind Evolving* (New York: Bantam, 1962). See also: Mark B. Adams, "From 'Gene Fund' to 'Gene Pool': On the Evolution of Evolutionary Language," *History of Biology* 3 (1979), pp. 241–85, and "The Founding of Population Genetics: Contributions of the Chetvevikov School 1924-1934," *Journal of the History of Biology* 1 (1968), pp. 23–39; Alfred Sturtevant, *A History of Genetics* (New York: Harper & Row, 1965); and James Crow, "Population Genetics History: A Personal View," *Annual Review of Genetics* 21 (1987), pp. 1–22.

[13]Extending Mendel's Laws by deducing the mathematical consequences of a nonblending system of heredity (i.e., that genes are discrete and do not blend), G. H. Hardy and W. Weinberg laid the cornerstone for population genetics; in turn, the modern science of statistics was born in the study of quantitative genetics. G. H. Hardy, "Mendelian Proportions in Mixed Populations," *Science* 28 (1908), pp. 49–50; W. Weinberg, "Über den Nachweis der Vererbung beim Menschen," *Jh. Ver. Vaterl. Naturk. Wurttemb.* 64 (1908), pp. 368–82.

In the first decade of this century, this revolution in population genetics was not well integrated with Darwinian views on natural selection as the mechanism of evolution. Curiously, as concepts in genetics emerged, they were often considered to represent an alternative explanation to Darwinian natural selection. Indeed, during the first part of this century, a major anti-Darwinian movement rejecting natural selection as the force behind evolution emerged with opposition so negative that one noted scholar could write: "We are now standing at the death bed of Darwinism. . . ." Indeed, by the 1920s even major textbooks on evolution carried forth an antiselectionist position; for many scholars it was already a foregone conclusion that the Darwinian revolution had passed and that ". . . a new generation has grown up that knows not Darwin."[14]

It is from the revival of Darwinian notions of natural selection and their eventual coupling with genetics that the modern synthetic theory of evolution was to emerge. And it was at this point that the basic ideas of sociobiology were first articulated.

THE EMERGENCE AND CODIFICATION OF SOCIOBIOLOGY

In his *The Genetical Theory of Natural Selection*, R. A. Fisher was the first to draw out the implications of synthesizing genetics and Darwinian selection.[15] Fisher's first task was to refute the main competitor to natural selection as the force behind evolution. This competitor was "mutation theory," which argued that large mutations are the driving force behind evolution. Fisher demonstrated with elegant mathematical equations that the vast majority of mutations are harmful and doomed to extinction by natural selection. In particular, Fisher demonstrated that those large mutations that were posited as the key force behind evolution would be harmful and, hence, selected out of the gene pool. Instead, only small mutations offering slight advantages in promoting the fitness of organisms to an environment could be involved in evolution, but these modest mutations could not by themselves alter the gene pool. Rather, it is the power of natural selection to favor such mutations that drives evolution.

---

[14]See Eberhart Dennert, *At the Deathbed of Darwinism*, trans. E. G. O'Hara and John Peschges (Iowa: German Literary Board, 1904), p. 4. Also see J. H. Bennet, *Natural Selection, Heredity, and Eugenics* (Oxford: Clarendon Press, 1983), p. 1; Garland Allen, "Hugo de Vries and the Reception of the Mutation Theory," *Journal of the History of Biology* 2 (1969), pp. 56–87; and Sewall Wright, "Genetics and Twentieth-Century Darwinism," *American Journal of Human Genetics* 12 (1960).

[15]See footnote 4. For an overview of Fisher's contribution, see: J. H. Bennett, *Natural Selection, Heredity, and Eugenics* (Oxford: Clarendon Press, 1983). J. B. S. Haldane and Sewall Wright (see footnote 12) also laid the foundation for the reconciliation of Mendelian heredity and Darwinian selection, but it was Fisher's work that triggered this revitalization. Fisher was primarily interested in how an organism can increase its fitness, however it is achieved. This emphasis on fitness is summarized in his fundamental theorem: "the rate of increase in fitness of any organism at any time is equal to its genetic variance in fitness at that time." Fisher, *The Genetical Theory of Natural Selection*, p. 35.

With this line of argument, Fisher welded population genetics and natural selection while introducing an important concept: *fitness*. This concept was as old as Malthus, Spencer, and Darwin, but it now took on new meaning in Fisher's hands. The mean fitness of a population, Fisher argued, will usually be proportional to some component of the genetic variance. That is, gene pools revealing considerable variation provide a greater range of options for selection to work on, thereby increasing the mean fitness of that population to survive. Moreover, even with small variations in a gene pool, selection will still remain the significant force in determining gene frequencies.

In this way the concepts of selection, genetic variation, and fitness were linked together. Others in the 1930s extended this line of argument, but what is crucial for our purposes is the view of evolution as revolving around the power of natural selection to promote fitness by "selecting" those variations in genes that promote adaptation.

In the 1940s all of these leads were crystallized in the "modern synthesis," with natural selection hailed as the only directional force in evolution. But before this synthesis Fisher set the stage for modern human sociobiology. In the last one-third of *The Genetical Theory of Natural Selection*, Fisher quoted Herbert Spencer and turned to an analysis of "Man and Society," stressing the need for forms of eugenics to promote the "survival of the fittest genes." And although Fisher's ideas on sociology are not now important, they firmly planted in the minds of some biologists that human behavior and organization might be understood in terms of the same natural processes affecting other species—that is, variation in genes, natural selection, and fitness as the adaptive value of genes.

The First True Sociobiologists

Early instinct theories   The first "sociobiologists" were, curiously, not those working in the tradition established by Fisher and the synthesis of population genetics and natural selection. Rather, the initial thrust for a biological view of humans and society was a group of scholars who were throwbacks to the older "instinct theories" that had always existed in sociology and social philosophy. Perhaps the most influential figure was the father of ethology, Konrad Lorenz. In particular, Lorenz investigated the "aggression instinct," which he saw as becoming channeled by rituals (presumably through natural selection) in order to space members of an animal species for food gathering, mating, and mitigation of violent encounters.[16] Lorenz viewed human aggression as maladaptive, however, because natural selection had not equipped early human ancestors with ritualized mechanisms for inhibiting aggression in high-density situations. Other speculation on human instincts for aggression and domination followed—for example, Robert Ardrey's portrayal of the "killing

---

[16]Konrad Lorenz, *On Aggression* (New York: Harcourt Brace Jovanovich, 1960).

instinct" in *African Genesis* and *The Territorial Imperative*,[17] Desmond Morris' "naked ape,"[18] Lionel Tiger's and Robin Fox's "imperial animal,"[19] and the early sociobiological work of Pierre van den Berghe on age, sex, and domination.[20] We will examine this type of sociobiology later, but it should be recognized that this line of argument is not functional, whereas work following Fisher became increasingly functional.

Functional sociobiology   Sociobiology in its more functional incarnation emerged partly in response to what is known as the "group selectionist" argument. Perhaps V. C. Wynne-Edwards is the most closely associated with this group selection perspective.[21] For him the basic problem is this: how is altruistic behavior in animals to be explained? That is, how does natural selection, with its emphasis on the individual organism's effort to survive and reproduce, explain cooperative behavior in which organisms sacrifice their own fitness for the good of the group? For, despite the fact that it is a competitive world, Wynne-Edwards argued in 1962 that "the members of social groups cooperate in civilizing it and, so far as the competition is concerned, they enact according to rules. Everything the social code decrees is done for the common good. . . ."[22] Thus it may be that, in higher animals, the group is often the unit of selection rather than the individual. Variations in group structures are selected for in terms of their capacity to adapt and survive in an environment. Such a line of argument was not terribly different from Herbert Spencer's view that social groups, especially whole societies, often compete for existence in a given area, with the better organized society surviving (usually as a result of its superior military ability); so, evolution had, in Spencer's eyes, involved the successive competition and survival of ever more "fit" societies.[23] Whether the author was a Spencer or a Wynne-Edwards, the idea was that the individual organism is not the only unit of selection; groups or, in Spencer's terms, "superorganic" units composed of interacting and mutually dependent organisms can also constitute a "body" or "social organism" subject to selection pressures.

---

[17]Robert Ardrey, *African Genesis* (New York: Delta Books, 1961) and *The Territorial Imperative* (New York: Atheneum, 1966).

[18]Desmond Morris, *The Naked Ape* (New York: Dell Books, 1967).

[19]Lionel Tiger and Robin Fox, *The Imperial Animal* (New York: Holt, Rinehart & Winston, 1971).

[20]Pierre van den Berghe, *Age and Sex in Human Societies: A Biosocial Perspective* (Belmont, CA: Wadsworth, 1973). A comparison of this work with later ones hints at the changes that van den Berghe was to make. See: *Human Family Systems: An Evolutionary View* (New York: Elsevier, 1979).

[21]V. C. Wynne-Edwards, *Evolution through Group Selection* (Oxford: Blackwell, 1986); and *Animal Dispersion in Relation to Social Behavior* (New York: Hafner, 1962). For a review of the controversy surrounding group selection arguments, see: David Sloan Wilson, "The Group Selection Controversy: History and Current Status," *Annual Review of Ecological Systems* 14 (1983), pp. 159–87.

[22]Wynne-Edwards, *Evolution through Group Selection*, p. 9, outlining his original ideas in the course of writing *Animal Dispersion*.

[23]Spencer, *Principles of Sociology*.

Sociobiology in its current profile emerged as a reaction against such group selection arguments. In particular George C. Williams launched a critique of group selection, arguing that the real unit of selection is not the group or even the individual organism.[24] Rather, the unit of selection is "the gene," leading Williams to posit the concept of *genic selection*. Those genes, however temporarily housed in individuals and groups, that promote survival and reproduction in an environment—that is, "fitness"—will be retained. Whatever the effects of selection for promoting "groupness," it is selection at the level of the individual and the gene that is the operative mechanism. For "group-related adaptations do not, in fact, exist"; instead, the characteristics of groups—altruism, reciprocity, exchange, etc.—are the result of natural selection on individuals because "simply stated, an individual who maximizes his friendships and minimizes his antagonisms will have an evolutionary advantage, and selection should favor those characters that promote the optimization of personal friendships."[25] Thus, particular genes that promote those traits in individuals facilitating "groupness" will, in certain environments, promote fitness—that is, survival and reproduction. It is not necessary, Williams argued, to explain group processes in terms of group selection; "genic selection" can explain such group processes, for ". . . the fitness of a group will be high as a result of (the) summation of the adaptations of its members."

W. D. Hamilton took this kind of reasoning a step further by introducing the all-important concept of *inclusive fitness*.[26] This concept was intended to account for cooperation among relatives, and the argument goes something like this: Natural selection operates to promote "kin selection" in the sense that those who share genes will interact and cooperate to promote one another's fitness—or capacity to pass on their genes. Self-sacrifice for a biological relative is, in reality, not altruism at all but the selfish pursuit of fitness because, in helping a relative to survive and reproduce, one is passing on one's own genetic material (as stored in relatives' genotypes). Thus, from this point of view, self-sacrifice for, and cooperation with, relatives will be greater as the amount of shared genetic material increases. So, altruistic behaviors among parents and offspring or among siblings can be understood as behaviors that were "selected for" as a way to pass on one's genetic material, or to keep it in the gene pool. This is the process of inclusive fitness—"inclusive" in the sense that one shares identical genes with others and "fitness" in the sense that, in helping these

---

[24]George C. Williams, *Adaptation and Natural Selection: A Critique of Some Current Evolutionary Thought* (Princeton, NJ: Princeton University Press, 1966). For his defense of reductionism away from the group level, see: "A Defense of Reductionism in Evolutionary Biology," in *Oxford Surveys in Evolutionary Biology* 2, eds. R. Dawkins and M. Ridley (Oxford: Oxford University Press, 1985), pp. 1–27.

[25]Williams, *Adaptation and Natural Selection*, p. 95.

[26]W. D. Hamilton, "The Evolution of Altruistic Behavior," *American Naturalist* 97 (1963), pp. 354–56; "The Genetical Theory of Social Behavior I and II," *Journal of Theoretical Biology* 7 (1964), pp. 1–52; "Innate Social Aptitudes of Man: An Approach from Evolutionary Genetics," in *Biosocial Anthropology*, ed. R. Fox (New York: John Wiley, 1984), pp. 135–55; "Geometry for the Selfish Herd," *Journal of Theoretical Biology* 31 (1971), pp. 295–311.

others, one is also assuring that shared genetic material will remain in the gene pool. This kind of argument takes "the altruism out of altruism" among family members and sees such behaviors as simple matters of self-interest: to maximize the amount of one's genetic material that stays in the gene pool. Hence the "goal" of genes is to preserve themselves, and it is "rational" for them to help preserve the bodies of those individuals who carry common genetic material. Of course, genes do not "think," but blind natural selection has operated in the distant past to promote behaviors in organisms, such as altruism among relatives, that increased "fitness" in ways that *maximize* the passing on of particular sets of genes.

Although Hamilton's notions of "kin selection" and "inclusive fitness" might be seen to account for cooperation among relatives, the question was soon raised: how can such arguments explain altruism and cooperation among nonrelatives who do not share genetic material? Robert Trivers sought to overcome this objection with the concept of *reciprocal altruism*.[27] In a series of modeling procedures he presents the following scenario: natural selection can operate to produce organisms that will incur the "costs" of helping a nonrelative because at some later time this nonrelative can "reciprocate" and help "altruistic" organisms (thereby increasing the latter's fitness, or ability to survive and pass on genes). Thus, for species that live a long time and congregate, natural selection can operate to promote reciprocal altruism and increase all individuals' fitness, whereas those that would "cheat" and fail to reciprocate others' altruism will be selected out (since eventually, without signs of reciprocity, others would not come to their aid). And so, once again what seems like altruism is, in reality, "selfishness" on the part of individual organisms, each of which is trying to maximize its capacity to enhance its genes in the pool.

The last major conceptual development in sociobiology has been the use of "game theory" to describe the process of fitness. Here the key figure has been J. Maynard-Smith,[28] although Trivers had started his analysis with the classic "Prisoner's Dilemma" game to show how selfish individuals can cooperate in order to increase their fitness beyond what it would be without cooperation.[29] In game theory it is assumed that actors are rational decision-

---

[27]Robert L. Trivers, "The Evolution of Reciprocal Altruism," *Quarterly Review of Biology* 46 (4, 1971), pp. 35–57; "Parental Investment and Sexual Selection," in *Sexual Selection and the Descent of Man, 1871–1971*, ed. B. Campbell (Chicago: Aldine, 1972); and "Parent-Offspring Conflict," *American Zoologist* 14 (1974), pp. 249–64.

[28]J. Maynard Smith, "The Theory of Games and the Evolution of Animal Conflicts," *Journal of Theoretical Biology* 47 (1974), pp. 209–21; "Optimization Theory in Evolution," *Annual Review of Ecological Systems* 9 (1978), pp. 31–56; *Evolution and the Theory of Games* (London: University of Cambridge Press, 1982). See also: Susan E. Riechert and Peter Hammerstein, "Game Theory in the Ecological Context," *Annual Review of Ecological Systems* 14 (1983), pp. 377–409.

[29]The basic format for the "Prisoner's Dilemma" is this: Two criminals are caught together and accused of a crime; they are taken to separate rooms for questioning, with each being offered some leniency for telling on the other. If both refuse to talk, the police have no real evidence; yet, if one talks and the other does not, then the latter is at a disadvantage. Thus the dilemma is to "talk" or "keep quiet" under conditions in which each partner in crime does not know what the other will do. The maximizing strategy is for both to "keep quiet," but each can get less than the maximum benefit (and far more than the worst outcome) by telling on the other.

making units who seek to get the best possible payoff, under particular conditions imposed by the game, by adopting a particular behavioral strategy. The payoffs in game theory are typically some unit of subjective value. In contrast, unlike classical utilitarianism, in which actors are assumed to be conscious, rational, and payoff maximizing, game theory as it is applied to evolutionary theory cannot assume rational consciousness of its players (genes), and the payoffs are always a measure of fitness (capacity to pass on genes). The process of selection is presumed to have "decided" (unconsciously) the strategy that maximizes fitness in a given environment; the investigator's task is then to determine what behavioral strategy for a particular species in an environment would best assure maximal payoffs in terms of fitness, or the capacity to survive, reproduce, and pass on genes. This is what Hamilton did with the concept of inclusive fitness: helping one's biological relatives is the best strategy, as "decided" by the forces of natural selection, for passing on one's genetic material.

Maynard-Smith went a step further and developed the concept of *evolutionary stable strategy*, or ESS, to describe the stabilization of behavioral strategies among the individuals of a population. Without outlining the mathematical and statistical details, the ESS enables Maynard-Smith to calculate an equilibrium point, around which the relative amounts of various behaviors, or strategies, for fitness will stabilize. In this way it is possible to show that all members of a population do not have to adopt the same strategy; rather, each potential strategy affects the payoffs of the others, and over time the relative frequencies of various strategies for survival will, as a result of natural selection, reach equilibrium, or the ESS.

This application of game theory gave sociobiologists a well-developed and powerful set of mathematical tools for making predictions about how natural selection will produce behavioral strategies maximizing fitness and how varying configurations of such strategies can reach equilibrium. Such configurations can, sociobiologists argued, explain patterns of social organization.

Richard Dawkins popularized this emerging sociobiological approach in his well-known work *The Selfish Gene*.[30] His argument is that genes are "replicator" or "copy" machines that try to reproduce themselves. Natural selection favored those replicators that could find a "survival machine" to live in—initially, in the distant past, a cell wall, then a grouping of cells, then an organism, and eventually a grouping of organisms. As Dawkins notes:[31]

> What weird machines of self-preservation would the millennia bring forth. . . . They (replicators) did not die out, for they are past masters of the survival arts. . . . Now they swarm in huge colonies, safe inside gigantic lumbering robots, sealed off from the outside world, communicating with it by tortuous indirect routes, manipulating it by remote control. They are in you and me; they created us, body and mind; and their preservation is the ultimate rationale for our existence. They have come a long way, those replicators. Now they go by the name of *genes*, and we are their survival machines.

---

[30]Richard Dawkins, *The Selfish Gene* (Oxford: Oxford University Press, 1976).
[31]Ibid., p. 21.

Such a rich metaphor captures the essence of modern sociobiology because the unit of selection becomes the gene, and evolution is the result of genes competing and adopting strategies in order to leave their DNA in the gene pool. Evolution is not an effort of individuals or species to survive; these are only vehicles for the real driving force of evolution: genes that are "ruthlessly selfish" in adopting strategies to maximize their fitness. At times it serves the genes' interest to foster limited forms of altruism and other social behaviors that are often considered to be the exclusive domain of the social sciences. But from a sociobiological perspective, many of the behaviors, strategies, and organizational traits of animals, including humans, are simply the genes' way of coping with an unpredictable environment. Indeed, even the human capacity for thinking and learning, Dawkins avers, can be viewed as the genes' way to construct a better survival machine; cooperation can similarly be seen as one survival machine making use of another survival machine in an effort to further assure its fitness; and various patterns of social organization can thus be conceptualized as nothing more than more complex and inclusive survival machines for genes.

Yet Dawkins hedges in his last chapter, as have many contemporary sociobiologists in recent years. Dawkins posits a "new replicator," which he terms *memes*. The basic tenets of sociobiology—genic selection, inclusive fitness, and reciprocal altruism, all producing strategies and "survival machines" for genes—can explain how humans came to exist, but culture begins to supplement and supplant biology as the major replicating mechanism. Memes are those new cultural units that exist inside brains and that, via socialization, are passed on and preserved in a "meme pool." Meme evolution will now begin to accelerate, for "once genes have provided their survival machines with brains which are capable of rapid imitation, the memes will automatically take over." And it might even be possible for memes to rebel against their creators, the selfish genes. Similarly, other sociobiologists have begun to talk in terms of "co-evolution," operating at both the genetic and the cultural levels.

Although these metaphors are colorful, sociobiology is highly technical—involving extensive use of mathematics, game theory, and computer simulations. It changes the image of natural selection as a process working on passive individuals and posits, instead, active actors driven by their genes to maximize their reproductive success by any strategy available (which can be modeled and simulated with various game theoretic approaches). The challenge of this perspective is that many processes considered by sociologists to be explicable only by sociological laws are seen by sociobiologists to be understandable in terms of biological processes derived from the laws of the synthetic theory of evolution. In mounting this challenge, sociobiology has created a new kind of functionalism.

WHY IS SOCIOBIOLOGY FUNCTIONAL?

Sociobiology has been termed, here and there in this chapter, as *biological functionalism*. Why is this label used? The answer is this: from a sociobiological perspective, behavioral and perhaps organizational traits are analyzed with respect to their functions for one master need or requisite, fitness. That is, a behavior or structure is seen to function to meet the need of genes to pass themselves on and remain in the gene pool.

Like most functional arguments, those in sociobiology reveal the problems of illegitimate teleology and tautology. Let us take the problem of illegitimate teleology first. Sociobiology begins with the very utilitarian assumption that genes seek to maximize fitness by adopting strategies; then a behavioral strategy or structure is analyzed in terms of how it maximizes reproductive fitness for genes. For example, family bonds are seen to maximize fitness by assuring that those organisms that share genetic material will help one another out. The function of the family, therefore, is to maximize the fitness of genes. In such arguments the goal of the family—fitness for genes—is its cause. Behind this argument, or course, is a "selection" mechanism: sometime in the past those genes that encapsulated themselves in organisms and adopted strategies that created family structures were more likely to survive than those that could not do so. But this speaks only to some kind of ultimate causation without specifying how particular types of families evolved and operated—the sociologically interesting questions. Such explanations are a big gloss on matters of interest to sociologists; moreover, they are not testable or, hence, refutable. In fact, any human structure can be analyzed in this way: structure x exists because it meets the needs of genes for inclusive fitness; structure x meets this need in the following way (now use your creative intellect to make up a story or scenario about how the needs of genes are served); and this function of structure x is the result of selection pressures in the distant past. What has really been explained by this kind of argument? The need to be met—fitness—is seen to cause those structures—x, y, z, etc.—meeting this need, but none of the detail as to how, when, and why this occurred is ever provided. Rather, some global pronouncements—selection in the distant past produced structure x to maximize fitness—are made.

With respect to tautology, sociobiological arguments are often circular. For example, to state the matter simply: structure x exists; structure x must therefore promote fitness, because otherwise structure x would not exist. Many sociobiological arguments are of this tautological nature. For instance, to be more sociologically concrete: Family is a basic institution in human societies. It must therefore serve important functions for the purpose of meeting fitness needs of genes. How do we know that the family does so? Because the family always exists.

Thus, sociobiology reveals all of the problems of classic functionalism (see pp. 72–77 in Chapter 3).[32] And, despite efforts to make "natural selection" an unconscious force making "decisions" over strategies, sociobiology also evidences all of the problems of classic utilitarianism: goal-directed genes are seen as trying to maximize their fitness through varying strategies that are mys-

---

[32]See also: Turner and Maryanski, *Functionalism.*

teriously "selected for." Although models of selection processes seek to make this utilitarianism nonteleological (that is, selection favored those genes that, presumably by chance, hit upon this or that strategy), the models make assumptions that are reminiscent of utilitarian economic models: open competition, unlimited time, no existing structures to "load the competitive dice," and so on. Moreover, the basic assumption—maximization of fitness—is much like the same unquestioned and incorrect assumption in utilitarianism—maximization of utilities. And the mathematical models that are produced with these assumptions are very much like those in economics—elegant but so impregnated with unrealistic assumptions as to make them nonisomorphic with the real world.

In a sense, then, sociobiology blends the worst of utilitarianism and functionalism. Yet we can ask: why has it caught on in biology? Part of the answer is that it extends population genetics into new territory—sociology—and couples biological analysis with game theory. And since biologists do not know very much sociology, such explanations seem "reasonable" or at least "intriguing." Indeed, most sociobiologists admit to not knowing very much sociology, even conceding that "in the future" they must learn more. The naiveté of sociobiologists in sociological matters and their willingness to fall into the old traps of functional theorizing have enabled most sociologists to dismiss sociobiology as ridiculous. But several sophisticated sociologists have adopted this perspective, and so we should not reject this biological challenge out of hand. We should first examine what a sociologist would find appealing about biological functionalism.

A SOCIOLOGICAL APPROACH TO BIOLOGICAL FUNCTIONALISM: PIERRE VAN DEN BERGHE

Although several sociologists have adopted elements of sociobiology,[33] Pierre van den Berghe has consistently been the most prolific advocate of a biological perspective on human affairs. For van den Berghe, it "is high time that we seriously look at ourselves as merely one biological species among many";[34] until this shift in perspective is accomplished, sociology will remain stagnant, for there can be no doubt that biological forces shape and constrain patterns of human organization. In his early advocacy, van den Berghe proposed a "biological" approach that corresponded to the instinct approach of some early sociologists and several contemporary thinkers, whereas in his more recent works, van den Berghe has become a full-blown sociobiologist in the tradition

---

[33]Probably the most tempered among these figures is Joseph Lopreato, *Human Nature and Biocultural Evolution* (Boston: Allen and Unwin, 1984). This book also offers an excellent summary, overview, and assessment of sociobiology while at the same time developing the author's own perspective.

[34]van den Berghe, *Age and Sex in Human Societies*, p. 2. See also his *Man in Society: A Biosocial View* (New York: Elsevier, 1975); "Territorial Behavior in a Natural Human Group," *Social Science Information* 16 (1977), pp. 421-30; "Bringing Beasts Back In: Toward a Biosocial Theory of Aggression," *American Sociological Review* 39 (December 1974), pp. 777-88.

of Fisher, Williams, Trivers, Wilson, and Dawkins.[35] Let us first mention the early instinct approach and then turn to van den Berghe's more functional biological arguments.

The Early Biosocial Instinct Approach

In this early phase, van den Berghe simply postulated the following: humans are a type of mammal or, more specifically, a primate, and evolution must surely have produced biological propensities in our mammalian and primate ancestors that, no doubt, have been elaborated by culture and social structure. For, in van den Berghe's mind, "there is such a thing as human nature, just as there is a chimp nature, or an elephant nature."[36] Thus there are predispositions among humans to behave in certain ways and hence to organize themselves into particular patterns of social structure.

In one of his most polemical pieces,[37] van den Berghe contended that humans are aggressive and that such aggressive tendencies are a consequence of competition for resources; in turn, this competition is the result of population pressures and constantly escalating needs (which the large human brain can dream up). In particular, the use of hunting and predation as competitive strategies of adaptation created selection pressures for aggressive behavior among humans. Such aggression became channeled into two biological propensities for regulating the aggression in ways that promoted the survival of humans: (1) territoriality, or the biological tendency to defend and control space in which resources are located; and (2) hierarchy, or the propensity to create stable rankings that give preferred access to resources. Of course, once patterns of hierarchy and territoriality exist, they feed back and often escalate the very aggression and competition that they evolved to regulate; but, without hierarchy and territoriality, constant aggression would exist and destroy the species. Thus there is a biological basis for war, stratification, politics, and other features of human societies, and in van den Berghe's view we will not fully understand human societies until we recognize this biological fact. For, although these biological tendencies become culturally elaborated and although they vary in different environmental contexts, there is still a biological basis underpinning the organizational features of human society.

In longer works, van den Berghe explored specific aspects of social structure from this general perspective. For example, in *Age and Sex in Human Soci-*

---

[35]For example, see: Pierre van den Berghe, "Bridging the Paradigms," *Society* 15 (1977–78), pp. 42–49; *The Ethnic Phenomenon* (New York: Elsevier, 1981); "Skin Color Preference, Sexual Dimorphism and Sexual Selection: A Case of Gene Cultural Co-Evolution," *Ethnic and Racial Studies* 9 (1, 1986), pp. 87–113; Pierre van den Berghe and David P. Barash, "Inclusive Fitness and Human Family Structure," *American Anthropologist* 79 (1977), pp. 809–23.

[36]van den Berghe, "Bringing Beasts Back In," p. 778.

[37]van den Berghe, "Bringing Beasts Back In." See, for a critique of this piece, Sandra R. Turner, Jonathan H. Turner, and Alan Fix, "Throwing the Beast Back Out," *American Sociological Review* 41 (June 1976), pp. 551–55.

eties,[38] the differentiation of human populations by age and sex is seen as the first axis for domination—of women by men and of the young by the old. These power dynamics are the building blocks for more elaborated features in society—politics, age strata, inheritance, kinship structures, etc. These features are often elaborated upon by culture and social structure, but they are nonetheless tied to biological processes revolving around sexual dimorphism and age.

In another example, van den Berghe wrote an introductory sociology text from a "biosocial" viewpoint, which argued that the behavioral repertoire of every species is determined, in part, by biological parameters and that, although the "complex interplay of biogenetical and environmental factors" complicates our capacity to see and sort out these parameters, there is still "the strong possibility that fundamental aspects of our social organization, such as hierarchy, family, and marriage are biologically predisposed."[39] For although "Biology does not explain everything . . . it is foolish to pretend that it does not explain anything."[40] Accordingly, family, sexual relations, religion, inequality, arts and sciences, fun and games, and other features of human organization are analyzed in terms of biological predispositions.

In all of this work, a cultural and structural feature of society is explained by a biological predisposition. This predisposition is seen as a result of past selection processes as humans' primate ancestors adjusted to the environment. And although the feature to be explained is seen to vary and to be augmented by nonbiological forces, it is still considered to have a biological basis. In particular, if a sociocultural feature is universal in human societies, then it is a good candidate for biosocial inquiry. Although suggestive and provocative, this early work was vague. It typically consisted of finding a "bioprogramer" for each universal structural feature and then constructing an evolutionary scenario as to why this bioprogramer should exist in the first place.

This line of inquiry emerged about the time sociobiology was becoming established in the 1970s and was not informed by sociobiology. But soon van den Berghe switched his mode of explanation to the conceptual vocabulary of sociobiology. The basic biological thrust of his analysis did not change, but the explanations for sociocultural features of human society underwent a significant recasting.

The Switch to Biological Functionalism

Criticisms of sociology   Van den Berghe has been one of the most persistent and truculent critics of sociology. One line of criticism is the failure of sociology to be a science. In van den Berghe's view, sociologists are increasingly likely to be scholastics who are

> . . . very good at quoting classics, seeking academic ancestors, and establishing intellectual pedigrees. Their textbooks are tiresome commentaries on the gos-

---

[38]van den Berghe, *Age and Sex in Human Societies*.

[39]van den Berghe, *Man in Society: A Biosocial View*, pp. 28, 42.

[40]Ibid., p. 42.

pels according to Saints Marx, Durkheim, Weber, and Pareto . . . (and) during their half century of lofty isolation from the natural sciences, the social sciences have become, in short, a scholastic tradition rather than an evolving scientific discipline.[41]

Moreover, in addition to these scholastic tendencies in sociological theory, numerous barriers prevent sociology's reintegration with the sciences.[42] For openers, sociologists are resistant to reductionist arguments in which phenomena are explained in terms of the properties of their constituent elements. In science, van den Berghe claims, this "is the only game in town," for the goal is always to explain phenomena in terms of the dynamics of their parts. Second, and related to this first obstacle, is sociology's persistent "reification of the group." Here emphasis is on the social structure rather than on the individual, whereas in science an effort is always made to explain "emergent phenomena" in terms of more elemental processes. Third, sociologists begin analysis with a dogmatic "environmental determinism," viewing individual behavior and emergent structures as the products of present social and cultural conditions. Such dogmatism prevents sociologists from recognizing the selection processes in the distant past that have produced biological propensities in humans to behave and organize in certain ways. Fourth, sociologists rely on "dualistic thinking" or artificial "dichotomization" when talking about human biology. For example, distinctions like "nature/nurture" and "heredity/environment" keep sociologists from viewing sociocultural phenomena as having continuity with biological and evolutionary processes. Fifth, sociologists have become rabid anti-evolutionists in their overreaction to models of societal evolution (e.g., see Parsons' work in Chapter 3). This overreaction has produced a general bias against any evolutionary thinking—an obviously flawed conclusion in light of the fact that humans as a species evolved by natural selection. Sixth, in van den Berghe's opinion there has been far too much emphasis in sociology on conscious motivation and voluntarism. Humans are seen by sociologists as symbolic, cerebral, verbal, and plastic; as a consequence, unconscious biological constraints and limitations on human behavior and organization are given insufficient attention. Seventh, sociologists have relied upon data—typically statistically aggregated verbal responses of individuals to questionnaires—that are far removed from the actual behavior of people and the biological forces that guide such behavior. And, finally, sociologists are insecure and afraid of being "gobbled up" by other sciences, and so they continually proclaim the unique and emergent qualities of their domain of inquiry. But, in reality, there should be no real problem linking the biological and social sciences, or even seeing sociology as a subdiscipline within the natural sciences. Only our insecurities keep us from reintegrating sociology with biology.

These kinds of criticisms frame van den Berghe's adoption of sociobiology. Perhaps by stating the critique in such extreme form, van den Berghe has felt

---

[41]van den Berghe, "Bridging the Paradigms," p. 42.

[42]Ibid. Also see: "Response to Lee Ellis' 'The Decline and Fall of Sociology' ", *The American Sociologist* 12 (1977), pp. 75–76.

justified in adopting extreme sociobiology. At any rate, he adopts the conceptual thrust of sociobiology, adding only a few minor extensions.

Conceptualizing sociobiological processes   Van den Berghe initiates his turn to sociobiology by posing the question: why are humans "social" in the first place? His answer is functional: "sociality," or the banding together of animals in cooperative groups, functions to increase their reproductive fitness by (1) protecting them against predators and (2) providing advantages in locating, gathering, and exploiting resources. Such fitness allows the alleles of those who are social to stay in the gene pool.

There are three sociobiological mechanisms producing the sociality that promotes reproductive fitness:[43] (1) kin selection (or what Hamilton termed "inclusive fitness"), (2) reciprocity (or what Trivers labeled "reciprocal altruism"), and (3) coercion (which appears to be a holdover from van den Berghe's early emphasis on aggression and its regulation through hierarchy and territoriality). Since each of these mechanisms promotes reproductive fitness of individuals, they lie at the base of sociocultural phenomena. Let us examine each of these mechanisms separately and then see how van den Berghe generates theoretical explanations with them.

(1) In van den Berghe's view, *kin selection* is the oldest mechanism behind sociality. And, as with all modern sociobiological arguments, he sees this mechanism as operating at the genetic level, for "the gene is the ultimate unit of natural selection, (although) each gene's reproduction is dependent on what Robert Dawkins terms its 'survival machine,' the organism in which it happens to be at any given time."[44]

This "survivor machine" carrying genetic material in its genotype need not be consciously aware of how natural selection in the distant past has created "replicators" or genes in its body that seek to survive and to be immortal. But selection clearly favored those replicators that could pass themselves on to new and better survival machines. One early survivor machine was the body of individual organisms, and a later one has been the development of cooperative arrangements among bodies sharing alleles in their genotypes. In van den Berghe's words:

> Since organisms are survival machines for genes, by definition those genes that program organisms for successful reproduction will spread. To maximize their reproduction, genes program organisms to do two things: successfully compete against ... organisms that carry alternative alleles ..., and successfully cooperate with (and thereby contribute to the reproduction of) organisms that share the same alleles of the genes.[45]

So, in van den Berghe's terms, individuals are *nepotistic* and favor kin over nonkin, and close kin over distant kin, for the simple reason that close kin will share genetic material (in their genotypes):

---

[43]van den Berghe, "Bridging the Paradigms" and *The Ethnic Phenomenon.*
[44]van den Berghe, "Bridging the Paradigms," p. 46.
[45]van den Berghe, *The Ethnic Phenomenon*, p. 7.

. . . each individual reproduces its genes directly through its own reproduction and indirectly through the reproduction of its relatives, to the extent that it shares genes with them. In simple terms, each organism may be said to have a 100 percent genetic interest in itself, a 50 percent interest in its parents, offspring, and full-siblings, a 25 percent interest in half-siblings, grandparents, grandchildren, uncles, aunts, nephews, and nieces, a 12½ percent interest in first cousins, half-nephews, great-grandchildren, and so on.[46]

The degree of "altruism" for kin or "nepotism" will vary in terms of the amount of genetic material shared with a relative *and* the ability of that relative to reproduce this material (this latter statement, van den Berghe argues, helps explain why parents, who are past their period of reproduction, will reveal more altruism toward their children than vice versa; children can still pass on parental genetic material, whereas parents cannot pass on the children's genetic material once they cease bearing children). For van den Berghe, then, "blood is thicker than water" for a simple reason: reproductive fitness, or the capacity to help those "machines" carrying one's genetic material to survive and reproduce.

Van den Berghe is careful, however, to emphasize that these biologically based propensities for nepotism, or kin selection, are elaborated upon and modified by cultural processes; yet there is clearly a biological factor operating in the overwhelming tendency of humans to be nepotistic. One cannot, van den Berghe insists, visualize such nepotism as a purely cultural process.

(2) When individuals exchange assistance, they create a bond of *reciprocity*. Such reciprocity serves the function of increasing the "fitness" of genes carried by the organisms (and organizations of organisms) involved; that is, if organisms help each other or can be relied upon to reciprocate for past assistance, the survival of genetic material is increased for both organisms. Such exchange, or what sociobiologists sometimes term "reciprocal altruism,"[47] greatly extends cooperation beyond nepotism, or "inclusive fitness" and "kin selection" in sociobiological jargon. Thus, reciprocal exchange is not a purely social product; it is a behavioral tendency programed by genes. Such programing occurred in the distant past as natural selection favored those genes that could create new kinds of survivor machines beyond an individual's physical body and social groupings of close kin. Those genes that could lodge themselves in nonkin groupings organized around bonds of exchange and reciprocity were more likely to survive; today the descendants of these genes provide a biological push for creating and sustaining such bonds of reciprocity.

At this point, van den Berghe reveals the affinity of such arguments with utilitarianism (see pp. 53–54 of Chapter 3) by introducing the problem of "free-riding" or, as he terms it, "free-loading." That is, what guarantee does an individual (and its genotype) have that others will indeed reciprocate favors and assistance? Here van den Berghe appears to argue that this problem of

---

[46]van den Berghe, "Bridging the Paradigms," pp. 46–47.
[47]See earlier discussion of Trivers and footnote 27.

free-riding created selection pressures for greater intelligence so that our pro-tohuman ancestors could remember and monitor whether or not others recip-rocated past favors. But, ironically, intelligence would also generate greater capacities for sophisticated deceit, cheating, and free-riding, which in turn would escalate selection pressures for the extended intelligence to "catch" and "detect" such concealed acts of nonreciprocation.

There is, then, a kind of selection cycle operating: reciprocity increases fitness, but it also leads to cheating and free-riding; hence, once reciprocity is a mechanism of cooperation, it can generate its own selection pressures for greater intelligence to monitor free-riding; but, ironically, increased intelligence enables individuals to engage in more subtle and sophisticated deceptions to hide their free-riding; in order to combat this tendency, there is increased pressure for greater intelligence; and so on. At some point culture and structure supplant this cycle by creating organizational mechanisms and cultural ideas to limit free-riding (see Chapter 17 on "rational choice" theory for a discussion of what these sociocultural mechanisms might be).

Van den Berghe does not pursue this discussion, however, except to point out that this cycle eventually produces "self-deceit." For van den Berghe, the best way to deceive is to believe one's own lies and deceptions, and in this way one can sincerely make and believe verbal pronouncements that contradict, at least in some ways, one's actual behavior. Religion and ideology, van den Berghe posits, are "the ultimate forms of self-deceit," because religion "denies mor-tality" and ideology facilitates "the transmission of credible, self-serving lies."[48] This conceptual leap from free-riding to religion and ideology is, to say the very least, rather vague, but it does give us a sense of what sociobiology tries to do: connect what might be considered purely "cultural" processes—religion and ideology—to a fundamental biological process—in this case, reproductive fitness through reciprocity.

(3) There are limits to social organization in terms of reciprocal exchanges, van den Berghe argues, because each party has to perceive that it receives benefits in the relationship.[49] Such perceptions of benefit can, of course, be manipulated by ideologies and other forms of deceit that hide the asymmetry of the relationship, but there are probably limits to the manipulation of per-ceptions. Power is an alternative mechanism to both kin selection and recip-rocal exchanges, because its mobilization allows some organisms to dominate others in terms of their access to resources that promote fitness. *Coercion* thus enables some to increase their fitness at a cost to others. Although this mech-anism is hardly unique to our species, humans "hold pride of place in their ability to use to good effect conscious, collective, organized, premeditated coer-cion in order to establish, maintain and perpetuate systems of intraspecific parasitism."[50] Coercion allows for the elaboration of the size and scale of human

---

[48]van den Berghe, *The Ethnic Phenomenon*, p. 9. See also "Bridging the Paradigms," p. 48.
[49]van den Berghe, "Bridging the Paradigms."
[50]van den Berghe, *The Ethnic Phenomenon*, p. 10.

organization (states, classes, armies, courts, etc.), but it is nonetheless tied to human biology as this was molded by selection. That is, genes that could use coercion to create larger and more elaborate social structures as their survival machines were more likely to survive and reproduce.

In sum, then, sociality or cooperation and organization are the result of natural selection as it preserved genes that could produce better survival machines through nepotism (kin selection and inclusive fitness), reciprocity (or reciprocal altruism), and coercion (or territorial and hierarchical patterns of dominance). These mechanisms are the result of the self-interest of genes in maximizing their fitness. Van den Berghe is one of the few who recognizes that "sociobiology is a utilitarian model of behavior that sees organisms as blindly selected to maximize their reproductive success,"[51] although he does not appear to recognize the implicit functionalism in this model (structures and processes exist to meet needs for fitness).[52] But the critical insight of this perspective, van den Berghe insists, is the recognition that there is a biological basis of human behavior and organization.

The linkages between biology, on the one side, and culture and society, on the other, are complex and often indirect, even after we recognize that these linkages occur along the three basic dimensions or axes of nepotism, reciprocity, and coercion. For there have certainly been complex interactions between ecological factors and genetic ones to produce patterns of human (and many other animal) organization; once initiated, selection processes producing greater intelligence allow for culture as "an impressive bag of tricks" to operate as a force of human evolution.

Conceptualizing cultural processes   From van den Berghe's vantage point, culture is created and transmitted in humans through mechanisms fundamentally different from those involved in genetic natural selection. In fact, van den Berghe portrays cultural evolution in Lamarckian rather than Darwinian terms: "acquired cultural characteristics, unlike . . . genetic evolution, can be transmitted, modified, transformed or eliminated through social learning."[53] Yet culture is not an emergent reality, *sui generis*, because, like all sociobiologists, van den Berghe views society as the additive sum of self-interested individuals (driven by "selfish genes") trying to maximize their fitness, with one consequence of such individual efforts being the creation of culture. And so culture is not a *separate* entity; it is yet one more product and process of biological evolution driven by natural selection as it produced genes trying to maximize their fitness by nesting themselves in better and better survival machines. And, despite the majestic forms that culture can take, it is still a maximizing strategy for genes; whereas culture can transform itself, it cannot do so in violation of the needs of genes to promote their fitness.

---

[51]Ibid., p. xii.

[52]Indeed, he sees sociobiology as an alternative to flawed functional analyses but does not, apparently, recognize how functional his perspective is.

[53]van den Berghe, *The Ethnic Phenomenon*, p. 6.

Such is van den Berghe's argument in general theoretical terms. As is clear, he has adopted the sociobiology perspective as it emerged from Fisher's initial blending of natural selection and genetics and as it was extended by the work of Hamilton, Williams, Trivers, Wilson, and others.[54] Van den Berghe has also moved beyond a general portrayal of sociobiology and sought to use this perspective to explain particular phenomena; so, to appreciate fully van den Berghe's work, as well as the challenge posed by sociobiology in general, we should examine several of these more concrete explanations.

Explanations of social phenomena with sociobiology   At various times van den Berghe has used the concepts of sociobiology to explain such empirical phenomena as kinship systems, incest taboos, ethnicity, skin color and sexual selection, and classes. Probably his two most detailed empirical analyses are on (1) kinship systems and (2) ethnicity. Each of these is briefly summarized below.

(1) In one of his early sociobiological articles, van den Berghe and David Barash sought to use the concept of "inclusive fitness" or "kin selection" to explain human family structure.[55] Much of this analysis involves explaining various features of kinship systems in terms of behavioral strategies of males and females to maximize their fitness (i.e., keep their alleles in the gene pool). Because the details of their suggestive arguments can be complex, just a few examples of the argument will be summarized.

For example, the widespread preference in human societies for polygyny (males with the option for multiple wives), hypergamy (females marrying males of higher-ranking kin groups), and double standards of sexual morality (males being given more latitude than females) can be explained in terms of reproductive fitness strategies. Women have comparatively few eggs to offer in their lifetime, have periods when they are infertile (during lactation, for example), and, even in the most liberal/egalitarian societies, have to spend more time than males raising their children. Hence a female will seek a reproductive strategy that will assure the survival of her less abundant genetical material, and the most maximizing strategy is to marry males with the resources and capacity to assure the survival of her children (and, thereby, one-half of her genotype). Thus women will seek to "marry up" (hypergamy) in the sense of securing a male who has more resources than she or male kin.

On the male side, men produce an uninterrupted and large supply of sperm and have fewer child-care responsibilities; so, they can be more promiscuous with little cost (and, in fact, they will derive some benefit in terms of the fitness that comes with spreading their genotype). Although men have an interest in assuring that as many women as possible bear children with one-half their genes, a male cannot know if a female is bearing his child when women are promiscuous; so, men also have an interest in restricting female sexual activity

---

[54]See earlier discussion.
[55]van den Berghe and Barash, "Inclusive Fitness and Family Structure."

through polygyny in order to assure that their genes are indeed contained in the genotype of the children born to a female (creating limits of female sexuality outside marriage). Thus, what are commonly viewed as purely cultural phenomena—the preference for polygyny in human societies (or monogamy with promiscuity), hypergamy ("marrying up"), and sexual double standards (favoring male promiscuity)—can be explained as the result of varying fitness strategies for males and females.

To take another example, van den Berghe and Barash attempt to explain the patrilineal descent (lineages ordered through the male line) over matrilineality (lineages through the female line). In their view, matrilineal societies create a conflict between the father of the children and the mother's brother (uncle), because in such systems of descent a man is supposed to favor his sister's children over his own. This kind of system goes against maximizing fitness because a male's nieces and nephews will have only one-fourth of his genetic material, whereas his own children will on average have one-half of his genotype. In contrast, patrilineal and bilateral (where neither side of the family is favored) systems do not produce these obligations decreasing fitness, and it is for this reason that matrilineality is rare (emerging under special circumstances) and can shift to a bilateral or patrilineal system (whereas the reverse is rare). Thus, simple "calculations" of fitness maximization can explain patterns of unilineal descent in human kinship systems.

(2) Turning to another empirical example, van den Berghe has sought to apply sociobiology to what has always been his main area of research, ethnicity.[56] Again, kin selection is his starting point, but he extends this idea beyond family relatives helping one another (in order to maximize their fitness) to a larger subpopulation. Historically, larger kin groups (composed of lineages) constituted a breeding population of close and distant kin who would sustain trust and solidarity with one another while mistrusting other breeding populations. Van den Berghe coins the term *ethny* for "ethnic group" and views an ethny as an extension of these more primordial breeding populations. An ethny is a cluster of kinship circles and is created by endogamy (intermarriage of its members) and territoriality (physical proximity of its members and relative isolation from nonmembers). An ethny represents a reproductive strategy for maximizing fitness beyond the narrower confines of kinship, because, by forming an ethny—even a very large one of millions of people—individuals create bonds with those who can help preserve their fitness, whether by actually sharing genes or, more typically, by reciprocal acts of altruism to fellow "ethnys." An ethny is, therefore, a manifestation of more basic "urges" for helping "those like oneself." Whereas ethnys become genetically diluted as their size increases and become subject to social and cultural definitions, the very tendency to form and sustain ethnys is the result of natural selection as it produced nepotism. Van den Berghe's argument is, of course, much more complicated and sophisticated, but we can get at least a general sense for how a

---

[56]van den Berghe, *The Ethnic Phenomenon.*

supposedly emergent phenomenon—ethnic groups—is explained by reduction to a theoretical perspective built upon the principles of genetic evolution.

SOCIOBIOLOGY: A CONCLUDING COMMENT

The great virtue of sociobiology is its recognition that humans are the product of natural selection and possess bioprogramers in their genes that limit and circumscribe behavior and social organization. Van den Berghe and others are clearly correct in their assertion that sociologists tend to forget the simple fact that humans are animals and that their behavior is, to some extent, explicable in terms of their ancestral past. And to the degree that advocates of sociobiology can reconnect sociology to biology—as both Spencer and Comte sought to do over 100 years ago—this radical challenge to sociological theory will have beneficial consequences.

Having said this, the question remains as to whether or not the specific theoretical thrust of sociobiology is useful. As we observed earlier, sociobiology is functional: all phenomena are analyzed with respect to meeting one master requisite, fitness. Such a mode of analysis easily slips into illegitimate teleologies: the need for fitness "causes" a given phenomenon, with such "causes" being assumed to have operated in the distant past through natural selection. This kind of reasoning simply uses the concept of natural selection, as it operated in the distant past, as a gloss that does not indicate exactly how and in what precise way "fitness" caused a particular phenomenon. Sociobiological explanations are also prone to being tautologies: phenomena exist to promote fitness, and we know that such is the case because phenomena exist.

But probably the most damaging criticism is that sociobiological explanations are easily "*ad hoced.*" It is possible to construct a story or scenario explaining any phenomenon in terms of maximization of fitness. However, exactly opposite phenomena can also be explained by fitness. For example, if matrilineal kinship systems exist, some story about how they promote fitness under certain conditions can be concocted; if patrilineal systems dominate, another story can be developed, ad hoc, to explain this fact. Thus, contrary to pronouncements that sociobiology constitutes a rigorous scientific explanation of social phenomena, the reality is that sociobiological explanations are rather vague and ad hoc when applied to human organization. These explanations are, however, creative, and we should not dismiss this perspective out of hand, especially for being functional, vague, and ad hoc (since, by these criteria, we would have to reject a good deal of sociological theory). Rather, we should remain skeptical; perhaps even more importantly, we should seek to develop alternative ways of linking the biological and sociological that need not rely on the reductionistic assumptions of sociobiology.

PART 2

Conflict Theorizing

CHAPTER 9

◆

The Origins of Conflict and Critical Theorizing

THE CONFLICT ALTERNATIVE

During the 1950s and 1960s functional theory in sociology—especially the Parsonian variety—was seen as underemphasizing the conflictual nature of social reality. Soon, attacks along these lines became ceremonial rituals for sociologists who sought theoretical redemption for past sins and who now held that conflict theory was to carry sociology out of its theoretical morass.

As early as 1956, for example, David Lockwood argued that, in continually assuming for analytical purposes a system in equilibrium, Parsons had created a fictionalized conception of the social world.[1] From this world of fantasy, as Lockwood phrased the matter, it was inevitable that analysis would emphasize mechanisms that maintained social order rather than those that systematically generated disorder and change. Furthermore, by assuming order and equilibrium, the ubiquitous phenomena of instability, disorder, and conflict too easily became viewed as deviant, abnormal, and pathological. For in reality, Lockwood insisted, there are mechanisms in societies that make conflict inevitable and inexorable. For example, power differentials assure that some groups will exploit others, thereby constituting a built-in source of tension and conflict in social systems. Additionally, the existence of scarce resources in societies will inevitably generate fights over the distribution of these resources. Finally, the fact that different interest groups pursue different goals and hence vie with one another assures that conflict will erupt. These forces, Lockwood contended, represent mechanisms of social disorder that should be as analytically significant to the understanding of social systems as Parsons' mechanisms of socialization and social control.

---

[1]David Lockwood, "Some Remarks on 'The Social System,' " *British Journal of Sociology* 7 (June 1956), pp. 134–46.

Ralf Dahrendorf crystallized this line of argument toward the end of the 1950s by comparing functional theory to a utopia.[2] Utopias usually have few historical antecedents, much like Parsons' hypothesized equilibrium; utopias display universal consensus on prevailing values and institutional arrangements, in a vein remarkably similar to Parsons' concept of institutionalization; and utopias always display processes that operate to maintain existing arrangements, much like the mechanisms of Parsons' social system. Hence utopias and the social world, when viewed from a functional perspective, do not change very much, since they do not concern themselves with history, dissension over values, or conflict in institutional arrangements.

At the same time, another body of conflict criticism was emerging. This criticism was mounted against all positivistic social science, especially the functional variety but also any system of theory that proclaimed itself as objective and neutral.[3] All theory that seeks to understand the social world without also exposing the patterns of oppression and domination is, in effect, an ideology supporting the status quo. For, in studying the world "as it is," there is an implicit assumption that this is how the social order *must be*. Theoretical knowledge cannot, therefore, merely describe events. It must expose exploitive social arrangements and, at the same time, suggest alternative ways to organize humans in less oppressive ways. Social theory cannot be neutral; rather, it must be emancipatory.

To some extent the critiques of Parsonian functionalism and positivism owe their inspiration to what I see as contradictory strains in the work of Karl Marx. Those who were to criticize Parsons for his failure to examine conflict tended to accept the positivistic strains in Marx and to propose a conflict sociology that tried to develop abstract propositions explaining the conflictual nature of social reality.[4] In contrast, those who criticized positivism in general

---

[2] Ralf Dahrendorf, "Out of Utopia: Toward a Reorientation of Sociological Analysis," *American Journal of Sociology* 744 (September 1958), pp. 115–27.

[3] Herbert Marcuse, *Reason and Revolution* (Boston: Beacon Press, 1960) and *Negations* (Boston: Beacon Press, 1969; originally published in 1965); Theodor Adorno, *Negative Dialectics* (New York: Seaburg Press, 1973); Max Horkheimer, *Critical Theory* (New York: Herder and Herder, 1972); Trent Schroyer, "Toward a Critical Theory for Advanced Industrial Society," in *Recent Sociology*, ed. H. P. Dreitzer (New York: Macmillan, 1970); and *The Critique of Domination: The Origins and Development of Critical Theory* (New York: Braziller, 1973); Jurgen Habermas, *Knowledge and Human Interest* (Boston: Beacon Press, 1971).

[4] For a diverse sampling of these efforts, see William Gamson, *The Strategy of Social Protest* (Belmont, CA: Wadsworth, 1990); Louis Kriesberg, *The Sociology of Social Conflict* (Englewood Cliffs, NJ: Prentice-Hall, 1973); Anthony Oberschall, *Social Conflict and Social Movements* (Englewood Cliffs, NJ: Prentice-Hall, 1973); Robin M. Williams, Jr., *Mutual Accommodation: Ethnic Conflict and Cooperation* (Minneapolis: University of Minnesota Press, 1977) and *The Reduction of Intergroup Tensions* (New York: Social Science Research Council, 1947); Randall Collins, *Conflict Sociology* (New York: Academic Press, 1975); Ted Gurr, "Sources of Rebellion in Western Societies: Some Quantitative Evidence," *The Annals* 391 (September 1970), pp. 128–44; A. L. Jacobson, "Intrasocietal Conflict: A Preliminary Test of a Structural Level Theory," *Comparative Political Studies* 6 (1973), pp. 62–83; David Snyder, "Institutional Setting and Industrial Conflict," *American Sociological Review* 40 (June 1975), pp. 259–78; David Britt and Omer R. Galle, "Industrial Conflict and Unionization," *American Sociological Review* 37 (February 1972), pp. 46–57; E. McNeil, ed., *The Nature of Human Conflict* (Englewood Cliffs, NJ: Prentice-Hall, 1965);

drew inspiration from the more moralistic and anti-science portions in Marx and, surprisingly, have not been so hostile to Parsons or functionalism in general.

But I hasten to add that Marx was not the only source of inspiration for this emerging range of alternatives. The other prominent German sociologists of the late 19th and early 20th centuries—Max Weber and Georg Simmel—were also decisive in the development of the various conflict sociologies. Simmel was to inspire a more positivistic and functional view of conflict processes than Marx, and his criticisms of Marx's analysis of capitalism have influenced critical theorists who dislike positivism and who emphasize that social theory must be emancipatory. Similarly, Max Weber was to stimulate both the positivistic and critical versions of conflict theorizing. For positivistically inclined theorists, his analysis of stratification has provided a crucial corrective for errors in Marx's analysis of class conflict. And for critical theorists, his historical and evolutionary account of the rise of capitalism and his pessimistic assessment of the process of rationalization in the modern world have forced a reassessment of Marx's naive faith in the coming emancipation of humans.

Thus, out of this ferment among the three great German sociologists— Marx, Weber, and Simmel—emerged a variety of approaches that share little else in common than their use of the works of Marx, Simmel, and Weber to analyze elements of inequality, power, domination, and conflict in human societies. I have grouped this diverse range of approaches under the rubric of *conflict theory,* but I do so only with the important qualification that this label embraces very different types of theoretical activity—ranging from dialectical and functional theories of conflict processes to more encompassing philosophical schemes about the state of human oppression and domination in the modern world. My purpose in this chapter, therefore, is to examine the origins of this eclectic mix of approaches by summarizing the diverse strains in the work of Marx, Simmel, and Weber that have been used to build these very different types of conflict theorizing.

Jessie Bernard, *American Community Behavior* (New York: Dryden Press, 1949), "Where Is the Modern Sociology of Conflict?" *American Journal of Sociology* 56 (1950), pp. 111–16, and "Parties and Issues in Conflict," *Journal of Conflict Resolution* 1 (June 1957), pp. 111–21; Kenneth Boulding, *Conflict and Defense: A General Theory* (New York: Harper & Row, 1962); Thomas Carver, "The Basis of Social Conflict," *American Journal of Sociology* 13 (1908), pp. 628–37; James Coleman, *Community Conflict* (Glencoe, IL: Free Press, 1957); James C. Davies, "Toward a Theory of Revolution," *American Journal of Sociology* 27 (1962), pp. 5–19; Charles P. Loomis, "In Praise of Conflict and Its Resolution," *American Sociological Review* 32 (December 1967), pp. 875–90; Raymond Mack and Richard C. Snyder, "The Analysis of Social Conflict," *Journal of Conflict Resolution* 1 (June 1957), pp. 388–97; John S. Patterson, *Conflict in Nature and Life* (New York: Appleton-Century-Crofts, 1883); Anatol Rapoport, *Fights, Games and Debates* (Ann Arbor: University of Michigan Press, 1960); Thomas C. Schelling, *The Strategy of Conflict* (Cambridge, MA: Harvard University Press, 1960); Pitirim Sorokin, "Solidary, Antagonistic, and Mixed Systems of Interaction," in *Society, Culture, and Personality* (New York: Harper & Row, 1947); Nicholas S. Timasheff, *War and Revolution* (New York: Sheed and Ward, 1965); and Clinton F. Fink, "Some Conceptual Difficulties in the Theory of Social Conflict," *Journal of Conflict Resolution* 12 (December 1968), pp. 429–31.

KARL MARX AND THE ORIGINS OF CONFLICT AND CRITICAL THEORY

As with any scholar whose work evolves over time, there are contradictions and inconsistencies in Marx's (1818–1883) ideas. I do not see this as a cause for great dismay, as many commentators have. Indeed, any creative scholar's thought will change and evolve over time. Thus, Marx's and Friedrich Engels' first work, *The German Ideology*,[5] is vastly different from Marx's later works, such as *Capital.*[6] In between these early and late works is Marx's and Engels' *The Communist Manifesto.*[7] There are, of course, other important works to consider, but I feel that these have been the most influential in the development of critical theory, which stresses the emancipatory themes in Marx's thought, and dialectical conflict theory, which has followed the more positivistic hints in Marx's work.

Critical Strains in Marx's Thought

In 1846 Marx and Engels completed *The German Ideology,* which was initially turned down by the publisher.[8] Much of this work is an attack on the "Young Hegelians," who were advocates of the German philosopher Georg Hegel, and is of little interest today. Yet in this attack are certain basic ideas that, I feel, have served as the impetus behind "critical theory," or the view that social theory must be critical of oppressive arrangements and propose emancipatory alternatives. This theme exists, of course, in all of Marx's work, but it is in this first statement by Marx that the key elements of contemporary critical theory are most evident.

Marx criticized the Young Hegelians severely because he had once been one of them and was now making an irrevocable break with them. He saw the Hegelians as hopeless idealists, in the philosophical sense. That is, they saw the world as reflective of ideas, with the dynamics of social life revolving around consciousness and other cognitive processes by which "ideal essences" work their magic on humans. Marx saw this emphasis on the "reality of ideas" as nothing more than a conservative ideology that supports people's oppression by the material forces of their existence. His alternative was "to stand Hegel on his head," but in this early work there is still an emphasis on the relation between consciousness and self-reflection, on the one hand, and social reality, on the other. This dualism becomes central to contemporary critical theory.

Marx saw humans as being unique by virtue of their conscious awareness of themselves and their situation. They are capable of self-reflection and, hence,

---

[5]Karl Marx and Friedrich Engels, *The German Ideology* (New York: International Publishers, 1947; originally written in 1846).

[6]Karl Marx, *Capital: A Critical Analysis of Capitalist Production*, vol. 1 (New York: International Publishers, 1967; originally published in 1867).

[7]Karl Marx and Friedrich Engels, *The Communist Manifesto* (New York: International Publishers, 1971; originally published in 1848).

[8]Marx and Engels, *The German Ideology.*

assessment of their positions in society. Such consciousness arises out of people's daily existence and is not a realm of ideas that is somehow independent of the material world, as much German philosophy argued. For Marx, people produce their ideas and conceptions of the world in light of the social structures in which they are born, are raised, and live.

The essence of people's lives is the process of production, since, for Marx, human "life involves before anything else eating and drinking, a habitation, clothing, and many other material things."[9] To meet these contingencies of life, production is necessary; but, as production satisfies one set of needs, new needs arise and encourage alterations in the ways that productive activity is organized. The elaboration of productive activity creates a division of labor, which, in the end, is alienating because it increasingly deprives humans of their capacity to determine their productive activities. Moreover, as people work, they are exploited in ways that generate private property and capital for those who enslave them. Thus, as people work as alienated cogs in the division of labor, they produce that which enslaves them: private property and profits for those who control the modes and means of production. Marx provided a more detailed discussion of the evolution of productive forces to this capitalist stage, but, for my purposes, his analysis of consciousness is more important.[10]

Marx argued that the capacity to use language, to think, and to analyze allows humans to alter their environment. People do not merely have to react to their material conditions in some mechanical way; they can also use their capacities for thought and reflection to construct new material conditions and corresponding social relations. Indeed, the course of history involved such processes as people actively restructured the material conditions of their existence. The goal of social theory, Marx implicitly argues, is to use humans' unique facility to expose those oppressive social relations and to propose alternatives. Marx's entire career was devoted to this goal, and it is this emancipatory aspect of Marx's thought that forms the foundation for critical theory, which I will examine in Chapter 13.

Positivistic Strains in Marx's Thought

In developing this emancipatory project, Marx produced a formal theory of conflict and change—one that he might disavow as a positivistic theory but that has been used nonetheless in developing contemporary conflict theory. In elaborating his model of revolutionary class conflict and social change, Marx delineated an image of social organization that still influences a major portion of contemporary sociological theory. Marx began with a simple—and I think simplistic—assumption: economic organization, especially the ownership of property, determines the organization of the rest of a society. The class struc-

---

[9]Ibid., p. 15.

[10]Indeed, Marx was as much an evolutionist as any functionalist, and, in fact, there is much functional reasoning in his arguments. For illustrations, see Arthur L. Stinchcombe, *Constructing Social Theories* (New York: Harcourt, Brace & World, 1968).

ture and institutional arrangements, as well as cultural values, beliefs, religious dogmas, and other idea systems, are ultimately a reflection of the economic base of a society. He then added another assumption: inherent in the economic organization of any society—save the ultimate communistic society—are forces inevitably generating revolutionary class conflict. Such revolutionary class conflict is seen as dialectical and conceptualized as occurring in epochs, with successive bases of economic organization sowing the seeds of their own destruction through the polarization of classes and subsequent overthrow of the dominant by the subjugated class. Hence, there is a third assumption: conflict is bipolar, with exploited classes, under conditions created by the economy, becoming aware of their true interests and eventually forming a revolutionary political organization that stands against the dominant, property-holding class.

I see the criticisms that can be leveled against these assumptions as self-evident: (1) societies are more than mere reflections of economic organization and patterns of property ownership; (2) social conflict is rarely bipolarized across an entire society; (3) interests in a society do not always cohere around social class; (4) power relations in a society are not always direct reflections of ownership of property; and (5) conflict does not always cause social change, dialectical or otherwise. In addition to a whole series of incorrect predictions—such as the formation of the modern proletariat into a revolutionary class during the present "capitalistic" epoch, the subsequent overthrow of capitalist economic systems, and the formation of a communist society—the wisdom of following Marx's lead can, I believe, be seriously questioned.

With abstraction above the specifics of Marx's economic determinism and excessive polemics, however, I and other theorists see a set of assumptions that directly challenge those imputed to functionalism and that can serve as an intellectual springboard for a conflict alternative in sociological theorizing:

1. Although social relationships display systemic features, these relationships are rife with conflicting interests.
2. This fact reveals that social systems systematically generate conflict.
3. Conflict is therefore an inevitable and pervasive feature of social systems.
4. Such conflict tends to be manifested in the opposition of interests.
5. Conflict most frequently occurs over the distribution of scarce resources, most notably power and material wealth.
6. Conflict is the major source of change in social systems.

In addition to these assumptions, I think that the form and substance of Marx's causal analysis have been equally influential in the development of modern conflict theory. This analysis takes the general form of assuming that conflict is an inevitable and inexorable force in social systems and is activated under certain specified conditions. Some of these conditions are viewed as allowing for the transformation of latent class interests (lying in a state of "false consciousness") into manifest class interests ("class consciousness"), which, under additional conditions, lead to the polarization of society into classes joined in conflict. Thus, for Marx, a series of conditions are cast into the role of intervening variables that accelerate or retard the inevitable transformation of class interests into revolutionary class conflict.

In addition to the form of the argument (which I will discuss more thoroughly in the next chapter), the substance of the Marxian model is of great importance in understanding modern sociological theory. Contrary to most contemporary Marxist theorists, I believe that this substantive contribution can best be seen if the propositions of his theoretical scheme are stated in a highly abstract form and divorced from his polemics and rhetoric about social class and revolution. I admit that much of the flavor of Marx's analysis is lost in such an exercise, but the indebtedness of modern sociological theory to Marxian propositions becomes more evident.[11] I have summarized these abstract propositions in Table 9-1.

In Table 9-1, Marx's assumptions about the nature of the social world and the key causal connections in this world are stated propositionally. For it is in a propositional form that Marx's contribution to conflict theory has been most frequently used by contemporary theorists (see following Chapters 10, 11, and 13). In Proposition I of Table 9-1, the degree of inequality in the distribution of resources is held by Marx to influence the extent to which segments of a social system will reveal conflicts of interest. Proposition II then documents some of the conditions that would make members of deprived or subordinate segments of a population aware of their conflict of interest with those holding the largest share of scarce resources. For, once segments of a population become aware of their true interests, they will begin to question the legitimacy of a system in which they come out on the short end of the distribution of scarce resources. Propositions II-A, B, C, and D deal, respectively, with the disruption in the social situation of deprived populations, the amount of alienation people feel as a result of their situation, the capacity of members of deprived segments to communicate with one another, and their ability to develop a unifying ideology that codifies their true interests. Marx sees these conditions as factors that increase and heighten awareness of subordinates' collective interests and, hence, decrease their willingness to accept as legitimate the right of superordinates to command a disproportionate share of resources.

In turn, some of these forces heightening awareness are influenced by such structural conditions as ecological concentration (II-C-1), educational opportunities (II-C-2), the availability of ideological spokespeople (II-D-1), and the control of socialization processes and communication networks by superordinates (II-D-2). In Proposition III, Marx hypothesizes that the increasing awareness by deprived classes of their true interests and the resulting questioning of the legitimacy of the distribution of resources serve to increase the likelihood that the disadvantaged strata will begin to organize collectively their opposition against the dominant segments of a system. This organization is seen as especially likely under several conditions: the more disorganized the dominant

---

[11]For criticism of my efforts along these lines, see Richard P. Appelbaum, "Marx's Theory of the Falling Rate of Profit: Towards a Dialectical Analysis of Structural Social Change," *American Sociological Review* 43 (February 1978), pp. 64–73.

TABLE 9–1 Marx's Key Propositions

I. The more unequal the distribution of scarce resources in a system, the greater the conflict of interest between dominant and subordinate segments in a system.

II. The more subordinate segments become aware of their true collective interests, the more likely they are to question the legitimacy of the existing pattern of distribution of scarce resources.

  A. The more social changes wrought by dominant segments disrupt existing relations among subordinates, the more likely the latter are to become aware of their true interests.

  B. The more practices of dominant segments create alienative dispositions among subordinates, the more likely the latter are to become aware of their true collective interests.

  C. The more members of subordinate segments can communicate their grievances to one another, the more likely they are to become aware of their true collective interests.

    1. The more ecologically concentrated members of subordinate groups, the more likely communication of grievances.

    2. The greater the educational opportunities of subordinate group members, the more diverse the means of their communication and the more likely the communication of grievances.

  D. The more subordinate segments can develop unifying ideologies, the more likely they are to become aware of their true collective interests.

    1. The greater the capacity to recruit or generate ideological spokespeople, the more likely ideological unification.

    2. The less the ability of dominant groups to regulate the socialization processes and communication networks in a system, the more likely ideological unification.

III. The more subordinate segments of a system are aware of their collective interests and the greater their questioning of the legitimacy of the distribution of scarce resources, the more likely they are to join in overt conflict against dominant segments of a system.

  A. The less the ability of dominant groups to make manifest their collective interests, the more likely subordinate groups are to join in conflict.

  B. The more the deprivations of subordinates move from an absolute to a relative basis, the more likely they are to join in conflict.

  C. The greater the ability of subordinate groups to develop a political leadership structure, the more likely they are to join in conflict.

IV. The greater the ideological unification of members of subordinate segments of a system and the more developed their political leadership structure, the more likely dominant and subjugated segments of a system are to become polarized.

V. The more polarized the dominant and subjugated, the more violent their conflict.

VI. The more violent the conflict, the greater the structural change of the system and the greater the redistribution of scarce resources.

---

segments with respect to understanding their true interests (III-A), the more the subordinates' deprivations escalate as they begin to compare their situation with that of the privileged (III-B), and the more the ease with which the deprived can develop political leadership to carry out the organizational tasks of mobilizing subordinates (III-C). In Proposition IV, Marx emphasizes that, once deprived groups possess a unifying ideology and political leadership, their true interests begin to take on clear focus and their opposition to superordinates begins to increase. As polarization increases, the less possibility there is for reconciliation, compromise, or mild conflict, since now the deprived are sufficiently alienated, organized, and unified to press for a complete change in the pattern of resource distribution (V)—thus making violent confrontation the only way to overcome the inevitable resistance of superordinates. Finally (VI), Marx notes that, the more violent the conflict, the greater the change in patterns of organization in a system, especially its distribution of scarce resources.

Thus, for those who have tried to use these ideas to build positivistic theory, I see the Marxian legacy as consisting of a set of conflict-oriented assumptions, a particular form of causal analysis that stresses the importance of intervening conditions for accelerating or retarding inexorable conflict processes, and a series of abstract propositions. My sympathies reside with the more positivistic approach in Marx's work, but there are many who would disagree. Indeed, I would guess that most contemporary Marxists disavow positivism and the search for universal and timeless laws of human organization. And they certainly do not condone my and others' efforts to translate Marx's ideas into the language of positivism (as I did in Table 9-1). Such is clearly the case, as I will document in Chapter 13, for most critical theorists, but it is also true for the vast majority of Marxist sociologists. But, in terms of actual theory development, I think that the more positivistic efforts to translate Marx's ideas have been more dominant than the work of Marxists and critical theorists. And even the most prominent critical theorist, Jurgen Habermas (who is the topic of Chapter 13) has increasingly sought to make such theory more objective. All this will become more evident, I think, as we proceed into the next chapters. But for the present let me return to examining the origins of contemporary conflict theory.

GEORG SIMMEL AND CONFLICT THEORIZING

Functional Strains in Simmel's Thought

Georg Simmel (1858–1918) was committed to developing a body of theoretical statements that captured the *form of basic social processes*, an approach he labeled *formal sociology*. Primarily on the basis of his own observations, he sought to abstract the essential properties from processes and events in a wide variety of empirical contexts. In doing so, Simmel hoped to develop abstract statements that depicted the most fundamental social processes of social organization. Nowhere is his genius in such activity more evident than in a short essay on conflict, which serves as a major source of insight for contemporary conflict theory in sociology.[12]

Much like Marx, Simmel viewed conflict as ubiquitous and inevitable in society. Unlike Marx, however, he did not view social structure as a domination and subjugation, but rather as an inseparable mingling of associative and dissociative processes, which are separable only in analysis:

---

[12]All subsequent references to this work are taken from Georg Simmel, *Conflict and the Web of Group Affiliation*, trans. K. H. Wolff (Glencoe, IL: Free Press, 1956).

> The structure may be *sui generis*, its motivation and form being wholly self-consistent, and only in order to be able to describe and understand it, do we put it together, *post factum*, out of two tendencies, one monistic, the other antagonistic.[13]

Part of the reason for this emphasis, I think, lies in Simmel's "organismic" view of the social world. In displaying formal properties, social processes evidence a systemic character—a notion apparently derived from the organismic doctrines dominating the sociology of his time. This subtle organicism led Simmel to seek out the consequences of conflict for social continuity rather than for change:

> Conflict is thus designed to resolve dualisms; it is a way of *achieving some kind of unity*, even if it be through the annihilation of one of the conflicting parties. This is roughly parallel to the fact that it is the most violent *symptom of a disease* which represents the effort of the *organism* to free itself of disturbances and damages caused by them. [Italics added.][14]

In apparent contradiction to the harmony implied by this organicism, Simmel postulated an innate "hostile impulse" or a "need for hating and fighting" among the units of organic wholes, although this instinct is mixed with others for love and affection and is circumscribed by the force of social relationships. Therefore Simmel viewed conflict as a reflection not only of conflicts of interest but also of hostile instincts. Such instincts can be exacerbated by conflicts of interest or mitigated by harmonious relations as well as by instincts for love. But, in the end, Simmel still viewed one of the ultimate sources of conflict as lying in the innate biological makeup of human actors.

In what I see as an effort to reconcile his assumptions about the nature of the social organism with notions of hating and fighting instincts, Simmel devoted considerable effort to analyzing the positive consequences of conflict for the maintenance of social wholes and their subunits. In this way, hostile impulses were seen not so much as a contradiction or cancer to the organic whole but as one of many processes maintaining the "body social." Thus, although Simmel recognized that an overly cooperative, consensual, and integrated society would show "no life process," his analysis of conflict is still loaded in the direction of how conflict promotes solidarity and unification.

It is this aspect of Simmel's work on conflict that reveals an image of social organization decidedly different from that emphasized by Marx:

1. Social relationships occur within systemic contexts that can be typified only as an organic intermingling of associative and dissociative processes.
2. Such processes are a reflection of both the instinctual impulses of actors and the imperatives dictated by various types of social relationships.

---

[13]Ibid., p. 23. However, with his typical caution, Simmel warns: "This fact should not lead us to overlook the numerous cases in which contradictory tendencies really co-exist in separation and can thus be recognized at any moment in the overall situation" (pp. 23–24).

[14]Ibid., p. 13.

3. Conflict processes are therefore a ubiquitous feature of social systems, although they do not necessarily, in all cases, lead to breakdown of the system and/or to social change.
4. In fact, conflict is one of the principal processes operating to preserve the social whole and/or some of its subparts.

These assumptions are reflected in a large number of specific propositions, which Simmel apparently developed from direct observations of events occurring around him and from readings of historical accounts of conflict.[15] In these propositions, Simmel views conflict as a *variable* that manifests different states of intensity or violence. The polar ends of this variable continuum are "competition" and the "fight." Competition involves the more regulated and parallel strivings of parties toward a mutually exclusive end, and fight denotes the less regulated and more direct combative activities of parties against one another.[16] Although he does not elaborate extensively on the variable properties of conflict or consistently employ his labels, Simmel's distinctions have inspired a long debate among contemporary sociologists on what is and what is not conflict.[17] My sense is that this debate has often degenerated into terminological quibbling, but at its heart is the important issue of clarifying the concepts to be employed in propositions on conflict processes—a theoretical issue that Simmel clearly recognized as crucial.

Simmel's organicism probably was critical in forcing this conceptualization of conflict as a variable phenomenon. Unlike Marx, who saw conflict as ultimately becoming violent and revolutionary and leading to the structural change of the system, Simmel was quite often led to the analysis of the opposite phenomena—less intense and violent conflicts that promoted the solidarity, integration, and orderly change of the system.[18] Yet, within the apparent constraints of his subtle organicism, Simmel enumerated a number of suggestive propositions on the intensity of conflict—that is, the degree of direct action and violence of parties against one another. As with Marx, I think that the full impact of Simmel's analysis on modern theory can be seen more readily when his propositions are stated formally and abstractly. I have listed these propositions in Table 9-2.

---

[15]Simmel was not concerned with developing scientific theory; rather, he was interested in inducting social forms from interaction processes. This emphasis on forms makes many of Simmel's analytical statements rather easily converted into propositions. I should emphasize, however, that transforming Simmel's analytical statements into propositions involves some risk of misinterpretation.

[16]Simmel, *Conflict*, p. 58.

[17]For an excellent summary of this debate, see C. F. Fink, "Some Conceptual Difficulties in the Theory of Social Conflict." See also my closing remarks in Chapter 8.

[18]Pierre van den Berghe has argued that a dialectical model of conflict is ultimately one in which unification, albeit temporary, emerges out of conflict. But, as I will examine extensively in the next chapter, the differences between Marx and Simmel have inspired vastly different theoretical perspectives in contemporary sociology. See Pierre van den Berghe, "Dialectic and Functionalism: Toward a Theoretical Synthesis," *American Sociological Review* 28 (October 1963), pp. 695–705.

TABLE 9–2 Simmel's Propositions on Conflict Intensity

I. The greater the degree of emotional involvement of parties to a conflict, the more likely the conflict is to be violent.
- A. The greater the respective solidarity among members of conflict parties, the greater the degree of their emotional involvement.
- B. The greater the previous harmony among members of conflict parties, the greater the degree of their emotional involvement.

II. The more that conflict is perceived by members of conflict groups to transcend individual aims and interests, the more likely the conflict is to be violent.

III. The more that conflict is a means to a clearly specified end, the less likely the conflict is to be violent.

Propositions I, I-A, and I-B overlap somewhat with those developed by Marx. In a vein similar to Marx, Simmel emphasized that violent conflict is the result of emotional arousal. Such arousal is particularly likely when conflict groups possess a great deal of internal solidarity and when these conflict groups emerge out of previously harmonious relations. Marx postulated a similar process in his contention that polarization of groups previously involved in social relations (albeit exploitive ones) leads to violent conflict. In Proposition II, Simmel indicated that, coupled with emotional arousal, the extent to which members see the conflict as transcending their individual aims increases the likelihood of violent conflict. Proposition III is Simmel's most important, I think, because it contradicts Marx's hypothesis that objective consciousness of interests will lead to organization for violent conflict. In this proposition, Simmel argued that, the more clearly articulated their interests, the more focused are the goals of conflict groups. With clearly articulated goals, it becomes possible to view violent conflict as only one of many means for their achievement, since other, less combative, conflicts, such as bargaining and compromise, can often serve to meet the specific objectives of the group. Thus, for Simmel, consciousness of common interests (Proposition II) can, under unspecified conditions, lead to highly instrumental and nonviolent conflict. In the context of labor/management relations, for example, I think that Simmel's proposition is more accurate than Marx's, for violence has more often accompanied labor/management disputes in the initial formation of unions, when interests and goals are not well articulated. As interests become clarified, violent conflict has been increasingly replaced by less violent forms of social interaction.[19]

Simmel then turned his attention to the consequences of conflict for (1) the conflict parties and (2) the systemic whole in which the conflict occurs. Simmel first analyzed how violent conflicts *increase* solidarity and internal organization of the conflict parties, but, when he shifted to an analysis of the functions of conflict for the social whole, he drew attention primarily to the

---

[19]Admittedly, Marx's late awareness of the union movement in the United States forced him to begin pondering this possibility, but he did not incorporate this insight into his theoretical scheme.

TABLE 9–3 Simmel's Propositions on the Functions of Conflict for the Respective Parties Involved

I. The more violent are intergroup hostilities and the more frequent is conflict among groups, the less likely group boundaries are to disappear.

II. The more violent the conflict and the less integrated the group, the more likely is despotic centralization of conflict groups.

III. The more violent the conflict, the greater will be the internal solidarity of conflict groups.

    A. The more violent the conflict and the smaller the conflict groups, the greater will be their internal solidarity.

        1. The more violent the conflict and the smaller the conflict groups, the less will be the tolerance of deviance and dissent in each group.

    B. The more violent the conflict and the more a group represents a minority position in a system, the greater will be the internal solidarity of that group.

    C. The more violent the conflict and the more a group is engaged purely in self-defense, the greater will be its internal solidarity.

fact that conflict promotes system integration and adaptation. I think it reasonable to ask, however: how can violent conflicts promoting increasing organization and solidarity of the conflict groups suddenly have these positive functions for the systemic whole in which the conflict occurs? For Marx such a process is seen to lead to polarization of conflict groups and then to the violent conflicts, which radically alter the systemic whole. But for Simmel the increased level of organization within conflict groups enables them to realize many of their goals without overt violence (but perhaps with a covert threat of violence), and such partial realization of clearly defined goals cuts down internal system tension and hence promotes integration. Let me document Simmel's reasoning with his propositions.

In Table 9–3, I have listed Simmel's propositions on the functions of violent conflict for the parties to the conflict. In these propositions, violent conflict is seen, under certain conditions, to increase the degree of centralization and the level of internal solidarity of groups. Unlike Marx, however, Simmel does not assume that conflict begets increasingly violent conflicts between ever more organized and polarized segments that, in the end, will cause radical change in the system. This difference between their analyses is most clear when Simmel's propositions on the consequences of conflict for the systemic whole are reviewed. The most notable feature of several key propositions, which I have listed in Table 9–4, is that Simmel was initially concerned with less violent conflicts and with their integrative functions for the social whole; and only later did he turn to more violent conflicts. And even then he emphasized their integrative consequences for the social whole.

I think that Proposition I in Table 9–4 provides an important qualification to Marx's analysis. Marx visualized mild conflicts as intensifying as the combatants become increasingly polarized; in the end the resulting violent conflict would lead to radical social change in the system. In contrast, Simmel argued that conflicts of low intensity and high frequency in systems of high degrees

TABLE 9–4   Simmel's Propositions on the Functions of Conflict for the Systemic Whole

---

I. The less violent the conflict among groups of different degrees of power in a system, the more likely the conflict is to have integrative consequences for the social whole.

   A. The less violent and more frequent the conflict, the more likely the conflict is to have integrative consequences for the social whole.

      1. The less violent and more frequent the conflict, the more members of subordinate groups can release hostilities and have a sense of control over their destiny and thereby maintain the integration of the social whole.
      2. The less violent and more frequent the conflict, the more likely that norms regularizing the conflict will be created by the conflicting parties.

   B. The less violent the conflict and the more the social whole is based on functional interdependence, the more likely the conflict is to have integrative consequences for the social whole.

      1. The less violent the conflict in systems with high degrees of functional interdependence, the more likely it is to encourage the creation of norms regularizing the conflict.

II. The more violent and the more prolonged the conflict relations between groups, the more likely the formation of coalitions among previously unrelated groups in a system.

III. The more prolonged the threat of violent conflict between groups, the more enduring the coalitions of each of the conflict parties.

---

of interdependence do not necessarily intensify or lead to radical social change. On the contrary, they release tensions and become normatively regulated, thereby promoting stability in social systems. Further, Simmel's previous propositions on violent conflict present the possibility that, with the increasing organization of the conflicting groups, the degree of violence of their conflict will decrease as their goals become better articulated. The consequence of such organization and articulation of interests will be a greater disposition to initiate milder forms of conflict, involving competition, bargaining, and compromise. What I see as critical for developing a sociology of conflict is that Simmel's analysis provides more options than do Marx's propositions on conflict outcomes. First, conflicts do not necessarily intensify to the point of violence; when they do not, they can have, under conditions that need to be further explored, integrative outcomes for the social whole. Marx's analysis precludes exploration of these processes. Second, Simmel's propositions allow for inquiry into the conditions under which initially violent conflicts can become less intense and thereby have integrative consequences for the social whole. This insight dictates a search for the conditions under which the level of conflict violence and its consequences for system parts and the social whole can shift and change over the course of the conflict process. I see this expansion of options as representing a much broader and firmer foundation for building a theory of conflict.

Finally, Propositions II and III in Table 9–4 note the functions of violent conflict for integrating systemic wholes. These propositions could represent a somewhat different way to state Marx's polarization hypothesis, since conflict was seen by Simmel as drawing together diverse elements in a system as their respective interests become more clearly recognized. But Simmel was not ideologically committed to dialectical assumptions. Thus, unlike Marx, he appeared to be arguing only that violent conflicts pose threats to many system units that, depending upon calculations of their diverse interests, will unite to form larger social wholes. Such unification will persist as long as the threat of violent conflict remains. Should violent conflict no longer be seen as necessary, with increasing articulation of interests and the initiation of bargaining relations, then Simmel's Propositions I-A and I-B on the consequences of conflict for the social whole would become operative.

Simmel's abstract propositions on conflict processes represent, I feel, an important qualification of Marx's reasoning. Although various conflict theorists have focused on either Marx or Simmel as their principal resource, there has been sufficient cross-fertilization to knock off the extremes in their respective analyses—for Marx, the overemphasis on organization and polarization and, for Simmel, the unmitigated functionalism and analysis of positive consequences.

Simmel's Implicit Attack on Marx's Emancipatory Project

There is yet another sense in which Simmel's ideas represent an important qualification to Marx's reasoning. Marx's more emancipatory side saw modern capitalism, especially as it creates a division of labor and inequality, as oppressing individuals. His ideas about conflict are, of course, a reflection of his view that capitalism produces the conditions that will lead to a revolutionary conflict ushering in a new form of human organization in which individuals are freed from the capitalists' domination. Thus capitalism expands the division of labor, makes workers appendages to machines, concentrates them in urban areas, quantifies social relations through money and markets, and forces workers to be mere role players (as opposed to fully involved participants) in social relations. In so doing, it creates the personal alienation and resentments as well as the social structural conditions that will lead subordinates to become aware of their domination and to organize in an effort to change their plight (the propositions in Table 9-1 simply represent more abstract statements of these ideas).

Simmel called into question much of Marx's analysis in his *The Philosophy of Money*,[20] in which, as I will document in Chapter 14, he also produced an important theory of social exchange. In fact, Marx's theory as represented in Table 9-1 is also an exchange theory, emphasizing what occurs when the exchange of valued resources is unequal (see Chapter 14). It is this implicit theory of exchange behind much conflict sociology that led me to view Randall Collins' approach, which, although drawing heavily from Max Weber and others, is an explicit exchange theory (see Chapter 12). For the present, I want to stress the critique of Marx contained in the exchange theory developed by Simmel.

---

[20]Georg Simmel, *The Philosophy of Money*, trans. T. Bottomore and D. Frisbie (Boston: Routledge & Kegan Paul, 1978; originally published in 1907).

This critique revolves around one of the themes in Marx's writing: capitalism quantifies social life with money and, in so doing, makes exchanges in markets paramount; the result is that human social relations are increasingly quantified, as is personified in the labor market, where workers sell themselves as a commodity; moreover, the growing division of labor makes workers mere cogs in an impersonal organizational machine. Such processes would, Marx believed, be so oppressive as to initiate revolutionary pressures for their elimination. Simmel, however, looked at these forces much differently. Although a certain level of alienation from work and "commodification" of relations through the use of money is inevitable with increasing differentiation and expansion of productive forces and markets, people are also freed from traditional constraints. They have more options as to how they spend their money and what they do; they can move about with more freedom and form new and varied social relations; they can live lifestyles that reflect their tastes and values; and, in general, they are more liberated than their counterparts in less complex, traditional societies.

This critique of Marx was, however, to be rejected by the early critical theorists who did not want to visualize modern societies as liberating. And yet they were confronted with the failure of Marx's predictions about the coming emancipation of society with the communist revolution. In an attempt to reconstruct Marx's vision of humans' capacity to make history, they were forced to accept Weber's highly pessimistic view of the constraints of modern society and to reject Simmel's more benign views. But, in so doing, they became trapped in a dilemma: if capitalism is structurally not self-transforming in terms of Marx's revolutionary model, if modern life is not so liberating as Simmel felt, and if Weber's analysis of increasing constraint in societies must therefore be accepted as true, then how is liberation to occur? What force is to drive people's emancipation from domination? Early critical theorists would not accept Simmel's judgment—that is, people are more *free* than in traditional societies—and so they conceptually retreated into a contemplative subjectivism. They viewed the liberating force as somehow springing from human nature and its capacity for conscious reflection. And, as I will document in Chapter 13, they moved away from Marx, who had "turned Hegel on his head," and put "Hegel back on his feet."

Thus Weber becomes a critical link in the reinterpretation of Marx in this century by critical theorists. But Weber also presents, I feel, an important corrective to Marx's more formal theory of revolutionary conflict. Weber's analysis of stratification and social change presents some important propositions that consistently reappear in the more positivistic conflict theoretic literature. And so, to understand the development of either critical theory or positivistically oriented conflict theory, we need to examine some of the works of Max Weber.

MAX WEBER'S THEORY OF CONFLICT

Positivistic Strains in Weber's Thought

Max Weber (1864–1920) did not believe that sociology could be a natural science, as positivists claim. Instead, he devoted his efforts to historical analyses, especially of the transition to industrial/bureaucratic social orders. Yet, despite his misgivings over timeless laws about invariant properties of the social universe, I see his work as filled with more abstract generalizations that imply more enduring and invariant social processes. Nowhere is this "implicit positivism" more evident than in Weber's analysis of stratification, conflict, and change. For, in his seminal ideas on these topics, he developed a number of important conflict principles, which are similar to those espoused by Marx but which, at the same time, subtly shift points of emphasis. I believe that much contemporary conflict theory implicitly uses these principles, although this debt to Weber frequently goes unacknowledged.

Most of these principles can be found in his discussion of the transition from societies based on traditional authority to those organized around rational/legal authority.[21] In systems where the sanctity of traditions legitimates political and social activity, there are three conditions that encourage the emergence of charismatic leaders who organize conflict groups that challenge such traditional authority. One condition is a situation in which there is a high degree of correlation among power, wealth, and prestige or, in his terms, incumbency in positions of political power (party), occupancy in advantaged economic positions (class), and membership in high-ranking social circles (status groups). That is, when economic elites, for example, are also social and political elites, and vice versa, then those who are excluded from power, wealth, and prestige become resentful and receptive to conflict alternatives.

Another condition is dramatic discontinuity in the distribution of rewards, or the existence of divisions in social hierarchies that give privilege to some and very little to others. When only a few hold power, wealth, and prestige and the rest are denied these rewards, then tensions and resentments exist. Such resentments become a further inducement for those without power, prestige, and wealth to engage in conflict with those who hoard these resources.

A final condition encouraging conflict is low rates of social mobility. When those of low rank have little chance to move up social hierarchies or to enter a new class, party, or status group, then resentment accumulates. Those denied opportunities to increase their access to resources become restive and willing to challenge the system of traditional authority.

The critical force that galvanizes the resentments inhering in these three conditions is charisma. Weber felt that whether or not charismatic leaders emerge is, to a great extent, a matter of historical chance. But if such leaders do emerge to challenge traditional authority and to mobilize resentments over the hoarding of resources by elites and the lack of opportunities to gain access to wealth, power, or prestige, then structural change would ensue.

When successful, such leaders confront organizational problems of consolidating their gains. One result is that charisma becomes routinized, as leaders create formal rules, procedures, and structures for organizing followers after their successful mobilization to pursue conflict. If routinization takes a tra-

---

[21]For a fuller discussion, see Jonathan H. Turner, Leonard Beeghley, and Charles Powers, *The Emergence of Sociological Theory* (Belmont, CA: Wadsworth, 1989). For original sources, see Max Weber, *Economy and Society* (New York: Bedminster Press, 1968).

TABLE 9–5 Weber's Propositions on Inequality and Conflict

I. The greater the degree of withdrawal of legitimacy from political authority, the more likely is conflict between superordinates and subordinates.

   A. The greater the correlation of membership in class, status group, and party (or, alternatively, access to power, wealth, and prestige), the more intense the level of resentment among those denied membership (or access) and hence, the more likely they are to withdraw legitimacy.

   B. The greater the discontinuity in social hierarchies, the more intense the level of resentment among those low in the hierarchies and, hence, the more likely they are to withdraw legitimacy.

   C. The lower the rates of mobility up social hierarchies of power, prestige, and wealth, the more intense the level of resentment among those denied opportunities and, hence, the more likely they are to withdraw legitimacy.

II. The more charismatic leaders can emerge to mobilize resentments of subordinates in a system, the greater will be the level of conflict between superordinates and subordinates.

   A. The more conditions I–A, I–B, and I–C are met, the more likely the emergence of charismatic leadership.

III. The more effective are charismatic leaders in mobilizing subordinates in successful conflict, the greater the pressures to routinize authority through the creation of a system of rules and administrative authority.

IV. The more a system of rules and administrative authority increases conditions I–A, I–B, and I–C, the greater will be the withdrawal of legitimacy from political authority and the more likely is conflict between superordinates and subordinates.

---

ditional form, thus creating a new system of traditional authority, renewed conflict can be expected as membership in class, status, and party becomes highly correlated, as the new elites hoard resources, and as mobility is blocked. However, if rational/legal routinization occurs, then authority is based upon equally applied laws and rules; performance and ability become the basis for recruitment and promotion in bureaucratic structures. Under these conditions, conflict potential will be mitigated.

I have, of course, ripped Weber's discussion out of its historical context, but this is just what contemporary conflict theorists do when using Weber's work to develop principles about social conflict. In Table 9–5, I have abstracted these ideas even further and presented them as a series of propositions. When stated in this way, I think their similarity to and differences with those by Marx become more evident. For me the unique feature of Weber's Proposition I is the recognition that inequality exists along several dimensions and that the level of correlation among incumbents along these dimensions is critical. Moreover, the degree of discontinuity in the distribution of resources and the rates of social mobility are also crucial.

Unlike Marx, who tended to overemphasize the economic basis of inequality and to argue for a simple polarization of societies into propertied and nonpropertied (exploited) classes, Weber's Proposition I allows more theoretical options. He tells us to look at variations in the distribution of power, wealth, and prestige and the extent to which holders of one resource control the other resources. He tells us to examine the degree of discontinuity in the distribution of these resources—in other words, the extent to which there are clear lines demarking privilege and nonprivilege. Finally, he advises us to examine the degree of mobility—the chance to gain access to power, wealth, and prestige—in order to understand the resentments and tensions that make people prone to conflict. Marx's scheme denotes the same processes but with a much heavier and polemical hand. I think that Weber's proposition encourages theorists to explore more variations along more dimensions than does Marx's scheme.

In Propositions II, III, and IV in Table 9-5, Weber emphasizes the importance of political leadership and organization. Political leaders emerge to galvanize resentments, and their effectiveness determines the course of conflict. But for Weber leadership is not inevitable, nor is it necessarily liberating. Indeed, it can restore a new system of inequality and privilege that will initiate a new wave of escalating resentments and potential conflict. Marx, on the other hand, makes similar points but with much more optimism about leaders' capacity to further the evolutionary progress of societies toward his utopian end state, communism.

In addition to the propositions in Table 9-5, which pertain primarily to *intra*societal conflict processes, Weber developed theoretical ideas on *inter*societal processes.[22] Since conflict between societies is, as Spencer recognized early in his work, a basic condition of human societies that have settled in territories and developed political leadership, it is not surprising that Weber would also analyze intersocietal conflict, or the geopolitics between societies. Such a line of emphasis has been one theme in the dramatic revival of historical sociology in both its neo-Marxian[23] and neo-Weberian[24] forms. And for Weber the degree of legitimacy accorded political authority within a system is very much dependent upon its capacity to generate prestige in the wider geopolitical system, or what today we might term "world system." Thus withdrawal of legitimacy is not just the result of conditions I-A, I-B, and I-C in Table 9-5; legitimacy also depends upon the "success" and "prestige" of a state in relation to other states.[25]

In fact, political legitimacy is a precarious situation because it relies upon the capacity of political authority to sustain needs among system members for defense and attack against external enemies, even during periods of relative peace. For without this sense of "threat" and a corresponding "success" in dealing with this threat, legitimacy lessens. Weber did not argue that legitimacy is always necessary for superordinates to dominate, for there can be periods of

---

[22]Weber, *Economy and Society*, pp. 901–1372, especially pp. 901–20.

[23]For example, see: Immanuel Wallerstein, *The Modern World System*, 3 volumes (New York: Academic Press, 1974–1989).

[24]For example, see: Randall Collins, *Weberian Sociological Theory* (Cambridge, England: Cambridge University Press, 1986); Michael Mann, *The Sources of Social Power*, vol. 1 (New York: Cambridge University Press, 1986); Theda Skocpol, *States and Social Revolutions* (New York: Cambridge University Press, 1979). The last work combines the analysis of internal revolution and geopolitics, seeing the former as a potential consequence of failed policies in the latter.

[25]Ibid., p. 904.

apathy, supported by tradition and routine. And there can also be periods of coercive force by superordinates to quell potential rebellion. Nor does Weber argue that "external enemies" must always be present to keep legitimacy revved up; rather, internal conflicts that pose threats can also give legitimacy to political authority. Thus the very processes that might lead some to withdraw legitimacy and initiate conflict under charismatic leadership can sometimes work to bolster the legitimacy of political authority, *if* enough other groupings in a society feel threatened. Indeed, Weber argued, political authorities often stir up internal or external "enemies" as a ploy for increasing their legitimacy and power to control the distribution of resources.

But the attention of those with political authority to the external system is not always political. Prestige, per se, can motivate some groupings to encourage military and other forms of contact with other systems (in Weber's terms, status groups can often encourage military and other forms of external contact). More important, however, are economic interests (classes, in Weber's terms). Those economic interests—colonial booty capitalists, privileged traders, financial interests, arms dealers, and the like—who rely upon the state to sustain their viability encourage foreign military expansion, whereas those economic interests who rely upon market dynamics and free trade will usually resist military expansionism (because it can hurt domestic productivity, or profits from unregulated activity in external markets) and encourage cooptive efforts through trade relations and market dependencies of external systems on commodities and services provided by these interests.

Table 9-6 presents Weber's argument in more abstract terms; it leads into the propositions in Table 9-5 in which less legitimacy is seen by Weber as increasing the likelihood of conflict. The essential point is not so much that Weber developed a mature theory but, rather, that he stimulated a conflict approach that examined the relationship between internal and external conflict processes, or, phrased differently, geopolitical processes.

Weber's Pessimism and the Dilemma for the Emancipation Project of Critical Theorists

As I indicated earlier, Weber was concerned with the historical transition from traditional societies to modern capitalist societies. In his description and explanation of this transition, as it occurred in the Western European nations, is a devastating critique of Marx's optimism that the conditions for the revolutionary transition to a new utopian society were being created. Weber's analysis is complex, and the historical detail that he presents to document his case is impressive, but his argument is captured by the word *rationalization*. Weber argued that the rationality that defines modern societies is "means/ends rationality." The nature of such rationality involves selecting the best means to achieve a defined end. The process of rationalization involves, Weber felt, the increasing penetration of means/ends rationality into ever more spheres of life and the consequent destruction of traditions. For, as bureaucracies expand in the economic and governmental sphere, and as markets allow individuals to

TABLE 9-6 Weber's Propositions on Geopolitics and Conflict

I. The greater the legitimacy of political authority, the greater its capacity to dominate other groupings in a system.

  A. The more those with power can sustain a sense of prestige and success in relations with external systems, the greater their capacity to be viewed as legitimate.

    1. The more productive sectors of a system depend upon political authority for their viability, the more they encourage political authority to engage in military expansion to augment their interests; and, when successful, such expansion increases prestige.

    2. The less productive sectors depend upon the state for their viability, the less likely they are to encourage political authority to engage in military expansion and the more likely they are to rely upon cooptation; and, when successful, such cooptation increases prestige.

  B. The more those with power can create a sense of threat from external forces, the greater their capacity to be viewed as legitimate.

  C. The more those with power can create a sense of threat among the majority by internal conflict with a minority, the greater their capacity to be viewed as legitimate.

II. The less political authority can sustain a sense of legitimacy, the more vulnerable it becomes to outbreaks of internal conflict.

  A. The more a political authority loses prestige in the external system, the less able it is to remain legitimate.

    1. The less successful a political authority in external conflict, the greater the loss of prestige.

    2. The less successful a political authority in cooptive efforts in the external system, the greater the loss of prestige.

pursue their personal ends rationally, the traditional moral fabric is broken. Weber agreed with Simmel that this rationalization of life brings individuals a new freedom from domination by religious dogmatism, community, class, and other traditional forces; but in their place it creates a new kind of domination by impersonal economic forces, such as markets and corporate bureaucracies, and by the vast administrative apparatus of the ever-expanding state. Human options were, in Weber's view, becoming every more constrained by the "iron cage" of rational/legal systems. And, unlike Marx, he did not see such a situation as rife with revolutionary potential; on the contrary, he saw the social world as ever more administered by impersonal bureaucratic forces.

This pessimistic view seemed, by the early 1930s, to be a far more reasonable assessment of modern society than was Marx's utopian dream. Indeed, the communist revolution in Russia had degenerated into Stalinism and bureaucratic totalitarianism by "the Party"; in the West, particularly the United States, workers seemed ever more willing to sell themselves in markets and work in large-scale organizations; and political fascism in Germany and Italy seemed to be increasing as dictators created large authoritarian bureaucracies. How, then, was the first generation of critical theorists to reconcile Weber's more accurate assessment of empirical trends with Marx's optimistic emancipatory vision? Such is the central question of all critical theory. And so, just as Weber's ideas forced revision of Marx's more formal propositions among modern positivists, so his analysis of the process of rationalization required critical theorists to reformulate the emancipatory dream of Marx.

THE PROLIFERATION OF CONFLICT SOCIOLOGY

From the reexamination of Marx, Weber, and Simmel came a powerful critique of positivist sociology in general and of Parsons' conceptual scheme in particular. As the Parsonian scheme and functionalism receded, conflict sociology gained greater prominence. Although not as dominant as the Parsonian scheme once was, conflict sociology has been the most conspicuous successor to functionalism. Aside from self-conscious conflict theories, the basic tenets of conflict sociology—inequality, tension, conflict—are now incorporated into many other theoretical orientations in sociology.[26]

Within the narrower confines of conflict sociology itself, there is an enormous diversity of activity. All that I can possibly hope to do with such a proliferating perspective is examine the range of activity encompassed by the label *conflict theory*. Of course, I will examine in later sections other approaches that incorporated elements of the more self-conscious conflict theorists. But in the chapters to follow I will explore four very different conflict theories: the dialectical approach of Ralf Dahrendorf, the conflict functionalism of Lewis Coser, the exchange conflict approach of Randall Collins, and the critical theory of Jurgen Habermas. I will miss much theoretical activity by focusing on these four approaches, but my sense is that these four are the most central and, thus, worthy of detailed analysis.

---

[26]For example, there is the structuralist Marxism of Louis Althusser and Maurice Godelier (who will not be examined in this book), the exchange theory of Peter Blau (whose blending of conflict ideas into exchange theories is examined in Chapter 16), and the structuration theory of Anthony Giddens (who is examined in Chapter 26).

CHAPTER 10

◆

Dialectical Conflict Theory

Ralf Dahrendorf

RITUALIZED CRITICISM OF FUNCTIONALISM

In the late 1950s, Ralf Dahrendorf persistently argued that the Parsonian scheme, and functionalism in general, presents an overly consensual, integrated, and static vision of society. In Dahrendorf's view, society has two faces—one of consensus, the other of conflict. And it is time to begin analysis of society's ugly face and abandon the utopian image created by functionalism. To leave utopia, Dahrendorf offered the following advice:

> Concentrate in the future not only on concrete problems but on such problems as involve explanations in terms of constraint, conflict, and change. This second face of society may aesthetically be rather less pleasing than the social system—but, if all sociology had to offer were an easy escape to Utopian tranquility, it would hardly be worth our efforts.[1]

To escape utopia, therefore, requires that a one-sided conflict model be substituted for the one-sided functional model. Although this conflict perspective was not considered by Dahrendorf to be the only face of society, it is a necessary supplement that will make amends for the past inadequacies of functional theory.[2] The model that emerges from this theoretical calling is a dialectical conflict perspective, which, I feel, still represents one of the best efforts to incorporate the insights of Marx and (to a lesser extent) Weber into

---

[1]Ralf Dahrendorf, "Out of Utopia: Toward a Reorientation of Sociological Analysis," *American Journal of Sociology* 64 (September 1958), p. 127.

[2]As Dahrendorf emphasizes: "I do not intend to fall victim to the mistake of many structural-functional theorists and advance for the conflict model a claim to comprehensive and exclusive applicability. . . . it may well be that in a philosophical sense, society has two faces of equal reality; one of stability, harmony, and consensus and one of change, conflict and constraint" (ibid.). Such disclaimers are, in reality, justifications for arguing for the primacy of conflict in society. By claiming that functionalists are one-sided, it becomes fair game to be equally one-sided in order to "balance" past one-sidedness.

a coherent set of theoretical propositions. I have my doubts that this dialectical conflict theory is more isomorphic than functional theory with what occurs in the actual world. I do think that it represents an important corrective to Parsonian functionalism, which tended to overemphasize social integration.

In his analysis, Dahrendorf is careful to note that processes other than conflict are evident in social systems and that even the conflict phenomena he examines are not the only kinds of conflict in societies. Having said this, however, my sense is that Dahrendorf believes his conflict model represents a more comprehensive theory of society and provides a more adequate base for theorizing about human social organization than either functionalism or other alternatives.

DAHRENDORF'S IMAGE OF THE SOCIAL ORDER

For Dahrendorf the process of institutionalization involves the creation of "imperatively coordinated associations" (hereafter referred to as *ICAs*) that, in terms of criteria not specified, represent a distinguishable organization of roles. This organization is characterized by power relationships, with some clusters of roles having power to extract conformity from others. I see Dahrendorf as somewhat vague on the point, but it appears that any social unit— from a small group or formal organization to a community or an entire society— can be considered for analytical purposes an ICA if an organization of roles displaying power differentials exists. Furthermore, although power denotes the coercion of some by others, these power relations in ICAs tend to become legitimated and can therefore be viewed as authority relations in which some positions have the "accepted" or "normative right" to dominate others. Dahrendorf thus conceives the social order as maintained by processes creating authority relations in the various types of ICAs existing throughout all layers of social systems.[3]

At the same time, however, power and authority are the scarce resources over which subgroups within a designated ICA compete and fight. They are thus the major sources of conflict and change in these institutionalized patterns. This conflict is ultimately a reflection of where clusters of roles in an ICA stand in relation to authority, since the "objective interests" inherent in any role are a direct function of whether that role possesses authority and power over other roles. However, even though roles in ICAs possess varying degrees of authority, any particular ICA can be typified in terms of just two basic types of roles, ruling and ruled. The ruling cluster of roles has an interest in preserving the status quo, and the ruled cluster has an interest in redistributing power, or authority. Under certain specified conditions, awareness of these contradictory interests increases, with the result that ICAs polarize into two conflict

---

[3]Ralf Dahrendorf, "Toward a Theory of Social Conflict," *Journal of Conflict Resolution* 2 (June 1958), pp. 170–83; *Class and Class Conflict in Industrial Society* (Stanford, CA: Stanford University Press, 1959), pp. 168–69; *Gesellschaft un Freiheit* (Munich: R. Piper, 1961); *Essays in the Theory of Society* (Stanford, CA: Stanford University Press, 1967).

groups, each now aware of its objective interests, which then engage in a contest over authority. The resolution of this contest or conflict involves the redistribution of authority in the ICA, thus making conflict the source of change in social systems. In turn, the redistribution of authority represents the institutionalization of a new cluster of ruling and ruled roles that, under certain conditions, polarize into two interest groups that initiate another contest for authority. Social reality is thus typified in terms of this unending cycle of conflict over authority within the various types of ICAs the social world comprises. Sometimes the conflicts within diverse ICAs in a society overlap, leading to major conflicts cutting across large segments of the society; at other times and under different conditions, these conflicts are confined to a particular ICA.

As I think is clear, this image of social organization represents a revision of Marx's portrayal of social reality:

1. Social systems are seen by both Dahrendorf and Marx as in a continual state of conflict.
2. Such conflict is presumed by both authors to be generated by the opposed interests that inevitably inhere in the social structure of society.
3. Opposed interests are viewed by both Marx and Dahrendorf as reflections of differences in the distribution of power among dominant and subjugated groups.
4. Interests are seen by both as tending to polarize into two conflict groups.
5. For both, conflict is dialectical, with resolution of one conflict creating a new set of opposed interests that, under certain conditions, will generate further conflict.
6. Social change is thus seen by both as a ubiquitous feature of social systems and the result of inevitable conflict dialectics within various types of institutionalized patterns.

Much like Marx, this image of institutionalization as a cyclical or dialectic process has led Dahrendorf into the analysis of only certain key causal relations: (1) conflict is assumed to be an inexorable process arising out of opposing forces within social/structural arrangements; (2) such conflict is accelerated or retarded by a series of intervening structural conditions or variables; (3) conflict resolution at one point in time creates a structural situation that, under specifiable conditions, inevitably leads to further conflict among opposed forces.

For Marx the source of conflict ultimately lies beneath cultural values and institutional arrangements, which represent edifices constructed by those with power. In reality the dynamics of a society are found in its substructure, where the differential distribution of property and power inevitably initiates a sequence of events leading to revolutionary class conflict. While borrowing much of Marx's rhetoric about power and coercion in social systems, Dahrendorf actually ends up positing a much different source of conflict: the institutionalized authority relations of ICAs. Such a position is very different from that of Marx, who viewed such authority relations as simply a superstructure erected by the dominant classes, which in the long run would be destroyed by the conflict dynamics occurring below institutional arrangements. Although Dahrendorf acknowledges that authority relations are imposed by the dominant groups in ICAs and frequently makes reference to "factual substrates," the source of conflict becomes, upon close examination, the legitimated authority role relations of ICAs. I think that this drift away from Marx's emphasis on the institutional substructure forces Dahrendorf to seek the source of conflict in those very relations that integrate, albeit temporarily, an ICA. By itself, I see this shift in emphasis as desirable, since Dahrendorf clearly recognizes that not all power is a reflection of property ownership—a fact Marx's polemics tended to underemphasize. But, as I will later observe, viewing power only in terms of authority can lead to analytical problems as severe as those in Marx's model and, in fact, somewhat reminiscent of those in Parsons' "social systems."

Although emphasizing different sources of conflict, Dahrendorf's and Marx's models reveal similar causal chains of events leading to conflict and the reorganization of social structure: relations of domination and subjugation create an "objective" opposition of interests; awareness or consciousness by the subjugated of this inherent opposition of interests occurs under certain specifiable conditions; under other conditions this newfound awareness leads to the political organization and then polarization of subjugated groups, which then join in conflict with the dominant group; the outcome of the conflict will usher in a new pattern of social organization; this new pattern of social organization will have within it relations of domination and subjugation that set off another sequence of events leading to conflict and then change in patterns of social organization.

The intervening conditions affecting these processes are outlined by both Marx and Dahrendorf only with respect to the formation of awareness of opposed interests by the subjugated, the politicization and polarization of the subjugated into a conflict group, and the outcome of the conflict. The intervening conditions under which institutionalized patterns generate dominant and subjugated groups and the conditions under which these can be typified as having opposed interests remain unspecified—apparently because they are in the nature of institutionalization, or ICAs, and do not have to be explained.

In Figure 10–1, I have outlined the causal imagery of Marx and Dahrendorf. At the top of the figure are Marx's analytical categories, stated in their most abstract form. The other two rows specify the empirical categories of Marx and Dahrendorf, respectively. Separate analytical categories for the Dahrendorf model are not enumerated because they are the same as those in the Marxian model. As I think is clear, the empirical categories of the Dahrendorf scheme differ greatly from those of Marx. But the form of analysis is much the same, since each considers as nonproblematic and not in need of causal analysis the empirical conditions of social organization, the transformation of this organization into relations of domination and subjugation, and the creation of opposed interests. The causal analysis for both begins with an elaboration of the conditions leading to growing class consciousness (Marx) or awareness among quasi groups (Dahrendorf) of their objective interests; then analysis shifts to the creation of a politicized class "for itself" (Marx) or a true "conflict group" (Dahrendorf); finally, emphasis focuses on the emergence of conflict

FIGURE 10–1 The Dialectical Causal Imagery

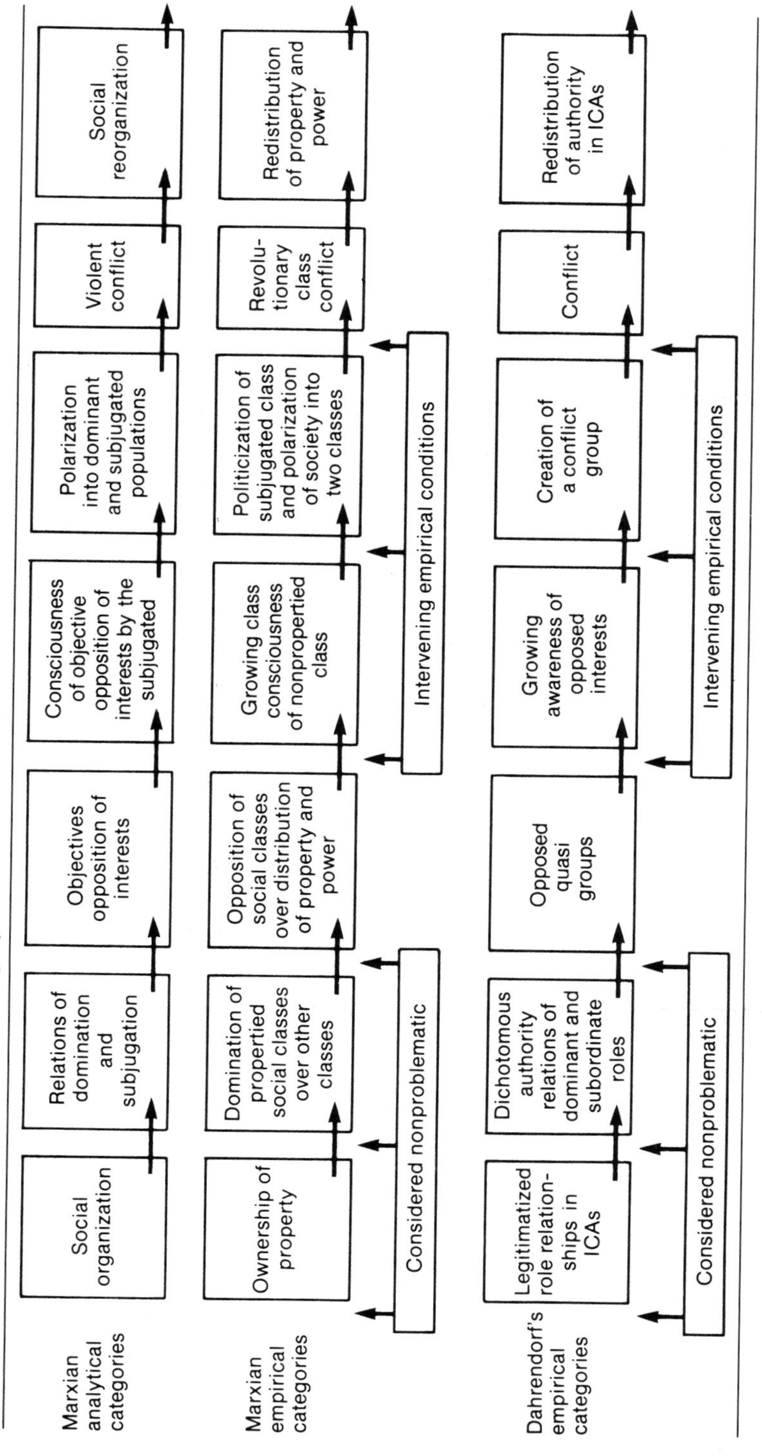

| Marxian analytical categories | Social organization | → | Relations of domination and subjugation | → | Objectives opposition of interests | → | Consciousness of objective opposition of interests by the subjugated | → | Polarization into dominant and subjugated populations | → | Violent conflict | → | Social reorganization |

| Marxian empirical categories | Ownership of property | → | Domination of propertied social classes over other classes | → | Opposition of social classes over distribution of property and power | → | Growing class consciousness of nonpropertied class | → | Politicization of subjugated class and polarization of society into two classes | → | Revolutionary class conflict | → | Redistribution of property and power |

Considered nonproblematic | Intervening empirical conditions

| Dahrendorf's empirical categories | Legitimatized role relationships in ICAs | → | Dichotomous authority relations of dominant and subordinate roles | → | Opposed quasi groups | → | Growing awareness of opposed interests | → | Creation of a conflict group | → | Conflict | → | Redistribution of authority in ICAs |

Considered nonproblematic | Intervening empirical conditions between polarized and politicized classes (Marx) or conflict groups (Dahrendorf).

CRITICISMS OF THE DIALECTICAL CONFLICT MODEL

Problems in the Causal Analysis

The most conspicuous criticism of Dahrendorf's causal imagery comes from Peter Weingart. He has argued that, in deviating from Marx's conception of the "substructure of opposed interests" existing below the cultural and institutional edifices of the ruling classes, Dahrendorf forfeits a genuine causal analysis of conflict and, therefore, an explanation of how patterns of social organization are changed.[4] This criticism asks questions reminiscent of Dahrendorf's portrayal of Parsonian functionalism: How is it that conflict emerges from legitimated authority relations among roles in an ICA? How is it that the same structure that generates integration also generates conflict? Although for the Marxian scheme there are, I think, rather severe analytical and empirical problems, the causal analysis is clear, since the source of conflict—the opposition of economic interests—is clearly distinguished from the institutional and cultural arrangements maintaining a temporary order—the societal superstructure. Dahrendorf, however, has failed to make explicit this distinction and thus falls into the very analytical trap he has imputed to functional theory: change-inducing conflict must mysteriously arise from the legitimated relations of the social system.

In an attempt to escape this analytical trap, I think that Dahrendorf's causal analysis often becomes confusing. One tack Dahrendorf employs is to assert that many roles also have a nonintegrative aspect because they represent fundamentally opposed interests of the incumbents. These opposed interests are reflected in role conflict, which seemingly reduces the issues of role strain and conflict to dilemmas created by objectively opposed interests—surely a dubious assertion that is correct only some of the time. In equating interests and role expectations, Dahrendorf would seemingly have to hypothesize that all institutionalized patterns, or ICAs, display two mutually contradictory sets of role expectations—one to obey, the other to revolt—and that actors must decide which set they will follow. Presumably, actors wish to realize their objective interests and hence revolt against the role expectations imposed upon them by the dominant group. This approach forces the Dahrendorf model to reduce the origins of conflict to the wishes, wills, and sentiments of a person or group—a reductionist imperative that I am sure Dahrendorf would reject but one that his causal imagery would seemingly dictate.[5]

---

[4]Peter Weingart, "Beyond Parsons? A Critique of Ralf Dahrendorf's Conflict Theory," *Social Forces* 48 (December 1969), pp. 151–65. See also Jonathan H. Turner, "From Utopia to Where: A Strategy for Reformulating the Dahrendorf Conflict Model," *Social Forces* 52 (December 1973), pp. 236–44.

[5]Weingart, "Beyond Parsons."

Many of these problems might be overcome, I feel, if Dahrendorf had provided a series of theoretical statements that would indicate the conditions under which legitimated role relationships in ICAs create dichotomous authority relations of domination and subjugation. To simply assume that this is the case is to avoid what I see as the critical causal link in his analytical scheme. I believe that these propositions—or, as Dahrendorf describes them, "intervening empirical conditions"—are a necessary part of the model. Without them it is unclear how the types of authority, coercion, and domination that lead to conflict ever emerge in the first place. Assuming that they just emerge or are an endemic part of social structure is to define away the theoretically important question about what types of authority in what types of ICAs lead to what types of domination and subjugation that, in turn, lead to what types of opposed interests and what types of conflict. These are all phenomena that must be conceptualized as variables and incorporated into the causal chains of the dialectical conflict model. From Figure 10–1, this task would involve stating the "intervening empirical conditions" at each juncture of all of Dahrendorf's empirical categories. What is now considered nonproblematic would become as problematic as subsequent empirical conditions.

Initiating this task is difficult,[6] but to do so would enable Dahrendorf to avoid some of the more standard criticisms of this causal imagery: (1) not only does conflict cause change of social structure, but changes of structure also cause conflict (under conditions that need to be specified in greater detail than Simmel's initial analysis); and (2) conflict can inhibit change (again, under conditions that need to be specified).[7] Unless these conditions are part of the causal imagery, conflict theory merely states the rather obvious fact that change occurs, without answering the theoretical questions of why, when, and where such change occurs.

Despite the vagueness of Dahrendorf's causal analysis, I see the great strength of his approach residing in the formulation of explicit propositions. These state the intervening empirical conditions that cause quasi groups to become conflict groups, as well as the conditions affecting the intensity (involvement of group members) and violence (degree of regulation) of the conflict and the degree and rate of structural change caused by it. More formally, Dahrendorf outlines three types of intervening empirical conditions: (1) conditions of organization that affect the transformation of latent quasi groups into manifest conflict groups; (2) conditions of conflict that determine the form and intensity of conflict; and (3) conditions of structural change that influence the kind, speed, and depth of the changes in social structure.[8]

Thus the variables in the theoretical scheme are (1) the degree of conflict-group formation; (2) the degree of intensity of the conflict; (3) the degree of

---

[6]For my best effort on this score, see Jonathan H. Turner, "A Strategy for Reformulating the Dialectical and Functional Theories of Conflict," *Social Forces* 53 (March 1975), pp. 433–44.

[7]For a convenient summary of these conditions, see Percy Cohen, *Modern Sociological Theory* (New York: Basic Books, 1968), pp. 183–91.

[8]Dahrendorf, "Toward a Theory of Social Conflict."

violence of the conflict; (4) the degree of change of social structure; and (5) the rate of such change. I think it significant, for criticisms to be delineated later, that concepts such as ICAs, legitimacy, authority, coercion, domination, and subjugation are not explicitly characterized as variables requiring statements on the conditions affecting their variability. Rather, these concepts are simply defined and interrelated to one another in terms of definitional overlap or stated as assumptions about the nature of social reality.

For those phenomena that are conceptualized as variables, Dahrendorf's propositions appear to be an elaboration of those developed by Marx. I have reworked them a bit, as shown in Table 10-1.[9]

Like Marx, Dahrendorf sees conflict as related to subordinates' growing awareness of their interests and formation into conflict groups (Proposition I). Such awareness and group formation are a positive function of the degree to which (a) the technical conditions (leadership and unifying ideology), (b) the political conditions (capacity to organize), and (c) the social conditions (ability to communicate) are met. These ideas clearly come from Marx's discussion (see Table 9-1). However, Proposition II borrows from Simmel and contradicts Marx. It emphasizes that, if groups are *not* well organized—that is, if the technical, political, and social conditions are not met—then conflict is likely to be emotionally involving (see Table 9-2). Then Dahrendorf borrows from Weber in Proposition III by stressing that the superimposition of rewards—that is, the degree of correlation among those who enjoy privilege with respect to power, wealth, and prestige—also increases the emotional involvement of subordinates who pursue conflict (see Table 9-5). Proposition IV also takes as much from Weber as from Marx because it sees the lack of mobility into positions of authority as escalating the emotional involvement of subordinates. Proposition V is clearly from Simmel and contradicts Marx, in that the violence of conflict is related to the lack of organization and clear articulation of interests. But Proposition VI returns to Marx's emphasis that sudden escalation in people's perception of deprivation—that is, relative deprivation—increases the likelihood of violent conflict. In Proposition VII, however, Dahrendorf returns to Simmel and argues that violence is very much related to the capacity of a system to develop regulatory procedures for dealing with grievances and releasing tensions. And in Propositions VIII and IX Dahrendorf moves again to Marx's emphasis on how conflict produces varying rates and degrees of structural change in a social system.

I feel that Dahrendorf must be commended for placing the propositions in a reasonably systematic format—a difficult task too infrequently performed by theorists in sociology. However, even though this propositional inventory is highly suggestive, I see a number of criticisms that need to be addressed.

One of the most obvious criticisms of the Dahrendorf perspective is the failure to visualize crucial concepts as variables. Indeed, my rephrasing has

---

[9]The propositions listed in the table differ from those in a list provided by Dahrendorf, *Class and Class Conflict*, pp. 239-40, in two respects: (1) they are phrased consistently as statements of covariance, and (2) they are phrased somewhat more abstractly without reference to "class," which in this particular work was Dahrendorf's primary concern.

TABLE 10–1 Dahrendorf's Abstract Propositions

I. The more members of quasi groups in ICAs can become aware of their objective interests and form a conflict group, the more likely conflict is to occur.
   A. The more the "technical" conditions of organization can be met, the more likely the formation of a conflict group.
      1. The more a leadership cadre among quasi groups can be developed, the more likely are the technical conditions of organization to be met.
      2. The more a codified idea system, or charter, can be developed, the more likely are the technical conditions of organization to be met.
   B. The more the "political" conditions of organization can be met, the more likely the formation of a conflict group.
      1. The more the dominant groups permit organization of opposed interests, the more likely are the political conditions of organization to be met.
   C. The more the "social" conditions of organization can be met, the more likely the formation of a conflict group.
      1. The more opportunity for members of quasi groups to communicate, the more likely are the social conditions of organization to be met.
      2. The more recruiting is permitted by structural arrangements (such as propinquity), the more likely are the social conditions to be met.
II. The less the technical, political, and social conditions of organization are met, the more intense the conflict.
III. The more the distribution of authority and other rewards are associated with each other (superimposed), the more intense the conflict.
IV. The less the mobility between super- and subordinate groups, the more intense the conflict.
V. The less the technical, political, and social conditions of organization are met, the more violent the conflict.
VI. The more the deprivation of the subjugated in the distribution of rewards shifts from an absolute to a relative basis, the more violent the conflict.
VII. The less the ability of conflict groups to develop regulatory agreements, the more violent the conflict.
VIII. The more intense the conflict, the more structural change and reorganization it will generate.
IX. The more violent the conflict, the greater the rate of structural change and reorganization.

converted some of Dahrendorf's concepts into variables. But still, most conspicuous for their nominal character are such central concepts as authority, domination/subjugation, and interest. Since it is from legitimated authority relations that conflict ultimately springs, I find it somewhat surprising that this concept is not viewed as a variable, varying at a minimum in terms of such properties as intensity, scope, and legitimacy. Rather, Dahrendorf has chosen to define away the problem:

> No attempt will be made in this study to develop a typology of authority. But it is assumed throughout that the existence of domination and subjugation is a common feature of all possible types of authority and, indeed, of all possible types of association and organization.[10]

[10]Ibid., p. 169.

A typology of authority might offer some indication of the variable states of authority and related concepts—a fact Dahrendorf ignores by simply arguing that authority implies domination and subjugation, which in turn gives him the structural dichotomy necessary for this dialectical theory of conflicting interests. He refuses to speculate on what types of authority displaying what variable states lead to what types of variations in domination and subjugation that, in turn, cause what variable types of opposed interests leading to what variable types of conflict groups. Thus I see Dahrendorf as linking only by assumption and definition crucial variables that causally influence each other as well as the more explicit variables of his scheme: the degree of conflict, the degree of intensity of conflict, the degree of violence in conflict, the degree of change, and the rate of social change. In fact, it is likely that these unstated variable properties of authority, domination, and interests have as much influence on the scheme's explicit variables as the "intervening empirical conditions" that Dahrendorf chooses to emphasize. Furthermore, as I noted earlier, when viewed as variables, the concepts of authority, domination/subjugation, and interests require their own intervening empirical conditions. These conditions may in turn influence other subsequently intervening conditions in much the same way as the "conditions of organization" also influence the subsequent intensity and violence of conflict in the scheme.

My criticisms suggest an obvious solution: to conceptualize ICAs, legitimacy, authority, domination/subjugation, and interests as variable phenomena and to attempt a statement of the intervening empirical conditions influencing their variability. Expanding the propositional inventory in this way would, I feel, reduce the vagueness of the causal imagery in the present scheme. Such an alteration would also cut down on the rather protracted set of dialectical assumptions—which are of dubious isomorphism in reality—and address a theoretical (rather than philosophical) question: under what conditions do ICAs create legitimated authority relations that generate clear relations of domination and subjugation leading to strongly opposed interests?

Methodological Problems

To his credit, Dahrendorf provides formal definitions of major concepts and suggests operational clues about their application in concrete empirical settings, as is evident in his analysis of class conflict in industrial societies.[11] Furthermore, the incorporation of at least *some* concepts into an explicit propositional inventory—albeit an incomplete one—makes the scheme appear more testable and amenable to refutation.

A number of methodological problems remain, however. One of these concerns the extremely general definitions given to concepts. Although these definitions are stated formally, they are often so general that they can be used in an ad hoc and ex post facto fashion to apply to such a wide variety of phenomena

---

[11]Ibid., pp. 241–318.

that their current utility for the development and testing of theory can be questioned. For example, I think that the concepts of power, legitimacy, authority, interests, domination, and even conflict are defined so broadly that instances of these concepts can be found in almost any empirical situation. Thus Dahrendorf can rather easily confirm his assumption that social life is rife with conflict. But how is one to measure these vaguely and globally defined concepts? I noted this problem earlier when discussing Dahrendorf's reluctance to view crucial concepts, such as authority and domination, as variables. If these concepts were so conceptualized, I feel that it would be easier for investigators to put empirical handles on them, since definitional statements about their variable states would specify more precisely the phenomena denoted by the concepts. Dahrendorf rarely does this service, preferring to avoid typologies; even when concepts are defined as variables, Dahrendorf avoids the issue with such statements as the following: "The intensity of class conflict varies on a scale (from 0 to 1)." Coupled with the formal definition of conflict intensity as the "energy expenditure and degree of involvement of conflicting parties," few operational guidelines are provided about how such a concept might be measured. Were these definitions supplemented with, at a minimum, a few examples of prominent points along the 0-to-1 scale, then the concepts and propositions of the scheme would be more amenable to empirical investigation. As the definitions stand now, Dahrendorf does the very thing for which he has so resoundingly criticized Parsons: he uses vague concepts in a way that will inevitably confirm his overall scheme. More attention to precise definitions would give the concepts more power as theoretical constructs and as guidelines for investigators.

In sum, then, the Dahrendorf scheme presents a number of problems for empirical investigators. Such a statement can be made for almost all theoretical perspectives in sociology and by itself is not a unique indictment. For the Dahrendorf scheme, however, it appears that methodological problems could be minimized with just a little additional work. To the extent that my suggested corrections are made, it is likely that the dialectical conflict perspective will offer a fruitful strategy for developing sociological theory.

FROM UTOPIA TO WHERE? A CONCLUDING COMMENT

As I emphasized at the beginning of this chapter, Dahrendorf has been one of the harshest critics of functional forms of theorizing, likening them to an ideological utopia. In order to set sociological theorizing on the road out of utopia, Dahrendorf felt compelled to delineate a dialectical conflict scheme that presumably mirrors more accurately than Parsonian functionalism the real character of the social world. In so doing, Dahrendorf presumably views his theoretical perspective and strategy as providing a more adequate set of theoretical guidelines for understanding the nature and dynamics of human social organization.

What I find curious about Dahrendorf's approach is that, upon close examination, it appears quite similar to the one he imputed to Parsonian functionalism. For example, a number of commentators[12] have noted that both Parsons and Dahrendorf view the social world in terms of institutionalized patterns—for Parsons, the "social systems"; for Dahrendorf, "imperatively coordinated associations."[13] Both view societies as composed of subsystems involving the organization of roles in terms of legitimate normative prescriptions. For Dahrendorf these legitimated normative patterns reflect power differentials in a system; and, despite his rhetoric about the coercive nature of these relations, I see this vision of power as remarkably similar to Parsons' conception of power as the legitimate right of some status-roles to regulate the expectations attendant upon other statuses.[14] Furthermore, in Dahrendorf's model, deviation from the norms established by status-roles will lead dominant groups to attempt to employ negative sanctions—a position that is very close to Parsons' view that power exists to correct for deviations within a system.[15]

The apparent difference between Dahrendorf's and Parsons' emphasis with respect to the functions of power in social systems (or ICAs) is that Dahrendorf argues explicitly that power differentials cause both integration (through legitimated authority relations) and disintegration (through the persistence of opposed interests). However, to state that conflict emerges out of legitimated authority is nothing more than to state, a priori, that opposed interests exist and cause conflict. The emergence of conflict follows from vague assumptions about such processes as the "inner dialectic of power and authority" and the "historical function of authority,"[16] rather than from carefully documented causal sequences. Thus I think that the genesis of conflict in Dahrendorf's model remains as unexplained as it does in his portrayal of the inadequacies of the functional utopia, primarily because its emergence is set against a background of unexplained conceptions of system norms and legitimated authority.

Dahrendorf's problem in explaining why and how conflict groups can emerge from a legitimated authority structure is partly a reflection of hidden assumptions about functional requisites. In a subtle and yet consistent fashion, he assumes that authority is a functional requisite for system integration and that the conflict which somehow emerges from authority relations is a functional requisite for social change. As Dahrendorf argues, "the historical function of authority" is to generate conflict and thereby maintain the vitality of social systems. From this notion of the requisite for change, it is all too easy to assert that conflict exists to meet the system's needs for change—an illegitimate teleology that echoes Marx's teleological assumption that cycles of dialectical change are necessary to create the communist utopia.

---

[12]For the best of these critiques, see Weingart, "Beyond Parsons."

[13]Ibid.; Pierre L. van den Berghe, "Dialectic and Functionalism: Toward a Theoretical Synthesis," *American Sociological Review* 28 (October 1963), pp. 695–705.

[14]For example, see Talcott Parsons and Edward Shils, eds., *Toward a General Theory of Action* (New York: Harper & Row, 1962), pp. 197–205.

[15]Ibid., p. 230.

[16]Weingart, "Beyond Parsons," p. 161.

More fundamental, however, is Dahrendorf's inability to explain the organization of ICAs. To assert that they are organized in terms of power and authority defines away a theoretically interesting problem of how, why, and through what processes the institutionalized patterns generating both integration and conflict come to exist. On the other hand, Parsons' analysis does attempt—however inadequately—to account for how institutionalized patterns, or social systems, become organized: by actors adjusting their various orientations, normative prescriptions emerge, which affect the subsequent organization of action; such organization is maintained by various mechanisms of social control—interpersonal sanctions, ritual activity, safety-valve structures, role segregation, and, on some occasions, power—and by mechanisms of socialization—the internalization of relevant values and the acquisition of critical interpersonal skills. Quite naturally, because of his commitment to developing systems of concepts instead of formats of propositions, Parsons gives only a vague clue about the variables involved in the process of institutionalization by which the types of opposed interests that lead to the organization of conflict groups and social change are created. However, Parsons at least attempts to conceptualize the variables involved in creating and maintaining the very social order that Dahrendorf glosses over in his formulation of the ICA concept. Yet it is from the institutionalized relations in ICAs that conflict-ridden cycles of change are supposed to emerge. Asserting one's way out of utopia, as I think Dahrendorf does, will not obviate the fundamental theoretical question facing sociological theory: how is social organization, in all its varied and changing forms, possible?

In sum, then, I think that Dahrendorf has used the rhetoric of coercion, dialectics, domination, subjugation, and conflict to mask a vision of social reality that is very close to the utopian image he has imputed to Parsons' work. In Dahrendorf's ICA is Parsons' social system; in his concepts of roles and authority is Parsons' concern with social control; in his portrayal of conflict, the origin of conflict is just as unclear as he assumes it to be in Parsons' work; and even in the analysis of social change, conflict is considered, in a way reminiscent of Parsons, to meet the functional need for change. Thus we can at least be suspicious about Dahrendorf's claim that we are on the road out of utopia. Yet, in making this harsh judgment, I still think that the propositions in Table 10-1 will be very useful in understanding some conflict processes. They will not, I feel, be as fundamental to understanding the nature of social reality as Dahrendorf seems to imply. But I find them very useful for helping to increase our knowledge of conflict processes in social systems. However, there is more to social reality than dialectical conflict, and this is where Dahrendorf and most conflict theorists go wrong: they assume—or, should I say, presume—too much of their more limited and delimited schemes about conflict.

CHAPTER 11

Conflict Functionalism

Lewis A. Coser

THE STANDARDIZED CRITIQUE

In the 1960s and 1970s, I see the criticisms of functionalism as looking much the same. They all berated Parsons and other functionalists for viewing society as overly institutionalized and equilibrating. At the same time the conflict schemes offered as alternatives revealed considerable diversity. The divergence in conflict theory is particularly evident when the conflict functionalism of Lewis Coser is compared with Ralf Dahrendorf's dialectical conflict perspective. Although Coser consistently criticized Parsonian functionalism for its failure to address the issue of conflict, he has also been sharply critical of Dahrendorf and other dialectical theorists for underemphasizing the positive functions of conflict for maintaining social systems.

In his first major work on conflict, Coser launched what became the standard polemic against functionalism: conflict is not given sufficient attention, and related phenomena such as deviance and dissent are too easily viewed as "pathological" for the equilibrium of the social system.[1] Parsons, in his concern for developing a system of concepts denoting the process of institutionalization, underemphasized conflict in his formal analytical works, seemingly viewing conflict as a disease that needs to be treated by the mechanisms of the body social.[2] I think that this rather one-sided portrayal of Parsons' work allows Coser to posit the need for redressing the sins of Parsonian functionalism with a one-sided conflict scheme. Apparently such analytical compensation was to be carried out for well over a decade, since after the tenth anniversary of his first polemic Coser was moved to reassert his earlier claim that it was "high time to tilt the scale in the direction of greater attention to social conflict."[3]

---

[1]Lewis A. Coser, *The Functions of Social Conflict* (London: Free Press of Glencoe, 1956).

[2]Ibid., pp. 22–23.

[3]Lewis A. Coser, "Some Social Functions of Violence," *Annals of the American Academy of Political and Social Science* 364 (March 1966), p. 10.

Yet, although Coser has consistently maintained that functional theorizing "has too often neglected the dimensions of power and interest," he does not follow either Marx's or Dahrendorf's emphasis on the disruptive consequences of violent conflict.[4] On the contrary, Coser seeks to correct Dahrendorf's analytical excesses by emphasizing the integrative and "adaptability" functions of conflict for social systems. Thus, Coser justifies his efforts by criticizing functionalism for ignoring conflict and by criticizing conflict theory for underemphasizing the functions of conflict.[5]

IMAGES OF SOCIAL ORGANIZATION

As I emphasized in Chapter 2, Émile Durkheim can be considered one of the fathers of functionalism. Thus I think it is interesting to note that a "conflict functionalist" is critical of Durkheim's approach.[6] In particular, Coser views Durkheim as taking a conservative orientation to the study of society, an orientation that "prevented him from taking due cognizance of a variety of societal processes, among which social conflict is the most conspicuous." Furthermore, this abiding conservatism forced Durkheim to view violence and dissent as deviant and pathological to the social equilibrium, rather than as opportunities for constructive social changes. Although Coser appears intent on rejecting the organicism of Durkheim's sociology, I find his own work filled with organismic analogies. For example, in describing the "functions of violence," Coser likens violence to pain in the human body, since both can serve as a danger signal that allows the body social to readjust itself.[7] To take another example, in his analysis of the "functions of dissent," Coser rejects the notion that dissent is explainable in terms of individual sickness and embraces the assumption that "dissent may more readily be explained as a reaction to what is perceived as a sickness in the body social."[8] I do not think that this form of analogizing fatally wounds Coser's analysis, but it does reveal that he has not rejected organicism. Apparently Coser has felt compelled to criticize Durkheim's or-

---

[4]Lewis A. Coser, *Continuities in the Study of Social Conflict* (New York: Free Press, 1967), p. 141.

[5]A listing of some of Coser's prominent works, to be utilized in subsequent analysis, reveals the functional flavor of his conflict perspective: *Functions of Social Conflict*; "Some Social Functions of Violence"; "Some Functions of Deviant Behavior and Normative Flexibility," *American Journal of Sociology* 68 (September 1962), pp. 172–81; and "The Functions of Dissent," in *The Dynamics of Dissent* (New York: Grune & Stratton, 1968), pp. 158–70. Other prominent works with less revealing titles but critical substance include: "Social Conflict and the Theory of Social Change," *British Journal of Sociology* 8 (September 1957), pp. 197–207; "Violence and the Social Structure," in *Science and Psychoanalysis*, ed. J. Masserman, vol. 7 (New York: Grune & Stratton, 1963), pp. 30–42. These and other essays are collected in Coser's *Continuities in the Study of Social Conflict*. One should also consult his *Masters of Sociological Thought* (New York: Harcourt Brace Jovanovich, 1977).

[6]Lewis Coser, "Durkheim's Conservatism and Its Implications for His Sociological Theory," in *Émile Durkheim, 1858–1917: A Collection of Essays*, ed. K. H. Wolff (Columbus: Ohio State University Press, 1960); also reprinted in Coser's *Continuities*, pp. 153–80.

[7]Coser, "Some Functions of Violence," pp. 12–13.

[8]Coser, "The Functions of Dissent," pp. 159–60.

ganicism because it did not allow the analysis of conflict as a process that could promote the further adaptation and integration of the body social.[9]

In rejecting the analytical constraints of Durkheim's analogizing, Coser embraces Georg Simmel's organicism (see Chapter 9). Conflict is viewed as a process that, under certain conditions, functions to maintain the body social or some of its vital parts. From this vantage point, Coser develops an image of society that stresses the following ideas:

1. The social world can be viewed as a system of variously interrelated parts.
2. All social systems reveal imbalances, tensions, and conflicts of interest among variously interrelated parts.
3. Processes within and among the system's constituent parts operate under different conditions to maintain, change, and increase or decrease a system's integration and adaptability.
4. Many processes, such as violence, dissent, deviance, and conflict, which are typically viewed as disruptive to the system, can also be viewed, under specifiable conditions, as strengthening the system's basis of integration as well as its adaptability to the environment.

From these assumptions, Coser articulates a rather extensive set of propositions about the functions (and, to a limited extent, the dysfunctions) of conflict for social systems. Coser offers some propositions about the conditions under which conflict leads to disruption and malintegration of social systems, but I see the main thrust of his analysis as revolving around statements on how conflict maintains or reestablishes system integration and adaptability to changing conditions. Thus I see Coser's analysis as follows: (1) imbalances in the integration of system parts lead to (2) the outbreak of varying types of conflict among these parts, which in turn causes (3) temporary reintegration of the system, which causes (4) increased flexibility in the system's structure, increased capability to resolve future imbalances through conflict, and increased capacity to adapt to changing conditions.

The causal imagery presents, I feel, a number of obvious problems. The most important of these is that processes are too frequently viewed as contributing to system integration and adaptation. Such an emphasis on the positive functions of conflict may reveal hidden assumptions of system needs that can be met only through the functions of conflict. Although Coser is careful to point out that he is simply correcting for analytical inattention to the positive consequences of conflict, the strategy is nonetheless one-sided.[10] Despite these

---

[9]I should note that such an emphasis on the "positive functions" of conflict could be construed as the pinnacle of conservative ideology—surpassing that attributed to Parsons. Even conflict promotes integration rather than disruption, malintegration, and change. Such a society, as Dahrendorf would argue, no longer has an ugly face and is as utopian as that of Parsons. For Coser's reply to this kind of charge, see *Continuities*, pp. 1 and 5.

[10]It is somewhat tragic for theory building in sociology that the early promising lead of Robin M. Williams, Jr., in his *The Reduction of Intergroup Tensions: A Survey of Research on Problems of Ethnic, Social, and Religious Group Relations* (New York: Social Science Research Council,

shortcomings, I think that Coser mitigates his functionalism by translating his image of social organization into a series of abstract propositions about conflict in social systems. As a result, the scheme takes on considerably more clarity than when stated as a cluster of assumptions. Equally significant, the scheme becomes more testable and amenable to reformulation on the basis of empirical findings. Let me now turn to these propositions, since they are the great strength of Coser's conflict functionalism.

PROPOSITIONS ON CONFLICT PROCESSES

Using both the substance and the style of Georg Simmel's provocative analysis, Coser has expanded the scope of Simmel's initial insights, incorporating propositions not only from Marx but also from Weber and the contemporary literature on conflict. Although the scheme reveals a large number of problems, stemming from his primary concern with the functions of conflict, I feel that Coser's conflict perspective remains one of the most comprehensive in the current literature. This comprehensiveness can, I think, be made evident by a partial list of the phenomena covered by his propositions: (1) the causes of conflict, (2) the violence of conflict, (3) the duration of conflict, and (4) the functions of conflict. Under each of these headings a variety of specific variables are incorporated into propositions.

One drawback to Coser's propositional inventory is that it has not been presented in a systematic or ordered format. Rather, the propositions appear in a number of discursive essays on substantive topics and in his analysis of Simmel's essay on conflict. Although each discrete proposition is usually stated quite clearly, I have had to extract and order the propositions. There is, of course, some danger of misinterpretation in this kind of exercise. Yet I feel that Coser's approach is much stronger when stated as a series of formal propositions.

The Causes of Conflict

In Table 11-1, I have formalized Coser's discursive analysis on the causes of conflict.[11] Much like Weber (see Table 9-5), Coser emphasizes in Proposition I that the withdrawal of legitimacy from an existing system of inequality is a critical precondition for conflict. In contrast, dialectical theorists such as Dah-

---

1947) was not consistently followed. Most of the propositions developed by Dahrendorf and Coser were summarized in this volume ten years prior to their major works. More important, the propositions are phrased more neutrally, without an attempt to reveal society's "ugly face" or to correct for inattention to the "functions of conflict."

[11]Again, it should be emphasized that it is dangerous and difficult to pull from diverse sources discrete propositions and attempt to relate them systematically without doing some injustice to the theorist's intent. However, unless this kind of exercise is performed, the propositions will contribute little to the development of sociological theory. The propositions in Table 11-1 were extracted from: *Functions of Social Conflict*, pp. 8-385; "Social Conflict and the Theory of Social Change," pp. 197-207; "Internal Violence as a Mechanism for Conflict Resolution"; and "Violence and Social Structure."

TABLE 11–1 Coser's Propositions on the Causes of Conflict

I. The more subordinate members in a system of inequality question the legitimacy of the existing distribution of scarce resources, the more likely they are to initiate conflict.
  A. The fewer the channels for redressing grievances over the distribution of scarce resources by subordinates, the more likely they are to question legitimacy.
     1. The fewer the internal organizations segmenting emotional energies of subordinates, the more likely they are to be without grievance alternatives and, as a result, to question legitimacy.
     2. The greater the ego deprivations of those without grievance channels, the more likely they are to question legitimacy.
  B. The more membership in privileged groups is sought by subordinates and the less mobility is allowed, the more likely they are to withdraw legitimacy.
II. The more deprivations of subordinates are transformed from absolute to relative, the greater will be their sense of injustice and, hence, the more likely they are to initiate conflict.
  A. The less the degree to which socialization experiences of subordinates generate internal ego constraints, the more likely they are to experience relative deprivation.
  B. The less the external constraints applied to subordinates, the more likely they are to experience relative deprivation.

---

rendorf tend to view the causes of conflict as residing in "contradictions" or "conflicts of interest." As subordinates become aware of their interests, they pursue conflict; hence the major theoretical task is to specify the conditions raising levels of awareness. But Coser is arguing that conflicts of interest are likely to be exposed only after the deprived withdraw legitimacy from the system. Coser emphasizes that the social order is maintained by some degree of consensus over existing arrangements and that "disorder" through conflict occurs when conditions decreasing this consensus or legitimacy over existing arrangements are present. Two such conditions are specified in Propositions I-A and I-B, both of which owe their inspiration more to Weber than to Marx. Proposition I-A argues that, when channels for expressing grievances do not exist, the withdrawal of legitimacy is likely, especially if there are few organizations to deflect and occupy people's energy (I-A-1) and if there are felt ego deprivations (I-A-2). Proposition I-B specifies that, if the deprived desire membership in higher ranks or if they have been led to believe that some mobility is possible, a withdrawal of legitimacy will be likely when little mobility is allowed.

Proposition II indicates, however, that the withdrawal of legitimacy, in itself, is not likely to result in conflict. People must first become emotionally aroused. The theoretical task then becomes one of specifying the conditions that translate the withdrawal of legitimacy into emotional arousal, as opposed to some other emotional state such as apathy and resignation. Here Coser draws inspiration from Marx's notion of relative deprivation. For, as Marx observed and as a number of empirical studies have documented, absolute deprivation does not always foster revolt.[12] When people's expectations for a better future

---

[12]James Davies, "Toward a Theory of Revolution," *American Journal of Sociology* 27 (1962),

TABLE 11-2 Coser's Propositions on the Violence of Conflict

I. The more groups engage in conflict over realistic issues (obtainable goals), the more likely they are to seek compromises over the means to realize their interests and, hence, the less violent the conflict.

II. The more groups engage in conflict over nonrealistic issues, the greater the level of emotional arousal and involvement in the conflict and, hence, the more violent the conflict.

   A. The more conflict occurs over core values, the more likely it is to be over nonrealistic issues.

   B. The more a realistic conflict endures, the more likely it is to become increasingly nonrealistic.

III. The less functionally interdependent the relations among social units in a system, the less the availability of institutional means for absorbing conflicts and tensions and, hence, the more violent the conflict.

   A. The greater the power differentials between super- and subordinates in a system, the less functionally interdependent are relations.

   B. The greater the level of isolation of subpopulations in a system, the less functionally interdependent are relations.

suddenly begin to exceed perceived avenues for realizing these expectations, only then do they become sufficiently aroused to pursue conflict. The level of arousal will, in turn, be influenced by their commitments to the existing system, by the degree to which they have developed strong internal constraints, and by the nature and amount of social control in a system. Such propositions, for example, lead to predictions that, in systems with absolute dictators who ruthlessly repress the masses, revolt by the masses is less likely than in systems where some freedoms have been granted and where the deprived have been led to believe that things will be getting better. Under these conditions the withdrawal of legitimacy can be accompanied by released passions and emotions.

The Violence of Conflict

In Table 11-2, I have extracted and listed Coser's most important propositions on the level of violence in a conflict.[13] I find Coser somewhat vague in his definition of conflict violence, but he appears to be denoting the degree to which conflict parties seek to injure or eliminate each other. As most functional theorists are likely to emphasize, Coser's Proposition I is directed at specifying the conditions under which conflict will be less violent. In contrast, dialectical theorists, such as Marx, often pursue just the opposite: specifying the conditions under which conflict will be more violent. Yet the inverse of Coser's first proposition can indicate a condition under which conflict will be violent. The key

pp. 5-19; Ted Robert Gurr, *Why Men Rebel* (Princeton, NJ: Princeton University Press, 1970), and "Sources of Rebellion in Western Societies: Some Quantitative Evidence," *The Annals* 38 (1973), pp. 495-501.

   [13]These propositions are taken from Coser's *Functions of Social Conflict*, pp. 45-50. Again, I have made them more formal than Coser's more discursive text.

concept in this proposition is "realistic issues." For Coser, realistic conflict involves the pursuit of specific aims against real sources of hostility, with some estimation of the costs to be incurred in such pursuit. As I noted in Chapter 9, Simmel recognized that, when clear goals are sought, compromise and conciliation are likely alternatives to violence. Coser restates this proposition but adds Proposition II on conflict over "nonrealistic issues," such as ultimate values, beliefs, ideology, and vaguely defined class interests. When nonrealistic, then, the conflict will be violent. Such nonrealism is particularly likely when conflict is over core values, which emotionally mobilize participants and make them unwilling to compromise (Proposition II-A). Moreover, if conflict endures for a long period of time, then it becomes increasingly nonrealistic as parties become emotionally involved, as ideologies become codified, and as "the enemy" is portrayed in increasingly negative terms (Proposition II-B).

Proposition III adds a more structural variable to the analysis of conflict violence. In systems in which there are high degrees of functional interdependence among actors—that is, where there are mutual exchanges and cooperation—conflict is less likely to be violent. However, if there is great inequality in power among units (Proposition III-A) or isolation of subpopulations (Proposition III-B), functional interdependence decreases; hence, when conflict occurs, it will tend to be nonrealistic and violent.

The Duration of Conflict

Despite the recognition that conflict is a process, unfolding over time, I find it surprising that few theorists have incorporated the variable of time in their work. I emphasize, however, that this oversight is true not just of conflict theory but also of all theory in sociology, although some recent theoretical efforts have made time an explicit property of social structure (see Chapters 12 and 26). Coser's incorporation of the time variable is extremely limited. He views time in terms of the duration of conflict and as a dependent variable, when it can also be an independent variable (see Proposition II-B in Table 11-2, for example). Just how the duration of conflict operates as an independent variable, influencing such variables as conflict intensity, violence, or functions, is never specified. Thus Coser's analysis is confined to the more limited, yet I think important, question: what variables influence the length of conflict relations? In Table 11-3, I have extracted some of these propositions on conflict duration.[14]

In Propositions I and II, Coser underscores the fact that conflicts with a broad range of goals or with vague ones will be prolonged. When goals are limited and articulated, it is possible to know when they have been attained. With perception of attainment, the conflict can be terminated. Conversely, with a wide variety or long list of goals, a sense of attainment is less likely to occur—thus prolonging the conflict. In Proposition III, Coser emphasizes that

---

[14]These propositions come from Coser, "The Termination of Conflict," in *Continuities*, pp. 37–52; and *Functions of Social Conflict*, pp. 20, 48–55, 59, 128–33.

TABLE 11–3 Coser's Propositions on the Duration of Conflict

I. The less limited the goals of the opposing parties to a conflict, the more prolonged the conflict.

II. The less the degree of consensus over the goals of conflict, the more prolonged the conflict.

III. The less the parties in a conflict can interpret their adversary's symbolic points of victory and defeat, the more prolonged the conflict.

IV. The more leaders of conflicting parties can perceive that complete attainment of goals is possible only at very high costs, the less prolonged the conflict.

    A. The more equal the power between conflicting groups, the more likely leaders are to perceive the high costs of complete attainment of goals.

    B. The more clear-cut the indexes of defeat or victory in a conflict, the more likely leaders are to perceive the high costs of complete attainment of goals.

V. The greater the capacity of leaders of each conflict party to persuade followers to terminate conflict, the less prolonged the conflict.

    A. The more centralized the conflict parties, the greater a leader's capacity to persuade followers.

    B. The fewer the internal cleavages within conflict parties, the greater a leader's capacity to persuade followers.

---

knowledge of what would symbolically constitute victory and defeat will influence the length of conflict. Without the ability to recognize defeat or victory, conflict is likely to be prolonged to a point where one party destroys the other. Propositions IV and V deal with the role of leadership in conflict processes. The more leaders can perceive that complete attainment of goals is not possible and the greater their ability to convince followers to terminate conflict, the less prolonged the conflict will be.

Thus I see Coser's overall image of conflict duration as follows. Where the goals of conflict parties are extensive, where there is dissent over goals, where conflict parties cannot interpret symbolic points of victory and defeat, where leaders cannot assess the costs of victory, and where leaders cannot effectively persuade followers, the conflict will be of longer duration than when the converse conditions hold true. In turn, as I think is evident, these variables are interrelated. For example, clear and limited goals help members to determine symbolic points of victory or defeat and leaders to assess the costs of victory. Although these interrelations are not fully or systematically developed, Coser has provided an important lead in the study of the time dimension in the conflict process.

The Functions of Social Conflict

As I emphasized in the chapters on functionalism, the concept of "function" presents a number of problems. If some process or structure has functions for some other feature of a system, there is often an implicit assumption about what is good and bad for a system. If this implicit evaluation is not operative, how does one assess when an item is functional or dysfunctional? Even seem-

TABLE 11–4 Coser's Propositions on the Functions of Conflict for the Respective Parties

---

I. The more violent or intense the conflict, the more clear-cut the boundaries of each respective conflict party.

II. The more violent or intense the conflict and the more internally differentiated the conflict parties, the more likely each conflict party is to centralize its decision-making structure.

III. The more violent or intense the conflict and the more it is perceived to affect the welfare of all segments of the conflict parties, the more conflict promotes structural and ideological solidarity among members of each conflict party.

IV. The more violent or intense the conflict, the more conflict leads to the suppression of dissent and deviance within each conflict party as well as to forced conformity to norms and values.

V. The more conflict between parties leads to forced conformity, the greater is the accumulation of hostilities and the more likely is internal group conflict to surface in the long run.

---

ingly neutral concepts, such as survival or adaptability, merely mask the implicit evaluation that is taking place. Sociologists are usually not in a position to determine what is survival and adaptation. To say that an item has more survival value or increases adaptation is frequently a way to mask an evaluation of what is "good."

In Coser's propositions on the functions of conflict, I think that this problem exists. Conflict is good when it promotes integration based on solidarity, clear authority, functional interdependence, and normative control. In Coser's terms, it is more *adaptive*. Other conflict theorists might argue that conflict in such a system is bad because integration and adaptability in this specific context could be exploitive. And so, with these qualifications, let me now turn to Coser's propositions.

Coser divides his analysis of the functions of conflict along lines similar to those by Simmel: the functions of conflict for (1) the respective parties to the conflict and (2) the systemic whole in which the conflict occurs. In Table 11-4, I have delineated the propositions on the functions of conflict for the parties.[15]

In the propositions listed in Table 11-4, the intensity of conflict—that is, people's involvement in and commitment to pursue the conflict—and its level of violence increase the demarcation of boundaries (Proposition I), centralization of authority (Proposition II), ideological solidarity (Proposition III), and suppression of dissent and deviance (Proposition IV) within each of the conflict parties. Conflict intensity is presumably functional because it increases integration, although Proposition V indicates that centralization of power as well as the suppression of deviance and dissent create malintegrative pressures in the long run. Thus there appears to be an inherent dialectic in conflict-

---

[15]These propositions are taken from Coser, *Functions of Social Conflict*, pp. 37–38, 45, 69–72, 92–95.

TABLE 11–5 Coser's Propositions on the Functions of Conflict for the Social Whole

---

I. The more differentiated and functionally interdependent the units in a system, the more likely is conflict to be frequent but of low degrees of intensity and violence.

II. The more frequent are conflicts, the less is their intensity, and the lower is their level of violence, then the more likely are conflicts in a system to (a) increase the level of innovation and creativity of system units, (b) release hostilities before they polarize system units, (c) promote normative regulation of conflict relations, (d) increase awareness of realistic issues, and (e) increase the number of associative coalitions among social units.

III. The more conflict promotes (a), (b), (c), (d), and (e) above, the greater will be the level of internal social integration of the system and the greater will be its capacity to adapt to its external environment.

---

group unification—one that creates pressures toward disunification. Unfortunately, Coser does not specify the conditions under which these malintegrative pressures are likely to surface. In focusing on positive functions—that is, forces promoting integration—the analysis ignores a promising area of inquiry. This bias becomes even more evident when Coser shifts attention to the functions of conflict for the systemic whole within which the conflict occurs. I have listed these propositions in Table 11–5.[16]

I have not presented Coser's propositions in their full complexity in Table 11–5, but the essentials of his analysis are clear.[17] In Proposition I, complex systems that have a large number of interdependencies and exchanges are more likely to have frequent conflicts that are less emotionally involving and violent than those systems that are less complex and in which tensions accumulate. It is in the nature of interdependence, Coser argues, for conflicts to erupt frequently; but, since they emerge periodically, emotions do not build to the point where violence is inevitable. Conversely, systems in which there are low degrees of functional interdependence will often polarize into hostile camps; when conflict does erupt, it will be intense and violent. In Proposition II, frequent conflicts of low intensity and violence are seen to have certain positive functions. First, they will force those in conflict to reassess and reorganize their actions (Proposition IIa). Second, such conflicts will release tensions and hostilities before they build to a point where adversaries become polarized around nonrealistic issues (Proposition IIb). Third, frequent conflicts of low intensity and violence encourage the development of normative procedures—laws, courts, mediating agencies, and the like—to regulate tensions (Proposition IIc). Fourth, these kinds of conflicts also increase a sense of realism over what the conflict is about. That is, frequent conflicts in which intensity and violence are kept

---

[16]Ibid., pp. 45–48; "Internal Violence as a Mechanism for Conflict Resolution"; "Social Conflict and the Theory of Social Change"; "Some Social Functions of Violence"; "The Functions of Dissent"; and "Social Conflict and the Theory of Social Change."

[17]See third edition of Jonathan H. Turner, *The Structure of Sociological Theory* (Homewood, IL: Dorsey Press, 1982).

under control allow conflict parties to articulate their interests and goals, thereby allowing them to bargain and compromise (Proposition IId). Fifth, conflicts promote coalitions among units that are threatened by the action of one party or another. If conflicts are frequent and of low intensity and violence, such coalitions come and go, thereby promoting flexible alliances (Proposition IIe). However, if conflicts are infrequent and emotions accumulate, coalitions often polarize threatened parties into ever more hostile camps, with the result that, when conflict does occur, it is violent. And Proposition III simply asserts Coser's functional conclusion that, when conflicts are frequent and when violence and intensity are reduced, conflict will promote flexible coordination within the system and increased capacity to adjust and adapt to environmental circumstances. This increase in flexibility and adaptation is possible because of the processes listed in Proposition II.

COSER'S FUNCTIONAL APPROACH: AN ASSESSMENT

I think that Coser's approach has done much to correct for the one-sidedness of Dahrendorf's analysis while reintroducing Simmel's ideas into conflict theory. Yet Coser's scheme represents an analytical one-sidedness that, if followed exclusively, would produce a skewed vision of the social world. Coser begins with statements about the inevitability of force, coercion, constraint, and conflict, but his analysis quickly turns to the integrative and adaptive consequences of such processes. This emphasis could rather easily transform the integrative and adaptive functions of conflict into functional needs and requisites that necessitate, or even cause, conflict to occur. Such teleology was inherent in Marx's work, where revolutionary conflict was viewed as necessary to meet the need for a communist society. But Coser's teleological inspiration appears to have come more from Simmel's organic model than from Marx's dialectical scheme. Once he documents how conflict contributes to the systemic whole, or body social, Coser inadvertently implies that the body social causes conflict in order to meet its integrative needs. Although conflict is acknowledged by Coser to cause change in social systems, it is still viewed primarily as a crucial process in promoting integration and adaptation.

I think that Coser, like so many conflict theorists, creates a problem when he tries to correct for past weaknesses in other approaches. For, in trying to compensate for the one-sidedness of dialectical theory and functionalism, Coser presents yet one more skewed approach. In my presentation of propositions in Tables 11-1 through 11-5, I have sought to mitigate against this overemphasis on the positive functions of conflict, but the actual scheme itself is heavily functional. I have not listed the many other functional statements here,[18] primarily because I have emphasized the strong points in Coser's scheme—that is, the propositions on the causes, violence, and duration of conflict (see Tables 11-1, 11-2, and 11-3). Tables 11-4 and 11-5 on the functions of conflict could,

---

[18]Ibid.

in fact, be longer than the others combined, but I decided to limit my summary to only the strong points. As a consequence, I have distorted Coser's ideas somewhat.

Thus the major substantive problem in Coser's scheme is its functionalism. To correct this problem, I see little need to redirect his propositions on the causes, violence, and duration of conflict. These propositions address, I feel, important questions neutrally and do not attempt to balance or correct for past theoretical one-sidedness with another kind of one-sidedness. Indeed, they display an awareness of key aspects of conflict in social systems, and with supplementation and reformulation, they offer an important theoretical lead. The substantive one-sidedness in the scheme comes with Coser's borrowing and then supplementing Simmel's functional propositions, and it is here that drastic changes in the scheme must come. One corrective strategy, which does not smack of another form of one-sidedness, is to ask the more neutral theoretical question: under what conditions can what kinds of outcomes of conflict for what types of systems and subsystems be expected? Although this is not a startling theoretical revelation, I think that it keeps assessments of conflict processes away from what ultimately must be evaluative questions of functions and dysfunctions. If the question of outcomes of conflicts is more rigorously pursued, the resulting propositions will present a more balanced and substantively accurate view of social reality.

Because of the long and unfortunate organic connotations of words such as *function*, I believe that it is wise to drop their use in sociology. They all too frequently create logical and substantive problems of interpretation. Coser appears well aware of these dangers, but he has invited misinterpretation by continually juxtaposing notions of "the body social" and the "functions" of various conflict processes and related phenomena, such as dissent and violence. Had he not done this, he would have better achieved his goal of correcting for the inadequacies of functional and conflict theorizing in sociology. Instead, he has created a heavily functional scheme.

In sum, then, I think that it makes little sense to have new perspectives that correct for the deficiencies of either dialectical or functional conflict theory. Sociological theory has far too long engaged in this kind of activity, and it is far more appropriate, I feel, to visualize conflict as one of many important processes in the social universe and to develop some abstract principles about this process that avoid problems inherent in all forms of functional analysis.

CHAPTER 12

Exchange Conflict Theory

Randall Collins

Most conflict theories initiate analyses at the macro level, focusing on an entire society and exploring how social classes and other collective actors become mobilized to pursue conflict. In contrast to this tendency, Randall Collins develops a conflict sociology that begins at the micro level of reality, examining what individuals in face-to-face interaction actually do. Moreover, he adopts ideas from scholars, such as Émile Durkheim, Erving Goffman, and Harold Garfinkel, who are not associated with the conflict tradition. Indeed, he has been criticized for calling his work "conflict sociology." Yet, when one looks closely, the basic interactive processes that undergird his theory are the exchange of resources and the emission of rituals—in particular, the exchange of resources among unequals and the effort to mitigate the inherent tension in this situation with rituals.

As I will argue in Part III, on exchange theory,[1] there is a basic conflict dynamic in this approach: actors with unequal resources will always engage in exchanges that generate potential conflict. In fact, much of conflict theory might be seen as a subset of ideas that can be derived from exchange theory. Thus there is considerable utility, I believe, in approaching conflict from an exchange perspective, although Collins appears to work rather hard at keeping this portion of his theory implicit. But, as I will emphasize, his analysis makes the most sense when the exchange assumptions are made explicit, and even highlighted.

Collins' conflict credentials, assuming that I need to present them, become even more evident when he begins to move from the micro analysis of exchange and ritual to more macro processes revolving around stratification, organizations, the state, war, and geopolitics. Here his inspiration is clearly Max Weber, who, as I noted in Chapter 9, articulated a conflict theory that converges with

---

[1]See Chapters 14, 15, 16, and 17.

that developed by Karl Marx. But Collins does more than develop an abstract conflict theory in the Weberian tradition; instead, he reveals the same eclectic interests as Weber in the analysis of conflict in many different substantive arenas—scientific organizations, interpersonal encounters, deference and demeanor rituals, classes, political parties, military operations, world empires, status groups, class cultures, and the like. And he does what Weber and other conflict theorists failed to do: develop a theory about the underlying interpersonal processes that create, sustain, and change the macrostructures where conflict occurs. Let me begin with Collins' critique of the way sociologists view "social structure"; then we can move to his conceptualization of micro processes as "interaction ritual chains" (I will argue that he should be more honest and say "exchange and ritual chains"); finally, we can review some of the more macro contexts in which he has applied his micro-level theory.

THE CRITIQUE OF MACROSTRUCTURAL ANALYSIS

Collins' approach is *microstructural* because of his assertion that most sociological conceptions of structure are reifications and metaphors. He does not deny the existence of social structures and macro reality; rather, he argues that macro theories often disconnect structures from the very processes by which they are constituted. For Collins asks: what is structure, and what does it really look like? His answer in early works is somewhat metaphorical but nonetheless suggestive:[2]

> Imagine the view of human society from the vantage point of an airplane. What we can observe are buildings, roads, vehicles, and—if our senses were keen enough—people moving back and forth and talking to each other. Quite literally, this is all there is; all of our explanations and all of our subjects to be explained must be grounded in such observations. "Social structure" could be brought into such a picture if we understand that men live by anticipating future encounters and remembering past ones. Structure is recurring sorts of encounters. An imaginary aerial time-lapse photograph, then, would render social structure as a set of light streaks showing the heaviness of social traffic. If we go on to imagine different colored streaks corresponding to the emotional quality of contacts—perhaps gray for purely formal relations, brown for organizational relations infused with more personal commitments, yellow for social relations, and red for close personal friendships—we would have an even more significant map.

More recent work has gone on to conceptualize social structure as "chains of interaction rituals" that are stretched out in time through their repetition by individuals in concrete settings.[3] That is, if encounters among individuals

---

[2]Randall Collins, *Conflict Sociology: Toward an Explanatory Science* (New York: Academic Press, 1975), p. 56.

[3]Randall Collins, "On the Micro-Foundations of Macro-Sociology," *American Journal of Sociology* 86 (March 1981), pp. 984–1014; "Micro-translation as a Theory Building Strategy," in *Advances in Social Theory and Methodology: Toward an Integration of Micro- and Macro-So-*

FIGURE 12–1 Micro and Macro: Time, Space, and Number

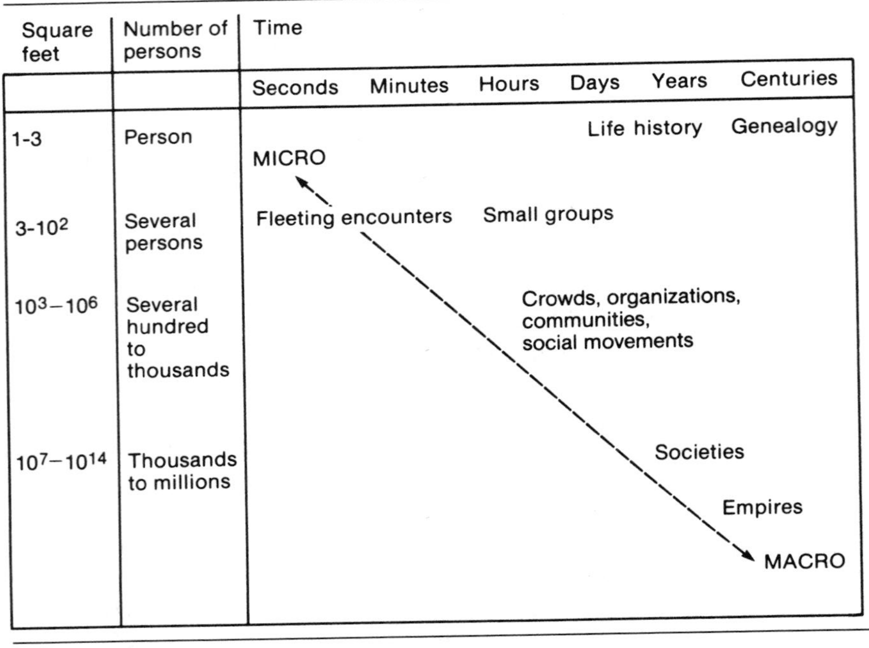

Square feet	Number of persons	Time					
		Seconds	Minutes	Hours	Days	Years	Centuries
1-3	Person					Life history	Genealogy
		MICRO					
3-10²	Several persons	Fleeting encounters		Small groups			
10³–10⁶	Several hundred to thousands				Crowds, organizations, communities, social movements		
10⁷–10¹⁴	Thousands to millions					Societies	
							Empires
							MACRO

are repeated again and again over time, then there is a structure to their interaction. Social structure becomes ever more macro when there are an increasing number of persons involved in social encounters and an extension of the physical space in which these encounters occur.[4] Thus, for Collins macrostructure consists of only three dimensions: (1) the sheer numbers of persons and encounters involved, (2) the amount of time consumed by an encounter and the degree to which it is connected to previous encounters, and (3) the amount and pattern of physical space used to emit the interaction rituals that typify an encounter. The more encounters are repeated across time, the more people who are involved, and the more space that is consumed, then the more macro a social structure will be.

This is, I think, a very extreme line of argument, for macrostructure is defined simply by space, number, and time. In Figure 12–1, I have simplified a figure used by Collins to illustrate this point.[5] Across the top of the figure is listed a crude time scale, ranging from seconds to centuries. And down the left

ciology, ed. K. Knorr-Cetina and A. V. Cicourel (London: Routledge & Kegan Paul, 1981); "Interaction Ritual Chains, Power and Property: The Micro-Macro Connection as an Empirically-Based Theoretical Problem," in The Micro-Macro Problem, ed. J. C. Alexander et al. (Berkeley: University of California Press, 1987); and Theoretical Sociology (San Diego: Academic Press, 1988), pp. 187–228.

[4]Collins, "Interaction Ritual Chains."

[5]Collins, "On the Micro-Foundations of Macro-Sociology."

side I have indicated the number of people involved and the space utilized in encounters. Within the body of the figure, illustrative structures are mentioned. As one moves from the upper-left to lower-right portions of the figure, one also goes from micro- to macroanalysis.

With respect to the time dimension, one critical issue is the length of time consumed by an encounter in which interaction occurs; another more important issue is the degree to which each encounter is a link in a chain of encounters. Thus a one-time interaction that lasts only a few seconds or minutes is sociologically much less interesting than encounters involving various interactive rituals that are repeated again and again. In Collins' eyes, then, sociological theory should be most interested in "chains of interaction rituals" that extend across time. Such chains represent a life history for an individual when enacted over years; if enacted over generations, they are part of a genealogy. More interesting sociologically, however, are chains of interaction among several people in an encounter, especially if interaction rituals are repeated over relatively long time periods. And if there are many individuals linked together, directly and indirectly, through chains of interaction rituals and using ever larger amounts of space, then social structures become increasingly macro. Micro and macro are, therefore, on a continuum that, at the extreme micro end, involves few people using little space in short-term interactions and, at the macro end, involves very complex and long-term chains of direct and indirect interaction among large numbers of people extended across physical space.

As is evident, Collins does not reject macro reality as a phenomenon worth studying. But there is an important implication in his position: the macro world is constructed from chains of interaction. There can be variation with respect to the number of people involved, the length of time consumed, and the amount of space utilized, but the one invariant property of structure is interaction among people. Social structure is ultimately chains of interaction rituals, and so sociological theory must emphasize what individuals actually do, for "structures never do anything; it is only persons in real situations who act."[6] Thus a theory of social structure must seek the "energizing processes" of social structure in the interactions of individuals. Theory must explain macrostructures in terms of principles that help us understand how micro interactions are stretched over time, expanded in number, and extended across space.

Unlike other theorists who have made similar assertions, Collins does not advocate a microreduction of macro processes. Explanation of macrostructural processes is not simply a matter of asserting that the only reality is the individual in interaction. We cannot, Collins would argue, merely outline how people interact and leave the matter settled. Instead, theory must provide principles that inform us about how and through what processes interactions are stretched across time in chains of encounters and aggregated in space among increasing numbers of individuals. Thus a theory of social structure must ad-

---

[6]Collins, "Interaction Ritual Chains."

dress the macro variables of time, space, and number if it is to say anything about those structures of most interest to sociologists—groups, organizations, crowds, social classes, societies, empires, and the like.

I think that this attention to macrostructural processes makes Collins' approach unique among those who stress the interactive foundations of structure. Moreover, he emphasizes that many interactive encounters occur within an existing macrostructure (built, of course, from past chains of interaction) that determines the number of people present, their respective resources, how long they will interact, and where they will position themselves in space. As they interact, they reproduce the macrostructure or potentially change it. And so there is almost always a reciprocity between the purely micro, interactive dimensions of social reality and the macro context in which these micro processes occur.

Yet the "energizer" behind social reality is the interaction ritual. Macrostructures are created, sustained, and changed at the level of what people in interaction do in their encounters. And thus the central task of sociological theory is to develop some principles about interaction rituals. Surprisingly, Collins' early work left many of these interaction processes somewhat implicit, moving rather quickly away from microanalysis of interaction to macrostructural concerns about organizations, the state, geopolitics, and the like. It is only recently that he has sought to explicate in more detail a theory of "interaction ritual chains."[7] Let me review the tentative and provisional outlines of this latest theory and then return to Collins' earlier work devoted explicitly to conflict.

INTERACTION RITUAL CHAINS

The Nature of Rituals

For Collins the basic micro unit of analysis is the *encounter* of at least two people who confront each other and interact. What transpires in such encounters is mediated by the exchange of resources and rituals.[8] For the present let me focus on rituals, and then we can turn to the underlying exchange dynamic. In Collins' broad portrayal, a ritual contains the following elements:[9]

1. a physical assembly of co-present individuals;
2. a common focus of attention, and mutual awareness of each other's attention, among co-present individuals;
3. a common emotional mood among co-present individuals;
4. a symbolic representation of this common focus and mood with objects, persons, gestures, words, and ideas among interacting individuals.

[7]Ibid.

[8]Jonathan H. Turner and Randall Collins, "Toward a Microtheory of Structuring," in *Theory Building in Sociology*, ed. J. H. Turner (Newbury Park, CA: Sage, 1989).

[9]Collins, *Theoretical Sociology*, p. 193. In this portrayal the ideas of Émile Durkheim are central.

FIGURE 12–2 Collins' Conceptualization of Rituals

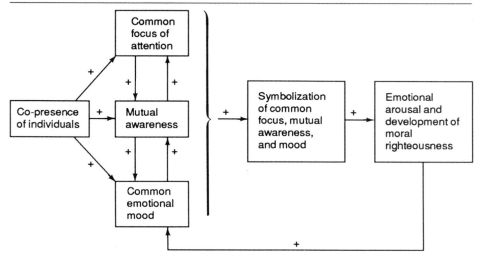

+ = Positive effects on

For Collins the ceremonial quality of a ritual, such as a certain stereotyped sequence of gestures, is not the entire ritual, as some would contend. Rather, gestural sequences are only one mechanism for focusing attention; it is also necessary for individuals to develop a common mood and to symbolize this mood in some way.

When this occurs, people will receive enhanced energy and feel negatively toward those who do not honor symbols.[10] Thus rituals produce emotional energy and a sense of mortality, and it is these outcomes of rituals that give them the power to "energize" encounters and to string them together across time. The key elements of this view of rituals are diagramed in Figure 12-2.[11]

Rituals are what mobilize individuals in encounters, and, ultimately, they are what enable long chains of interactions that produce and reproduce macrostructures to exist. Some rituals are macro in focus and invoke as objects of attention larger groupings and the entire society, whereas others are local and pertain to small groupings of people.[12] Rituals also vary in terms of whether they are intentional and institutionalized or natural and unintended. In institutionalized rituals (a graduation, funeral, political rally, wedding, etc.), participants are usually aware of the effort to focus attention and symbolize a common mood, whereas in more natural rituals (greetings and talk among

---

[10]Thus Collins includes the consequences of a ritual *as part of* the ritual itself, a most inclusive and, I think, illusive conceptualization.

[11]I have redrawn somewhat a figure presented by Collins in *Theoretical Sociology*, p. 194.

[12]Ibid., pp. 197–201.

friends, for example), participants are likely to be less conscious of what transpires.

Whether micro or macro in focus and intended or unintended, rituals all reveal the same basic elements, as listed above, but their form and consequences for structuring the micro encounter, as well as for larger macrostructures that are sustained by linked encounters, will vary depending on certain conditions:[13] (1) the degree of *inequality* in resources among interactants, (2) the level of *social density* of interactants (how many people are co-present), and (3) the degree of *social diversity* of interactants (the number and extent of alternative encounters and networks available to interactants). Inequality denotes the "power" or "vertical" dimension of interaction in which some interactants have more material, symbolic, and coercive resources than others and, as a consequence, can give orders and demand deference. Social density and diversity are the "network" dimensions of interaction in which the location of individuals both in the immediate situation of co-presence and in larger networks of relations outside the immediate situation influences their ritual behavior. These vertical and network dimensions become essential, as I will outline shortly, to understanding the nature of the rituals and their effects on social structure.

The Underlying Exchange Dynamic[14]

For Collins the encounter is a "shared conversational reality" revolving around not only rituals but also the negotiation and exchange of resources. Collins visualizes two basic types of resources as crucial to understanding exchanges and rituals:[15] (1) cultural capital and (2) emotional energy. *Cultural capital* consists of such resources as stored memories of previous conversations, vocal styles, special types of knowledge or expertise, the prerogatives to make decisions, and the right to receive honor. The concept of *generalized cultural capital* denotes those impersonal symbols that mark general classes of resources (for example, knowledge, positions, authority, and groupings), whereas *particularized cultural capital* refers to the memories that individuals have of the particular identities, reputations, and network/organizational positions of specific persons. *Emotional energy* is composed of the level and type of affect, feeling, and sentiment that individuals can, or will, mobilize in a situation.

Interaction consists of individuals using their cultural capital and emotional energy to talk with one another. Such conversations involve an investment of capital and energy, with each individual attracted to situations that bring the best available payoff in cultural capital and emotional energy. In particular, people augment their cultural capital and emotional energy whenever they carry on an interaction that results in a sense of group membership. That is, people spend cultural capital and emotional energy to receive in exchange with others

---

[13]Ibid., p. 224, although this is a restatement of his earlier position in *Conflict Sociology*.
[14]Jonathan H. Turner and Randall Collins, "Toward a Microtheory of Structuring."
[15]Ibid.

a feeling of group involvement and inclusion. Indeed, they will spend additional capital and energy if they can increase their position and rights to control the flow of group activity (thereby increasing their cultural capital). Moreover, people try to achieve membership in a group that is powerful and high ranking in the larger society. These processes of negotiating group inclusion may take place by conscious calculation, but more usually they happen unconsciously by emotional attraction to persons who emit certain kinds of symbols. Thus, to the extent that individuals can increase their cultural capital in interaction, they will expend emotional energy; reciprocally, to the degree that individuals can produce positive feelings and emotions in their conversations, they will invest cultural capital.

The critical variables determining the flow of interaction in the encounter are (1) levels of inequality in the respective cultural and emotional resources of individuals, (2) degrees of social density among individuals, and (3) degree of social diversity, or the number of alternative persons with whom they might otherwise interact (i.e., their network position). In situations involving power or property, these network relationships also include one's "enforcement coalition," or allies who will back up one's claims.

In general, when the participants' resources are similar, especially with respect to their positions as order givers or takers and cosmopolitans or locals, their "conversational rituals" will be personal, flexible, and long-term; these rituals will lead to strong, positive emotions and a willingness to renew the encounter in the future.[16] Conversely, when there is inequality in resources, it becomes less likely that strong personal ties will be created or sustained. In fact, under these conditions those with fewer resources may withhold cultural capital and emotional energy because they fear rejection or domination by the other person. On the other side of such inequalities, persons with greater resources in emotional energy and cultural capital are likely to feel little attraction to continuing a shared reality-constructing situation with those who bring them little symbolic status, or no advantage in network alliances. As a result, conversations will be highly ritualized, formal, impersonal, and short-term. Social density variables intersect with these processes revolving around inequality. The more people who are co-present and can observe one another, the more likely they are to interact, even under conditions of inequality. If people can leave the interaction and have other options (through increased social diversity associated with cosmopolitan networks), however, the inequalities are less likely to control the flow of the interaction, since people will simply take their emotional energy and cultural capital elsewhere.

As people assess the inequalities in their resources, the level of density, and the degree of diversity, they also determine the substantive nature or content of a situation as *work-practical* (designed to get something done), *ceremonial* (intended to mark and signify some occasion with rituals), or *social*

---

[16]Collins, *Conflict Sociology*, pp. 73–79.

(created for mutual enjoyment and pleasure).[17] As with inequality, density, and diversity, the content of the encounter is often circumscribed by larger macrostructures (which, of course, were created by previous chains of interaction). For example, a funeral, a work station, and an office party are all "prepackaged" by existing structures in terms of what is expected and are not, therefore, spontaneous occurrences. Situations can, of course, have mixtures of work-practical, ceremonial, and social content, but generally one of these prevails and dominates the other two.

In sum, then, the mechanics of interaction dynamics involve assessing situations to determine their general nature as "work-practical," "ceremonial," or "social" and then assessing densities, inequalities, and relative network positions in order to determine how much energy and capital should be invested in the conversational exchange and rituals. If densities are high, then some capital and energy must be spent to achieve a minimal sense of ritual involvement, but just how much is to be invested will depend upon an individual's assessment of whether or not a profitable return on this investment is likely—that is, whether or not he or she will receive positive emotions and increased cultural capital from conversational exchanges and rituals. High-density situations, though, are socially coercive on the individual; they tend to generate a flow of emotional energy, a shared mood among the participants. Hence they charge symbols with significance for local membership. If people live in, or just episodically encounter, local communities or networks that are omnipresent and cannot be evaded (usually as the result of macrostructural constraints), they will, like it or not, become imbued with the local culture, even if they also possess access to cosmopolitan networks.

In Figure 12-3, I have diagrammed the key elements of Collins' conceptualization of the encounter. The arrows specify the key causal connections for each actor in the encounter. Of particular importance are the feedback loops because they determine the extent to which interaction produces sufficient emotional energy and cultural capital for individuals to continue the relationship, stringing it together over time in an interaction exchange and ritual chain. Three interrelated forces guide the flow of the encounter: (1) expectations for group membership, (2) level of cultural capital, and (3) level of emotional energy. Collins appears to posit expectations for group membership as a primal force behind people's emotional energy, at least in this sense:[18] energy levels are greatly influenced by whether or not people feel involved in, and part of, ongoing group activity. They expect this to happen—indeed, they want it to occur—and their level of energy is a result of these expectations *and* the actual emission of rituals (note feedback loops *p* and *q* at the bottom of Figure 12-3). Additionally, emotional energy is related to how much cultural capital one has in a situation (causal arrow *c*): the more one has, the greater will be the level of emotional energy mobilized in the encounter, especially if one is able to gain

---

[17]Ibid., pp. 114–31. I have collapsed somewhat a longer list. See also: Jonathan H. Turner, *A Theory of Social Interaction* (Stanford, CA: Stanford University Press, 1988).

[18]Collins does not, however, ever make this "need" or "force" explicit.

FIGURE 12-3 Collins' Exchange Model of Interaction Rituals

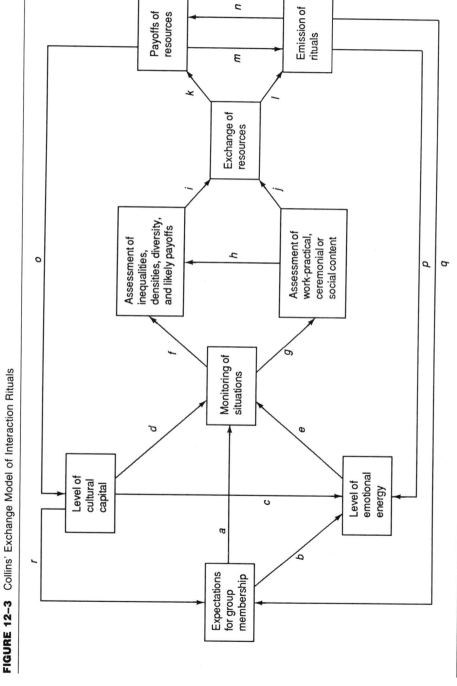

capital in the situation (note feedback loop o across the top of Figure 12–3). Such augmented cultural capital also raises expectations for enhanced standing in the group and, in this way, indirectly raises emotional energy (via feedback path o-r-b). This cluster of forces leads individuals to monitor situations to determine if they should invest their resources in exchanges and rituals. It should be emphasized, of course, that such monitoring is greatly circumscribed by macrostructural conditions (hence one might draw arrows from outside the model in order to denote that there are external constraints on encounters). Moreover, much monitoring is implicit and involves a "sense" or "feeling" about what transpires. Once initiated, monitoring tends to revolve around:

1. assessments of one's own resources and those possessed by other actors in order to determine the level of inequality, assessment of the level of social density (number of others co-present) in the situation, assessment of the level of social diversity (one's own, as well as others', options outside the encounter), and, as a result of these assessments, calculation of the likely payoff to be derived from exchanges of resources and the emission of rituals.
2. assessment of the degree to which a situation is to be work-practical, ceremonial, or social.

On the basis of these two classes of assessment, individuals exchange resources, or withdraw from the encounter if they perceive and feel that they will lose emotional energy and cultural capital. Of course, oftentimes they must exchange resources because of macrostructural constraints, even when they are in a situation that does not present very good possibilities (for example, when density is high, when diversity is low and one has no other place to go, and when inequality is high and others have power, one must go through with the exchange).

As actors exchange resources, they receive payoffs and emit rituals. When payoffs are high, rituals will be more enthusiastically emitted (causal arrow m); reciprocally, when rituals are focusing attention, symbolizing group involvement, and arousing emotions, then payoffs are likely to be positive (causal arrow n). There is, then, a reciprocity between payoffs and rituals. Rituals provide payoff, per se, especially those critical resources revolving around group membership; the degree of payoff determines the nature of the rituals emitted, as was noted above and as is delineated in Table 12–2 on deference and demeanor (see discussion of class cultures in later section).

The feedback loops in the model are crucial because they determine, first of all, whether an interaction is likely to be repeated and, second, the form and pattern that rituals will take when repeated (whether voluntarily or involuntarily by virtue of macrostructural constraints). Thus, if conversational exchanges produce positive feelings and augment capital (feedback loops p and o), actors will use their capital and energy during the course of the interaction rituals and will be likely to repeat the exchange over time. As a consequence, individuals will create a "chain" of interaction rituals. Ultimately, one's emotional feelings and sense of augmenting cultural capital are tied to the feedback arrows (q and r) flowing into expectations for group membership. Augmenting cultural capital often allows individuals to feel that they have a favorable position in a group context, and escalating positive emotions leads individuals to emit those rituals that produce group solidarity and enhance one's expectations for group involvement.

MICRO AND MACRO LINKAGES

The repetition of exchanges and rituals is what produces and reproduces social structure; as these ritualized chains of exchange embrace an increasingly larger number of people and are extended in space and time (see Figure 12-1), macrostructures are created and sustained. There is an important methodological corollary to this line of theoretical argument.[19] If macrostructures are produced and sustained by chains of exchange and ritual among individuals using cultural capital and expending emotional energy, then one should study structure by sampling encounters across time. By sampling encounters over time, it becomes possible to have a sense for what people are doing—how they talk, gesture, and position themselves as well as how they feel about what is occurring. Such a strategy for studying social structure involves what Collins calls *micro-translation* of macrostructures. That is, if macrostructures are nothing more than complex chains of exchange and ritual stretched across time and space among large numbers of people, then, to understand such structures, we need to "translate" them into those interactive processes from which they are constituted—exchange and ritual chains. Research projects should not, therefore, sample at one point in time with questionnaires; rather, they should select encounters over time during which people can be observed talking to one another, exchanging resources, and emitting rituals (that is, engaging in those processes modeled in Figure 12-3).

Having said this, however, there is always a problem of how to reconcile micro-level processes of exchange and ritual among individual people and the large-scale, macro-level processes of whole societies and empires (see bottom right of Figure 12-1). At some point, assertions about the micro bases of these large macrostructures become vague and metaphorical (see, for example, quote on page 229), and caveats about micro-translations of macrostructures are not likely to explain very much about these macrostructures. This micro versus macro question is, of course, not unique to Collins' approach; it plagues all of sociological theory. Collins attempts to resolve the "gap" between micro and macro by positing, at least implicitly, a *meso* level of reality revolving around organizational structures (complex organizations and bureaucracies) and stratification systems (classes, status/prestige groupings, and power groupings).[20] Interaction rituals and exchanges occur in encounters, typically lodged in groups; and groups are, in turn, generally embedded in organizations of various

---

[19]Collins, "Micro-translation as a Theory Building Strategy" as well as "Micro-methods as a Basis for Macro-sociology," *Urban Life* 12 (1983), pp. 184–202.

[20]Collins, *Theoretical Sociology*, pp. 373–489.

kinds and aspects of the stratification system, such as social classes. In turn, a society is an assemblage of many varied organizations, power networks, classes, and status groupings. And really large-scale macrostructures, such as political empires, are built from the merger or conquest of different societies (which can be successively decomposed down to organizations, classes, and status groupings and then down to encounters involving rituals and exchanges among individual people).

It is imagery such as this, borrowed from Max Weber, that has guided Collins' efforts at theorizing. Although he has theorized at the group and encounter levels on the micro side and at the societal and empire levels on the macro side, most of his theorizing is about stratification and organizational processes. That is, he has concentrated the application of his theory on *meso* structures. In his first and still best theoretical work, *Conflict Sociology*,[21] this line of emphasis is clearly evident. He begins with an earlier version of his analysis of interaction rituals[22] but quickly jumps to the meso level, examining successively deference and demeanor, class cultures, sexual stratification, age stratification, organization control systems (especially those in scientific organizations), and finally the state (as a coercive organization). Elsewhere he has analyzed intersocietal systems, such as empires.[23] Thus the bulk of theorizing is at the meso level, with periodic efforts at clarification of micro-level processes of exchange and ritual, on the one side, and mega-macro analyses of historical and contemporary geopolitical processes, on the other side. Let me now turn to a sampling of these theoretical efforts in order to give a flair for how Collins uses his micro theory of exchange and ritual to construct a conflict theory.

CONFLICT SOCIOLOGY

Collins' general argument about the exchange and ritual foundations of social structure and the effects of the macro dimensions of space, time, and size on interaction were anticipated in earlier works. In *Conflict Sociology*, Collins proposed the following steps for building social theory.[24] First, examine typical real-life situations in which people encounter one another. Second, focus on the material arrangements that affect interaction—the physical layout of situations, the means and modes of communication, the available tools, weapons, and goods. Third, assess the relative resources that people bring to, use in, or extract from encounters. Fourth, entertain the general hypotheses that those

---

[21]Collins, *Conflict Sociology*. Most of Collins' subsequent theoretical work has focused on fine-tuning and amplifying the insights in this ground-breaking work.

[22]I will not repeat these, because they have been considerably updated, as is summarized in Figures 12–2 and 12–3.

[23]For example, see Randall Collins, *Weberian Sociological Theory* (Cambridge, England: Cambridge University Press, 1986); "Long-term Social Change and the Territorial Power of States," in his *Sociology Since Midcentury: Essays in Theory Cumulation* (New York: Academic Press, 1981).

[24]Collins, *Conflict Sociology*, pp. 60–61.

with resources press their advantage, that those without resources seek the best deal they can get under the circumstances, and that stability and change are to be explained in terms of the lineups and shifts in the distribution of resources. Fifth, assume that cultural symbols—ideas, beliefs, norms, values, and the like—are used to represent the interests of those parties who have the resources to make their views prevail. Sixth, look for the general and generic features of particular cases so that more abstract propositions can be extracted from the empirical particulars of a situation.

Thus, as with his more recent work on interaction rituals, there is a concern with the encounter, with the distribution of individuals in physical space, with their respective capital or resources to use in exchanges, and with inequalities in resources. Also as with his recent work, the respective resources of individuals are critical: "power" is the capacity to coerce or to have others do so on one's behalf; "material resources" are wealth and the control of money as well as property or the capacity to control the physical setting and people's place in it; and "symbolic resources" are the respective levels of linguistic and conversational resources as well as the capacity to use cultural ideas, such as ideologies, values, and beliefs, for one's purposes.

And, as emphasized above, a central consideration in all of Collins' propositions is "social density," or the number of people co-present in a situation where an encounter takes place. Social density is, of course, part of the macrostructure since it is typically the result of past chains of interaction. But it can also be a "material resource" that some individuals can use to their advantage. Thus the interaction in an encounter will be most affected by the participants' relative resources and the density or number of individuals co-present. These variables influence the two underlying micro dynamics in Collins' scheme, talk and ritual.

Talk and Ritual

For Collins, talk is the emission of verbal and nonverbal gestures that carry meaning and that are used to communicate with others and to sustain (or create) a common sense of reality.[25] Since talk is one of the key symbolic resources of individuals in encounters, much of what transpires among interacting individuals is talk and the use of this cultural capital to develop their respective lines of conduct. In Table 12–1, I have listed some of Collins' propositions on conversations, which, I should add, have been altered somewhat in order to make them more parsimonious and to reconcile them with his more recent terminology. As can be seen from Proposition I in Table 12–1, the likelihood that people will talk is related to their sheer co-presence: if others are near, one is likely to strike up a conversation. More important sociologically are conversations that are part of a "chain" of previous encounters. If people felt good about a past conversation, they will usually make efforts to have

---

[25]Ibid., pp. 114–31.

TABLE 12–1   Key Propositions on the Conditions Producing Talk and Conversation

---

I.   The likelihood of talk and conversational exchanges among individuals is a positive and additive function of (*a*) the degree of their physical co-presence; (*b*) the emotional gratifications retained from their previous conversational exchanges; (*c*) the perceived attractiveness of their respective resources; and (*d*) their level of previous ritual activity.

II. The greater the degree of equality and similarity in the resources of individuals, the more likely conversational exchanges are to be (*a*) personal, (*b*) flexible, and (*c*) long-term.

III. The greater the level of inequality in the resources of individuals, the more likely conversational exchanges are to be (*a*) impersonal, (*b*) highly ritualized, and (*c*) short-term.

IV. The greater the amount of talk among individuals, especially among equals, the more likely are (*a*) strong, positive emotions; (*b*) sentiments of liking; (*c*) common agreements, moods, outlooks, and beliefs; and (*d*) strong social attachments sustained by rituals.

---

another; if they perceive each other's resources, especially symbolic or cultural but also material ones, as desirable, then they will seek to talk again. And if they have developed ritualized interaction that affirms their common group membership, they will be likely to enact those rituals again. As Proposition II indicates, conversations among equals who share common levels of resources will be more personal, flexible, and long-term because people feel comfortable with such conversations. As a result, the encounter raises their levels of emotional energy and increases their cultural capital. That is, they are eager to talk again and to pick up where they left off. However, the nature of talk in an encounter changes dramatically when there is inequality in the resources of the participants. As Proposition III emphasizes, subordinates will try to avoid wasting or losing emotional energy and spending their cultural capital by keeping the interaction brief, formal, and highly ritualized with trite and inexpensive words. Yet, as Proposition IV argues, even under conditions of inequality and even more when equality exists, people who interact and talk in repeated encounters will tend, over time, to develop positive sentiments and will have positive emotional feelings. Moreover, they will also converge in their definitions of situations and develop common moods, outlooks, beliefs, and ideas. And, finally, they will be likely to develop strong attachments and a sense of group solidarity, which is sustained through rituals.

Thus the essence of interaction is talk and ritual as mediated by an exchange dynamic; as chains of encounters are linked together over time, conversations take on a more personal and also a ritualized character that results from and at the same time reinforces the growing sense of group solidarity among individuals. Such is the case because the individuals have "invested" their cultural capital (conversational resources) and have derived positive feelings from being defined as group members. I do not think that Collins has developed adequately the underlying exchange theory behind these ideas, but we can see his intent: to view social structure as the linking together of encounters through talk and ritual. This basic view of the micro reality of social life pervades all of Collins' sociology, especially as he begins to move to the analysis of inequalities in social life.

Deference and Demeanor

Inequality and stratification are structures only in the sense of being temporal chains of interaction rituals and exchanges among varying numbers of people with different levels of resources. Thus, to understand these structures, it is necessary to examine what people actually do across time and in space. One thing that they do in interaction is exhibit deference and demeanor. Collins and coauthor Joan Annett define *deference* as the process of manipulating gestures to show respect to others; or, if one is in a position to command respect, the process of manipulation of gestures is to elicit respect from others.[26] The actual manipulation of gestures is termed *demeanor*. Deference and demeanor are, therefore, intimately connected to each other. They are also tied to talk and rituals, since talk involves the use of gestures and since deference and demeanor tend to become ritualized. Hence deference and demeanor can be visualized as one form of talk and ritual activity—a form that is most evident in those interactions that create and sustain inequalities among people.

As would be expected, Collins visualizes several variables as central to understanding deference and demeanor:

1. Inequality in resources, particularly wealth and power.
2. Social density variables revolving around the degree to which behaviors are under the "surveillance of others" in a situation.
3. Social diversity variables revolving around the degree to which communications networks are "cosmopolitan" (i.e., unrestricted to others who are co-present in a situation).

In Table 12-2 these variables are incorporated into a few abstract propositions that capture the essence of Collins' and Annett's numerous propositions and descriptions of the history of deference and demeanor.[27] In these propositions, Collins and Annett argue that rituals and talk revealing deference and demeanor are most pronounced between people of unequal status, especially when their actions are observable and when communication outside the situation is restricted. Such density and surveillance are, of course, properties of the macrostructure as it distributes varying numbers of people in space. As surveillance decreases, however, unequals avoid contact or perform deference and demeanor rituals in a perfunctory manner. For example, military protocol will be much more pronounced between an officer and enlisted personnel in public on a military base than in situations where surveillance is lacking (e.g., off the base). Moreover, Collins and Annett stress that inequalities and low

---

[26]Randall Collins and Joan Annett, "A Short History of Deference and Demeanor," in *Conflict Sociology*, pp. 161-224.

[27]Ibid., pp. 216-19.

TABLE 12–2 Key Propositions on Deference and Demeanor

I. The visibility, explicitness, and predictability of deference and demeanor rituals and talk among individuals are a positive and additive function of:

  A. The degree of inequality in resources among individuals, especially with respect to:

    1. Material wealth.

    2. Power.

  B. The degree of surveillance by others of behaviors emitted by individuals, with surveillance being a positive function of:

    1. The extent to which others are co-present.

    2. The degree of homogeneity in outlook of others.

  C. The restrictiveness of communication networks (low cosmopolitanism), with restrictiveness being a negative function of:

    1. The degree of complexity in communications technologies.

    2. The degree of mobility of individuals.

II. The greater the degree of inequality among individuals and the lower the level of surveillance, the more likely behaviors are to be directed toward:

  A. Avoidance of contact and emission of deference and demeanor by individuals.

  B. Perfunctory performance of deference and demeanor by individuals when avoidance is not possible.

III. The greater the degree of inequality among individuals and the lower the level of cosmopolitanism among individuals, the more likely behaviors are to be directed toward simplified but highly visible deference and demeanor.

IV. The greater the degree of inequality among individuals, and the less the degree of mobility among groups with varying levels of resources, the more visible, explicit, and predictable are deference and demeanor rituals and talk within these groups.

V. The greater the equality among individuals, and the greater the degree of cosmopolitanism and/or the less the level of surveillance, the less compelling are deference and demeanor talk and rituals.

---

mobility between unequal groups create pressures for intragroup deference and demeanor rituals, especially when communications outside the group are low (for example, between new army recruits and their officers or between prison inmates and guards). But as communication outside the group increases and/ or as surveillance by group members decreases, then deference and demeanor will decrease.

Class Cultures

These exchange processes revolving around talk, ritual, deference, and demeanor explain what are often seen as more macro processes in societies. One such process is variation in the class cultures. That is, people in different social classes tend to exhibit diverging behaviors, outlooks, and interpersonal styles. These differences are accountable in terms of two main variables:

1. The degree to which one possesses and uses the capacity to coerce, to materially bestow, and to symbolically manipulate others so that one can give orders in an encounter and have these orders followed.

TABLE 12–3 Key Propositions on Class Cultures

I. Giving orders to others in a situation is a positive and additive function of the capacity to mobilize and use coercive, material, and symbolic resources.

II. The behavioral attributes of self-assuredness, the initiation of talk, positive self-feelings, and identification with the goals of a situation are a positive function of the capacity to give orders to others in that situation.

III. The behavioral attributes of toughness, courage, and action in a situation are a positive function of the degree of physical exertion and danger in that situation.

IV. The degree of behavioral conformity exhibited in a situation is a positive function of the degree to which people can communicate only with others who are physically co-present in that situation and is a negative function of the degree to which people can communicate with a diversity of others who are not physically co-present.

V. The outlook and behavioral tendencies of an individual are an additive function of those spheres of life—work, politics, home, recreation, community—where varying degrees of giving/receiving orders, physical exertion, danger, and communication occur.

2. The degree to which communication is confined to others who are physically co-present in a situation or, conversely, the degree to which communication is diverse, involving the use of multiple modes of contact with many others in different situations.

Utilizing these two general classes of variables, which are part of the macrostructure that has been built up in past chains of interaction as well as several less central ones such as wealth and physical exertion on the job, Collins describes the class cultures of American society. More significantly for theory building, he also offers several abstract propositions that stipulate certain important relationships among power, order giving, communication networks, and behavioral tendencies among individuals. I have restated these relationships in somewhat altered form in Table 12–3.[28] With these principles, Collins explains variations in the behaviors, outlooks, and interpersonal styles of individuals in different occupations and status groups. For example, those occupations that require order giving, that reveal high co-presence of others, and that involve little physical exertion will generate behaviors that are distinctive and that circumscribe other activities, such as whom one marries, where one lives, what one values, and what activities one pursues in various spheres of life. Different weights to these variables would cause varying behavioral tendencies in individuals. Thus it is from the processes delineated in the propositions of Table 12–3 that understanding of such variables as class culture, ethnic cultures, lifestyles, and other concerns of investigators of stratification is to be achieved. But such understanding is anchored in the recognition that these class cultures are built up and sustained by interaction chains in which deference and demeanor rituals have figured prominently. Thus a class culture is not mere internalization of values and beliefs or simple socialization (al-

---

[28]Collins, *Conflict Sociology*, pp. 49–88.

TABLE 12–4 Key Propositions on Sex Stratification

I. Control over sexual activities between males and females as well as talk and ritual activities is a positive and additive function of:

   A. The degree of one sex's control over the means of coercion, which is a negative function of:

      1. The existence of coercive powers outside sexual partners and family groupings (such as the state).

      2. The presence of relatives of the subordinate sex.

   B. The degree to which one sex controls material resources, which is a positive and additive function of:

      1. The level of economic surplus in a population.

      2. The degree to which key economic activities are performed by one sex.

      3. The degree to which resources are inherited rather than earned.

II. The greater the degree of control of sexual relations and related activities by one sex, the more likely sexual relations are to be defined as property relations, and the more likely they are to be normatively regulated by rules of incest, exogamy, and endogamy.

III. The greater the degree of control of sexual relations and related activities by one sex, the greater will be the efforts of the other sex to:

   A. Reduce sexual encounters.

   B. Regulate them through ritual.

---

though this is no doubt involved); rather, it is the result of repeated encounters among unequals under varying conditions imposed by the macrostructure as it has been built up from past chains of interaction.

Sexual Stratification

The propositions in Tables 12-1 through 12-3 represent, I think, Collins' theoretical approach to the analysis of stratification. The propositions in these tables provide the basic principles for the analysis of specific forms of stratification, such as those created by age and sexual categories. For as is clearly evident, stratification for Collins is a process that occurs in types of encounters among individuals in diverse spheres of life. One such sphere is encounters between males and females that become organized into family relations and elaborated into kinship systems. In Table 12-4, I have summarized some of Collins' propositions on sexual stratification. Deductions from these, Collins implicitly argues, can help explain the nature of talk and ritual between the sexes as well as the structuring of male/female encounters in kinship systems. These latter deductions are not produced in the table, but Collins provides numerous additional propositions[29] and discursive text[30] to illustrate the properties and dynamics of sexual stratification.

These propositions can, in Collins' eyes, explain the diversity in chains of sexual encounters and the resulting structure of family systems. For example,

---

[29]Ibid., pp. 281–84.

[30]Ibid., pp. 225–59; see also the relevant chapters in his *Family Sociology*, 2nd ed. (Chicago: Nelson-Hall, 1988).

TABLE 12–5 Key Propositions on Age Stratification

I. The degree of age stratification among individuals is a positive and additive function of the degree to which individuals in one age group control (*a*) the means of coercion; (*b*) material resources; (*c*) symbolic resources (e.g., capacity to evaluate actions in terms of moral values); and (*d*) sociability (e.g., access to peers, recreation, and related activities).

II. The form of control exercised by individuals in one age group over individuals in another age group is a direct function of the types of resources that are most controlled by the dominant groups (see I(*a*), (*b*), (*c*), (*d*) above).

III. The greater the degree of age stratification, the greater the level of ritual interaction among individuals in different age groups; conversely, the more equally balanced the resources among age groups, the less the level of ritual activity among individuals in different age groups.

IV. The greater the efforts by individuals of one age group to control individuals in another age group, but the greater the level of resources available to individuals of subordinate age groups, the greater the level of conflict among individuals of different age groups.

---

in societies where females perform much of the economic labor and where wealth is passed through the female line (as is the case in matrilineal kinship systems), male dominance is greatly mitigated. Or in American society, females who enter the labor force possess resources to counter the males' control of the means of coercion, thereby reducing male dominance (especially since the state rather than kin groups is the ultimate source of coercive power in the society). Conversely, women who do not work are often forced to counter the economic and coercive power of males by limiting sexual encounters and by ritualizing such encounters when they do occur. Many other specific forms of male/female stratification can, in Collins' view, be explained by making deductions from these abstract propositions to specific empirical cases. But such explanations do not rely upon concepts of norms, values, and roles. Rather, they emphasize the respective resources that people have and how these determine what happens in an encounter as well as how structures are built up from chains of encounters among individuals with varying resources.

Age Stratification

Another universal form of stratification is by age, which, like sexual relations, involves control of resources by individuals in one age group and efforts by individuals in subordinate age groups to mitigate their situation in daily encounters. I have listed several of Collins' key propositions on this in Table 12–5. These propositions can, I think, explain variations in age stratification.[31] For example, among poor families in America, parents often possess only physical coercion in abundance (not having money, much leisure, or capacity to provide recreation), whereas youth often possess counterresources of their own, such as extensive peer relations and moral codes of peer evaluation that are

---

[31]Collins, *Conflict Sociology*, pp. 259–81.

built up from chains of encounters among themselves. The result is often high levels of conflict between parents and children. In contrast, affluent parents possess many resources, such as money, the capacity to schedule recreation, and the time to impose and sanction moral standards. As a result, they typically resort to coercion as a punishment of last resort, only after material incentives, moral shaming, or withholding of leisure-time privileges fails to generate conformity to parental dictates. Interaction in such families will, despite its seeming informality, involve the use of ritual and talk in clear deference and demeanor activities. Similar kinds of analyses can, I believe, be made of other societies using these principles as premises from which deductions to specific empirical cases can be made.

Organizational Processes

After examining stratification processes, Collins turns to an extensive analysis of organizations and develops a rather long inventory of propositions on their properties and dynamics.[32] These propositions overlap, to some degree, with those on stratification, since an organization is typically a stratified system with a comparatively clear hierarchy of authority. In Table 12-6, I have extracted only three groups of propositions from Collins' analysis. These revolve around processes of organizational control, the administration of control, and the general organizational structure. Other topic areas in Collins' inventory are either incorporated into these propositions or omitted. Moreover, I have rephrased considerably the propositions and have also stated them at a higher level of abstraction than in Collins' text. By placing the propositions at a high level of abstraction, I think that they can explain political control in more than complex organizations; in my view, they would seem relevant to understanding other patterns of social organization, such as entire societies.

Even as Collins begins to examine more macrostructures, like bureaucratic organizations, it is evident that the conceptual emphasis is on micro processes. Organization control, its administration, and the general structure of an organization are all created and sustained by people using resources in encounters. As these encounters are repeated, the chains of interactions develop a structure, whose profile depends on the respective levels and types of resources possessed by participants.

The State and Economy

Collins develops propositions on many topics in *Conflict Sociology,* but I will close my review of this work with an illustration of his more purely macroanalysis of the state and economy.[33] A cursory look at Table 12-7 reveals, I think, few references to individuals at the micro level. One might, as Collins would

---

[32]Ibid., pp. 286–347.
[33]Ibid., pp. 348–413.

TABLE 12–6 Key Propositions on Organizations

Processes of Organizational Control

I. The degree of control in patterns of social organization is a positive and additive function of the concentration among individuals of (*a*) coercive resources, (*b*) material resources, and (*c*) symbolic resources.

II. The form of control in patterns of social organization is a function of the type of resource that is concentrated in the hands of those individuals seeking to control others.

III. The more control is sought through the use of coercive resources, the more likely are those subject to the application of these resources to (*a*) seek escape, (*b*) fight back, if escape is impossible, (*c*) comply if the above are impossible and if material incentives exist, and (*d*) sluggishly comply if the above do not apply.

IV. The more control is sought through the use of material resources, the more likely are those subject to the manipulation of material incentives to (*a*) develop acquisitive orientations and (*b*) develop a strategy of self-interested manipulation.

V. The more control is sought through the use of symbolic resources, the more likely are those subject to the application of such resources to (*a*) experience indoctrination into values and beliefs, (*b*) be members of homogeneous cohorts of recruits, (*c*) be subject to efforts to encourage intraorganizational contact, (*d*) be subject to efforts to discourage extraorganizational contact, (*e*) participate in ritual activities, especially those involving rites de passage, and (*f*) be rewarded for conformity with upward mobility.

Administration of Control

VI. The more those in authority employ coercive and material incentives to control others, the greater the reliance on surveillance as an administrative device to control.

VII. The more those in authority use surveillance to control, the greater are (*a*) the level of alienation by those subject to surveillance, (*b*) the level of conformity in only highly visible behaviors, and (*c*) the ratio of supervisory to nonsupervisory individuals.

VIII. The more those in authority employ symbolic resources to control others, the greater their reliance on systems of standardized rules to achieve control.

IX. The greater the reliance on systems of standardized rules, the greater are (*a*) the impersonality of interactions, (*b*) standardization of behaviors, and (*c*) the dispersion of authority.

Organizational Structure

X. Centralization of authority is a positive and additive function of (*a*) the concentration of resources; (*b*) the capacity to mobilize the administration of control through surveillance, material incentives, and systems of rules; (*c*) the capacity to control the flow of information; (*d*) the capacity to control contingencies of the environment; and (*e*) the degree to which the tasks to be performed are routine.

XI. The bureaucratization of authority and social relations is a positive and additive function of (*a*) record-keeping technologies, (*b*) nonkinship agents of socialization of potential incumbents, (*c*) money markets, (*d*) transportation facilities, (*e*) nonpersonal centers of power, and (*f*) diverse centers of power and authority.

---

insist, make micro-translations of the processes incorporated into the propositions, but it is clear that Collins abandons discussions of micro-interactional ritual chains when he begins to talk about large-scale social processes. And even his insistence that the macro is no more than size, number, and time seems somewhat strained, although population size and extent of territory are crucial variables in the propositions.

TABLE 12–7 Key Propositions on the State, Economy, and Ideology

---

I. The size and scale of political organization are a positive function of the productive capacity of the economy.

II. The productive capacity of the economy is a positive and additive function of (a) level of technology, (b) level of natural resources, (c) population size, and (d) efficiency in the organization of labor.

III. The form of political organization is related to the levels of and interactive effects among (a) size of territories to be governed, (b) the absolute numbers of people to be governed, (c) the distribution and diversity of people in a territory, (d) the organization of coercive force (armies), (e) the distribution (dispersion or concentration) of power and other resources among a population, and (f) the degree of symbolic unification within and among social units.

IV. The stability of the state is a negative and additive function of:

    A. The capacity for political mobilization by other groups, which is a positive function of:

        1. The level of wealth.

        2. The capacity for organization as a status group.

    B. The incapacity of the state to resolve periodic crises.

Geopolitics

Borrowing from Weber but adding his own ideas, Collins argues that there are sociological reasons for the historical facts that only certain societies are able to form stable empires and that societies can extend their empires only to a maximal size of about 3 to 4 million square miles.[34] When a society has a resource (money, technology, population base) and marchland (no enemies on most of its borders) advantage, it can win wars, but eventually it will (a) extend itself beyond its logistical capacities, (b) bump up against another empire, (c) lose its marchland advantage as it extends its borders and becomes ever more surrounded by enemies, and (d) lose its technological advantages as enemies adopt them. The result of these forces is that empires begin to stall at a certain size, as each of these points of resistance is activated. These processes indicate that internal nation/states will not build long-term or extensive empires because they are surrounded by enemies, and increasingly so as they extend territory. Rather, it is marchland states, with oceans, mountains, or unthreatening neighbors at their back, that can move out and conquer others, since they have to fight a war on only one front. But eventually they overextend, confront another marchland empire, lose their technological advantage, and acquire enemies on a greater proportion of their borders (thereby losing the marchland advantage and, in effect, becoming an internal state that must now fight on several borders). Sea and air powers can provide a kind of marchland advantage, but the logistical loads of distance from home bases and maintenance of sophisticated technologies make such empires vulnerable. Only when

---

[34]Randall Collins, "Modern Technology and Geopolitics" and "The Future Decline of the Russian Empire," in *Weberian Sociological Theory*, pp. 167–212.

TABLE 12–8 Key Propositions on Geopolitics

I. The possibility of winning a war between nation/states is a positive and additive function of:

  A. The level of resource advantage of one nation/state over another, which is a positive function of:

    1. The level of technology.
    2. The level of productivity.
    3. The size of the population.
    4. The level of wealth formation.

  B. The degree of "marchland advantage" of one nation/state over another, which is a positive and additive function of:

    1. The extent to which the borders of a nation/state are peripheral to those of other nation/states.
    2. The extent to which a nation/state has enemies on only one border.
    3. The extent to which a nation/state has natural buffers (mountains, oceans, large lakes, etc.) on most of its borders.

II. The likelihood of an empire is a positive function of the extent to which a marchland state has resource advantages over neighbors and uses these advantages to wage war.

III. The size of an empire is a positive and additive function of the dominant nation/states' capacity to:

  A. Avoid showdown war with the empire of other marchland states.
  B. Sustain a marchland advantage.
  C. Maintain territories with standing armies.
  D. Maintain logistical capacity for communications and transportation, which is a positive function of levels of communication, transportation, and military technologies and a negative and additive function of:

    1. The size of a territory.
    2. The distances of borders from the home base.

  E. Diffusion of technologies to potential enemies.

IV. The collapse of an empire is a positive and additive function of:

  A. The initiation of war between two empires.
  B. The overextension of an empire beyond its logistical capacity.
  C. The adoption of its superior technologies by enemy nation/states.

they encounter little resistance can an empire be maintained across oceans and at great distances by air; as resistance mounts, the empire collapses quickly as its supply lines are disrupted. Table 12–8 summarizes these ideas more formally.

CONCLUSION

I think that Collins' work is among the most creative in social theory today. Its great strength resides in the effort to develop abstract propositions about fundamental properties of the universe. For, unlike so many in social theory today, Collins is not so antipositivist that he must retreat into an excessive relativism and solipsism.

A related strength in Collins' approach is his willingness to theorize. Metatheoretical pontification is minimal, and he gets right down to the business of developing propositions about the social universe. Moreover, he is willing to theorize about a wide range of phenomena in terms of just a few fundamental ideas. He thus achieves breadth of substantive analysis without an overly elaborate conceptual scheme, making deductions from a limited number of basic propositions.

Equally significant, I think, is that Collins' work is filled with creative insights. He takes ideas from diverse intellectual traditions and combines them with his own ideas to produce works that are stimulating. For, in the end, theory is not just mechanical induction or deduction; it involves creative leaps of imagination, and Collins' work is filled with them.

The substantiveness of Collins' theoretical approach reflects this creativity. He has sought to conceptualize the basic nature of interaction and how it produces and reproduces more macrostructures, and vice versa. I think that at times he argues too polemically as to the micro basis of reality, but his conceptualization of the micro realm in terms of exchange and ritual among actors using resources to their advantage and the macro realm in terms of chains of encounters that are strung out in time and stretched in space among various numbers of individuals is, I believe, a useful way to look at the social world. Simply stated in these terms, however, Collins' work would not be so interesting. But as he translates these metaphorical ideas into abstract propositions, the approach begins to come to life.

There are also, I think, some major problems with the approach. First, despite his assertion that much existing social theory is metaphorical in its conception of structure, I find his own formulations rather vague and metaphorical. He rarely defines concepts precisely, and he shifts his vocabulary in ways that often make it difficult to get a clear grasp of even such central concepts as resources, power, coercion, property, wealth, negotiations, encounters, rituals, and structures. Indeed, as I have presented his ideas in this chapter, I have had to make many inferences about his meaning. I think that Collins gets away with this vagueness in concept formation because he writes so well. Indeed, he is probably the best writer in academic sociology today, and, because he can be so engaging, he can also be conceptually very slippery and vague.

A second problem is that, in Collins' *Conflict Sociology* and elsewhere, there is a lack of deductive rigor in the propositions. There are simply too many propositions that overlap each other and that do not follow from well-articulated axioms, corollaries, and theorems. Naturally, it can be argued that, at the more preliminary stages of theory building, such deductive precision is not possible or even desirable. To lose richness of insight in the name of theoretical elegance and precision is obviously not desirable in an immature science like sociology. Yet, with more than 400 propositions in *Conflict Sociology*, Collins' inventory becomes unmanageable. In many ways my reformulations of Collins' propositions represent an attempt to make them more manageable. But in the end Collins will need to articulate some basic theorems if his theoretical strategy is to realize its full potential.

If this latter effort is made, one suspects that the "axioms" or highest-order propositions would look very much like those of exchange theory (see Chapters 14–17), and this is why I have labeled his approach "exchange conflict theory." For in the final analysis Collins' propositions typically deal with (1) the respective resources of individuals, (2) the effort to use resources to advantage and to avoid heavy costs in the process, (3) the differentiation of individuals in terms of their respective resources, and (4) the processes by which such differentiation is regularized, on the one hand, and is the source of conflict and change, on the other. Collins' ideas on "interaction ritual chains" do this to some extent, but only in a most imprecise and metaphorical sense. I believe that considerably more attention should be devoted to rigorous definitions of concepts and explicit formulation of his implicit exchange theoretic propositions. In this way the full potential of this exchange conflict approach can be realized.

CHAPTER 13

———————————◆———————————

Critical Theorizing

Jurgen Habermas

As I emphasized in Chapter 9, on the origins of conflict theorizing, the emancipatory theme in Karl Marx's thought has been carried over into this century in a number of guises, one of which is critical theory.[1] Such theorizing is explicitly evaluational in that it sees as its purpose the emancipation of individuals from domination. However, as I also mentioned in Chapter 9, the first generation of critical theorists confronted a modern world that did not appear rife with emancipatory potential. Indeed, Max Weber's analysis of rationalization and the extension of bureaucratic systems of control into ever more spheres of life seemed to be a more apt prognosis of the future than Marx's utopian dream.

Thus the first generation of critical theorists, who are frequently referred to as the Frankfurt School because of their location in Germany and their explicit interdisciplinary effort to interpret the oppressive events of the 20th century, confronted a real dilemma: how to reconcile Marx's emancipatory dream with the stark reality of modern society as conceptualized by Weber.[2]

---

[1]Some basic reviews and analyses of critical theory and sociology include: Paul Connerton, ed., *Critical Sociology* (New York: Penguin Books, 1970); Raymond Geuss, *The Idea of a Critical Theory* (New York: Cambridge University Press, 1981); David Held, *Introduction to Critical Theory* (Berkeley: University of California Press, 1980); Trent Schroyer, *The Critique of Domination: The Origins and Development of Critical Theory* (New York: Braziller, 1973); Albrecht Wellmer, *Critical Theory of Society* (New York: Seabury Press, 1974); Ellsworth R. Fuhrman and William E. Snizek, "Some Observations on the Nature and Content of Critical Theory," *Humboldt Journal of Social Relations* 7 (Fall–Winter 1979-80), pp. 33–51; Zygmunt Bauman, *Towards a Critical Society* (Boston: Routledge & Kegan Paul, 1976); Robert J. Antonio, "The Origin, Development, and Contemporary Status of Critical Theory," *The Sociological Quarterly* 24 (Summer 1983), pp. 325–51; Jim Faught, "Objective Reason and the Justification of Norms," *California Sociologist* 4 (Winter 1981), pp. 33–53.

[2]For descriptions of this activity, see Martin Jay, *The Dialectical Imagination* (Boston: Little, Brown, 1973), and "The Frankfurt School's Critique of Marxist Humanism," *Social Research* 39

Indeed, when the Frankfurt Institute for Social Research was founded in 1923 there seemed little reason to be optimistic about developing a theoretically informed program for freeing people from unnecessary domination. The defeat of the left-wing working-class movements, the rise of fascism in the aftermath of World War I, and the degeneration of the Russian Revolution into Stalinism had, by the 1930s, made it clear that Marx's analysis needed drastic revision. Moreover, the expansion of the state, the spread of bureaucracy, and the emphasis on means/ends rationality through the application of science and technology all signaled that Weber's analysis had to be confronted.

The members of the Frankfurt School wanted to maintain Marx's notion of praxis—that is, a blending of theory and action or the use of theory to stimulate action, and vice versa. And they wanted theory to expose oppression and to propose less constrictive options. Yet they were confronted with the spread of political and economic domination. Thus the development of modern critical theory in sociology was born in a time when there was little reason to be optimistic about realizing emancipatory goals.

Three members of the Frankfurt School are most central: George Lukács, Max Horkheimer, and Theodor Adorno.[3] Lukács' major work appeared in the 1920s,[4] whereas Horkheimer[5] and Adorno[6] were active well into the 1960s. In many ways I view Lukács as the key link in the transition from Marx and Weber to modern critical theory, for Horkheimer and Adorno were reacting to much of Lukács' analysis and approach.

All of these scholars are important for the purposes of this chapter because they directly influenced the intellectual development and subsequent work of Jurgen Habermas, the most prolific contemporary critical theorist. Thus, to appreciate fully the nature of critical theory in general and Habermas' work in particular, I will briefly examine Lukács', Horkheimer's, and Adorno's ideas

---

(1972), pp. 285–305; David Held, *Introduction to Critical Theory*, pp. 29–110; Robert J. Antonio, "The Origin, Development, and Contemporary Status of Critical Theory"; Phil Slater, *Origin and Significance of The Frankfurt School* (London: Routledge & Kegan Paul, 1977).

[3]Other prominent members included: Friedrich Pollock (economist), Erich Fromm (psychoanalyst, social psychologist), Franz Neumann (political scientist), Herbert Marcuse (philosopher), and Leo Loenthal (sociologist). During the Nazi years the school relocated to the United States, and many of its members never returned to Germany.

[4]George Lukács, *History and Class Consciousness* (Cambridge, MA: MIT Press, 1968; originally published in 1922).

[5]Max Horkheimer, *Critical Theory: Selected Essays* (New York: Herder and Herder, 1972), is a translation of essays written in German in the 1930s and 1940s; *Eclipse of Reason* (New York: Oxford University Press, 1947; reprinted by Seabury Press in 1974) was the only book by Horkheimer originally published in English. It takes a slightly different turn than earlier works, but it does present ideas that emerged from his association with Theodor Adorno; and *Critique of Instrumental Reason* (New York: Seabury Press, 1974). See David Held, *Introduction to Critical Theory*, pp. 489–91, for a more complete listing of Horkheimer's works in German.

[6]Theodor W. Adorno, *Negative Dialectics* (New York: Seabury Press, 1973; originally published in 1966); and, with Max Horkheimer, *Dialectic of Enlightenment* (New York: Herder and Herder, 1972; originally published in 1947). See Held, *Introduction to Critical Theory*, pp. 485–7, for a more complete listing of his works.

and how Habermas interpreted and reacted to them.[7] Then I will move directly into a more detailed review and analysis of Habermas' work as it has developed over the last decades.

EARLY CRITICAL THEORY: LUKÁCS, HORKHEIMER, ADORNO

In Habermas' eyes, George Lukács blended Marx and Weber together by seeing a convergence of Marx's ideas about commodification of social relations through money and markets with Weber's thesis about the penetration of rationality into ever more spheres of modern life. Borrowing from Marx's analysis of the "fetishism of commodities," Lukács employed the concept of "reification" to denote the process by which social relationships become "objects" that can be manipulated, bought, and sold. Then, reinterpreting Weber's notion of "rationalization" to mean a growing emphasis on the process of "calculation" of exchange values, Lukács combined Weber's and Marx's analysis. As traditional societies change, he argued, there is less reliance on moral standards and processes of communication to achieve societal integration; instead, there is more utilization of money, markets, and rational calculations. As a result, relations are coordinated by exchange values and by people's perceptions of one another as "things."[8]

As Habermas contends, Lukács painted himself into a conceptual corner: if indeed such is the historical process, how is it to be stopped? Lukács' answer is to resurrect Hegel and "stand him back on his feet" in opposition to Marx's effort to "stand Hegel on his head." That is, rather than look to contradictions in material conditions—economic and political forces—one must examine the dialectical forces inherent in human consciousness. For there are limits, Lukács argued, to how much reification and rationalization people will endure. There is an inner quality in human subjects that keeps rationalization from completely taking over.[9]

This emphasis on the process of consciousness is very much a part of critical theory that borrows much from the early Marx[10] and that, at the Frankfurt School, had a heavy dose of Freud and psychoanalytic theory. But, as Habermas was to emphasize, early critical theory was too subjectivist and failed to analyze intersubjectivity, or the ways people interact through mutually shared conscious

---

[7]Chapter 4, "From Lukács to Adorno: Rationalization as Reification," pp. 339-99 in Jurgen Habermas, *The Theory of Communicative Action*, vol. 1 (Boston: Beacon Press, 1984), contains Habermas' critique of Lukács, Horkheimer, and Adorno.

[8]Lukács, *History and Class Consciousness*.

[9]Ibid., pp. 89-102. In a sense, Lukács becomes another "Young Hegelian" whom Marx would have criticized. Yet, in Marx's own analysis, he sees alienation, per se, as producing resistance by workers to further alienation by the forces of production. This is the image that Lukács seems to take from Marx.

[10]Karl Marx and Friedrich Engels, *The German Ideology* (New York: International Publishers, 1947; originally written in 1846).

activity. Emphasizing the inherent resistance of subjects to their total reification, Lukács could only propose that the critical theorist's role is to expose reification at work by analyzing the historical processes that have dehumanized people. As a consequence, Lukács made critical theory highly contemplative, emphasizing that the solution to the problem of domination resides in making people more aware and conscious of their situation through a detailed, historical analysis of reification.

Both Horkheimer and Adorno were highly suspicious of Lukács' Hegelian solution to the dilemma of reification and rationalization. These processes do not imply their own critique, as Hegel would have suggested. Subjective consciousness and material reality cannot be separated. Consciousness does not automatically offer resistance to those material forces that commodify, reify, and rationalize. Critical theory must, therefore, actively seek to (1) describe historical forces that dominate human freedom and (2) expose ideological justifications of these forces. Such is to be achieved through interdisciplinary research among variously trained researchers and theorists who confront one another's ideas and use this dialogue to analyze concrete social conditions and to propose courses of ameliorative action. This emphasis on praxis—the confrontation between theory and action in the world—involves the development of ideas about what oppresses and what to do about it in the course of human struggles. As Horkheimer argued: "[The] value of theory is not decided alone by the formal criteria of truth ... but by its connection with tasks, which in the particular historical moment are taken up by progressive social forces."[11] Such critical theory is, Horkheimer claimed, guided by a "particular practical interest" in the emancipation of people from class domination.[12] Thus, critical theory is tied, in a sense that Marx might have appreciated, to people's practical interests.

As Adorno and Horkheimer interacted and collaborated, their positions converged (although by the late 1950s and early 1960s Horkheimer had seemingly rejected much of his earlier work). Adorno was, in my eyes, more philosophical and research oriented than Horkheimer; his empirical work on "the authoritarian personality" had a major impact on research in sociology and psychology, but I think that his theoretical impact came from his collaboration with Horkheimer and, in many ways, through Horkheimer's single-authored work.[13] Adorno was very pessimistic about the chances of critical theory making great changes, although his essays were designed to expose patterns of recognized and unrecognized domination of individuals by social and psychological forces. At best, his "negative dialectics" could allow humans to "tread water"

---

[11]Max Horkheimer, "Zum Rationalismusstreit in der gegenwartigen Philosophe," originally published in 1935; reprinted in *Kritische Theorie*, vol. 1, ed. A. Schmidt (Frankfurt: Fischer Verlag, 1968), pp. 146–7. This and volume 2, by the way, represent a compilation of many of Horkheimer's essays while at the Institute in Frankfurt.

[12]Habermas is to take up this idea, but he will extend it in several ways.

[13]Theodor W. Adorno, et al., *The Authoritarian Personality* (New York: Harper & Row, 1950).

until historical circumstances were more favorable to emancipatory movements. The goal of negative dialectics is to sustain a constant critique of ideas, conceptions, and conditions. This critique cannot by itself change anything, for it occurs only on the plane of ideas and concepts. But it can keep ideological dogmatisms from obscuring conditions that might eventually allow for emancipatory action.

I have not done justice to either Horkheimer or Adorno, but my goal is not to summarize their ideas in total but only those portions to which Jurgen Habermas reacted. Both Horkheimer and Adorno emphasized that humans' "subjective side" is restricted by the spread of rationalization. In conceptualizing this process, they created a kind of dualism between the subjective world and the realm of material objects, seeing the latter as working to oppress the former. And so, from their viewpoint, critical theory must expose this dualism, and it must see how this invasion of "instrumental reason" (means/ends rationality)[14] has invaded the human spirit. In this way some resistance can be offered to these oppressive forces.

Habermas sees this kind of philosophical analysis—even with the caveats to praxis—as terribly vague, excessively contemplative, and amazingly detached from what people actually do when they interact and use their conscious faculties.[15] Such critical theory is also fatalistic, seeing the forces of production as immobilizing the forces of subjective consciousness and ameliorative action. What Habermas proposes, then, is a shift in orientation away from the subjective consciousness of the individual to the processes by which humans create *inter*subjective understandings and to the processes by which they actually coordinate their actions. For, if there is oppression and domination, it can be relieved only by understanding the specific mechanisms that integrate societies. For to change a society requires detailed insight into what binds it together. And so it is in the analysis of these integrating processes that critical theory can propose realistic emancipatory solutions.

THE CRITICAL APPROACH OF JURGEN HABERMAS

Jurgen Habermas has been an enormously productive scholar in the past two decades. This fact makes it difficult to summarize his work in a simple way. Moreover, to be kind about the matter, Habermas' style of exposition is somewhat dense. Part of this denseness is his style of prose,[16] but much of it is Habermas' Germanic approach, which involves an effort to analyze all the relevant viewpoints on a topic before he presents his case. As a result, Habermas' main line of argument is often buried in tangential analyses of scholars

---

[14]Means/ends rationality is action designed to use the most efficient means to achieve stated ends or goals. For Max Weber's use of this concept, see Chap. 9.

[15]Habermas, *The Theory of Communicative Action*, vol. 1, Chap. 4.

[16]Indeed, the English translations of his work are difficult, but the German versions are even worse. Translators consistently improve Habermas' prose, but it still remains highly obscure.

and issues that he perceives to be relevant and essential to his point.[17] Consequently, there are varying interpretations of Habermas' views; I suspect that, in working through his work, I may misinterpret him. This danger of misinterpretation is particularly acute because I will try to extract the central themes from a conceptual scheme that has evolved and changed over the years. Yet I see a certain continuity in the argument, and it is this aspect of his work that I will emphasize.

The Central Problem of Critical Theory

Underlying Habermas' work, I believe, are a series of unarticulated questions. These include: (1) How is social theory to develop ideas that keep Karl Marx's emancipatory project alive and yet, at the same time, recognize the empirical inadequacy of his prognosis for advanced capitalist societies? (2) How is social theory to confront Max Weber's historical analysis of rationalization[18] in a way that avoids his pessimism and thereby keeps Marx's emancipatory goals at the center of theory? (3) How is social theory to avoid the retreat into subjectivism of earlier critical theorists, such as Lukács, Horkheimer, and Adorno, who increasingly focused on states of subjective consciousness *within individuals* and, as a consequence, lost Marx's insight that society is constructed from, and must therefore be emancipated in terms of, the processes that sustain *social relations among* individuals? (4) How is social theory to conceptualize and develop a theory that reconciles the forces of material production and political organization with the forces of intersubjectivity among reflective and conscious individuals in such a way that it avoids (*a*) Weber's pessimism about the domination of consciousness by rational economic and political forces, (*b*) Marx's naive optimism about inevitability of class consciousness and revolt, and (*c*) early critical theorists' retreat into the subjectivism of Hegel's dialectic, where oppression mysteriously mobilizes its negation through increases in subjective consciousnesses and resistance?

At different points in his career, Habermas focused on one or another of these questions, but my sense is that all four have always guided his approach, at least implicitly. Habermas has been accused of abandoning the critical thrust of his earlier works, but I think that this conclusion is incorrect. For, in trying to answer these questions, he has increasingly recognized that mere critique of oppression is not enough. Such critique becomes a "reified object itself"; although early critical theorists knew this, they never developed conceptual schemes that accounted for the underlying dynamics of societies. For critique to be useful in liberating people from domination, it is necessary, Habermas seems to say, for the critique to be couched in terms of the fundamental processes that integrate social systems. In this way the critique has some possibility

---

[17]In some ways I find these tangents more interesting than the main argument.
[18]That is, the spread of means/end rationality into ever more spheres of life.

for suggesting ways to create new types of social relations. Without theoretical understanding of how society works, critique is only superficial debunking and becomes an exercise in futility. It is this willingness to theorize about the underlying dynamics of society, to avoid the retreat into subjectivism, to reject superficial criticism and instead base critique on reasoned theoretical analysis, and to incorporate ideas from many diverse theoretical approaches that make Habermas' work theoretically significant.

In presenting his ideas, I will briefly summarize his earlier works, showing how they have culminated in the synthesis of his ideas in the two-volume work *The Theory of Communicative Action*.[19]

Habermas' Analysis of "The Public Sphere"[20]

In his first major publication, *Structural Transformation of the Public Sphere*, Habermas traces the evolution and dissolution of what he termed *the public sphere*.[21] This sphere is a realm of social life where people can bring up matters of general interest; where they can discuss and debate these issues without recourse to custom, dogma, and force; and where they can resolve differences of opinion by rational argument. To say the least, I find this conception of a public sphere to be rather romanticized, but the imagery of free and open discussion that is resolved by rational argumentation becomes a central theme in Habermas' subsequent approach. Increasingly throughout his career, Habermas has come to see emancipation from domination as possible through "communicative action," which is a reincarnation of the public sphere in more conceptual clothing.

In this early work, however, Habermas appears more interested in history and views the emergence of the public sphere as occurring in the 18th century, when various forums for public debate—clubs, cafés, journals, newspapers—proliferated. He concludes that these forums helped erode the basic structure of feudalism, which is legitimated by religion and custom rather than by agreements that have been reached through public debate and discourse. The public sphere was greatly expanded, Habermas argues, by the extension of market economies and the resulting liberation of the individual from the constraints of feudalism. Free citizens, property holders, traders, merchants, and members of other new sectors in society could now be actively concerned about the governance of society and could openly discuss and debate issues. But, in a vein similar to Weber's analysis of rationalization, Habermas argues that the public sphere was eroded by some of the very forces that stimulated its ex-

---

[19]Habermas, *The Theory of Communicative Action*.

[20]Some useful reviews and critiques of Habermas' work include: John B. Thompson and David Held, eds., *Habermas: Critical Debates* (London: Macmillan, 1982); David Held, *An Introduction to Critical Theory*, Chaps. 9-12.

[21]Jurgen Habermas, *Struckturwandel der Offentlichkeit* (Neuwied, Germany: Luchterhand, 1962).

pansion. As market economies experience instability, the powers of the state are extended in an effort to stabilize the economy; with the expansion of bureaucracy to ever more contexts of social life, the public sphere is constricted. And, increasingly, the state seeks to redefine problems as technical and soluble by technologies and administrative procedures rather than by public debate and argumentation.

The details of this argument are less important, I think, than the fact that this work established Habermas' credentials as a critical theorist. All the key elements of critical theory are there—the decline of freedom with the expansion of capitalism and the bureaucratized state as well as the seeming power of the state to construct and control social life. The solution to these problems is to resurrect the public sphere, but how is this to be done in light of the growing power of the state? Thus, in this early work, I see Habermas as having painted himself into the same conceptual corner as his teachers in the Frankfurt School. The next phase of his work extends this critique of capitalist society, but it also tries to redirect critical theory so that it does not have to retreat into the contemplative subjectivism of Lukács, Horkheimer, and Adorno. He begins this project in the late 1960s with an analysis of knowledge systems and a critique of science.

The Critique of Science

In *The Logic of the Social Sciences*[22] and *Knowledge and Human Interest*,[23] Habermas analyzes systems of knowledge in an effort to elaborate a framework for critical theory. The ultimate goal of this analysis is to establish the fact that science is but one type of knowledge that exists in order to meet only one set of human interests. To realize this goal, Habermas posits three basic types of knowledge that encompass the full range of human reason: (1) There is *empirical/analytic* knowledge, which is concerned with understanding the lawful properties of the material world. (2) There is *hermeneutic/historical* knowledge, which is devoted to the understanding of meanings, especially through the interpretations of historical texts. And (3) there is *critical* knowledge, which is devoted to uncovering conditions of constraint and domination. These three types of knowledge reflect three basic types of human interests: (1) a *technical* interest in the reproduction of existence through control of the environment, (2) a *practical* interest in understanding the meaning of situations, and (3) an *emancipatory* interest in freedom for growth and improvement. Such interests reside not in individuals but in more general imperatives for reproduction,

[22]Jurgen Habermas, *Zur Logik der Sozialwissenschaften* (Frankfurt: Suhrkamp, 1970).

[23]Jurgen Habermas, *Knowledge and Human Interest*, trans. J. Shapiro (London: Heinemann, 1970; originally published in German in 1968). The basic ideas in *Zur Logik der Sozialwissenschaften* and *Knowledge and Human Interest* were stated in Habermas' inaugural lecture at the University of Frankfurt in 1965 and were first published in "Knowledge and Interest," *Inquiry* 9 (1966), pp. 285–300.

meaning, and freedom that presumably are built into the species as it has become organized into societies. These three interests create, therefore, three types of knowledge. The interest in material reproduction has produced science or empirical/analytic knowledge, the interest in understanding of meaning has led to the development of hermeneutic/historical knowledge, and the interest in freedom has required the development of critical theory.

These interests in technical control, practical understanding, and emancipation generate different types of knowledge through three types of media: (1) "work" for realizing interests in technical control through the development of empirical/analytic knowledge, (2) "language" for realizing practical interests in understanding through hermeneutic knowledge, and (3) "authority" for realizing interests in emancipation through the development of critical theory. There is a kind of functionalism in this analysis: needs for "material survival and social reproduction," for "continuity of society through interpretive understanding," and for "utopian fulfillment" create interests. Then, through the media of work, language, and authority, these needs produce three types of knowledge: the scientific, hermeneutical, and critical.

This kind of typologizing is, of course, reminiscent of Weber and Parsons, and in many ways it is even more vague than Parsons' typologizing. Nonetheless, it is the vehicle through which Habermas makes the central point: positivism and the search for natural laws constitute only one type of knowledge, although the historical trend has been for the empirical/analytic to dominate the other two types of knowledge. Thus interests in technical control through work and the development of science have dominated the interests in understanding and emancipation. And so, if social life seems meaningless and cold, it is because technical interests in producing science have come to dictate what kind of knowledge is permissible and legitimate. Thus Weber's "rationalization thesis" is restated with the typological distinction among interest, knowledge, and media. There is, I should emphasize again, an implicit functionalism here: two sets of universal functional needs—for "continuity through interpretive understanding" and for "utopian fulfillment"—are not being met by virtue of the ascendancy of concern with practical interests. And, by implication, then, one of the underlying causes of the problems in advanced capitalism is the failure to meet the functional needs of Habermas' implicit functionalism. I will not pursue this point further, but it is worth considering. I have summarized Habermas' excessive typological argument in Figure 13–1. However, I have revised it somewhat in order to make the argument clear.

This typology allows Habermas to achieve several goals. First, he can attack the assumption that science is value free because, like all knowledge, it is attached to a set of interests. Second, he can revise the Weberian thesis of rationalization in such a way that it dictates a renewed emphasis on hermeneutics and criticism. For it is these other two types of knowledge that are being driven out by empirical/analytic knowledge, or science. Therefore it is necessary to reemphasize these neglected types of knowledge. Third, he can

FIGURE 13–1 Types of Knowledge, Interests, Media (and Functional Needs)

Functional Needs	Interests	Knowledge	Media
Material survival and social reproduction generate pressures for:	technical control of the environment, which leads to the development of:	empirical/analytic knowledge, which is achieved through:	work.
Continuity of social relations generates pressures for:	practical understanding through interpretations of others' subjective states, which leads to the development of:	hermeneutic and historical knowledge, which is achieved through:	language.
Desires for utopian fulfillment generate pressures for:	emancipation from unnecessary domination, which leads to the development of:	critical theory, which is achieved through:	authority.

justify certain topic areas that come to consume his work. Let me elaborate on this last point.

By viewing positivism in the social sciences as a type of empirical/analytic knowledge, Habermas can associate it with human interests in technical control. He can therefore visualize social science as a tool of economic and political interests. Science thus becomes an ideology; in fact, Habermas sees it as the underlying cause of the legitimation crises of advanced capitalist societies (more on this shortly). In dismissing positivism in this way, he can orient his own project to hermeneutics with a critical twist. That is, he can visualize the major task of critical theory as the analysis of those processes by which people achieve interpretative understanding of one another in ways that give social life a sense of continuity. Increasingly, Habermas has come to focus on the communicative processes among actors as the theoretical core for critical theorizing. Goals of emancipation cannot be realized without knowledge about how people interact and communicate. Such an emphasis represents a restatement in a new guise of Habermas' early analysis of the public sphere, but now the process of public discourse and debate is viewed as the essence of human interaction in general. Moreover, to understand interaction, it is necessary to analyze language and linguistic processes among individuals. Knowledge of these processes can, in turn, give critical theory a firm conceptual basis from which to launch a critique of society and to suggest paths for the emancipation of individuals. Yet, to justify this emphasis on hermeneutics and criticism, Habermas must first analyze the crises of capitalist societies in terms of the overextension of empirical/analytic systems of knowledge.

Legitimation Crises in Society

As Habermas' earlier work had argued, there are several historical trends in modern societies: (1) the decline of the public sphere, (2) the increasing intervention of the state into the economy, and (3) the growing dominance of science in the service of the state's interests in technical control. These ideas are woven together in *Legitimation Crisis,* which I see as a further critique of capitalist society and positivistic science.[24] But, in locating the crises of capitalist society in the ascendance of interests in technical control, Habermas can justify his own critical project, especially as it takes a hermeneutic turn— that is, as it moves toward an analysis of the processes by which actors interpret one another's subjective states.

The basic argument in *Legitimation Crisis* is that, as the state increasingly intervenes in the economy, it also seeks to translate political issues into "technical problems." Issues are not topics for public debate; rather, they represent technical problems that require the use of technologies by experts in bureaucratic organizations. As a result, there is a "depoliticization" of practical issues by redefining them as technical problems. To do this, the state propagates a "technocratic consciousness" that, to Habermas, represents a new kind of ideology. Unlike previous ideologies, however, it does not promise a future utopia; but, like other ideologies, it is seductive in its ability to veil problems, to simplify perceived options, and to justify a particular way of organizing social life. At the core of this technocratic consciousness is an emphasis on "instrumental reason," or what Weber termed *means/ends rationality.* That is, criteria of the efficiency of means in realizing explicit goals increasingly come to guide evaluations of social action and people's approach to problems. This emphasis on instrumental reason operates to displace other types of action, such as behaviors oriented to mutual understanding. This displacement occurs in a series of stages: science is first used by the state to realize specific goals; then, the criterion of efficiency is used by the state to reconcile competing goals of groupings; next, basic cultural values are themselves assessed and evaluated in terms of their efficiency and rationality; finally, in Habermas' version of *Brave New World,* decisions are completely delegated to computers, which seek the most rational and efficient course of action.

This reliance on the ideology of technocratic consciousnesses creates, Habermas argues, new dilemmas of political legitimation. For Habermas, capitalist societies can be divided into three basic subsystems: (1) the economic, (2) the politico-administrative, and (3) the cultural (what he later calls *lifeworld*). From this division of societies into these subsystems, Habermas then posits four points of crises: (1) an "economic crisis" occurs if the economic subsystem cannot generate sufficient productivity to meet people's needs; (2) a "rationality crisis" exists when the politico-administrative subsystem cannot generate a sufficient number of instrumental decisions; (3) a "motivation crisis" exists when actors cannot use cultural symbols to generate sufficient meaning for them to feel committed to participate fully in the society; and (4) a "legitimation

---

[24]Jurgen Habermas, *Legitimation Crisis,* trans. T. McCarthy (London: Heinemann, 1976; originally published in German in 1973).

crisis" arises when actors do not possess the "requisite number of generalized motivations" or diffuse commitments to the political subsystem's right to make decisions. Much of this analysis of crises is couched in Marxian terms but emphasizes that economic and rationality crises are perhaps less important than either motivational or legitimation crises. For, as technocratic consciousness penetrates all spheres of social life and creates productive economies as well as an intrusive state, the crisis tendencies of late capitalism shift from the inability to produce sufficient economic goods or political decisions to the failure to generate (*a*) diffuse commitments to political processes and (*b*) adequate levels of meaning among individual actors.

I have, of course, simplified Habermas' argument, but my main intent is only to highlight those aspects of his work that become prominent in his most recent theoretical synthesis. In *Legitimation Crisis* is an early form of what becomes an important distinction: "systemic" processes revolving around the economy and politico-administrative apparatus of the state must be distinguished from "cultural" processes. This distinction will later be conceptualized as *system* and *lifeworld*, respectively, but the central point is this: in tune with his Frankfurt School roots, Habermas is shifting emphasis from Marx's analysis of the economic crisis of production to crises of meaning and commitment; if the problems or crises of capitalist societies are in these areas, then critical theory must focus on the communicative and interactive processes by which humans generate understandings and meanings among themselves. For, if instrumental reason, or means/ends rationality, is driving out action based on mutual understanding and commitment, then the goal of critical theory is to expose this trend and to suggest ways of overcoming it, especially since legitimation and motivational crises make people aware that something is missing from their lives and, therefore, receptive to more emancipatory alternatives. And so the task of critical theory is to develop a theoretical perspective that allows for the restructuring of meaning and commitment in social life. This goal will be realized, Habermas argues, by further understanding of how people communicate, interact, and develop symbolic meanings.

Early Analyses of Speech and Interaction

In 1970 Habermas wrote two articles that marked a return to the idea of the public sphere, but with a new, more theoretical thrust. They also signaled an increasing emphasis on the process of speech, communication, and interaction. In his "On Systematically Distorted Communication," Habermas outlines the nature of undistorted communication.[25] True to Habermas' Weberian origins, this outline is an ideal type. The goal is to determine the essentials and essence of undistorted communication so that those processes that distort communication, such as domination, can be better exposed. What, then, are the features

---

[25]Jurgen Habermas, "On Systematically Distorted Communication," *Inquiry* 13 (1970), pp. 205–18.

of undistorted communication? Habermas lists five: (1) expressions, actions, and gestures are noncontradictory; (2) communication is public and conforms to cultural standards of what is appropriate; (3) actors can distinguish between the properties of language, per se, and the events and processes that are described by language; (4) communication leads to, and is the product of, intersubjectivity, or the capacity of actors to understand one another's subjective states and to develop a sense of shared collective meanings; and (5) conceptualizations of time and space are understood by actors to mean different things when externally observed and when subjectively experienced in the process of interaction.[26] The details of his analysis on the distortion of communication are less essential than the assertions about what critical theory must conceptualize. For Habermas the conceptualization of undistorted communication is used as a foil for mounting a critique against those social forces that make such idealized communication difficult to realize. Moreover, as his subsequent work testifies, Habermas emphasizes condition (4), or communication and intersubjectivity among actors.

This emphasis becomes evident in his other article published in 1970, "Toward a Theory of Communicative Competence."[27] Again, I do not think the details of this argument are as critical as the overall intent, especially since his ideas undergo subsequent modification. Habermas argues that, for actors to be competent, they must know more than the linguistic rules of how to construct sentences and to talk; they must also master "dialogue-constitutive universals," which are part of the "social linguistic structure of society." Behind this jargon is the idea that the meaning of language and speech is contextual and that actors use implicit stores or stocks of knowledge to interpret the meaning of utterances. Habermas then proposes yet another ideal type, "the ideal speech situation," in which actors possess all of the relevant background knowledge and linguistic skills to communicate without distortion.

Thus, in the early 1970s, Habermas began to view the mission of critical theory as emphasizing the process of interaction as mediated by speech. But such speech acts draw upon stores of knowledge—rules, norms, values, tacit understandings, memory traces, and the like—for their interpretation. I see these ideals of the speech process as a restatement of the romanticized public sphere, where issues were openly debated, discussed, and rationally resolved. What Habermas has done, of course, is to restate this view of "what is good and desirable" in more theoretical and conceptual terms, although it could be argued that there is not much difference between the romanticized portrayal of the public sphere and the ideal-typical conceptualization of speech. But with this conceptualization, the goal of critical theory must be to expose those conditions that distort communication and that inhibit realization of the ideal

---

[26]I am simplifying here. See also Jim Faught's discussion in his "Objective Reason and the Justification of Norms."

[27]Jurgen Habermas, "Toward a Theory of Communicative Competence," *Inquiry* 13 (1970), pp. 360-75.

speech situation. Habermas' utopia is thus a society where actors are able to communicate without distortion, to achieve a sense of one another's subjective states, and to openly reconcile their differences through argumentation that is free of external constraint and coercion. In other words, he wants to restore the public sphere but in a more encompassing way—that is, in people's day-to-day interactions.

Habermas moves in several different directions in trying to construct a rational approach for realizing this utopia. He borrows metaphorically from psychoanalytic theory as a way to uncover the distortions that inhibit open discourse,[28] but I think that this psychoanalytic journey is far less important than his growing concentration on the process of communicative action and interaction as the basis for creating a society that reduces domination and constraint. Thus, by the mid-1970s, he labeled his analysis *universal pragmatics*, whose centerpiece is the "theory of communicative action."[29] I will discuss this theory in more detail shortly, but let me briefly indicate its key elements. Communication involves more than words, grammar, and syntax; it also involves what Habermas terms *validity claims*. There are three types of claims: (1) those asserting that a course of action as indicated through speech is the most effective and efficient means for attaining ends; (2) those claiming that an action is correct and proper in accordance with relevant norms; and (3) those maintaining that the subjective experiences as expressed in a speech act are sincere and authentic. All speech acts implicitly make these three claims, although a speech act may emphasize one more than the other two. Those responding to communication can accept or challenge these validity claims; if challenged, then the actors contest, debate, criticize, and revise their communication. They use, of course, shared "stocks of knowledge" about norms, means/ends effectiveness, and sincerity to make their claims as well as to contest and revise them. This process (which, I should add, restates the public sphere in yet one more guise) is often usurped when claims are settled by recourse to power and authority. But if claims are settled by the "giving of reasons for" and "reasons against" the claim in a mutual give-and-take among individuals, then Habermas sees it as "rational discourse." Thus, built into the very process of interaction is the potential for rational discourse that can be used to create a more just, open, and free society. Such discourse is not merely means/ends rationality, for it involves adjudication of two other validity claims: those concerned with normative appropriateness and those concerned with subjective sincerity. Actors thus implicitly assess and critique one another in terms of effectiveness, normative appropriateness, and sincerity of their re-

---

[28]Habermas sometimes calls this aspect of his program "depth hermeneutics." And the idea is to create a methodology of inquiry for social systems that parallels the approach of psychoanalysis—that is, dialogue, removal of barriers to understanding, analysis of underlying causal processes, and efforts to use this understanding to dissolve distortions in interaction.

[29]For an early statement, see "Some Distinctions in Universal Pragmatics: A Working Paper," *Theory and Society* 3 (1976), pp. 155–67.

spective speech acts; and so the goal of critical theory is to expose those societal conditions that keep such processes from occurring for *all three types* of validity claims.

In this way Habermas has moved critical theory from Lukács', Horkheimer's, and Adorno's emphasis on subjective consciousness to a concern with *inter*subjective consciousness and the interactive processes by which intersubjectivity is created, maintained, and changed through the validity claims in each speech act. Moreover, rather than viewing the potential for liberating alternatives as residing in subjective consciousness, Habermas can assert that emancipatory potential inheres in each and every communicative interaction. And, since speech and communication are the basis of interaction and since society is ultimately sustained by interaction, the creation of less restrictive societies will come about by realizing the inherent dynamics of the communication process.

Habermas' Reconceptualization of Social Evolution

All critical theory is historical in the sense that it tries to analyze the long-term development of oppressive arrangements in society. Indeed, as I have emphasized, the central problem of critical theory is to reconcile Marx's and Weber's respective analyses of the development of advanced capitalism. It is not surprising, therefore, that Habermas produces a historical/evolutionary analysis, but, in contrast to Weber, he needs to see emancipatory potential in evolutionary trends. Yet at the same time he needs to avoid the incorrect prognosis in Marx's analysis and to retain the emancipatory thrust of Marx's approach. Habermas' first major effort to effect this reconciliation appears in his *The Reconstruction of Historical Materialism*,[30] parts of which have been translated and appear in *Communication and the Evolution of Society*.[31]

Habermas' approach to evolution pulls together many of the themes that I have discussed above, and so a brief review of his general argument can set the stage for an analysis of his most recent theoretical synthesis, *The Theory of Communicative Action*.[32] In many ways Habermas reintroduces traditional functionalism into Marx's and Weber's evolutionary descriptions, but with both a phenomenological and a structuralist emphasis (see Chapters 18, 23, 24, and 25). As with all functional theory, he views evolution as the process of structural differentiation and the emergence of integrative problems. He also borrows a page from Herbert Spencer, Talcott Parsons, and Niklas Luhmann when he argues that the integration of complex systems leads to an adaptive upgrading,

---

[30]Jurgen Habermas, *Zur Rekonstruktion des Historischen Materialismus* (Frankfurt: Suhrkamp, 1976).

[31]Jurgen Habermas, *Communication and the Evolution of Society*, trans. T. McCarthy (London: Heinemann, 1979).

[32]For an earlier statement, see Jurgen Habermas, "Towards a Reconstruction of Historical Materialism," *Theory and Society* 2 (3, 1975), pp. 84–98.

increasing the capacity of the society to cope with the environment.[33] That is, complex systems that are integrated are better adapted to their environments than are less complex systems. The key issue, then, is: what conditions increase or decrease integration? For, without integration, differentiation produces severe problems.

Habermas' analysis of system integration is both protracted and vague, and so I will offer only a cursory review. He argues that contained in the world views or stocks of knowledge of individual actors are learning capacities and stores of information that determine the overall learning level of a society. In turn, this learning level determines the society's steering capacity to respond to environmental problems. At times Habermas refers to these learning levels as *organization principles.* Thus, as systems confront problems of internal integration and external contingencies, the stocks of knowledge and world views of individual actors are translated into organization principles and steering capacities, which in turn set limits on just how a system can respond. For example, a society with only religious mythology will be less complex and less able to respond to environmental challenges than a more complex one with large stores of technology and stocks of normative procedures determining its organization principles. But societies can "learn"[34] that, when confronted with problems beyond the capacity of their current organization principles and steering mechanisms, they can draw upon the "cognitive potential" in the world views and stocks of knowledge of individuals who reorganize their actions. The result of this learning creates new levels of information that allow for the development of new organization principles for securing integration in the face of increased societal differentiation and complexity.

I have only skimmed the surface of Habermas' views; the details are perhaps best left out since they are ambiguous. Yet there is an implicit line of argument that I should emphasize: the basis for societal integration lies in the processes by which actors communicate and develop mutual understandings and stores of knowledge. To the extent that these interactive processes are arrested by the patterns of economic and political organization, the society's learning capacity is correspondingly diminished. One of the main integrative problems of capitalist societies is the integration of the material forces of production (economy as administered by the state), on the one side, and the cultural stores of knowledge that are produced by communicative interaction, on the other side. Societies that differentiate materially in the economic and political realms without achieving integration on a normative and cultural level (i.e., shared understandings) will remain unintegrated and experience crises.

Built into these dynamics, however, is their resolution. The processes of "communicative interaction" that produce and reproduce unifying cultural

---

[33]He borrows from Niklas Luhmann here (see Chapter 5), although much of Habermas' approach is a reaction to Luhmann.

[34]Habermas analogizes here to Piaget's and Kohlberg's analysis of the cognitive development of children, seeing societies as able to "learn" as they become more structurally complex.

symbols must be given equal weight with the "labor" processes that generate material production and reproduction. It is at this point that Habermas develops his more synthetic approach in *The Theory of Communicative Action.*

The Theory of Communicative Action

The two-volume *The Theory of Communicative Action* pulls together into a reasonably coherent framework various strands of Habermas' thought.[35] Yet, true to his general style of scholarship, Habermas wanders over a rather large intellectual landscape. In Thomas McCarthy's words, Habermas develops his ideas through "a somewhat unusual combination of theoretical constructions with historical reconstructions of the ideas of 'classical' social theorists."[36] Such thinkers as Marx, Weber, Durkheim, Mead, Lukács, Horkheimer, Adorno, and Parsons are, for Habermas, "still very much alive" and are treated as "virtual dialogue partners."[37] As a consequence, the two volumes meander through selected portions of various thinkers' work with an eye toward critiquing and yet utilizing key ideas. After the dust settles, however, the end result is a very creative synthesis of ideas into a critical theory.

Habermas' basic premise is summarized near the end of volume 1:

> If we assume that the human species maintains itself through the socially coordinated activities of its members and that this coordination is established through communication—and in certain spheres of life, through communication aimed at reaching agreement—then the reproduction of the species also requires satisfying the conditions of a rationality inherent in communicative action.[38]

In other words, intrinsic to the process of communicative action, where actors implicitly make, challenge, and accept one another's validity claims, is a rationality that can potentially serve as the basis for reconstructing the social order in less oppressive ways. The first volume of *The Theory of Communicative Action* thus focuses on action and rationality in an effort to reconceptualize both processes in a manner that shifts emphasis from the subjectivity and consciousness of the individual to the process of symbolic interaction. In a sense, volume 1 is Habermas' microsociology, whereas volume 2 is his macrosociology. In this second volume Habermas introduces the concept of system and tries to connect it to microprocesses of action and interaction through a reconceptualization of the phenomenological concept of lifeworld (see Chapters 18 and 23).

---

[35]Jurgen Habermas, *The Theory of Communicative Action*, 2 vols. The subtitle of volume 1, *Reason and the Rationalization of Society*, gives some indication of its thrust. The translator, Thomas McCarthy, has done an excellent service in translating what is very difficult prose. Also, his "Translator's Introduction" to volume 1, pp. v–xxxvii, is the best summary of Habermas' recent theory that I have come across.

[36]Thomas McCarthy, "Translator's Introduction," p. vii.

[37]Ibid.

[38]Jurgen Habermas, *The Theory of Communicative Action*, vol. 1, p. 397.

The overall project   Let me begin by briefly summarizing the overall argument, and then I will return to volumes 1 and 2 with a more detailed analysis. There are four types of action: (1) teleological, (2) normative, (3) dramaturgical, and (4) communicative. Only communicative action contains the elements whereby actors reach intersubjective understanding. Such communicative action—which is, in fact, interaction—presupposes a set of background assumptions and stocks of knowledge, or, in Habermas' terms, a *lifeworld*. Also operating in any society are "system" processes, which revolve around the material maintenance of the species and its survival. The evolutionary trend is for system processes and lifeworld processes to become internally differentiated *and* differentiated from each other. The integration of a society depends upon a balance between system and lifeworld processes. As modern societies have evolved, however, this balance has been upset as system processes revolving around the economy and the state (also law, family, and other reproductive structures) have "colonized" and dominated lifeworld processes concerned with mutually shared meanings, understandings, and intersubjectivity. As a result, modern society is poorly integrated.

These integrative problems in capitalist societies are manifested in crises concerning the "reproduction of the lifeworld"; that is, the acts of communicative interaction that reproduce this lifeworld are displaced by "delinguistified media," such as money and power, that are used in the reproduction of system processes (economy and government). The solution to these crises is a rebalancing of relations between lifeworld and system. This rebalancing is to come through the resurrection of the public sphere in the economic and political arenas and in the creation of more situations in which communicative action (interaction) can proceed uninhibited by the intrusion of system's media, such as power and money. The goal of critical theory, therefore, is to document those facets of society in which the lifeworld has been colonized and to suggest approaches whereby situations of communicative action (interaction) can be reestablished. Such is Habermas' general argument, and now I will fill in some of the details.

The reconceptualization of action and rationality   In volume 1 of *The Theory of Communicative Action*, Habermas undertakes a long and detailed analysis of Weber's conceptualization of action and rationalization. He does so because he wants to reconceptualize rationality and action in ways that allow him to view rational action as a potentially liberating rather than imprisoning force.[39] In this way, he feels, he can avoid the pessimism of Weber and the retreat into subjectivity of Lukács, Adorno, and Horkheimer.

Habermas argues by adding typological distinctions on top of one another, which, to say the least, makes the exposition somewhat ponderous. I will sim-

---

[39]Recall that its subtitle is *Reason and the Rationalization of Society*.

plify a bit and discuss only the critical distinctions. There are, he concludes, several different types of action:[40]

1. *Teleological action* is behavior oriented to calculating various means and selecting the most appropriate ones to realize explicit goals. Such action becomes strategic when other acting agents are involved in one's calculations. Habermas also calls this action "instrumental" because it is concerned with means to achieve ends. Most importantly, he emphasizes that this kind of action is too often considered to be "rational action" in previous conceptualizations of rationality. For, as he is to argue, this view of rationality is too narrow and forces critical theory into a conceptual trap: if teleological or means/ends rationality is what has taken over the modern world and what has, as a consequence, oppressed people, then how can critical theory propose rational alternatives? Would not such a rational theory be yet one more oppressive application of means/ends rationality? The answers to these questions lie in recognizing that there are several types of action and that true rationality resides not in teleological action but in communicative action.

2. *Normatively regulated action* is behavior that is oriented to common values of a group. Thus, normative action is directed toward complying with normative expectations of collectively organized groupings of individuals.

3. *Dramaturgical action* is action that involves conscious manipulation of oneself before an audience or public. It is ego-centered in that it involves actors mutually manipulating their behaviors to present their own intentions, but it is also social in that such manipulation is done in the context of organized activity.

4. *Communicative action* is interaction among agents who use speech and nonverbal symbols as a way of understanding their mutual situation and their respective plans of action in order to agree on how to coordinate their behaviors.

These four types of action presuppose different kinds of "worlds." That is, each action is oriented to a somewhat different aspect of the universe, which can be divided into the (1) "objective or external world" of manipulable objects; (2) "social world" of norms, values, and other socially recognized expectations; and (3) "subjective world" of experiences. Teleological action is concerned primarily with the objective world; normatively regulated action with the social; and dramaturgical with the subjective and external. But it is only with communicative action that actors "refer simultaneously to things in the objective, social, and subjective worlds in order to negotiate common definitions of the situation."[41]

---

[40]Jurgen Habermas, *The Theory of Communicative Action*, pp. 85–102.
[41]Ibid., p. 95.

Such communicative action is therefore potentially more rational than all of the others because it deals with all three worlds and because it proceeds in terms of speech acts that assert three types of validity claims. Such speech acts assert that (1) statements are true in "propositional content," or in reference to the external and objective world; (2) statements are correct with respect to the existing normative context, or social world; and (3) statements are sincere and manifest the subjective world of intention and experiences of the actor.[42] The process of communicative action in which these three types of validity claims are made, accepted, or challenged by others is inherently more rational than other types of action. For, if a validity claim is not accepted, then it is debated and discussed in an effort to reach understanding without recourse to force and authority.[43] The process of reaching understanding through validity claims, their acceptance, or their discussion takes place against:

> the background of a culturally ingrained preunderstanding. This background remains unproblematic as a whole; only that part of the stock of knowledge that participants make use of and thematize at a given time is put to the test. To the extent that definitions of situations are negotiated by participants *themselves*, this thematic segment of the lifeworld is at their disposal with the negotiation of each new definition of the situation.[44]

Thus, in the process of making validity claims through speech acts, actors use existing definitions of situations or create new ones that establish order in their social relations. Such definitions become part of the stocks of knowledge in their lifeworlds, and they become the standards by which validity claims are made, accepted, and challenged. Thus, in reaching an understanding through communicative action, the lifeworld serves as a point of reference for the adjudication of validity claims, which encompass the full range of worlds— the objective, social, and subjective. And so, in Habermas' eyes, there is more rationality inherent in the very process of communicative interaction than in means/ends or teleological action.[45] As Habermas summarizes in a rare moment of succinctness:

> We have . . . characterized the rational structure of the processes of reaching understanding in terms of (a) the three world-relations of actors and the corresponding concepts of the objective, social, and subjective worlds; (b) the validity claims of propositional truth, normative rightness, and sincerity or authenticity; (c) the concept of a rationally motivated agreement, that is, one based on the intersubjective recognition of criticizable validity claims; and (d)

---

[42]Ibid., p. 99.

[43]Recall Habermas' earlier discussion of nondistorted communication and the ideal speech act. This is his most recent reconceptualization of these ideas.

[44]Ibid., p. 100. Emphasis in original.

[45]Ibid., p. 302.

the concept of reaching understanding as the cooperative negotiation of common definitions of the situation.[46]

Thus, as people communicatively act (interact), they use and at the same time produce common definitions of the situation. Such definitions are part of the lifeworld of a society; if they have been produced and reproduced through the communicative action, then they are the basis for the rational and non-oppressive integration of a society. Let me now turn to Habermas' discussion of this lifeworld, which serves as the "court of appeals" in communicative action.

The lifeworld and system processes of society   For Habermas the lifeworld is a "culturally transmitted and linguistically organized stock of interpretative patterns." But what are these "interpretative patterns" about? What do they pertain to? His answer, as one comes to expect from Habermas, is yet another typology. There are three different types of interpretative patterns in the lifeworld: there are interpretative patterns with respect to culture, or systems of symbols; there are those pertaining to society, or social institutions; and there are those oriented to personality, or aspects of self and being. That is, (1) actors possess implicit and shared stocks of knowledge about cultural traditions, values, beliefs, linguistic structures and their use in interaction; (2) they also know how to organize social relations and what kinds and patterns of coordinated interaction are proper and appropriate; and (3) they understand what people are like, how they should act, and what is normal or aberrant.

These three types of interpretative patterns correspond, Habermas asserts, to the following functional needs for reproducing the lifeworld (and, by implication, for integrating society): (1) reaching understanding through communicative action serves the function of transmitting, preserving, and renewing cultural knowledge; (2) communicative action that coordinates interaction meets the need for social integration and group solidarity; and (3) communicative action that socializes agents meets the need for the formation of personal identities.[47]

Thus the three components of the lifeworld—culture, society, personality—meet corresponding needs of society—cultural reproduction, social integration, and personality formation—through three dimensions along which communicative action is conducted—reaching understanding, coordinating interaction, and effecting socialization. As Habermas summarizes in volume 2:

> In coming to an understanding with one another about their situation, participants in communication stand in a cultural tradition which they use and at the same time renew; in coordinating their actions via intersubjective

---

[46]Ibid., p. 137.

[47]We are now into volume 2, ibid., pp. 205–40, which is entitled *System and Lifeworld: A Critique of Functionalist Reason*, which is, I think, a somewhat ironic title because of the heavily functional arguments in volume 2. But, as I pointed out earlier, even Habermas' earlier work has always had an implicit functionalism.

recognition of criticizable validity claims, they rely upon their membership in groupings and at the same time reenforce their integration; through participating in interaction with competent persons, growing children internalize value orientations and acquire generalized capacities for action.[48]

These lifeworld processes are interrelated with system processes in a society. Action in economic, political, familial, and other institutional contexts draws upon, and reproduces, the cultural, societal, and personality dimensions of the lifeworld. Yet evolutionary trends are for differentiation of the lifeworld into separate stocks of knowledge with respect to culture, society, and personality and for differentiation of system processes into distinctive and separate institutional clusters, such as economy, state, family, and law. Such differentiation creates problems of integration and balance between the lifeworld and system.[49] And therein reside the dilemmas and crises of modern societies.

Evolutionary dynamics and societal crises   In a sense, Habermas blends traditional analysis by functionalists on societal and cultural differentiation with a Marxian dialectic whereby the seeds for emancipation are sown in the creation of an ever more rationalized and differentiated society. Borrowing from Durkheim's analysis of mechanical solidarity, Habermas argues that "the more cultural traditions predecide which validity claims, when, where, for what, from whom, and to whom must be accepted, the less the participants themselves have the possibility of making explicit and examining the potential groups in which their yes/no positions are based."[50] But "as mechanical solidarity gives way to organic solidarity based upon functional interdependence," then "the more the worldview that furnishes the cultural stock of knowledge is decentered" and "the less the need for understanding is covered *in advance* by an interpreted lifeworld immune from critique," and therefore "the more this need has to be met by the interpretative accomplishments of the participants themselves." That is, if the lifeworld is to be sustained and reproduced, it becomes ever more necessary with growing societal complexity for social actions to be based upon communicative processes. The result is that there is greater potential for rational communicative action because less and less of the social order is preordained by a simple and undifferentiated lifeworld. But system processes have operated to reduce this potential, and the task of critical theory is to document how system processes have worked to colonize the lifeworld and thereby arrest this potentially superior rationality inherent in the speech acts of communicative action.

How have system processes restricted this potential contained in communicative action? As the sacred/traditional basis of the lifeworld organization has dissolved and been replaced by linguistic interaction around a lifeworld

---

[48]Ibid., p. 208.

[49]This is the old functionalist argument of "differentiation" producing "integrative problems," which is as old as Spencer and which is Parsons reincarnated with a phenomenological twist.

[50]I have shifted now, back to portions of volume 1. All quotes here are from p. 70 of volume 1.

differentiated along cultural, social, and personality axes, there is a countertrend in the differentiation of system processes. System evolution involves the expansion of material production through the greater use of technologies, science, and "delinguistified steering mechanisms" such as money and power to carry out system processes.[51] These media do not rely upon the validity claims of communicative action; when they become the media of interaction in ever more spheres of life—markets, bureaucracies, welfare state policies, legal systems, and even family relations—the processes of communicative action so essential for lifeworld reproduction are invaded and colonized. Thus system processes use power and money as their media of integration, and in the process they "decouple the lifeworld" from its functions for societal integration.[52] There is an irony here because differentiation of the lifeworld facilitated the differentiation of system processes and the use of money and power,[53] and so "the rationalized lifeworld makes possible the rise of growth of subsystems which strike back at it in a destructive fashion."[54]

Through this ironical process, capitalism creates market dynamics using money, which in turn spawn a welfare state employing power in ways that reduce political and economic crises but that increase those crises revolving around lifeworld reproduction. For the new crises and conflicts "arise in areas of cultural reproduction, of social integration and of socialization."[55]

The goal of critical theory   Habermas has now circled back to his initial concerns and those of early critical theorists. He has recast the Weberian thesis by asserting that true rationality inheres in communicative action, not teleological (and strategic or instrumental) action, as Weber claimed. And he has redefined the critical theorist's view on modern crises; they are not ones of rationalization, but ones of colonization of those truly rational processes that inhere in the speech acts of communicative action, which reproduce the lifeworld so essential to societal integration. Thus, built into the integrating processes of differentiated societies (note: not the subjective processes of individuals, as early critical theorists claimed) is the potential for a critical theory that seeks to restore communicative rationality in the face of impersonal steering mechanisms. If system differentiation occurs in terms of delinguistified media, like money and power, and if these reduce the reliance upon communicative action, then crises are inevitable. The resulting collective frustration over the lack of meaning in social life can be used by critical theorists to mobilize people to restore the proper balance between system and lifeworld processes.

---

[51]Here Habermas is borrowing from Simmel's analysis in *The Philosophy of Money* (see Chapter 9) and from Parsons' conceptualization of generalized media (see Chapter 3).

[52]I am now back in volume 2 of *The Theory of Communicative Action*, pp. 256–76.

[53]Habermas appears in these arguments to borrow heavily from Parsons' analysis of evolution (see Chapter 3).

[54]Volume 2, p. 227.

[55]Ibid., p. 576.

Thus it is not crises of material production that will be the impetus for change, as Marx contended. It is the crises of lifeworld reproduction that will serve as the stimulus to societal reorganization. And returning to his first work, Habermas sees such reorganization as involving (1) the restoration of the public sphere in politics, where relinguistified debate and argumentation, rather than delinguistified power and authority, are used to make political decisions (thus reducing "legitimation crises"), and (2) the extension of communicative action back into those spheres—family, work, and social relations—that have become increasingly dominated by delinguistified steering media (thereby eliminating "motivational crises").

The potential for this reorganization inheres in the nature of societal integration through the rationality inherent in the communicative actions that reproduce the lifeworld. It is the purpose of critical theory to release this rational potential.

AN ASSESSMENT OF HABERMAS' CRITICAL THEORY AND PROJECT

Because it does not claim to be science or theory in the sense outlined in Chapter 1, critical theory is difficult to assess in other than its own terms.[56] Yet I have examined Habermas' project in detail because, unlike many in this area, it does try to develop a theoretical perspective for understanding the dynamics of human organization. In assessing Habermas' approach, therefore, I will focus only on those portions of his project that analyze human organization. I will leave to others the assessment of his ideological argument, which, I must confess, I find hopelessly naive. Naturally, Habermas' critical approach owes its guiding inspiration to the tradition of the Frankfurt School, and his formulation of concepts has thereby been biased by these ideological commitments. But, nonetheless, I will concentrate primarily on one central question: does Habermas' approach give us new insight and understanding into the processes of human action, interaction, and organization?

I admit that this question is somewhat unfair, because much of the creativity in Habermas' approach is the blending of theory and ideology. But, since my commitments are to the separation of science from ideology (of course, I recognize the problems involved), I would have few positive remarks for theory that is so explicitly ideological (of course, Habermas would counter that science is ideological because it serves certain interests). Thus, just as I did for Marx in Chapter 9, I will try to stress the merits of Habermas' approach *as theory*, divorced from his critical project.

Evaluating Habermas' Conception of Action

Habermas' distinction among teleological (plus its strategic and instrumental variants), normative, dramaturgical, and communicative action does direct our

---

[56]Indeed, it is both "ideological" and "scientific" in terms of the scheme drawn in Figure 1-1.

attention to different phases of behavior. The distinction between teleological and normative adds little beyond Parsons' formulation in 1937.[57] The dramaturgical does incorporate Goffman's ideas into action theory, but Habermas does not extend Goffman's conceptualization in any way.[58] The only unique portion of the approach, then, is the conceptualization of communicative action, which is, in reality, a perspective for the analysis of interaction.

Habermas' Conception of Interaction

In the analysis of communicative action, Habermas does blend elements of George Herbert Mead's behaviorist/interactionist approach and Alfred Schutz's phenomenological/interactionist ideas (see Chapter 18) with portions of ethnomethodology (see Chapter 23) and linguistic analysis. I have not discussed Habermas' detailed presentation of speech processes, but they figure prominently in his classification of types of action and the nature of validity claims in speech acts.[59] Moreover, the emphasis on the processes of reaching understanding and intersubjectivity through speech acts and implicit background assumptions contained in the lifeworld does, I think, represent a creative synthesis of linguistics, ethnomethodology, phenomenology, and symbolic interactionism. Speech acts are seen as contextual or indexical in that their meaning is shaped by implicit stocks of knowledge, and actors use these stocks to interpret one another's gestures and achieve a sense of intersubjectivity. There is nothing new in the formulation as I have just stated it, but the notion that speech acts make three types of validity claims—propositional truth, normative appropriateness, and subjective sincerity—is a very interesting idea. That is, Habermas is asserting that, in achieving understanding and intersubjectivity through the contextual interpretations of speech acts, humans make implicit validity claims along these dimensions. Furthermore, Habermas' contention that this sense of understanding and intersubjectivity depends upon such validity claims being made, accepted, or challenged without recourse to authority is, likewise, an interesting proposition. Thus, although I am not sure that Habermas' ideas denote how interaction actually occurs, he has given interactionist theory some new ideas that, I feel, need to be considered carefully.

Habermas' Macrosociology: System and Lifeworld

By viewing lifeworld as consisting of interpretative patterns with respect to culture, society, and personality, Habermas has done three things. First, he has tied Talcott Parsons' analytical distinctions among personality, social, and

---

[57]Talcott Parsons, *The Structure of Social Action* (New York: McGraw-Hill, 1937), does not go much beyond Weber's analysis.

[58]Erving Goffman, *The Presentation of Self in Everyday Life* (Garden City, NY: Doubleday, 1959). See also Chapter 22.

[59]Habermas, *The Theory of Communicative Action*, vol. 1, pp. 273–344.

cultural systems to the process of interaction.[60] That is, people use linguistically articulated patterns as well as more implicit and tacit stocks of knowledge about culture, society, and personality in their day-to-day interactions. These distinctions are not just those of analytical theory; they are also implicit "folk categories" of actual people and are part of the active lifeworld among interacting and communicating actors.

Second, Habermas has made less mysterious and vague the lifeworld as it has been conceptualized by phenomenologists and ethnomethodologists. It is not just an amorphous mass of implicit and tacit understandings; rather, it is a series of folk ideas about several classes of phenomena—culture (symbols), society (social organization), and personality (self and ego).

Third, Habermas has extended and refined functionalists' arguments about social integration as being dependent upon value consensus and other cultural forces, while he has synthesized them with the more processual arguments of interactionists. By viewing the lifeworld as interpretative patterns with respect to culture, society, and personality and as the background context for speech acts and validity claims, societal integration is reconceptualized as being sustained through the active processes of interaction that utilize and reproduce the interpretative patterns of the lifeworld. Culture is not some ex cathedra force external to actors, pushing and shoving them around. Instead, it is part of a more encompassing stock of knowledge and a crucial resource that is used in interaction. And, in being actively used, it is reproduced in ways that allow actors to understand one another's subjective points of view and to coordinate their actions.[61] As such, it has functions for sustaining the macrostructures of societies.

System processes are given a rather standard interpretation of institutional structures involved in material reproduction of the species through the institutionalization of economic, political, legal, and family activities. But there are creative points of synthesis in Habermas' argument when he views lifeworld and system processes as interdependent. That is, institutionalized structures like economy and polity function better when the behaviors that reproduce them are conducted in terms of communicative interactions utilizing and reproducing the interpretative patterns of the lifeworld. And Habermas' view of the integrative problems of modern societies as a decoupling of lifeworld and system is, I feel, an interesting way to reconceptualize the basic integrative problems of differentiated systems. For when Habermas talks about a decoupling, he is doing more than merely restating Durkheim's anomie thesis, Marx's alienation argument, or Parsons' discussion of imbalances of information or energy in his cybernetic hierarchy of control.[62] He is also specifying the particular *processes of interaction* that are being decoupled from institutionalized

---

[60]See Chapter 3. See also Talcott Parsons, *The Social System* (New York: Free Press, 1951).

[61]Habermas has also refined Mead's view of "role taking" with the "generalized other" in this conceptualization. See Chapter 18.

[62]See Chapter 3, Figure 3-7.

patterns. In so doing, Habermas is far less vague than are notions of malintegration so typical of functional theory. He is indicating more precisely what is being lost—certain types of speech acts involving validity claims and discourse that utilize lifeworld processes to create the mutual understandings that integrate the social order.

Habermas' View of Evolution

Habermas interprets evolution in highly functional terms as a dual process of (1) differentiation and (2) creation of integrative problems, especially those of sustaining common meanings among actors in differentiated niches in a complex society. He borrows from Durkheim's and Parsons' view of evolution as a process of differentiation between and within symbolic and structural realms, with the essential dilemma of modern societies revolving around how to put these two realms back together again so that actions in social structures are meaningful to individuals. But, unlike most functional theorists, he turns to interactionism (see Chapters 18-23) for his answer: restore the processes of communicative action, or symbolic interaction.

My sense is that he merely restated the problem in interactionist terms, without telling us *how* or *through what procedures* communicative action is to be restored. Moreover, I agree with Niklas Luhmann's critique that humans do in fact reach understanding in their interactions in complex systems dominated by the steering media of power and money.[63] Habermas has, I think, a rather romanticized view of what human interaction is really like and of what humans actually need in their social relations in order to be content. He assumes incorrectly, I believe, that the symbolic interactions that generate intersubjectivity do not occur in structures dominated by power and money; my sense is that they do, as is amply evidenced by studies of "the informal system" in bureaucratic structures.

And this line of argument brings me to my last observation of Habermas' approach. There is a kind of naive romanticism in all of his work, from his first discussion of the public sphere to the present. He employs a totally artificial yardstick—first the yardstick of undistorted communication, then one on the ideal speech situation, and finally the criterion of communicative action—for assessing what's wrong with modern societies. When one begins the analysis of human social organization with a set of standards that, I suspect, have never been met in human affairs, then one's interpretation of events will be rather dramatically obscured. Although Habermas has blended many diverse traditions in creative ways and offered, as I have indicated, a number of interesting conceptual leads, I do not think that these denote the operative dynamics of human social systems. And this is where all critical theory goes wrong—from

---

[63]Niklas Luhmann, *The Differentiation of Society* (New York: Columbia University Press, 1982). Indeed, Luhmann and Habermas have had a number of stimulating debates on the issue, with Luhmann emerging the victor in my opinion.

Marx's dream of the communist society to Habermas' desire for a society integrated by communicative action. When ideology and hope for a future society are the beginning points of analysis, I doubt if the end points will be as accurate as would be the case if these ideological hopes were never a part of the theoretical project. Yet it is a credit to both Marx's and Habermas' genius that, despite this great flaw, they have both produced a body of concepts that, when stripped of the ideological trappings, can be useful in building scientific theory.

PART 3

$\blacklozenge$

Exchange Theorizing

Early Forms of
Exchange Theorizing

The intellectual lineage culminating in modern exchange theory is very diverse. What typifies this heritage as much as its diversity are the frequently vague connections between contemporary exchange theorists and their predecessors. Indeed, I find current exchange theories to be a curious and unspecified mixture of utilitarian economics, functional anthropology, conflict sociology, and behavioral psychology. As a result, tracing the roots of exchange theory is an eclectic and uncertain enterprise. At best, I can only summarize various types of early exchange theory, leaving unanswered the question of how each modern theorist has drawn from this diverse legacy.

UTILITARIANISM: THE LEGACY OF
CLASSICAL ECONOMICS

The names of Adam Smith, David Ricardo, John Stuart Mill, and Jeremy Bentham loom large in the history of economic theorizing between 1770 and 1850. Each made a unique contribution to both economic and social thought, but I see several common assumptions in all of their works that enable their thought to be labeled *utilitarianism*. For these classical economists, humans are viewed as rationally seeking to maximize their material benefits, or utility, from transactions or exchanges with others in a free and competitive marketplace. As rational units in a free marketplace, people have access to all necessary information, can consider all available alternatives, and, on the basis of this consideration, rationally select the course of activity that will maximize material benefits. Entering into these rational considerations are calculations of the costs involved in pursuing various alternatives. Such costs must be weighed against material benefits in an effort to determine which alternative will yield the maximum payoff or profit (benefits less costs).[1]

---

[1]For interesting discussions of utilitarian thought as it bears on the present discussion, see Elie

With the emergence of sociology as a self-conscious discipline, there was considerable borrowing, revision, and reaction to this conception of humans. In fact, the debate between the intellectual descendants of utilitarianism and those reacting to this perspective has raged since sociology's inception. For example, as I outlined in Chapter 3, Talcott Parsons attempted to reformulate utilitarian principles and weld them to other theoretical traditions in an effort to develop a general perspective on social action.[2] Similarly, modern exchange theorists have attempted to reformulate utilitarian principles into various theories of social exchange.[3] This reformulation asserts the following alternative assumptions:

1. Humans do not seek to maximize profits, but they always attempt to make some profit in their social transactions with others.
2. Humans are not perfectly rational, but they do engage in calculations of costs and benefits in social transactions.
3. Humans do not have perfect information on all available alternatives, but they are usually aware of at least some alternatives, which form the basis for assessments of costs and benefits.
4. Humans always act under constraints, but they still compete with one another in seeking to make a profit in their transactions.
5. Humans always seek to make a profit in their transactions, but they are limited by the resources that they have when entering an exchange relation.

In addition to these alterations of utilitarian assumptions, exchange theory removes human interaction from the limitations of material transactions in an economic marketplace. Rather, these alternative assumptions are seen to apply to all social contexts, requiring that I add two more:

6. Humans do engage in economic transactions in clearly defined marketplaces in all societies, but these transactions are only a special case of more general exchange relations occurring among individuals in virtually all social contexts.
7. Humans do pursue material goals in exchanges, but they also mobilize and exchange nonmaterial resources, such as sentiments, services, and symbols.

Aside from this revised substantive legacy, I believe that some forms of modern exchange theory have also adopted the strategy of the utilitarians for constructing social theory. In assuming humans to be rational, utilitarians

---

Halévy, *The Growth of Philosophical Radicalism* (London: Farber & Farber, 1928); John Plamenatz, *Man and Society* (New York: McGraw-Hill, 1963).

[2]This effort continues, as we saw with Jurgen Habermas' work in the last chapter. And in both sociology and economics today, there is concern—excessive concern, in my view—with the notion of "rationality."

[3]The most blatant of these efforts, which I will not examine in the following chapters, is James Coleman's "Foundations for a Theory of Collective Decisions," *American Journal of Sociology* 71 (1966), pp. 615–27.

argued that exchanges among people could also be studied by a rational science, one in which the "laws of human nature" would stand at the top of a deductive system of explanation.[4] Thus utilitarians borrowed the early physical-science conception of theory as a logico-deductive system of axioms or laws and various layers of lower-order propositions that could be rationally deduced from the laws of "economic man." Most exchange theories are thus presented in a propositional format, typically in what I called a formal propositional scheme in Chapter 1.

In reviewing the tenets of utilitarianism, however, I think that only part of the historical legacy of exchange theory is revealed. For utilitarianism excited considerable debate and controversy in early anthropology. In fact, I sense that much of the influence of utilitarianism on current exchange theory has been indirect, passing through social anthropology around the turn of this century.

EXCHANGE THEORY IN ANTHROPOLOGY[5]

Sir James Frazer

In 1919 Sir James George Frazer's (1854–1941) second volume of *Folklore in the Old Testament* conducted what was probably the first explicit exchange-theoretic analysis of social institutions.[6] In examining a wide variety of kinship and marriage practices among primitive societies, Frazer was struck by the clear preference of Australian aborigines for cross-cousin over parallel-cousin marriages: "Why is the marriage of cross-cousins so often favored? Why is the marriage of ortho-cousins [that is, parallel cousins] so uniformly prohibited?"[7]

Although the substantive details of Frazer's descriptions of the aborigines' practices are fascinating in themselves (if only for their inaccuracy), I think that it is the form of his explanation that marks his theoretical contribution. In a manner clearly indebted to utilitarian economics, Frazer launched an economic interpretation of the predominance of cross-cousin marriage patterns. In this explanation Frazer invoked the "law" of "economic motives," since, in having "no equivalent in property to give for a wife, an Australian aborigine is generally obliged to get her in exchange for a female relative, usually a sister or daughter."[8] Thus the material or economic motives of individuals in society

---

[4]Jurgen Habermas would, of course, make this a much more complicated issue. See his *The Theory of Communicative Action*, vol. 1 (Boston: Beacon Press, 1981).

[5]For a similar treatment of these materials, see Peter Ekeh, *Social Exchange Theory and the Two Sociological Traditions* (Cambridge, MA: Harvard University Press, 1975).

[6]Sir James George Frazer, *Folklore in the Old Testament*, vol. 2 (New York: Macmillan, 1919); see also his *Totemism and Exogamy: A Treatise on Certain Early Forms of Superstition and Society* (London: Dawsons of Pall Mall, 1968; originally published in 1910); and his Preface to Bronislaw Malinowski's *Argonauts of the Western Pacific* (London: Routledge & Kegan Paul, 1922), pp. vii–xiv.

[7]Quote taken from Ekeh's *Social Exchange Theory*, pp. 41–42, discussion of Frazer (original quote in Frazer, *Folklore*, p. 199).

[8]Ibid., p. 195.

(lack of property and desire for a wife) explain various social patterns (cross-cousin marriages). What is more, Frazer went on to postulate that, once a particular pattern emanating from economic motives becomes established in a culture, it constrains other social patterns that can potentially emerge.

For Frazer, the social/structural patterns that come to typify a particular culture are a reflection of economic motives in humans, who, in exchanging commodities, attempt to satisfy their basic economic needs. Although Frazer's specific explanation was to be found sadly wanting by subsequent generations of anthropologists, especially Malinowski and Lévi-Strauss, modern exchange theory in sociology invokes a similar conception of social organization:

1. Exchange processes are the result of efforts by people to realize basic needs.
2. When yielding payoffs for those involved, exchange processes lead to the patterning of interaction.
3. Such patterns of interaction not only serve the needs of individuals but also constrain the kinds of social structures that can subsequently emerge.

In addition to anticipating the general profile of modern explanations on how elementary exchange processes create more complex patterns in a society, Frazer's analysis also foreshadowed another concern of contemporary exchange theory: the differentiation of social systems in terms of privilege and power. Much as Marx had done a generation earlier, Frazer noted that those who possess resources of high economic value can exploit those who have few such resources, thereby enabling them to possess high privilege and presumably power. Hence the exchange of women among the aborigines was observed by Frazer to lead to the differentiation of power and privilege in at least two separate ways. First, "Since among the Australian aboriginals women had a high economic and commercial value, a man who had many sisters or daughters was rich and a man who had none was poor and might be unable to procure a wife at all."[9] Second, "the old men availed themselves of the system of exchange in order to procure a number of wives for themselves from among the young women, while the young men, having no women to give in exchange, were often obliged to remain single or to put up with the cast-off wives of their elders."[10] Thus, at least implicitly, Frazer supplemented the conflict theory contribution with a fourth exchange principle:

4. Exchange processes operate to differentiate groups in terms of their relative access to valued commodities, resulting in differences in power, prestige, and privilege.

As provocative and seemingly seminal as Frazer's analysis appears, I doubt if it had a direct impact on modern exchange theory. Rather, it is to those in

---

[9]Frazer, *Folklore*, p. 198.
[10]Ibid., pp. 200–201.

anthropology who reacted against Frazer's utilitarianism that contemporary theory remains indebted.

Bronislaw Malinowski and Nonmaterial Exchange

Despite Malinowski's close ties with Frazer, he developed an exchange perspective that radically altered the utilitarian slant of Frazer's analysis of cross-cousin marriage. Indeed, Frazer himself, in his preface to Malinowski's *Argonauts of the Western Pacific*, recognized the importance of Malinowski's contribution to the analysis of exchange relations.[11] In his now-famous ethnography of the Trobriand Islanders—a group of South Seas Island cultures—Malinowski observed an exchange system termed the *Kula Ring,* a closed circle of exchange relations among tribal peoples inhabiting a wide ring of islands.[12] What was distinctive in this closed circle, Malinowski observed, was the predominance of exchange of two articles—armlets and necklaces—which the inhabitants constantly exchanged in opposite directions. Armlets traveling in one direction around the Kula Ring were exchanged for necklaces moving in the opposite direction around the ring. In any particular exchange between individuals, then, an armlet would always be exchanged for a necklace.

In interpreting this unique exchange network, Malinowski was led to distinguish material or economic from nonmaterial or symbolic exchanges. In contrast with the utilitarians and Frazer, who did not conceptualize nonmaterial exchange relations, Malinowski recognized that the Kula was not only an economic or material exchange network but also a symbolic exchange, cementing a web of social relationships: "One transaction does not finish the Kula relationship, the rule being 'once in the Kula, always in the Kula,' and a partnership between two men is a permanent and lifelong affair."[13] Although purely economic transactions did occur within the rules of the Kula, the ceremonial exchange of armlets and necklaces was observed by Malinowski to be the Kula's principal function.

The natives themselves, Malinowski emphasized, recognized the distinction between purely economic commodities and the symbolic significance of armlets and necklaces. However, to distinguish economic from symbolic commodities does not mean that the Trobriand Islanders failed to assign graded values to the symbolic commodities; indeed, they made gradations and used them to express and confirm the nature of the relationships among exchange partners as equals, superordinates, or subordinates. But, as Malinowski noted, "in all forms of [Kula] exchange in the Trobriands, there is not even a trace of gain, nor is there any reason for looking at it from the purely utilitarian and economic standpoint, since there is no enhancement of mutual utility through

---

[11]Frazer, Preface to *Argonauts* (see note 12).

[12]Bronislaw Malinowski, *Argonauts of the Western Pacific* (London: Routledge & Kegan Paul, 1922), p. 81.

[13]Ibid., pp. 82–83.

the exchange."[14] Rather, the motives behind the Kula were social psychological, for the exchanges in the ring were viewed by Malinowski to have implications for the needs of both individuals and society (recall from Chapter 2 that Malinowski was also a founder of functional theory). From his functionalist framework, he interpreted the Kula to mean "the fundamental impulse to display, to share, to bestow [and] the deep tendency to create social ties."[15] For Malinowski, then, an enduring social pattern such as the Kula Ring is considered to have positively functional consequences for satisfying individual psychological needs and societal needs for social integration and solidarity.

As Robert Merton and others were to emphasize (see Chapter 4), this form of functional analysis presents many logical difficulties. Nevertheless, I believe that Malinowski's analysis made several enduring contributions to modern exchange theory:

1. In Malinowski's words, "the meaning of the Kula will consist in being instrumental to dispel [the] conception of a rational being who wants nothing but to satisfy his simplest needs and does it according to the economic principle of least effort."[16]
2. Psychological rather than economic needs are the forces that initiate and sustain exchange relations and are therefore critical in the explanation of social behavior.
3. Exchange relations can also have implications beyond two parties, for, as the Kula demonstrates, complex patterns of indirect exchange can operate to maintain extended and protracted social networks.
4. Symbolic exchange relations are the basic social process underlying both differentiation of ranks in a society and the integration of society into a cohesive and solidary whole.

With these points of emphasis, Malinowski helped free exchange theory from the limiting confines of utilitarianism. By stressing the importance of symbolic exchanges for both individual psychological processes and patterns of social integration, he anticipated the conceptual base for two basic types of exchange perspectives, one emphasizing the importance of psychological processes and the other stressing the significance of emergent cultural and structural forces on exchange relations.

Marcel Mauss and the Emergence of Exchange Structuralism

Reacting to what he perceived as Malinowski's tendency to overemphasize psychological instead of social needs, Marcel Mauss reinterpreted Malinowski's

---

[14]Ibid., p. 175.

[15]Ibid.

[16]Ibid., p. 516.

analysis of the Kula.[17] In this effort he formulated the broad outlines of a "collectivistic," or structural-exchange, perspective.[18] For Mauss the critical question in examining an exchange network as complex as that of the Kula was: "In primitive or archaic types of societies what is the principle whereby the gift received has to be repaid? What force is there in the thing which compels the recipient to make a return?"[19] The "force" compelling reciprocity was, for Mauss, society or the group. As he noted: "It is groups, and not individuals, which carry on exchange, make contracts, and are bound by obligations."[20] The individuals actually engaged in an exchange represent the moral codes of the group. Exchange transactions among individuals are conducted in accordance with the rules of the group, thereby reinforcing these rules and codes. Thus, for Mauss, the overconcern with individuals' self-interests by utilitarians and the overemphasis on psychological needs by Malinowski are replaced by a conception of individuals as representatives of social groups. In the end, exchange relations create, reinforce, and serve a group morality that is an entity *sui generis*, to borrow a famous phrase from Mauss's mentor, Émile Durkheim. Furthermore, in a vein similar to that of Frazer, once such a morality emerges and is reinforced by exchange activities, it comes to regulate other activities in the social life of a group, above and beyond particular exchange transactions.

Mauss's work has received scant attention from sociologists, but I see him as the first to forge a reconciliation between the exchange principles of utilitarianism and the structural, or collectivistic, thought of Durkheim. In recognizing that exchange transactions give rise to and, at the same time, reinforce the normative structure of society, Mauss anticipated the structural position of some contemporary exchange theories. Mauss's influence on modern theory has been indirect, however, for I think that it is through the structuralism of Claude Lévi-Strauss that the French collectivist tradition of Durkheim and Mauss influenced the exchange perspectives of contemporary sociological theory.

---

[17]Marcel Mauss, *The Gift*, trans. I. Cunnison (New York: Free Press, 1954; originally published as *Essai sur le don en sociologie et anthropologie* [Paris: Presses universitaires de France, 1925]). It should be noted that Mauss rather consistently misinterpreted Malinowski's ethnography, but it is through such misinterpretation that he came to visualize a "structural" alternative to "psychological" exchange theories.

[18]In Peter Ekeh's excellent discussion of Mauss and Lévi-Strauss (*Social Exchange Theory*, pp. 55–122), the term *collectivist* is used in preference to *structural* and is posited as the alternative to *individualistic* or psychological exchange perspectives. I prefer the terms *structural* and *psychological*; thus, although I am indebted to Ekeh's discussion, these terms will be used to make essentially the same distinction. My preference for these terms will become more evident in subsequent chapters, since, in contrast with Ekeh's analysis, I consider Peter M. Blau and George C. Homans to have developed, respectively, structural and psychological theories. Ekeh considers the theories of both Blau and Homans to be individualistic, or psychological.

[19]Mauss, *The Gift*, p. 1.

[20]Ibid., p. 3.

Claude Lévi-Strauss and Structuralism

In 1949 Claude Lévi-Strauss launched an analysis of cross-cousin marriage in his classic work, *The Elementary Structures of Kinship*.[21] In restating Durkheim's objections to utilitarians, Lévi-Strauss took exception to Frazer's utilitarian interpretation of cross-cousin marriage patterns. And, in a manner similar to Mauss's opposition to Malinowski's emphasis on psychological needs, Lévi-Strauss developed a sophisticated structural-exchange perspective.[22]

In rejecting Frazer's interpretation of cross-cousin marriage, Lévi-Strauss first questions the substance of Frazer's utilitarian conceptualization. Frazer, he notes, "depicts the poor Australian aborigine wondering how he is going to obtain a wife since he has no material goods with which to purchase her, and discovering exchange as the solution to this apparently insoluble problem: 'men exchange their sisters in marriage because that was the cheapest way of getting a wife.'" In contrast, Lévi-Strauss emphasizes that "it is the exchange which counts and not the things exchanged." For Lévi-Strauss, exchange must be viewed in terms of its functions for integrating the larger social structure. Lévi-Strauss then attacks Frazer's and the utilitarians' assumption that the first principles of social behavior are economic. Such an assumption flies in the face of the fact that social structure is an emergent phenomenon that operates in terms of its own irreducible laws and principles.

Lévi-Strauss also rejects psychological interpretations of exchange processes, especially the position advocated by behaviorists (see later section). In contrast with psychological behaviorists, who see little real difference in the laws of behavior between animals and humans, Lévi-Strauss emphasizes that humans possess a cultural heritage of norms and values that separates their behavior and societal organization from that of animal species. Human action is thus qualitatively different from animal behavior, especially with respect to social exchange. Animals are not guided by values and rules that specify when, where, and how they are to carry out social transactions. Humans, however, carry with them into any exchange situation learned definitions of how they are to behave—thus assuring that the principles of human exchange will be distinctive.

Furthermore, exchange is more than the result of psychological needs, even those that have been acquired through socialization. Exchange cannot be understood solely in terms of individual motives, because exchange relations are a reflection of patterns of social organization that exist as an entity, *sui generis*. Exchange behavior is thus regulated from without by norms and values, re-

---

[21]Claude Lévi-Strauss, *The Elementary Structures of Kinship* (Boston: Beacon Press, 1969). This is a translation of Lévi-Strauss's 1967 revision of the original *Les structures élémentaires de la parenté* (Paris: Presses universitaires de France, 1949). The full impact of this work was probably never felt in sociology, since until 1969 it was not available in English. Yet, as will be noted in Chapter 15 on Homans, this work did have a profound impact on Homans' thinking, primarily because Homans felt compelled to reject Lévi-Strauss's analysis. I will examine Lévi-Strauss's work again in Chapter 24.

[22]See Ekeh's discussions for a more detailed analysis of Lévi-Strauss.

sulting in processes that can be analyzed only in terms of their consequences, or functions, for these norms and values.

In arguing this point of view, Lévi-Strauss posits several fundamental exchange principles. First, all exchange relations involve costs for individuals, but, in contrast with economic or psychological explanations of exchange, such costs are attributed to society—to those customs, rules, laws, and values that require behaviors incurring costs. Yet individuals do not assign the costs to themselves, but to the "social order." Second, for all those scarce and valued resources in society—whether material objects, such as wives, or symbolic resources, like esteem and prestige—their distribution is regulated by norms and values. As long as resources are in abundant supply or are not highly valued in a society, their distribution goes unregulated; but, once they become scarce and highly valued, their distribution is soon regulated. Third, all exchange relations are governed by a norm of reciprocity, requiring those receiving valued resources to bestow on their benefactors other valued resources. In Lévi-Strauss's conception of reciprocity there are various patterns of reciprocation specified by norms and values. In some situations norms dictate "mutual" and direct rewarding of one's benefactor, whereas in other situations the reciprocity can be "univocal," involving diverse patterns of indirect exchange in which actors do not reciprocate directly but only through various third (fourth, fifth, and so forth) parties. Within these two general types of exchange reciprocity—mutual and univocal—numerous subtypes of exchange networks can be normatively regulated.

For Lévi-Strauss, these three exchange principles offer a more useful set of concepts to describe cross-cousin marriage patterns, because they can now be viewed in terms of their functions for the larger social structure. Particular marriage patterns and other features of kinship organization no longer need be interpreted merely as direct exchanges among individuals but can be conceptualized as univocal exchanges between individuals and society. In freeing exchange from the analysis of only direct and mutual exchanges, Lévi-Strauss offers a tentative theory of societal integration and solidarity. His explanation extends Durkheim's provocative analysis and indicates how various subtypes of direct and univocal exchange both reflect and reinforce different patterns of societal integration and organization.

This theory of integration is, in itself, of theoretical importance. But I think it more significant for present purposes to stress Lévi-Strauss's impact on current sociological exchange perspectives. My sense is that two points of emphasis exerted strong influence on modern sociological theory.

1. Various forms of social structure, rather than individual motives, are the critical variables in the analysis of exchange relations.
2. Exchange relations in social systems are frequently not restricted to direct interaction among individuals but are protracted into complex networks of indirect exchange. On the one hand, these exchange processes are caused by patterns of social integration and organization; on the other hand, they promote diverse forms of such organization.

In looking back on this anthropological heritage, I believe that Lévi-Strauss's work represents the culmination of a reaction to economic utilitarianism as it was originally incorporated into anthropology by Frazer. Malinowski recognized the limitations of Frazer's analysis of only material or economic motives in direct exchange transactions. As the Kula Ring demonstrates, exchange can be generalized into protracted networks involving noneconomic motives that have implications for societal integration. Mauss drew explicit attention to the significance of social structure in regulating exchange processes and to the consequences of such processes for maintaining social structure. Finally, in this intellectual chain of events in anthropology, Lévi-Strauss began to indicate how different types of direct and indirect exchange are linked to different patterns of social organization. This intellectual heritage has influenced both the substance and the strategy of exchange theory in sociology, but it has done so only after considerable modification of assumptions and concepts by a particular strain of psychology: behaviorism.

PSYCHOLOGICAL BEHAVIORISM AND EXCHANGE THEORY

As a psychological perspective, behaviorism began from insights derived from observations of an accident. The Russian physiologist Ivan Petrovich Pavlov (1849–1936) discovered that experimental dogs associated food with the person bringing it.[23] He observed, for instance, that dogs on whom he was performing secretory experiments would salivate not only when presented with food but also when they heard their feeder's footsteps approaching. After considerable delay and personal agonizing, Pavlov undertook a series of experiments on animals to understand such "conditioned responses."[24] From these experiments he developed several principles that later were to be incorporated into behaviorism. These include: (1) A stimulus consistently associated with another stimulus producing a given physiological response will, by itself, elicit that response. (2) Such conditioned responses can be extinguished when gratifications associated with stimuli are no longer forthcoming. (3) Stimuli that are similar to those producing a conditioned response can also elicit the same response as the original stimulus. (4) Stimuli that increasingly differ from those used to condition a particular response will decreasingly be able to elicit this response. Thus Pavlov's experiments exposed the principles of conditioned responses, extinction, response generalization, and response discrimination. Although Pavlov clearly recognized the significance of these findings for human behavior, his insights came to fruition in America under the tutelage of Edward Thorndike and John B. Watson—the founders of behaviorism.[25]

---

[23]See, for relevant articles, lectures, and references, I. P. Pavlov, *Selected Works*, ed. K. S. Kostoyants, trans. S. Belsky (Moscow: Foreign Languages Publishing House, 1955); and *Lectures on Conditioned Reflexes*, 3rd ed., trans. W. H. Grant (New York: International Publishers, 1928).

[24]I. P. Pavlov, "Autobiography," in *Selected Works*, pp. 41–44.

[25]For an excellent summary of their ideas, see Robert I. Watson, *The Great Psychologists*, 3rd ed. (Philadelphia: Lippincott, 1971), pp. 417–46.

Edward Lee Thorndike conducted the first laboratory experiments on animals in America. In the course of these experiments he observed that animals would retain response patterns for which they were rewarded.[26] For example, in experiments on kittens placed in a puzzle box, Thorndike found that they would engage in trial-and-error behavior until emitting the response that allowed them to escape. And, with each placement in the box, the kittens would engage in less trial-and-error behavior, indicating that the gratifications associated with a response allowing the kittens to escape caused them to learn and retain this response. From these and other studies, which were conducted at the same time as Pavlov's, Thorndike formulated three principles or laws: (1) the "law of effect," which holds that acts in a situation producing gratification will be more likely to occur in the future when that situation recurs; (2) the "law of use," which states that situation/response connection is strengthened with repetitions and practice; and (3) the "law of disuse," which argues that the connection will weaken when practice is discontinued.[27]

These laws converge with those presented by Pavlov, but there is one important difference. Thorndike's experiments involved animals engaged in free trial-and-error behavior, whereas Pavlov's work was on the conditioning of physiological—typically glandular—responses in a tightly controlled laboratory situation. Thorndike's work could thus be seen as more directly relevant to human behavior in natural settings.

John B. Watson was one of only several thinkers to recognize the significance of Pavlov's and Thorndike's work, but he soon became the dominant advocate of what was becoming explicitly known as behaviorism.[28] Watson's opening shot for the new science of behavior was fired in an article entitled "Psychology as the Behaviorist Views It":

> Psychology as the behaviorist views it is a purely objective experimental branch of natural science. Its theoretical goal is the prediction and control of behavior. Introspection forms no essential part of its methods, nor is the scientific value of its data dependent upon the readiness with which they lend themselves to interpretation in terms of consciousness. The behaviorist, in efforts to get a unitary scheme of animal response, recognizes no dividing line between man and brute.[29]

---

[26]Edward L. Thorndike, "Animal Intelligence: An Experimental Study of the Associative Processes in Animals," *Psychological Review Monograph*, Supplement 2 (1989).

[27]See Edward L. Thorndike, *The Elements of Psychology* (New York: Seiler, 1905), *The Fundamentals of Learning* (New York: Teachers College Press, 1932), and *The Psychology of Wants, Interests, and Attitudes* (New York: D. Appleton, 1935).

[28]Others who recognized their importance included: Max F. Meyer, *Psychology of the Other-One* (Columbus, OH: Missouri Book, 1921); and Albert P. Weiss, *A Theoretical Basis of Human Behavior* (Columbus, OH: Adams Press, 1925).

[29]J. B. Watson, "Psychology as the Behaviorist Views It," *Psychological Review* 20 (1913), pp. 158–77. For other basic works by Watson, see *Psychology from the Standpoint of a Behaviorist*, 3rd ed. (Philadelphia: Lippincott, 1929); *Behavior: An Introduction to Comparative Psychology* (New York: Henry Holt, 1914).

Watson thus became the advocate of the extreme behaviorism against which many vehemently reacted.[30] For Watson, psychology is the study of stimulus/response relations, and the only admissible evidence is overt behavior. Psychologists are to stay out of the "Pandora's box" of human consciousness and to study only observable behaviors as they are connected to observable stimuli.[31]

In many ways, I see behaviorism as similar to utilitarianism, since it operates on the principle that humans are reward-seeking organisms pursuing alternatives that will yield the most reward and the least punishment. Rewards are simply another way of phrasing the economist's concept of "utility," and "punishment" is somewhat equivalent to the notion of "cost." For the behaviorist, reward is any behavior that reinforces or meets the needs of the organism, whereas punishment denies rewards or forces the expenditure of energy to avoid pain (thereby incurring costs).

Modern exchange theories have borrowed from behaviorists the notion of reward and used it to reinterpret the utilitarian exchange heritage. In place of utility, the concept of reward has been inserted, primarily because it allows exchange theorists to view behavior as motivated by psychological needs. However, the utilitarian concept of cost appears to have been retained in preference to the behaviorist's formulation of punishment, since the notion of cost allows exchange theorists to visualize more completely the alternative rewards that organisms forego in seeking to achieve a particular reward.

Despite these modifications of the basic concepts of behaviorism, its key theoretical generalizations have been incorporated with relatively little change into some forms of sociological exchange theory. Let me list these:

1. In any given situation, organisms will emit those behaviors that will yield the most reward and the least punishment.
2. Organisms will repeat those behaviors that have proved rewarding in the past.
3. Organisms will repeat behaviors in situations that are similar to those in the past in which behaviors were rewarded.
4. Present stimuli that on past occasions have been associated with rewards will evoke behaviors similar to those emitted in the past.
5. Repetition of behaviors will occur only as long as they continue to yield rewards.
6. An organism will display emotion if a behavior that has previously been rewarded in the same or similar situation suddenly goes unrewarded.
7. The more an organism receives rewards from a particular behavior, the less rewarding that behavior becomes (due to satiation) and the more

---

[30]For example, in his *Mind, Self, and Society* (Chicago: University of Chicago Press, 1934), Mead has 18 references to Watson's work.

[31]For a more detailed discussion of the emergence of behaviorism, see Jonathan H. Turner, Leonard Beeghley, and Charles Powers, *The Emergence of Sociological Theory* (Belmont, CA: Wadsworth, 1989).

likely the organism is to emit alternative behaviors in search of other rewards.

Since these principles were discovered in laboratory situations where experimenters typically manipulated the environment of the organism, it is difficult to visualize the experimental situation as interaction. The experimenter's tight control of the situation precludes the possibility that the animal will affect significantly the responses of the experimenter. This fact has forced modern exchange theories using behaviorist principles to incorporate the utilitarian's concern with transactions, or exchanges. In this way humans can be seen as mutually affecting one another's opportunities for rewards. In contrast to animals in a Skinner box or some similar laboratory situation, humans exchange rewards. Each person represents a potentially rewarding stimulus situation for the other.

As sociological exchange theorists have attempted to apply behaviorist principles to the study of human behavior, they have inevitably confronted the problem of the black box: humans differ from laboratory animals in their greater ability to engage in a wide variety of complex cognitive processes. Indeed, as the utilitarians were the first to emphasize, what is distinctly human is the capacity to abstract, to calculate, to project outcomes, to weigh alternatives, and to perform a wide number of other cognitive manipulations. Furthermore, in borrowing behaviorists' concepts, contemporary exchange theorists have also had to introduce the concepts of an introspective psychology and structural sociology. Humans not only think in complex ways; their thinking is emotional and circumscribed by many social and cultural forces (first incorporated into the exchange theories of Mauss and Lévi-Strauss). Once it is recognized that behaviorist principles must incorporate concepts denoting both internal psychological processes and constraints of social structure and culture, it is also necessary to visualize exchange as frequently transcending the mutually rewarding activities of individuals in direct interaction. The organization of behavior by social structure and culture, coupled with humans' complex cognitive abilities, allows protracted and indirect exchange networks to exist.

In looking back upon the impact of behaviorism on some forms of contemporary exchange theory, then, I see a convergence of concepts and principles. Although the vocabulary and general principles of behaviorism are clearly evident, the concepts have been redefined and the principles altered to incorporate the insights of the early utilitarians as well as the anthropological reaction to utilitarianism. The end result has been for proponents of an exchange perspective employing behaviorist concepts and principles to abandon much of what made behaviorism a unique perspective as they have dealt with the complexities introduced by human cognitive capacities and their organization into sociocultural groupings.

THE SOCIOLOGICAL TRADITION AND EXCHANGE THEORY

Thus far I have concentrated on the influence of conceptual work in economics, anthropology, and psychology on exchange theory. The vocabulary of exchange theory clearly comes from utilitarianism and behaviorism. Anthropological work forced the recognition that cultural and social dynamics need to be incorporated into exchange theory. When we look at early sociological work, however, the impact of early sociological theorists on modern exchange theory is difficult to assess. I think that this difficulty is the result of several factors. First, much sociological theory represented a reaction against utilitarianism and extreme behaviorism and, therefore, has been reluctant to incorporate concepts from these fields. Second, the most developed of the early exchange theories—that provided by Georg Simmel in his *The Philosophy of Money*—remained untranslated into English until recent years.[32] (Yet I suspect that German-reading theorists, such as Peter Blau and Talcott Parsons, were to some degree influenced by Simmel's ideas.) Third, the topics of most interest to many sociological exchange theorists—differentiations of power and conflict in exchanges—have more typically been conceptualized as conflict theory than as exchange theory. But, as I will document, sociological theories of exchange converge with those on conflict processes, and I have little doubt that Marx's ideas exerted considerable influence on modern exchange theories.

Thus, although I am not sure of the exact lines of influence, I think that Marx's and Simmel's work has been involved—perhaps only subliminally—in the development of sociological exchange theory. Let me draw out this line of argument for Marx and then for Simmel.

Marx's Theory of Exchange and Conflict

Most contemporary theories of exchange examine the dynamics of exchange that follow from the fact that actors typically have unequal levels of resources with which to bargain. Those with resources valued by others are in a position to strike a better bargain, especially if those others who value their resources do not possess equally valued resources to offer in exchange. This fact of social life is, I feel, the situation described in Marx's conflict theory.[33] Capitalists have the power to control the distribution of material rewards, whereas all that workers have is their labor to offer in exchange. Although labor is valued by the capitalist, it is in plentiful supply, and thus no one worker is in a position to bargain effectively with an employer. As a consequence, capitalists can get labor at a low cost and can force workers to do what they want. As capitalists press their advantage, they create the very conditions that allow workers to develop resources—political, organizational, ideological—that they can then use to strike a better bargain with capitalists and, in the end, to overthrow them.

Granted, I am simplifying Marx's implicit exchange theory, but my point is clear: dialectical conflict theory is, I think, a variety of exchange theory. Let me list some of these exchange dynamics more explicitly:

---

[32]Georg Simmel, *The Philosophy of Money*, trans. T. Bottomore and D. Frisby (Boston: Routledge & Kegan Paul, 1978; originally published in 1907).

[33]See Chapter 9.

1. Those who need scarce and valued resources that others possess but who do not have equally valued and scarce resources to offer in return will be dependent upon those who control these resources.
2. Those who control valued resources have power over those who do not. That is, the power of one actor over another is directly related to (a) the capacity of one actor to monopolize the valued resources needed by other actors and (b) the inability of those actors who need these resources to offer equally valued and scarce resources in return.
3. Those with power will press their advantage and will try to extract more resources from those dependent upon them in exchange for fewer (or the same level) of the resources that they control.
4. Those who press their advantage in this way will create conditions that encourage those who are dependent on them to (a) organize in ways that increase the value of their resources and, failing this, to (b) organize in ways that enable them to coerce those on whom they are dependent.

If the words *capitalist* and *proletarian* are inserted at the appropriate places in the above list, I think that Marx's exchange model becomes readily apparent. Dialectical conflict theory is thus a series of propositions about exchange dynamics in systems in which the distribution of resources is unequal. And, as I think will become evident in the next chapters, sociological exchange theories have emphasized these dynamics that inhere in the unequal distribution of resources. Such is, I feel, Marx's major contribution to exchange theory.

Georg Simmel's Exchange Theory

In Georg Simmel's *The Philosophy of Money*[34] is a critique of Marx's "value theory of labor"[35] and, in its place, a clear exposition of exchange theory. Since this important work remained untranslated until recently, I am not sure just how much influence it had on sociological exchange theory. Yet, since a number of exchange theorists are fluent in German, I suspect that Simmel's work has shaped modern exchange theory more than is commonly realized or acknowledged.

The Philosophy of Money is, as its title indicates, about the impact of money on social relations and social structure. I need not go into the details of Simmel's insightful analysis and critique of Marx for our present purposes. I only want to extract the purely exchange-theoretic principles from Simmel's discussion. For Simmel, social exchange involves the following elements:

1. The desire for a valued object that one does not have.
2. The possession of the valued object by an identifiable other.
3. The offer of an object of value to secure from another the desired object.

---

[34]Simmel, *The Philosophy.*

[35]Ibid. Simmel did not see value as inhering in "labor power," but in what people desired and the level of scarcity in what they desired.

4. The acceptance of this offer by the possessor of the valued object.[36]

Contained in this portrayal of social exchange are several additional points that Simmel emphasized. First, value is idiosyncratic and is, ultimately, tied to an individual's impulses and needs. Of course, what is defined as valuable is typically circumscribed by cultural and social patterns, but how valuable an object is will be a positive function of (a) the intensity of a person's needs and (b) the scarcity of the object. Second, much exchange involves efforts to manipulate situations so that the intensity of needs for an object is concealed and the availability of an object is made to seem less than what it actually is. Inherent in exchange, therefore, is a basic tension that can often erupt into other social forms, such as conflict. Third, to possess an object is to lessen its value and to increase the value of objects that one does not possess. Fourth, exchanges will occur only if both parties perceive that the object given is less valuable than the one received.[37] Fifth, collective units as well as individuals participate in exchange relations and hence are subject to the four processes listed above. Sixth, the more liquid the resources of an actor in an exchange— that is, the more that resources can be used in many types of exchanges—the greater will be that actor's options and power. For if an actor is not bound to exchange with any other and can readily withdraw resources and exchange them with another, then that actor has considerable power to manipulate any exchange.

Economic exchange involving money is only one case of this more general social form. But it is a very special case. For when money becomes the predominant means for establishing value in social relationships, the properties and dynamics of social relations are transformed. This process of displacing other criteria of value, such as logic, ethics, and aesthetics, with a monetary criterion is precisely the long-term evolutionary trend in societies. This trend is, as I mentioned earlier, both a cause and effect of money as the medium of exchange. Money emerged to facilitate exchanges and to realize even more completely humans' basic needs. But, once established, money has the power to transform the structure of social relations in society.

Thus the key insight in *The Philosophy of Money* is that the use of different criteria for assessing value has an enormous impact on the form of social relations. As money replaces barter and other criteria for determining values, social relations are fundamentally changed. But they are transformed in accordance with some basic principles of social exchange, which are never codified by Simmel but, as I have indicated above, are very clear. In Table 14-1, I have codified these ideas into abstract exchange principles, because I see them as *the* basic processes depicted in sociological analysis of exchange. At the risk

---

[36]Simmel, *The Philosophy*, pp. 85–88.

[37]Surprisingly, Simmel did not explore in any great detail the consequences of unbalanced exchanges, in which people are forced to give up a more valuable object for a less valuable one. Simmel simply assumed that, at the time of exchange, one party felt that redefinition might occur, but the exchange would not occur if at the moment people did not perceive that they had received more value than they had given up.

TABLE 14–1 Georg Simmel's Exchange Principles

I. *Attraction Principle:* The more actors perceive as valuable one another's respective resources, the more likely is an exchange relationship to develop among these actors.

II. *Value Principle:* The greater the intensity of an actor's needs for a resource of a given type, and the less available that resource, the greater the value of that resource to the actor.

III. *Power Principles:*
   A. The more an actor perceives as valuable the resources of another actor, the greater is the power of the latter over the former.

   B. The more liquid an actor's resources, the greater will be the exchange options and alternatives and, hence, the greater will be the power of that actor in social exchanges.

IV. *Tension Principle:* The more actors in a social exchange manipulate the situation in an effort to misrepresent their needs for a resource and/or conceal the availability of resources, the greater is the level of tension in that exchange and the greater is the potential for conflict.

---

of repeating myself, let me summarize again the ideas behind the propositions. Principle I states that interaction occurs because actors value one another's resources, and, in accordance with Principle II, value is a dual function of (*a*) actors' needs for resources and (*b*) the scarcity of resources. Principles III-A and III-B underscore the fact that power is a part of the exchange process. Actors who have resources that others value are in a position to extract compliance from those seeking these resources (Principle III-A), and actors who have liquid or generalized resources, such as money, will be more likely to have power, since liquid resources can be more readily exchanged with alternative actors. Principle IV states that, as actors seek to manipulate situations in order to conceal the availability of resources and their needs for resources, tensions are created, and these tensions can result in conflict.

MODERN EXCHANGE THEORY: A PREVIEW

Exchange theory is now one of the most prominent theoretical perspectives in sociology. A number of exchange perspectives has emerged in recent decades. Typically, they begin with inspiration from either the behaviorist tradition in psychology or the utilitarian heritage in economic theory. But, as we will observe, these two traditions merge in modern exchange theory, especially as they deal with the related issues of inequality, power, and conflict.

In the chapters to follow, I will begin with the behavioristic approach of George C. Homans, and then I will examine the more economic strategy of Peter M. Blau. As I believe will become clear, these two theorists begin at somewhat different places, and they advocate varying theoretical strategies, but I see their ideas as converging around the questions of tension and conflict in exchange systems. I will close with a review of rational choice theory—the most extreme of the utilitarian perspectives. The latter perspective has revived utilitarianism in sociology, accommodating it to the realities of social structure. Among prominent theorists working in this tradition, James Coleman and Michael Hechter stand out. I have decided to explore Hechter's approach in detail, although I could just as easily have focused on Coleman's related approach. But the key point is that utilitarian thinking has reemerged in sociology—making Talcott Parsons' earlier epitaph somewhat premature.[38]

---

[38]Talcott Parsons, *The Structure of Social Action* (New York: McGraw-Hill, 1937).

CHAPTER 15

◆

Exchange Behaviorism

George C. Homans

One of the most prominent theorists of this century is George C. Homans, who, before his recent death, advocated an exchange-theoretic approach that borrows key concepts and propositions from behavioristic psychology.[1] Homans' early work emphasized "what people do" and "how they behave" in concrete settings, and it was highly inductive. It stressed pulling from empirical contexts generalizations in an effort to tie theory to ongoing social processes. But, as I emphasized in Chapter 1, empirical generalizations are not theory; rather, they are what a theory is supposed to explain. And so, as Homans sought to explain a variety of empirical findings and generalizations, he shifted to a more deductive theoretical strategy. He labels this strategy *axiomatic*, but, as noted in Chapter 1, axiomatic theorizing in sociology is not really possible. A better name is *formal theorizing*, whereby the strict requirements of axiomatic theory are relaxed. But the basic idea is the same: to make deductions from abstract principles to empirical generalizations; and to the extent that the abstract statement subsumes or "covers" the empirical generalization, then the abstract statement explains the generalization (consult Figure 1–4).

I think that Homans' transition from empirical generalizations to deductive explanations is instructive. For, as I argued in Chapter 1, one does not mechanically generalize from empirical findings to abstract laws. Empirical generalizations always have contextual content, and it is not easy to simply raise the level of abstraction and create abstract theory. Rather, a creative leap in abstraction is necessary, and then, armed with more abstract principles, one makes deductions back down to empirical generalizations. Homans' "creative leap" involved borrowing from behavioristic psychology, and so perhaps I should view his exchange work as creative borrowing. Nonetheless, Homans

---

[1] I would like to thank George C. Homans for his constructive criticisms of this chapter as it appeared in an earlier edition of this book.

recognized that induction from empirical contexts can go only so far; at some point one must construct deductive propositional schemes to explain what one has induced from these contexts.

THE EARLY INDUCTIVE STRATEGY

By 1950, with the publication of *The Human Group*, Homans' work revealed a clear commitment to an inductive strategy for building sociological theory.[2] In studies ranging from the analysis of a work group in a factory and a street gang to the kinship system of a primitive society and the structure of an entire New England community, he stressed the importance of observing people's actual behaviors and activities in various types of groups. By observing what people actually do, it is possible to develop concepts that are attached to the ongoing processes of social systems. Such concepts are termed by Homans *first-order abstractions,* since they merely represent names that observers use to signify a single class of observations in a group. Homans chose these words carefully. He wanted to emphasize their difference from the "second-order abstractions" commonly employed by sociologists. As distinct from the abstractions he prefers, *second-order abstractions* refer to several classes of observations and are thereby somewhat detached from ongoing events in actual groups. For example, "status" and "role" are favorite concepts used by sociologists to denote processes in groups, but upon careful reflection it is evident that one does not observe directly a status or role; rather, they are highly abstract names that subsume numerous types and classes of events occurring in groups.

The typical practice of jumping to second-order abstractions in building theory was viewed by Homans as premature:

> [Sociologists should initially] attempt to climb down from the big words of social science, at least as far as common sense observation. Then, if we wish, we can start climbing up again, but this time with a ladder we can depend on.[3]

In constructing a firm bottom rung to the abstraction ladder, Homans introduced three first-order abstractions that provided labels or names to the actual events occurring in groups: *activities, interaction,* and *sentiments.* Activities pertain to what people do in a given situation, interaction denotes the process in which one unit of activity stimulates a unit of activity in another person, and sentiments refer to actions that signal the internal psychological states of people engaged in activities and interaction.

The three first-order abstractions, or *elements,* as Homans was fond of calling them, exist within an "external system." For Homans this external system represents the environmental parameters of a particular group under study. As he was to later label this analytical approach, the external system summarizes the "givens" of a particular situation, which, for the purposes at

---

[2]George C. Homans, *The Human Group* (New York: Harcourt Brace Jovanovich, 1950).
[3]Ibid., p. 13.

hand, are not examined extensively. Of more interest, however, is the "internal system," which operates within the constraints imposed by the external system and which is composed of the interrelated activities, interactions, and sentiments of group members. The fact that activities, interactions, and sentiments are interrelated is of great analytical significance: changes in one element lead to changes in the other elements.

Of critical importance in the analysis of internal systems is the process of "elaboration" in which new patterns of organization among activities, interaction, and sentiments are constantly emerging by virtue of their interrelatedness with one another and with the external system. A group thus "elaborates itself, complicates itself, beyond the demands of the original situation"; in so doing, it brings about new types of activities, forms of interaction, and types of sentiments.

In *The Human Group*, Homans' strategy was to present a descriptive summary of five case studies on diverse groups. From each summary, Homans incorporates the elements of the internal system into propositions that describe the empirical regularities he had observed in the case study.[4] As he proceeds in his summaries, Homans attempts to substantiate the generalizations in one study from the next, while using each successive study as a source for additional generalizations. With this strategy he hoped to generate a body of interrelated generalizations describing the various ways different types of groups elaborate their internal systems of activities, interactions, and sentiments. In this way the first rungs of the abstraction ladder used in sociological theorizing would be dependable, providing a firm base for subsequent theorizing at a more abstract level.

Let me offer an example from Homans' summary of the famous Bank Wiring Room in the Hawthorne Western Electric Plant. In this setting, Homans "observed" these regularities:[5] (1) If the frequency of interaction between two or more persons increases, the degree of their liking for one another will increase, and vice versa.[6] (2) Persons whose sentiments of liking for one another increase will express these sentiments in increased activity, and vice versa.[7] (3) The more frequently persons interact with one another, the more alike their activities and their sentiments tend to become, and vice versa.[8] (4) The higher the rank of a person within a group, the more nearly his activities conform to

---

[4]In his work, Homans rarely makes direct observations himself. Rather, he has tended to rely on the observations of others, making from their reports inferences about events. As would be expected, Homans has often been criticized for accepting too readily the imprecise and perhaps inaccurate observations of others as an inductive base for theorizing.

[5]This extended example is adapted from the discussion of Homans' propositions in M. J. Mulkay, *Functionalism and Exchange and Theoretical Strategy* (New York: Schocken Books, 1971), pp. 135–41. As was done by Mulkay, some of Homans' propositions are reworded for simplification.

[6]Homans, *Human Group*, p. 112.

[7]Ibid., p. 118.

[8]Ibid., p. 120.

the norms of the group, and vice versa.[9] (5) The higher the person's social rank, the wider will be the range of this person's interactions.[10]

These propositions describe some of the group-elaboration processes that Homans "observed" in the Bank Wiring Room. Propositions 1 and 2 summarize Homans' observations that, the more the workers interacted, the more they appeared to like one another, which, in turn, seemed to cause further interactions above and beyond the work requirements of the external system. Such elaboration of interactions, sentiments, and activities also results in the differentiation of subgroups that reveal their own levels of output, topics of conversation, and patterns of work assistance. This tendency is denoted by Proposition 3. Another pattern of differentiation in the Bank Wiring Room is described by Proposition 4, whereby the ranking of individuals and subgroups occurs when group members compare their activities with those of other members and the output norms of the group. Proposition 5 describes the tendency of high-ranking members to interact more frequently with all members of the group, offering more on-the-job assistance.

Having tentatively established these and other empirical regularities in the Bank Wiring Room, Homans then analyzes the equally famous Norton Street Gang, described in William Whyte's *Street Corner Society*. In this second case study, Homans follows his strategy of confirming his earlier propositions while inducing further generalizations: (6) The higher a person's social rank, the larger will be the number of persons who originate interaction for him or her, either directly or through intermediaries.[11] (7) The higher a person's social rank, the larger will be the number of persons toward whom interaction is initiated, either directly or through intermediaries.[12] (8) The more nearly equal in social rank a number of people are, the more frequently they will interact with one another.[13]

I see Propositions 6 and 7 as simply corollaries of Proposition 5. It follows, almost by definition, that those with a wide range of interactions will receive and initiate more interaction. However, Propositions 6 and 7 are induced separately from his observations of the Bank Wiring Room. For the Norton Street Gang, Homans observes that Doc, the leader of the gang, was the center of a complex network of communication. Such a pattern of communication was not observable in the Bank Wiring Room because stable leadership ranks had not emerged. Proposition 8 describes another process Homans sees in groups with clear leadership: the internal system tends to differentiate into super- and subordinate subgroups, whose members appear to interact more with one another than with those of higher or lower rank.

---

[9]Ibid., p. 141.
[10]Ibid., p. 145.
[11]Ibid., p. 182.
[12]Ibid.
[13]Ibid., p. 184.

After using the group-elaboration processes in the Norton Street Gang to confirm and supplement the propositions induced from the Bank Wiring Room, Homans then examines the Tikopia family, as described in Raymond Firth's famous ethnography. As with the Norton Street Gang, the Tikopia family system was used to confirm earlier propositions and as a field from which to induce further propositions: (9) The more frequently persons interact with one another, when no one of them originates interaction with much greater frequency than the others, the greater is their liking for one another and their feeling of ease in one another's presence.[14] (10) When two persons interact with each other, the more frequently one of the two originates interaction for the other, the stronger will be the latter's sentiment of respect (or hostility) and the more nearly will the frequency of interaction be kept to a minimum.[15]

These propositions reveal, I should emphasize, another facet of Homans' inductive strategy. They establish some conditions under which Proposition 1 will hold true. In Proposition 1, Homans notes that increased interaction between two persons increases their liking, but in the Tikopia family system Homans discovered that this generalization holds true only under conditions in which authority of one person over another is low. In the Tikopia family system, brothers revealed sentiments of liking as a result of their frequent interactions, primarily because they do not have authority over one another. However, frequent interaction with their father, who does have authority, was tense because the father initiated the interaction and because it usually involved the exercise of his authority.

Perhaps less critical than the substance of Propositions 9 and 10, I think, is the strategy they reveal. Homans has used an inductive technique to develop a large number of propositions that describe empirical regularities. By continually testing them in different types of groups, he can confirm them or, as is the case with Propositions 9 and 10, qualify these earlier propositions. By following this strategy, Homans argued in this early work that it is possible to develop a large body of empirical statements that reveal the form: y varies with x, under specifiable conditions. Stated in this form and with a clear connection to actual events in human groupings, such propositions encourage the development of more abstract concepts and theoretical statements. In addition, these statements are induced from a firm empirical footing that allows the abstract statements of sociology to be tested against the facts of ongoing group life. The next step in developing theory is to subsume these kinds of generalizations under more abstract laws. For, as I emphasized in Chapter 1, empirical generalizations by themselves are not explanations. They are regularities that require a theoretical explanation. And so it is not surprising that Homans eventually began to add another theoretical strategy to his general approach: formal theorizing, dressed up in the vocabulary of axiomatic theory (see Chapter 1 and Figure 1–4).

---

[14]Ibid., p. 243.
[15]Ibid., p. 247.

THE ADDITION OF A DEDUCTIVE STRATEGY

Homans visualizes the process of deduction and induction in the following way:

> The process of borrowing or inventing the more general propositions I call
> *induction*, whether or not it is the induction of the philosophers; the process
> of deriving the empirical propositions from the more general ones I call ex-
> planation, and this is the *explanation* of the philosophers.[16]

The Human Group was, therefore, an exercise in induction, and now the
task of the theorist is explanation through deduction. And it is to this deductive
approach—what I called formal theory in Chapter 1—that Homans turned in
the 1960s with the publication of *Social Behavior: Its Elementary Forms*.[17]
Such an approach emerged from Homans' long-standing criticism of most sociological theory, especially the functional strategy of Talcott Parsons (see
Chapter 3). For over the years Homans has mounted an increasingly pointed
criticism of sociological theorizing, with the hope that "we bring what we say
about theory into line with what we actually do, and so put an end to our
intellectual hypocrisy."[18] On the road out of "intellectual hypocrisy" is a rejection of Talcott Parsons' classificatory strategy of developing systems of concept and categorical schemes (what I termed "conceptual schemes" in Chapter
1):

> Some students get so much intellectual security out of such a scheme, be-
> cause it allows them to give names to, and to pigeonhole, almost any social
> phenomenon, that they are hesitant to embark on the dangerous waters of
> actually saying something about the relations between the phenomena—because
> they must actually take the risk of being found wrong.[19]

For Homans a more proper strategy is the construction of deductive systems
of propositions. At the top of the deductive system are the general axioms,
from which lower-order propositions are logically deduced. The lowest-order
propositions in the scheme are those composed of first-order abstractions that
describe actual events in the empirical world. Because these empirical generalizations are logically related to a hierarchy of increasingly abstract propositions, culminating in logical articulation with the axioms, the empirical generalizations are assumed to be explained by the axioms (consult Figures 1–4
and 1–7 in Chapter 1). Thus, for Homans, to have deduced logically an empirical

---

[16]George C. Homans, *Social Behavior: Its Elementary Forms* (New York: Harcourt Brace
Jovanovich, 1961), p. 10.

[17]Ibid.

[18]Ibid.

[19]George C. Homans, *The Nature of Social Science* (New York: Harcourt, Brace & World,
1967), p. 13. Parsons has replied that such systems of concepts can be theory: "I emphatically
dispute this [deductive theory] is all that can legitimately be called theory. In biology, for example,
I should certainly regard the basic classificatory schemes of taxonomy, for example in particular
the comparative anatomy of vertebrates, to be theoretical. Moreover very important things are,
with a few additional facts, explained on such levels as the inability of organisms with lungs and
no gills to live for long periods under water" (Talcott Parsons, "Levels of Organization and the
Mediation of Social Interaction,"*Sociological Inquiry* 34 [Spring 1964], pp. 219–20).

regularity from a set of more general propositions and axioms is to have explained the regularity.[20] Armed with this commitment to axiomatic theory, Homans articulates his exchange approach.

THE EXCHANGE MODEL

Sources of Homans' Psychological Exchange Perspective

I see Homans' exchange scheme first surfacing in a polemical reaction to Lévi-Strauss's structural analysis of cross-cousin marriage patterns. In collaboration with David Schneider, Homans previews what become prominent themes in his writings: (1) a skeptical view of any form of functional theorizing, (2) an emphasis on psychological principles as the axioms of social theory, and (3) a preoccupation with exchange-theoretic concepts.[21]

In their assessment of Lévi-Strauss's exchange functionalism, Homans and Schneider take exception to virtually all that made Lévi-Strauss's theory. First, the conceptualization of different forms of indirect, generalized exchange is rejected. In so conceptualizing exchange, Lévi-Strauss "thinned the meaning out of it." Second, Lévi-Strauss's position that different forms of exchange symbolically reaffirm and integrate different patterns of social organization is questioned, for an "institution is what it is because it results from the drives, or meets the immediate needs, of individuals or subgroups within a society."[22] The result of this rejection of Lévi-Strauss's thought is for Homans and Schneider to argue that exchange theory must initially emphasize face-to-face interaction, focus primarily on limited and direct exchanges among individuals, and recognize that social structures are created and sustained by the behaviors of individuals.[23]

With this critique of the anthropological tradition, which, I should emphasize, emerged as a reaction to utilitarianism, Homans resurrects the utili-

---

[20]Homans has championed this conception of theory in a large number of works; see, for example, Homans, *Social Behavior*; Homans, *Nature of Social Science*; George C. Homans, "Fundamental Social Processes," in *Sociology*, ed. N. J. Smelser (New York: John Wiley, 1967), pp. 27–78; George C. Homans, "Contemporary Theory in Sociology," in *Handbook of Modern Sociology*, ed. R. E. L. Faris (Skokie, IL: Rand McNally, 1964), pp. 251–77; and George C. Homans, "Bringing Men Back In," *American Sociological Review* 29 (December 1964), pp. 809–18. For an early statement of his position, see George C. Homans, "Social Behavior as Exchange," *American Journal of Sociology* 63 (August 1958), pp. 597–606. For a recent statement, see "Discovery and the Discovered in Social Theory," *Humboldt Journal of Social Relations* 7 (Fall–Winter 1979–80), pp. 89–102.

[21]George C. Homans and David M. Schneider, *Marriage, Authority, and Final Causes: A Study of Unilateral Cross-Cousin Marriage* (New York: Free Press, 1955). There are, however, hints of this interest in exchange theory in Homans' first major work, *English Villagers of the 13th Century* (New York: Russell & Russell, 1941).

[22]Homans and Schneider, *Marriage*, p. 15.

[23]In so doing, Homans advocates the principle of "methodological individualism," which sees social reality as the result of aggregated individual actions and behaviors. If social structure is aggregations of individual behavior, then the highest-order principles in the deductive schemes of sociology must be about individual behavior. For another version of this, see Chapter 12 on Randall Collins' exchange perspective.

tarian's concern with individual self-interest in the conceptual trappings of psychological behaviorism. For indeed, as Homans and Schneider emphasize: "We may call this an individual self-interest theory, if we remember that interests may be other than economic."[24] As becomes evident by the early 1960s, this self-interest theory is to be cast in the behaviorist language of B. F. Skinner. Given Homans' commitment to axiomatic theorizing and his concern with face-to-face interaction among individuals, it was perhaps inevitable that he would look toward Skinner and, indirectly, to the early founders of behaviorism—Pavlov, Thorndike, and Watson (see Chapter 14). But Homans borrows directly from Skinner's reformulations of early behaviorist principles.[25] Stripped of its subtlety, as Homans prefers, Skinnerian behaviorism states as its basic principle that, if an animal has a need, it will perform activities that in the past have satisfied this need. A first corollary to this principle is that organisms will attempt to avoid unpleasant experiences but will endure limited amounts of such experiences as a cost in emitting the behaviors that satisfy an overriding need. A second corollary is that organisms will continue emitting certain behaviors only as long as they continue to produce desired and expected effects. A third corollary of Skinnerian psychology emphasizes that, as needs are satisfied by a particular behavior, animals are less likely to emit the behavior. A fourth corollary states that, if in the recent past a behavior has brought rewards and if these rewards suddenly stop, the organism will appear angry and gradually cease emitting the behavior that formerly satisfied its needs. A final corollary holds that, if an event has consistently occurred at the same time as a behavior that was rewarded or punished, the event becomes a stimulus and is likely to produce the behavior or its avoidance.

These principles were derived from behavioral psychologists' highly controlled observations of animals, whose needs could be inferred from deprivations imposed by the investigators. Although human needs are much more difficult to ascertain than those of laboratory pigeons and mice and, despite the fact that humans interact in groupings that defy experimental controls, Homans believes that the principles of operant psychology can be applied to the explanation of human behavior in both simple and complex groupings. One of the most important adjustments of Skinnerian principles to fit the facts of human social organization involves the recognition that needs are satisfied by other people and that people reward and punish one another. In contrast with Skinner's animals, which only indirectly interact with Skinner through the apparatus of the laboratory and which have little ability to reward Skinner (except

---

[24]Homans and Schneider, *Marriage*, p. 15.

[25]Homans, *Social Behavior*, pp. 1–83. For an excellent summary of the Skinnerian principles incorporated into Homans' scheme, see Richard L. Simpson, "Theories of Social Exchange" (Morristown, NJ: General Learning Press, 1972), pp. 3–4. As an interesting aside, Peter P. Ekeh, *Social Exchange Theory and the Two Sociological Traditions* (Cambridge, MA: Harvard University Press, 1975), has argued that, had Homans not felt so compelled to reject Lévi-Strauss, he would not have embraced Skinnerian principles. Ekeh goes so far as to offer a hypothetical list of axioms that Homans would have postulated, had he not cast his scheme into the terminology of behaviorism.

perhaps to confirm his principles), humans constantly give and take, or exchange, rewards and punishments.[26]

The conceptualization of human behavior as exchange of rewards (and punishments) among interacting individuals leads Homans to incorporate, in altered form, the first principle of elementary economics: humans rationally calculate the long-range consequences of their actions in a marketplace and attempt to maximize their material profits in their transactions. I emphasize, however, that Homans qualifies this simplistic notion:

> Indeed we are out to rehabilitate the economic man. The trouble with him was not that he was economic, that he used resources to some advantage, but that he was antisocial and materialistic, interested only in money and material goods and ready to sacrifice even his old mother to get them.[27]

Thus, to be an appropriate explanation of human behavior, this basic economic assumption must be altered in four ways: (1) People do not always attempt to maximize profits; they seek only to make some profit in exchange relations. (2) Humans do not usually make either long-run or rational calculations in exchanges, for, in everyday life, "the Theory of Games is good advice for human behavior but a poor description of it." (3) The things exchanged involve not only money but also other commodities, including approval, esteem, compliance, love, affection, and other less materialistic goods. (4) The marketplace is not a separate domain in human exchanges, for all interaction involves individuals exchanging rewards (and punishments) and seeking profits.

The Basic Concepts

In an effort to meet the objections of his critics, Homans has altered his concepts somewhat since the original publication of his theory. Below, I have summarized the concepts of the theory in their most recent form:[28]

1. *Stimulus*—cues in the environment to which an organism responds with action.
2. *Action*—behaviors emitted by organisms directed at getting rewards and avoiding punishments.

---

[26]For a more detailed analysis of this point and the problems it presents, see Richard M. Emerson, "Social Exchange Theory," in *Annual Review of Sociology* 2 (Palo Alto, CA: Annual Reviews, 1976).

[27]Homans, *Social Behavior*, p. 79. Kenneth Boulding ("An Economist's View of 'Social Behavior: Its Elementary Forms,'" *American Journal of Sociology* 67 [January 1962], p. 458) has noted that in Homans' work "economic man is crossed with the psychological pigeon to produce what the unkind might call the Economic Pigeon theory of human interaction." For a more detailed and serious analysis of Homans' meshing of elementary economics and psychology, see Ekeh, *Social Exchange Theory*, pp. 162–71.

[28]George Caspar Homans, *Social Behavior: Its Elementary Forms* (New York: Harcourt Brace Jovanovich, 1974), pp. 15–47. In contrast to the first edition of *The Structure of Sociological Theory*, which concentrated on the earlier edition of Homans' book, I have relied almost exclusively on the revised edition of *Social Behavior*. For a discussion of Homans' original formulation of his exchange theory, the reader is referred to pp. 235–47 of my *The Structure of Sociological Theory* (Homewood, IL: Dorsey Press, 1974).

3. *Reward*—the capacity to bestow gratification or to meet the needs of an organism that a stimulus possesses.
4. *Punishment*—the capacity to harm, injure, or block the satisfaction of needs that a stimulus possesses.
5. *Value*—the degree of reward that a stimulus possesses.
6. *Cost*—rewards foregone or punishment incurred in engaging in one line of action.
7. *Perception*—the capacity to perceive, weigh, and assess rewards and costs.
8. *Expectation*—the level of rewards, punishments, or costs that an organism has come to associate with a particular stimulus.

These are the key concepts used in Homans' statement of his "elementary exchange principles."[29] He adds other concepts when applying these principles to human behavior. These new concepts, however, merely extend those listed here.

Elementary Principles of Social Behavior

Homans labels propositions in terms of the variables that each highlights.[30] These labels appear alongside the appropriate propositions as I have listed them in Table 15-1.

In Propositions I through III in Table 15-1, the principles of Skinnerian psychology are restated. The more valuable an activity (III), the more often such activity is rewarded (I); also, the more a situation approximates one in which activity has been rewarded in the past (II), the more likely a particular activity will be emitted. Proposition IV indicates the condition under which the first three fall into temporary abeyance. In accordance with the reinforcement principle of satiation or the economic law of marginal utility, humans eventually define rewarded activities as less valuable and begin to emit other activities in search of different rewards (again, however, in accordance with the principles enumerated in Propositions I-III). Proposition V introduces a more complicated set of conditions that qualify Propositions I through IV. From Skinner's observation that pigeons reveal anger and frustration when they do not receive an expected reward, Homans reasons that humans will probably reveal the same behavior.

In addition to these principles, Homans introduces a "rationality proposition," which summarizes the stimulus, success, and value propositions. I have also placed this proposition in Table 15-1 because it is so prominent in Homans' actual construction of deductive explanations. Let me translate the somewhat awkward vocabulary of Principle VI as Homans wrote it. People make calcu-

---

[29]The concept of *cost* does not appear in Homans' actual axioms, but it is so prominent in his scheme that it is listed here with his other important concepts.

[30]See Homans, *Social Behavior*, pp. 11–68. Masculine pronouns are maintained, since this is how Homans wrote the propositions.

TABLE 15-1   Homans' Exchange Propositions

 I. *Success Proposition:* For all actions taken by persons, the more often a particular action of a person is rewarded, the more likely the person is to perform that action.

 II. *Stimulus Proposition:* If in the past the occurrence of a particular stimulus or set of stimuli has been the occasion on which a person's action has been rewarded, then, the more similar the present stimuli are to the past ones, the more likely the person is to perform the action or some similar action now.

 III. *Value Proposition:* The more valuable to a person is the result of his action, the more likely he is to perform the action.

 IV. *Deprivation/Satiation Proposition:* The more often in the recent past a person has received a particular reward, the less valuable any further unit of that reward becomes for him.

 V. *Aggression/Approval Propositions:*

    A. When a person's action does not receive the reward he expected or receives punishment he did not expect, he will be angry and become more likely to perform aggressive behavior, and the results of such behavior become more valuable to him.

    B. When a person's action receives the reward expected, especially greater reward than expected, or does not receive punishment he expected, he will be pleased and become more likely to perform approving behavior, and the results of such behavior become more valuable to him.

 VI. *Rationality Proposition:* In choosing between alternative actions, a person will choose that one for which, as perceived by him or her at the time, the value of the result, multiplied by the probability of getting that result, is greater.

---

lations about various alternative lines of action. They perceive or calculate the value of the rewards that might be yielded by various actions. But they also temper this calculation in terms of the perceptions of how probable the receipt of rewards will be. Low probability of receiving highly valued rewards would lower their reward potential. Conversely, high probability of receiving a lower valued reward increases their overall reward potential. This relationship can be stated by the following formula:

$$\text{Action} = \text{Value} \times \text{Probability}$$

People are, Homans asserts, rational in the sense that they are likely to emit that behavior, or action, among alternatives in which value on the right side of the equation is largest. For example, if $Action_1$ is highly valued (say, 10) but the probability of getting it by emitting $Action_1$ is low (.20) and if $Action_2$ is less valued (say, 5) but the probability of receiving it is greater (.50) than $Action_1$, then the actor will emit $Action_2$ (because $10 \times .20 = 2$ yields less reward than $5 \times .50 = 2.5$).

This proposition was implicit in Homans' earlier version of his exchange theory, but he has now made it explicit. This proposition, as noted above, summarizes the implications of the stimulus, value, and success propositions, because it indicates why actors would choose one stimulus situation that has been rewarding over another. And, in fact, Homans had tended to use the rationality proposition in his explanations long before he made it explicit.[31] I see such a proposition as reevoking utilitarian notions of rational calculation but in the vocabulary of psychological behaviorism.

As summarized in Table 15-1, these basic principles or laws are seen by Homans to explain, in the sense of deductive explanation, patterns of human organization. As is obvious, they are psychological in nature. What is more, these psychological axioms constitute from Homans' viewpoint the only general sociological propositions, since "there are no general sociological propositions that hold good of all societies or social groups as such."[32]

However, the fact that psychological propositions are the most general does not make any less relevant or important sociological propositions stating the relationships among group properties or between group properties and those of individuals. On the contrary, these are the very propositions that are to be deduced from the psychological axioms. Thus, sociological propositions will be conspicuous in the deductive system emanating from the psychological principles. Homans stresses that sociology will finally bring what it says about theory into what it actually does when it arranges both abstract sociological statements and specific empirical generalizations in a deductive system with the psychological axioms at its top. For, as he continually emphasizes, to deduce propositions from one another is to explain them.

Homans' Construction of Deductive Systems

The fact that the basic axioms to be used in sociological explanation seem to be obvious truisms should not be a cause for dismay. Too often, Homans insists, social scientists assume that the basic laws of social organization will be more esoteric—and certainly less familiar—since for them the game of science involves the startling discovery of new, unfamiliar, and presumably profound principles. In reference to these social scientists, Homans writes:

> All this familiarity has bred contempt, a contempt that has got in the way of the development of social science. Its fundamental propositions seem so obvious as to be boring, and an intellectual, by definition a wit and a man of the world, will go to great lengths to avoid the obvious.[33]

However, if the first principles of sociology are obvious, despite the best efforts of scientists to the contrary, Homans suggests that sociology should cease its vain search for the esoteric and begin constructing deductive systems that recognize the fact that the most general propositions are not only psychological but familiar. In fact, if sociologists crave complexity, this task should certainly be satisfying, since the deductive systems connecting these simple principles to observed empirical regularities will be incredibly complex.

---

[31]See the first edition of *The Structure of Sociological Theory* and my "Building Social Theory: Some Questions about Homans' Strategy," *Pacific Sociological Review* 20 (April 1977), pp. 203–20.

[32]Homans, "Bringing Men Back In."

[33]Homans, *Nature of Social Science*, p. 73.

Unfortunately, Homans himself never offers a well-developed explanation. He has tended to simply invoke, in a rather ad hoc fashion, his axioms and to reconcile them in a very loose and imprecise manner with empirical regularities. Or he has constructed brief deductive schemes to illustrate his strategy. These limitations in themselves constitute an important criticism of Homans' work, because he has not fully implemented his own theoretical strategy. Yet I do not see Homans' failure to actually do what he says should be done as negating the utility of the strategy. Let me examine one of Homans' deductive schemes to assess its potential as well as its problems.

One of Homans' explanations is reproduced below. Homans recognizes that this is not a complete explanation, but he argues that it is as good as any other that exists. Moreover, although certain steps in the deductive scheme are omitted, it is the proper way to develop scientific explanations from his point of view. In this scheme, Homans seeks to explain Golden's Law that industrialization and the level of literacy in the population are highly correlated. As I emphasized in Chapter 1, such empirical generalizations are often considered to be theory, but Homans has correctly perceived that this "law" is only an empirical generalization that belongs at the bottom of the deductive scheme (see Figure 1–4). Propositions move from the most abstract statement, or the axiom(s),[34] to the specific empirical regularity to be explained. Golden's Law is thus explained in the following manner.

1. Men are more likely to perform an activity the more valuable they perceive the reward of that activity to be.
2. Men are more likely to perform an activity the more successful they perceive the activity to be in getting that reward.[35]
3. Compared with agricultural societies, a higher proportion of men in industrial societies are prepared to reward activities that involve literacy. (Industrialists want to hire bookkeepers, clerks, and persons who can make and read blueprints, manuals, and so forth.)
4. Therefore a higher proportion of men in industrial societies will perceive the acquisition of literacy as rewarding.
5. And [by (1)] a higher proportion will attempt to acquire literacy.
6. The provision of schooling costs money, directly or indirectly.
7. Compared with agricultural societies, a higher proportion of men in industrial societies are, by some standard, wealthy.
8. Therefore a higher proportion are able to provide schooling (through government or private charity), and a higher proportion are able to pay for their own schooling without charity.
9. And a higher proportion will perceive the effort to acquire literacy as apt to be successful.

---

[34]Note that I am using Homans' vocabulary, but, as I emphasized in Chapter 1, "axiomatic" theory for sociology is unrealistic.

[35]Note that "axioms" 1 and 2 here are simply an earlier version of Homans' rationality principle (VI) listed in Table 15–1.

10. And [by (2) as by (1)] a higher proportion will attempt to acquire literacy.

11. Since their perceptions are generally accurate, a higher proportion of men in industrial societies will in fact acquire literacy. That is, the literacy rate is apt to be higher in an industrial than in an agricultural society.[36]

Propositions 1 and 2 are an earlier statement of the rationality proposition that summarizes the success, stimulus, and value propositions. It is from these first two propositions, or axioms, that others are derived in an effort to explain Golden's Law (Proposition 11 in the scheme above). Examining some of the features of this explanation can provide insight into Homans' deductive strategy.

In this example, the transition between Propositions 2 and 3 ignores so many necessary variables as to simply describe in the words of behavioral psychology what Homans perceives to have occurred. Why are people in industrial societies prepared to reward literacy? This statement does not explain; it describes and thus opens a large gap in the logic of the deductive system. For Homans this statement is a given. It states a boundary condition, for the theory is not trying to explain why people are prepared to reward literacy. Another story is required to explain this event. Thus Homans argues that "no theory can explain everything" and that it is necessary to ignore some things and assume them to be givens for the purposes of explanation at hand. The issue remains, however: Has not Homans defined away the most interesting sociological issue—what makes people ready to reward literacy in a society's historical development? For Homans, people are just ready to do so.

Another problem in this scheme comes with the placement of the word *therefore*. This transitive is typically used immediately following a statement of the givens that define away important classes of sociological variables. For example, the "therefore" preceding key propositions (4 and 8) begs questions such as: Why do people perceive literacy as rewarding? What level of industrialization would make this so? What level of educational development? What feedback consequences does desire for literacy have for educational development? By ignoring the why and what of these questions, Homans can then in Propositions 5 and 10 reinsert the higher order axioms (1 and 2) of the explanation, thereby giving the scheme an appearance of deductive continuity. In fact, however, answers to the critical sociological questions have been avoided, such as why people perceive as valuable and rewarding certain crucial activities.

Homans' reply to such criticisms is, I think, important to note.[37] First, he emphasizes that this deductive scheme was just an example or illustrative

---

[36]George C. Homans, "Reply to Blain," *Sociological Inquiry* 41 (Winter 1971), pp. 19–24. This article was written in response to a challenge by Robert Blain for Homans to explain a sociological law: "On Homans' Psychological Reductionism," *Sociological Inquiry* 41 (Winter 1971), pp. 3–25.

[37]Personal communication to the author.

sketch, leaving out many details. Second, and more important, Homans contends that it is unreasonable for the critic to expect that a theory can and must explain every given condition (Propositions 3 and 7, for example). He offers the example of Newton's laws, which can help explain the movement of the tides (by virtue of gravitational forces of the moon relative to the axis of the earth), but these laws cannot explain why oceans are present and why the earth exists. I think that this latter argument is reasonable, in principle, since deductive theories explain only a delimited range of phenomena, as can be recalled by reference to Figure 1-6 in Chapter 1. But let me anticipate a point that I will emphasize shortly: although Homans does not need to explain the givens of a deductive scheme, all of the interesting sociological questions are contained in those givens. Moreover, these questions beg for the development of abstract *sociological* laws to explain them (Golden's Law is not an abstract principle, only an empirical generalization). And so I wonder why Homans devotes his time to the construction of quasi-axiomatic (formal) systems that contain no sociological laws. Why concentrate on the psychological laws from which sociological laws are to be ultimately deduced when the real task of sociology is to develop these sociological laws? But this is one of many criticisms leveled at Homans' scheme in particular and, to some extent, at all exchange theory. Let me now turn to these criticisms and discuss them in more detail.

CRITICISMS OF HOMANS' STRATEGY AND EXCHANGE THEORY

The Issue of Rationality

Homans' proposition on rationality could potentially open his exchange scheme to criticisms leveled against utilitarianism: do actors rationally calculate the costs and rewards? Homans partially meets this criticism by recognizing that people make calculations by weighing costs, rewards, and the probabilities of receiving rewards or avoiding punishments. But people do so in terms of value— that is, in terms of what bestows gratifications on them. Just what is rewarding is thus a personal matter and unique to all individuals. Depending on their past experiences, people establish their own values. Rationality, then, occurs in terms of calculations of personal value.[38]

There is, however, another implicit assumption that leaves Homans, and most exchange schemes, open to criticism. Do all human actions involve calculations? Do people always weigh and assess costs and rewards in all situa-

---

[38]Parsons, who similarly has dealt with the rationality issue ("Levels of Organization," in *Institutions and Social Exchange*, eds. H. Turk and R. L. Simpson [Indianapolis: Bobbs-Merrill, 1971], p. 219), notes: "History thus seems to become for Homans the ultimate residual category, recourse to which can solve any embarrassment which arises from inadequacy of the more specific parts of the conceptual scheme. The very extent to which he has narrowed his conceptualization of the variables, in particular adopting the atomistic conception of values, . . . increases the burden thrown upon history and with it the confession of ignorance embodied in the statement, 'things are as they are because of ways in which they have come to be that way.'"

tions?[39] Often, critics argue, people just receive rewards without prior calculations. For example, when a person receives a gift or becomes the beneficiary of another's desires to bestow rewards, prior calculations are not involved. Rewards or reinforcements are, of course, still involved, but the principle of rationality excludes much interaction. Thus Homans may have unduly limited his theory.

The Issue of Tautology

More fundamental to the exchange perspective is the problem of tautology. If one examines the definition of key concepts—value, reward, and action—they appear to be defined in terms of one another. Rewards are gratifications that have value. Value is the degree of reward, or reinforcement. Action is reward-seeking activity. The question arises, then, as to whether it is possible to build a theory from axioms that are tautological. For example, Homans' proposition that "The more valuable to a person is the result of his action, the more likely he is to perform the action" could be considered a tautology. Value is defined as the degree of reward, and action is defined as reward-seeking behavior.[40]

Homans acknowledges the circularity but views the problem as resolved by the use of deductive theory. Indeed, he argues that many axioms in deductive systems will be statements of equivalence, with concepts defined in terms of each other (for example, $f = ma$ and $E = mc^2$ in physics). For, if the axioms are viewed as part of a deductive system, the problem of tautology is soon obviated. Although value and action cannot be measured independently when stated so abstractly, the deductive system allows for their independent measurement at the empirical level. Thus, "a tautology can take part in the deductive system whose conclusion is not a tautology."[41] Thus Homans argues that the tautological nature of highly abstract axioms in a deductive system can be obviated when precise and clear derivations from the axioms are performed. In this process, independent definitions and indicators of key concepts can be provided. If these deductive steps are left out, however, and vague axioms are simply reconciled in an ad hoc fashion to empirical events, the problem of tautology will persist. And it is here that critics note: in virtually every explanation of social behavior in Homans' recent work, rigorous deductive systems are absent.[42] But, in fairness, I think that Homans has been only attempting

[39]Robert Bierstedt, "Review of Blain's Exchange and Power," *American Sociological Review* 30 (1965), pp. 789–90.

[40]For enlightening discussions of this problem, see Morton Deutsch, "Homans in the Skinner Box," in *Institutions and Social Exchange*," eds. H. Turk and R. L. Simpson, pp. 81–92; Bengt Abrahamsson, "Homans on Exchange: Hedonism Revived," *American Journal of Sociology* 76 (September 1970), pp. 273–85; Pitirim Sorokin, *Sociological Theories of Today* (New York: Harper & Row, 1966), especially Chapter 15, "Pseudo-Behavioral and Empirical Sociologies," and Mulkay, *Functionalism and Exchange*, pp. 166–69.

[41]Homans, *Social Behavior*, p. 35.

[42]Ronald Maris ("The Logical Adequacy of Homans' Social Theory," *American Sociological Review* 35 [December 1970], pp. 1069–81) came to a somewhat different conclusion about the

to suggest the utility of his concepts for future construction of deductive systems and to point out to sociologists that the ultimate explanatory principles of sociology are those about individual behavior.

The Issue of Reductionism

Periodically, old philosophical issues are resurrected and debated fervently. Homans' exchange perspective has rekindled one such debate: the issue of reductionism. Homans' statements on the issue are sometimes tempered and at other times polemic, but the thrust of his argument has been made amply clear. He writes:

> The institutions, organizations, and societies that sociologists study can always be analyzed, without residue, into the behavior of individual men. They must therefore be explained by propositions about the behavior of individual men.[43]

This position has been particularly disturbing to some sociologists, since it appears to pose a problem: if we accept the contention that sociological propositions are reducible to those about individuals, then those about individuals are reducible to physiological propositions, which in turn are reducible to biochemical propositions, and so on in a reductionist sequence ending in the basic laws of physical matter. Homans has not been very helpful in alleviating sociologists' concern with whether he is advocating this kind of reductionism. In fact, he advocates the position that, although the psychological axioms "cannot be derived from physiological propositions, . . . this condition is unlikely to last forever."[44] His tempered statements, however, are more revealing. Even if the laws of sociology are reducible to those of psychology and, at some future date, those of psychology to physiology, we will still need sociological laws. For example, to know that portions of Newtonian mechanics can be deduced from Einsteinian relativity theory does not stop us from using Newton's laws. Similarly, I think that Homans would argue: we need not reduce every sociological law to psychology each time we construct an explanation of an event.

This position argues against critics who have charged that Homans is a "nominalist" who asserts that society and its various collective forms (groups,

---

logical adequacy of Homans' theoretical manipulations. With the aid of symbolic logic and the addition of some assumptions, "Homans' theory of elementary social behavior has not been proven inadequate." But the criticisms of Maris' logical manipulation are sufficient to suggest that his analysis is not the definitive answer to the logical adequacy of Homans' deductions. For examples of these criticisms, see Don Gray, "Some Comments Concerning Maris on 'Logical Adequacy' "; Stephen Turner, "The Logical Adequacy of 'The Logical Adequacy of Homans' Social Theory' "; and Robert Price, "On Maris and the Logic of Time"; all in *American Sociological Review* 36 (August 1971), pp. 706–13. Maris' "Second Thoughts: Uses of Logic in Theory Construction" can also be found in this issue. For a more adequate construction of a logically sophisticated exchange perspective, see B. F. Meeker, "Decisions and Exchange," *American Sociological Review* 36 (June 1971), pp. 485–95.

[43]George C. Homans, "Commentary," *Sociological Inquiry* 34 (Spring 1964), p. 231.

[44]Homans, "Commentary," p. 225.

institutions, organizations, and so forth) are mere names sociologists arbitrarily assign to the only "really real" phenomenon, the individual. For Homans is very clear on the matter:

> I, for one, am not going to back into the position of denying the reality of social institutions. . . . The question is not whether the individual is the ultimate reality or whether social behavior involves something more than the behavior of individuals. The question is, always, how social phenomena are to be explained.[45]

I think that the thrust of the last phase has been underemphasized in the criticisms of Homans' reductionism. Critics have too often implied that his reductionism forces him to embrace a particular variety of nominalism. Yet, for Homans, the issue has always been one of how to explain with deductive— or axiomatic—systems the groups and institutions studied by sociologists.

Homans and the fallacy of "misplaced concreteness"   I see the most persistent criticism of Homans' reductionist strategy as revolving around charges that he has fallen into the fallacy of misplaced concreteness.[46] As originally conceived by the philosopher Alfred North Whitehead, scientists had at one time fallen into the trap of thinking that they could analyze the universe into its constituent parts and thereby eventually discover the basic elements or building blocks of all matter.[47] Once *the* basic building block had been found, it would only then be necessary to comprehend the laws of its operation for an understanding of everything else in the universe. In the eyes of Whitehead and others, these scientists had erroneously assumed that the basic parts of the universe were the only reality. In so doing, they had "misplaced" the concreteness of phenomena. In reality, the *relationships* among parts forming a whole are just as real as the constituent parts. The organization of parts is not the sum of the parts but, rather, the constitution of a new kind of reality.

Has Homans fallen into this fallacy? Numerous critics think that he has when he implies that the behaviors of persons or "men" are the basic units, whose laws need only be understood to explain more complex sociocultural arrangements. I think that these critics are overreacting to Homans' reductionism, perhaps confusing his reductionist strategy with the mistaken assumption that Homans is a nominalist in disguise. Actually, Homans has never denied the importance of sociological laws describing complex sociocultural processes; on the contrary, they are critical propositions in any deductive system that attempts to explain these processes. In my view, all that Homans asserts is: these sociological laws are not the most general; they are subsumable under more general psychological laws (his axioms), which, with more knowledge and sophisticated intellectual techniques, will be subsumable under a still

---

[45]Homans, *Nature of Social Science*, pp. 61–62.

[46]For examples of this line of criticism, see Parsons, "Levels of Organization"; and Blain, "On Homans' Psychological Reductionism," *Sociological Inquiry* 41 (Winter 1971), pp. 10–19.

[47]Alfred North Whitehead, *Science and the Modern World* (New York: Macmillan, 1925).

more general set of laws. At no point in this reductionist philosophy is Homans asserting that the propositions subsumed by a more general set of laws are irrelevant or unimportant. Thus Homans has not misplaced the concreteness of reality. He has not denied the existence of emergent properties such as groups, organizations, and institutions, nor the theoretical significance of the laws describing these emergent phenomena. Homans is not a nominalist in disguise but a sociological realist, who is advocating a particular strategy for understanding sociocultural phenomena.

The utility of Homans' reductionist strategy   Once it becomes evident that Homans' reductionism is a theoretical strategy that does not deny the metaphysical or ontological existence of emergent phenomena, the next question becomes: is this strategy useful in explaining phenomena? Some critics have emphasized that a reductionist strategy will affect the kinds of theoretical and research questions sociologists will ask.[48] If one is concerned primarily with psychological laws as explanatory principles, it is likely that research questions and theoretical generalizations will begin to revolve around psychological and social-psychological phenomena, because these phenomena are the most readily derived from psychological axioms. Thus, despite a recognition that complex sociological phenomena are real, the adoption of a reductionist strategy for building theory will inadvertently result in the avoidance of the macro patterns of social organization studied by many sociologists. To the extent that such one-sided research and theory building are the result of a reductionist strategy, this strategy is undesirable and can be questioned on these grounds alone. However, critics charge that there are more fundamental grounds on which to reject Homans' strategy: adherence to his strategy at the present time will lead to logically imprecise and empirically empty theoretical formulations.

This indictment is, of course, severe and needs to be documented. I sense that Homans may be correct in holding that a deductive axiomatic strategy necessitates reductionism, for the goal of such a strategy is to subsume under ever more general axioms what we previously considered the most general ones. Such a process of subsumption may indeed lead first to the subsumption of sociological axioms under psychological axioms and then to the subsequent subsumption of these latter axioms under physiological, biochemical, and physical laws. Just as many of the laws of chemistry can be subsumed under the laws of physics, so sociological laws may be subsumed by the laws of psychology. However, I believe that deduction of sociological laws from psychological axioms should occur in a two-step process: (1) A series of well-established sociological laws from which it is possible to deduce a wide variety of sociological propositions that have received consistent empirical support must be developed. (2) Then, and only then, a clear body of psychological axioms, which are amendable to similar reductions and which have received consistent empirical support,

---

[48]For example, see Blain, "On Homans' Psychological Reductionism"; and Walter Buckley, *Sociology and Modern Systems Theory* (Englewood Cliffs, NJ: Prentice-Hall, 1967), pp. 109–11.

can be used to explain the sociological laws.[49] I think that Step 1 must occur prior to Step 2, as it typically has in the physical sciences. Homans has recognized the fact that the social sciences have not achieved Step 1 when he notes that the "issue for the social sciences is not whether we should be reductionists, but rather, if we were reductionists, whether we could find any propositions to reduce."[50]

Yet I believe that Homans fails to realize the full implications of his statement. Without well-established sociological laws to subsume, the critics can correctly ask: What is the utility of attempting to subsume what does not exist? Would it not be far wiser to expend our efforts in developing sociological laws and let the issue of reductionism take care of itself when these laws are established? To attempt prematurely to develop psychological axioms and then deduce sociological propositions from them in the absence of well-established sociological laws is likely to generate logically imprecise deductions, as I illustrated with Homans' explanation of Golden's Law. What Homans typically does in his deductions is to take as givens all the interesting sociological questions, answers to which would lead to the development of the sociological laws needed to fill out properly his deductive system. The end result is that an empirical generalization—say, Golden's Law—may be explained without any of the desirable components of a deductive system—clear sociological laws.

Such deductive systems will ultimately boil down to statements such as: things are as they are because they are rewarding. What this does is to repeat the empirical generalization to be explained in the words of behavioral psychology, without logically deducing the generalization from sociological laws, which in turn are deduced (if one is so inclined) from psychological laws.[51]

HOMANS' IMAGE OF SOCIETY

What has always intrigued me about Homans' approach is the substantive image of society that he presents. I share his commitment to deductive theory, although I think that he is premature in his overconcern with psychological laws. But these kinds of criticisms aside, Homans has approached the sub-

---

[49]Turner, "Building Social Theory."

[50]Homans, *Nature of Social Science*, p. 86.

[51]The concepts of behavioral psychology do not have to muddle empirical generalizations, so long as one does not prematurely try to deduce sociological propositions to crude psychological propositions. In fact, operant principles can be used quite fruitfully to build (note: not reduce) more complex exchange principles that pertain to sociological processes (note: not psychological). For an impressive attempt at employing operant principles as a starting point and then changing them to fit the facts of emergent properties, see Richard M. Emerson's various works, especially "Power-Dependence Relations," *American Sociological Review* 27 (February 1962), pp. 31–41; "Power-Dependent Relations: Two Experiments," *Sociometry* 27 (September 1964), pp. 282–98; "Operant Psychology and Exchange Theory," in *Behavior Sociology: The Experimental Analysis of Social Process*, eds. R. L. Burgess and D. Bushell, Jr. (New York: Columbia University Press, 1969), pp. 379–405; and "Exchange Theory, Part I: A Psychological Basis for Social Exchange" and "Exchange Theory, Part II: Exchange Relations and Network Structures," in *Sociological Theories in Progress*, eds. J. Berger, M. Zelditch, Jr., and B. Anderson (Boston: Houghton Mifflin, 1972), pp. 38–87.

stantive issues that divide sociologists in a very creative way. The most prominent of these substantive issues is the micro versus macro controversy that I outlined in Chapter 1. That is, how do we reconcile the microprocesses of behavior, action, and interaction of individuals with the macrostructures and institutions of society? How do we explain their relationship? One answer is through deductive reduction; that is, if sociological laws about social structures can be deduced from those about individual behavior, then the micro/macro schism is resolved by the logic of deductive reasoning. I find this answer less compelling than the substantive image that Homans discursively develops in both *The Human Group* and *Social Behavior*. For I think that the *substantive* vision of the social world first communicated in *The Human Group* and expanded upon in the later exchange works is provocative and likely to be the most enduring feature of Homans' theoretical perspective. Thus I will close this chapter on Homans' approach with a review of this more substantive image of society.

In *The Human Group*, Homans' numerous empirical generalizations describe the processes of group elaboration and disintegration. Groups are observed to differentiate into subgroups, to form leadership ranks, to codify norms, to establish temporary equilibriums, and then, in his last case study of a dying New England town, to reveal the converse of these processes. In the later exchange works, the concepts of *activities, interactions,* and *sentiments,* which are incorporated into the propositions describing such group elaboration, are redefined in order to give Homans the opportunity to explain why these processes should occur. Human activity becomes action directed toward the attainment of rewards and avoidance of punishments. Interaction becomes social behavior whereby the mutual actions of individuals have cost and reward implications for the parties to the interaction. People are now seen as emitting those activities that increase the likelihood of profits—rewards less costs—measured against a set of expectations. Such rewarding and costly exchanges of activities are not viewed as necessarily involving the exchange of material rewards and punishments, but more frequently "psychic profits."

Just as he did in *The Human Group*, Homans enumerates concepts that enable him to denote processes of group elaboration in his more recent work. In *Social Behavior*, the particular explanations are of less interest, I feel, than Homans' descriptions of how vital group processes—interaction, influence, conformity, competition, bestowal of esteem, justice, ranking, and innovation—ebb and flow as actors seek psychic profits in their exchanges of rewards and punishments. In these descriptions, considerable intuitive insight into the basic processes of human interaction is evident. It was these insights that made *The Human Group* appealing, and I think that it is this same feature of *Social Behavior* that makes it an important work.

Despite the suggestiveness of his descriptions of basic processes in earlier chapters, the most theoretically interesting section of *Social Behavior* is the closing chapter, "The Institutional and the Subinstitutional." Introduced apologetically as a last-gasp "orgy," Homans nevertheless returns to an issue first raised in *The Human Group*: the relationship of processes in groups to the

FIGURE 15–1   Homans' Image of Social Organization

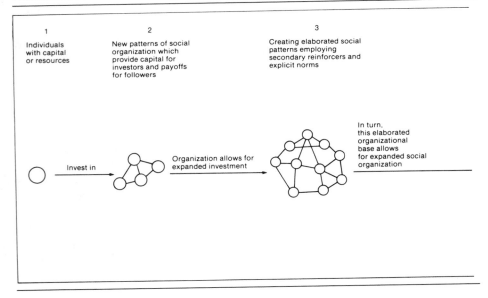

structures of larger societies, or "civilizations," as he phrased the issue then. As he emphasized in the last paragraph of *The Human Group*, the development of civilizations is ultimately carried out by persons in groups:

> At the level of the small group, society has always been able to cohere. We infer, therefore, that if civilization is to stand, it must maintain, in the relation between the groups that make up society and the central direction of society, some of the features of the small group itself.[52]

In *Social Behavior*, Homans has a more sophisticated answer to why it appears that society coheres around the small group: society is elaborated and structured from fundamentally the same exchange processes that cause the elaboration of the small group. All social structures are thus built up from basically the same exchange processes. In the explication of why this should be so, Homans provides an interesting image of how patterns of social organization are created, maintained, changed, and broken down. This image is not developed into what can be considered adequate theory, as has already been shown. But it does provide a vision of the social world that can perhaps initiate a more useful exchange-theoretic perspective on the processes underlying various patterns of social organization.

To explicate the relationship between elementary exchange processes and more complex patterns of social organization, Homans—much like Parsons a decade earlier—provides a sketch of the process of institutionalization, which

---

[52]Homans, *Human Group*, p. 468.

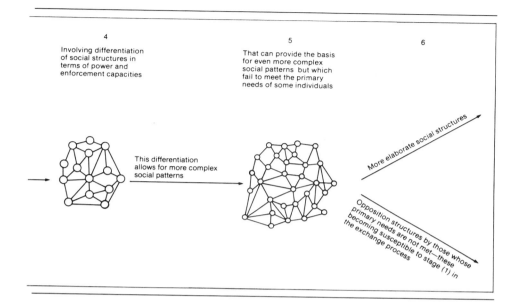

4

Involving differentiation of social structures in terms of power and enforcement capacities

5

That can provide the basis for even more complex social patterns but which fail to meet the primary needs of some individuals

6

This differentiation allows for more complex social patterns

More elaborate social structures

Opposition structures by those whose primary needs are not met—these becoming susceptible to stage (1) in the exchange process

I have illustrated in Figure 15-1.[53] At points in history, some people have the "capital" to reinforce or provide rewards for others, whether it comes from their possessing a surplus of food, money, a moral code, or valued leadership qualities. With such capital, "institutional elaboration" can occur, since some can invest their capital by trying to induce others (through rewards or threats of punishments) to engage in novel activities. These new activities can involve an "intermeshing of the behavior of a large number of persons in a more complicated or roundabout way than has hitherto been the custom." Whether this investment involves conquering territory and organizing a kingdom or creating a new form of business organization, those making the investment must have the resources—whether it be an army to threaten punishment, a charismatic personality to morally persuade followers, or the ability to provide for people's subsistence needs—to keep those so organized in a situation in which they derive some profit. At some point in this process, such organization can become more efficient and hence rewarding to all when the rewards are clearly specified in terms of generalized reinforcers, such as money, and when the activities expended to get their rewards are more clearly specified, such as when explicit norms and rules emerge. In turn, this increased efficiency allows

---

[53]Homans, *Social Behavior*, Chapter 16. The reader should find interesting a comparison of this model of institutionalization and that provided by Parsons, since Homans implicitly sees this model as an alternative to that presented by functionalists such as Parsons. But a careful reading of Talcott Parsons, *The Social System* (New York: Free Press, 1951), pp. 1–91, and of his more recent work on evolution would reveal a remarkable similarity between his conceptualization of this basic process and that of Homans.

for greater organization of activities. This new efficiency increases the likelihood that generalized reinforcers and explicit norms will be used to regulate exchange relations and hence increase the profits to those involved. Eventually the exchange networks involving generalized reinforcers and an increasingly complex body of rules require differentiation or subunits—such as a legal and banking system—that can maintain the stability of the generalized reinforcers and the integrity of the norms.

Out of this kind of exchange process, then, social organization—whether at a societal, group, organizational, or institutional level—is constructed. The emergence of most patterns of organization is frequently buried in the recesses of history, but such emergence is typified by these accelerating processes: (1) People with capital (reward capacity) invest in creating more complex social relations that increase their rewards and allow those whose activities are organized to realize a profit. (2) With increased rewards, these people can invest in more complex patterns of organization. (3) Increasingly complex patterns of organization require, first of all, the use of generalized reinforcers and then the codification of norms to regulate activity. (4) With this organizational base, it then becomes possible to elaborate further the pattern of organization, creating the necessity for differentiation of subunits that assure the stability of the generalized reinforcers and the integrity of norms. (5) With this differentiation, it is possible to expand even further the networks of interaction, since there are standardized means for rewarding activities and codifying new norms as well as enforcing old rules.

However, these complex patterns of social organization employing formal rules and secondary or generalized reinforcers can never cease to meet the more primary needs of individuals.[54] Institutions first emerged to meet these needs, and, no matter how complex institutional arrangements become and how many norms and formal rules are elaborated, these extended interaction networks must ultimately reinforce humans' more primary needs. When these arrangements cease meeting the primary needs from which they ultimately sprang, an institution is vulnerable and apt to collapse if alternative actions, which can provide primary rewards, present themselves as a possibility. In this situation, low- or high-status persons—those who have little to lose by nonconformity to existing prescriptions—will break from established ways to expose to others a more rewarding alternative. Institutions may continue to extract conformity for a period, but they will cease to do so when they lose the capacity to provide primary rewards. Thus, complex institutional arrangements must ultimately be satisfying to individuals, not simply because of the weight of culture or norms but because they are constructed to serve people:

> Institutions do not keep going just because they are enshrined in norms, and it seems extraordinary that anyone should ever talk as if they did. They keep going because they have pay-offs, ultimately pay-offs for individuals. Nor is society a perpetual-motion machine, supplying its own fuel. It cannot keep

---

[54]Unfortunately, Homans never defines "primary need."

itself going by planting in the young a desire for these goods and only those goods that it happens to be in shape to provide. It must provide goods that men find rewarding not simply because they are sharers in a particular culture but because they are men.[55]

The fact that institutions of society must also meet primary needs sets the stage for a continual conflict between institutional elaboration and the primary needs of humans. As one form of institutional elaboration meets one set of needs, it may deprive people of other important rewards—opening the way for deviation and innovation by those presenting the alternative rewards that have been suppressed by dominant institutional arrangements. In turn, the new institutional elaborations that may ensue from innovators who have the capital to reward others will suppress other needs, which, through processes similar to its inception, will set off another process of institutional elaboration.

In sum, this sketch of how social organization is linked to elementary processes of exchange represents an interesting perspective for analyzing how patterns of social organization are built up, maintained, altered, and broken down. Although there are obvious conceptual problems—for example, the difficulty of distinguishing primary rewards from other types—the image of society presented by Homans remains creative and provocative.

---

[55]Homans, *Social Behavior*, p. 366.

CHAPTER 16

◆

Structural Exchange Theory

Peter M. Blau

Peter M. Blau has been one of the most productive social theorists over the last three decades. In his early work on informal processes within bureaucracies, he noted how frequently employees exchanged assistance with their work for respect, information for social approval, and other processes of giving and receiving nonmaterial rewards. Yet, in looking back on this early empirical work, he recently remarked: "I was not aware, or did not remember, that conceptions of social exchange had been used by many others before, from Aristotle to Mauss."[1] In some ways I think that this ignorance may have been an advantage because, in rediscovering exchange processes, Blau created a constructive blend of exchange, functional, and dialectical conflict theories. And although Blau has in recent years abandoned his exchange approach for another theoretical strategy (see Chapter 28),[2] I think that his exchange orientation is still very important and worthy of a more detailed analysis.[3]

BLAU'S THEORETICAL STRATEGY

In contrast to Homans' concern with developing deductive explanations, Blau offers what he terms a *theoretical prolegomenon*—or a conceptual sketch that

---

[1]Peter M. Blau, "Contrasting Theoretical Perspectives," paper delivered at the German-American Theory Conference, 1984, p. 2.

[2]Peter M. Blau, *Inequality and Heterogeneity* (New York: Free Press, 1977).

[3]Peter M. Blau's major exchange work is *Exchange and Power in Social Life* (New York: John Wiley, 1964). This formal and expanded statement on his exchange perspective was anticipated in earlier works. For example, see Peter M. Blau, "A Theory of Social Integration," *American Journal of Sociology* 65 (May 1960), pp. 545–56; and Peter M. Blau, *The Dynamics of Bureaucracy*, 1st and 2nd eds. (Chicago: University of Chicago Press, 1955, 1963). It is of interest to note that George C. Homans in *Social Behavior: Its Elementary Forms* (New York: Harcourt, Brace & World, 1961) makes frequent reference to the data summarized in this latter work. For a more recent statement of Blau's position, see Peter M. Blau, "Interaction: Social Exchange," in *International Encyclopedia of the Social Sciences*, vol. 7 (New York: Macmillan, 1968), pp. 452–58.

can serve as a preliminary to more mature forms of theorizing. In many ways Blau's strategy resembles Talcott Parsons', for he appears less concerned with developing a rigorous system of propositions than with enumerating concepts that can capture in loosely phrased and related propositions the fundamental processes occurring at diverse levels of social organization. Although there is less categorization than in Parsons' conceptual efforts, Blau is concerned with developing an initial bundle of concepts and propositions that can provide insight into the operation of a wide range of sociological processes, from the behavior of individuals in small-group contexts to the operation of whole societies.

Yet, as his more recent theorizing reveals and as he was led to remark in a retrospective look at his exchange approach, sociologists should be concerned with exchange analysis because "it is one of the few subject matters, outside of mathematical sociology, that lends itself to the development of systematic axiomatic theory."[4] Thus, although he does not develop his exchange ideas into an axiomatic format as does George Homans (see Chapter 15), he clearly had this goal in mind. And, as we will see in Chapter 28, he has tried in recent years to implement a more formal deductive approach. And so, as I proceed with Blau's approach, I will convert his ideas into formal principles. Such conversion is, I feel, in the spirit of his underlying theoretical strategy.

But at this stage in his theoretical work on exchange, Blau tries to use a bundle of exchange concepts and implicit principles to bridge the micro/macro gap. For he thought that exchange theory could provide a means for analyzing individual interactions as well as more structural relations in terms of the same basic framework. Although he is no longer so convinced that this can be the case, he sought in *Exchange and Power in Social Life* to (1) conceptualize some of the simple and direct exchange processes occurring in relatively small interaction networks and (2) then expand the conceptual edifice to include some of the complexities inherent in less direct exchange processes in larger social systems. In a vein similar to Homans' analysis, Blau first examines "elementary" forms of social exchange with an eye to how they help in the analysis of "subinstitutional" behavior. However, unlike Homans, who terminates his analysis by simply presenting a conceptual "orgy" in *Social Behavior*,[5] Blau begins to supplement the exchange concepts describing elementary processes in an effort to understand more complex processes of institutionalization.

Thus, in a manner reminiscent of Parsons' analysis of the process of institutionalization in *The Social System*[6] (see Chapter 3), Blau begins with a conceptualization of basic interactive processes; then, utilizing and supplementing the concepts developed in this analysis, he shifts to the analysis of more elaborate institutional complexes.

---

[4]Blau, "Contrasting Theoretical Perspectives."

[5]Homans, "The Institutional and the Subinstitutional," *Social Behavior*, rev. ed., Chapter 16.

[6]Talcott Parsons, *The Social System* (New York: Free Press, 1951), especially pp. 1–200.

BASIC EXCHANGE PRINCIPLES

I find that Blau does not define the variables in his exchange scheme as explicitly as Homans does. Rather, considerably more attention is devoted to defining exchange as a particular type of association, involving "actions that are contingent on rewarding reactions from others and that cease when these expected reactions are not forthcoming."[7] For Blau, exchange occurs only among those relationships in which rewards are expected and received from designated others. Much like Parsons' conception of voluntarism and Homans' rationality proposition, Blau conceptualizes as exchange activities only those behaviors that are oriented to specified goals, or rewards, and that involve actors selecting from various potential alternatives, or costs, a particular line of action that will yield an expected reward. In pursuing rewards and selecting alternative lines of behavior, actors are conceptualized as seeking a profit (rewards less costs) from their relations with others. Thus, Blau employs the basic concepts of all exchange theories—reward, cost, and profit—but he limits their application to relations with others from whom rewards are expected and received. This definition of exchange is considerably more limited than Homans' definition, which encompasses all activity as exchange, regardless of whether rewards are expected or received.

In common with Homans, however, Blau recognizes that, in focusing on associations involving "an exchange of activity, tangible or intangible, and more or less rewarding and costly, between two persons," an elementary economic model is being employed.[8] Indeed, social life is conceived to be a marketplace in which actors negotiate with one another in an effort to make a profit. But Blau shares the skepticism that led Homans to reject the theory of games as good advice but a poor description of human behavior and that induced Parsons earlier, in *The Structure of Social Action*, to discard the extremes of utilitarianism.[9] Blau recognizes that, unlike the simple "economic man" of classical economics (and of more recent rationalistic models of human behavior), humans (1) rarely pursue one specific goal to the exclusion of all others, (2) are frequently inconsistent in their preferences, (3) virtually never have complete information on alternatives, and (4) are never free from social commitments limiting the available alternatives. Furthermore, in contrast with a purely economic model of human transactions, social associations involve the exchange of rewards whose value varies from one transaction to another without a fixed market value and whose value cannot be expressed precisely in terms of a single, accepted medium of exchange (such as money). In fact, the vagueness of the values exchanged in social life is a "substantive fact, not simply a methodological problem."[10] As Blau emphasizes, the values people hold are inherently diffuse and ill defined.[11]

---

[7]Blau, *Exchange and Power*, p. 6.

[8]Ibid., p. 88.

[9]Talcott Parsons, *The Structure of Social Action* (New York: McGraw-Hill, 1937).

[10]Blau, *Exchange and Power*, p. 95.

[11]As I noted for Homans' scheme, this "fact" creates both methodological and logical problems.

TABLE 16-1 Blau's Implicit Exchange Principles

I. *Rationality Principle:* The more profit people expect from one another in emitting a particular activity, the more likely they are to emit that activity.

II. *Reciprocity Principles:*

A. The more people have exchanged rewards with one another, the more likely are reciprocal obligations to emerge and guide subsequent exchanges among these persons.

B. The more the reciprocal obligations of an exchange relationship are violated, the more disposed are deprived parties to sanction negatively those violating the norm of reciprocity.

III. *Justice Principles:*

A. The more exchange relations have been established, the more likely they are to be governed by norms of "fair exchange."

B. The less norms of fairness are realized in an exchange, the more disposed are deprived parties to sanction negatively those violating the norms.

IV. *Marginal Utility Principle:* The more expected rewards have been forthcoming from the emission of a particular activity, the less valuable is the activity and the less likely is its emission.

V. *Imbalance Principle:* The more stabilized and balanced are some exchange relations among social units, the more likely are other exchange relations to become imbalanced and unstable.

As I just indicated, Blau does not develop explicit exchange axioms or principles. But in Table 16-1, I have listed his basic ideas in propositional form. Proposition I, which I have termed the *rationality principle,* combines Homans' Axioms 1, 2, and 3, whereby rewarding stimulus situations, the frequency of rewards, and the value of rewards increase the likelihood that actions will be emitted. Actually, as we saw in the last chapter, Homans' own deductive systems—for example, the one he constructed to explain Golden's Law—collapses his first three axioms into a similar rationality principle: "In choosing between alternative actions, a person will choose that one for which, as perceived by him at the time, the value of the result, multiplied by the probability of getting the result, is greater."[12] In practice, then, Homans and Blau utilize the same basic principle. Blau's use of the concept of "reward expectation" would encompass the same phenomena denoted by Homans' "perception of reward" and "probability" of getting a reward.

Propositions II-A and II-B on reciprocity borrow from Malinowski's and Lévi-Strauss's initial discussion as reinterpreted by Alvin Gouldner.[13] Blau postulates that "the need to reciprocate for benefits received in order to con-

---

If value cannot be precisely measured, how is it possible to discern just how value influences behavior? If value cannot be measured independently of the behavior it is supposed to regulate, then propositions will be tautologous and of little use in building sociological theory.

[12]Homans, *Social Behavior,* p. 43.

[13]Alvin W. Gouldner, "The Norm of Reciprocity," *American Sociological Review* 25 (April 1960), pp. 161-78.

tinue receiving them serves as a 'starting mechanism' of social interaction."[14] Equally important, once exchanges have occurred, a "fundamental and ubiquitous norm of reciprocity" emerges to regulate subsequent exchanges. Thus, inherent in the exchange process, per se, is a principle of reciprocity. Over time, and as the conditions of Principle I are met, a social "norm of reciprocity," whose violation brings about social disapproval and other negative sanctions, emerges in exchange relations. And I should emphasize here in anticipation of later discussion that violations of the norm of reciprocity become significant in Blau's subsequent analysis of opposition and conflict.

Much like Homans, Blau recognizes that people establish expectations about what level of reward particular exchange relations should yield. Unlike Homans, however, Blau recognizes that these expectations are normatively regulated. These norms are termed *norms of fair exchange* since they determine what the proportion of rewards to costs should be in a given exchange relation. And, like Homans, Blau asserts that aggression is forthcoming when these norms of fair exchange are violated. I have incorporated these ideas into Propositions III-A and III-B, terming them the *justice principles.*[15]

Following economists' analyses of transactions in the marketplace, Blau introduces a principle on "marginal utility" (Proposition IV). The more a person has received a reward, the more satiated he or she is with that reward and the less valuable are further increments of the reward.[16] I should note that this marginal utility principle is the same as Homans' axiom on "deprivation/satiation." For both Blau and Homans, then, actors will seek alternative rewards until their level of satiation declines.

Blau's exchange model is vitally concerned with the conditions under which conflict and change occur in social systems, and so these justice principles become crucial generalizations. As I will document shortly, the deprivations arising from violating the norms of fair exchange can lead to retaliation against violators.

Proposition V on imbalance completes my listing of Blau's abstract laws. For Blau, as for all exchange theorists, established exchange relations are seen to involve costs or alternative rewards foregone. Since most actors must engage in more than one exchange relation, the balance and stabilization of one exchange relation is likely to create imbalance and strain in other necessary exchange relations. For Blau, social life is thus filled with dilemmas in which people must successively trade off stability and balance in one exchange relation for strain in others as they attempt to cope with the variety of relations they must maintain. In his last chapter on institutionalization, Homans hinted at this principle when he emphasized that, in satisfying some needs, institutional arrangements deny others and thereby set into motion a perpetual dialectic

---

[14]Blau, *Exchange and Power*, p. 92.

[15]See Peter M. Blau, "Justice in Social Exchange," in *Institutions and Social Exchange: The Sociologies of Talcott Parsons and George C. Homans*, eds. H. Turk and R. L. Simpson (Indianapolis: Bobbs-Merrill, 1971), pp. 56–68; see also: Blau, *Exchange and Power*, pp. 156–57.

[16]Blau, *Exchange and Power*, p. 90.

between dominant institutions and change-oriented acts of innovation and deviance.[17] It is from this concluding insight of Homans into the dialectical nature of relationships between established social patterns and forces of opposition that Blau is to begin his analysis of exchange in social life. Indeed, we might even term Blau's approach a *dialectical exchange theory* because inherent in exchange relations are potentials for conflict. As I have indicated in the last chapter, Marxian conflict theory and exchange theory have much in common since both are concerned with the conditions under which actors feel deprived of expected rewards and are mobilized to redress their grievances.

BASIC EXCHANGE PROCESSES IN SOCIAL LIFE

Elementary Systems of Exchange

Blau initiates his discussion of elementary exchange processes with the assumption that people enter into social exchange because they perceive the possibility of deriving rewards (Principle I). Blau labels this perception *social attraction* and postulates that, unless relationships involve such attraction, they are not relationships of exchange. In entering an exchange relationship, each actor assumes the perspective of another and thereby derives some perception of the other's needs. Actors then manipulate their presentation of self so as to convince one another that they have the valued qualities others appear to desire. In adjusting role behaviors in an effort to impress others with the resources that they have to offer, people operate under the principle of reciprocity, for, by indicating that one possesses valued qualities, each person is attempting to establish a claim on others for the receipt of rewards from them. All exchange operates under the presumption that people who bestow rewards will receive rewards in turn as payment for value received.

Actors attempt to impress one another through competition in which they reveal the rewards they have to offer in an effort to force others, in accordance with the norm of reciprocity, to reciprocate with an even more valuable reward. Social life is thus rife with people's competitive efforts to impress one another and thereby extract valuable rewards. But, as interaction proceeds, it inevitably becomes evident to the parties to an exchange that some people have more valued resources to offer than others, putting them in a unique position to extract rewards from all others who value the resources that they have to offer.

It is at this point in exchange relations that groups of individuals become differentiated in terms of the resources they possess and the kinds of reciprocal demands they can make on others. Blau then asks an analytical question: what generic types or classes of rewards can those with resources extract in return for bestowing their valued resources upon others? Blau conceptualizes four general classes of such rewards: money, social approval, esteem or respect, and

---

[17]Homans, *Social Behavior*, pp. 390–98 (1961 ed.); pp. 366–73 (1974 ed.).

TABLE 16–2 Blau's Conditions for the Differentiation of Power in Social Exchange

---

I. The fewer services people can supply in return for the receipt of particularly valued services, the more those providing these particularly valued services can extract compliance.

II. The fewer alternative sources of rewards people have, the more those providing valuable services can extract compliance.

III. The less those receiving valuable services from particular individuals can employ physical force and coercion, the more those providing the services can extract compliance.

IV. The less those receiving the valuable services can do without them, the more those providing the services can extract compliance.

---

compliance. Although Blau does not make full use of his categorization of rewards, he offers some suggestive clues about how these categories can be incorporated into abstract theoretical statements. Let me elaborate upon his argument.

Blau first ranks these generalized reinforcers in terms of their value in exchange relations. In most social relations, money is an inappropriate reward and hence is the least valuable reward. Social approval is an appropriate reward, but for most humans it is not very valuable, thus forcing those who derive valued services to offer with great frequency the more valuable reward of esteem or respect to those providing valued services. In many situations the services offered can command no more than respect and esteem from those receiving the benefit of these services. At times, however, the services offered are sufficiently valuable to require those receiving them to offer, in accordance with the principles of reciprocity and justice, the most valuable class of rewards— compliance with one's requests.

When people can extract compliance in an exchange relationship, they have power. They have the capacity to withhold rewarding services and thereby punish or inflict heavy costs on those who might not comply. To conceptualize the degree of power possessed by individuals, Blau formulates four general propositions that determine the capacity of powerful individuals to extract compliance. I have listed and somewhat reformulated them in Table 16-2.[18]

These four propositions list the conditions leading to differentiation of members in social groups in terms of power. To the extent that group members can supply some services in return, seek alternative rewards, potentially use physical force, or do without certain valuable services, individuals who can provide valuable services will be able to extract only esteem and approval from group members. Such groups will be differentiated in terms of prestige rankings but not power. Naturally, as Blau emphasizes, most social groups reveal complex patterns of differentiation of power, prestige, and patterns of approval, but of

---

[18]Blau, *Exchange and Power*, pp. 118–19.

particular interest to him are the dynamics involved in generating power, authority, and opposition.

In focusing almost exclusively on the questions of power, authority, and opposition, I think that Blau fails to complete his analysis of how different types of social structures are influenced by the exchange of different classes of rewards. The logic of Blau's argument would, I believe, require additional propositions that would indicate how various types of rewards lead to the differentiation of groups, not only in terms of power and authority but also with respect to prestige rankings and networks of social approval. Interesting theoretical questions left unanswered include: What are the conditions for the emergence of different types of prestige rankings? What are the conditions for the creation of various types of approval networks? Presumably, answers to these questions are left for others to provide. Blau chooses to focus primarily on the problem of how power is converted into authority and how, in accordance with his basic exchange principles listed in Table 16-1, various patterns of integration and opposition become evident in human groupings.

For Blau, power differentials in groups create two contradictory forces: (1) strains toward integration and (2) strains toward opposition and conflict.

Strains toward integration Differences in power inevitably create the potential for conflict. However, such potential is frequently suspended by a series of forces promoting the conversion of power into authority, in which subordinates accept as legitimate the leaders' demands for compliance. Principles II and III in Table 16-1 denote two processes fostering such group integration: exchange relations always operate under the presumption of reciprocity and justice, forcing those deriving valued services to provide other rewards in payment. In providing these rewards, subordinates are guided by norms of fair exchange, in which the costs that they incur in offering compliance are to be proportional to the value of the services that they receive from leaders. Thus, to the extent that actors engage in exchanges with leaders and to the degree that the services provided by leaders are highly valued, subordination must be accepted as legitimate in accordance with the norms of reciprocity and fairness that emerge in all exchanges. Under these conditions, groups elaborate additional norms specifying just how exchanges with leaders are to be conducted in order to regularize the requirements for reciprocity and to maintain fair rates of exchange. Leaders who conform to these emergent norms can usually assure themselves that their leadership will be considered legitimate. In fact, Blau emphasizes that, if leaders abide by the norms regulating exchange of their services for compliance, norms carrying negative sanctions typically emerge among subordinates stressing the need for compliance to leaders' requests. Through this process, subordinates exercise considerable social control over one another's actions and thereby promote the integration of super- and subordinate segments of groupings.

Authority, therefore, "rests on the common norms in a collectivity of subordinates that constrain its individual members to conform to the orders of a superior."[19] In many patterns of social organization, these norms simply emerge out of the competitive exchanges among collective groups of actors. Frequently, however, in order for such "normative agreements" to be struck, participants in an exchange must be socialized into a common set of values that define not only what constitutes fair exchange in a given situation but also the way such exchange should be institutionalized into norms for both leaders and subordinates. Although it is quite possible for actors to arrive at normative consensus in the course of the exchange process itself, an initial set of common values facilitates the legitimation of power. Actors can now enter into exchanges with a common definition of the situation, which can provide a general framework for the normative regulation of emerging power differentials. Without common values, the competition for power is likely to be severe. In the absence of guidelines about reciprocity and fair exchange, considerable strain and tension will persist as definitions of these are worked out. For Blau, then, legitimation "entails not merely tolerant approval but active confirmation and promotion of social patterns by common values, either preexisting ones or those that emerge in a collectivity in the course of social interaction."[20]

With the legitimation of power through the normative regulation of interaction, as confirmed by common values, the structure of collective organization is altered. One of the most evident changes is the decline in interpersonal competition, for now actors' presentations of self shift from a concern of impressing others with their valuable qualities to an emphasis on confirming their status as loyal group members. Subordinates come to accept their status and manipulate their role behaviors to assure that they receive social approval from their peers as a reward for conformity to group norms. Leaders can typically assume a lower profile, since they must no longer demonstrate their superior qualities in each and every encounter with subordinates—especially since norms now define when and how they should extract conformity and esteem for providing their valued services. Thus, with the legitimation of power as authority, the interactive processes (involving the way group members define the situation and present themselves to others) undergo a dramatic change, reducing the degree of competition and thereby fostering group integration.

With these events the amount of direct interaction between leaders and subordinates usually declines, since power and ranking no longer must be constantly negotiated. This decline in direct interaction marks the formation of distinct subgroupings as members seek to interact with those of their own social rank, avoiding the costs of interacting with either their inferiors or their superiors.[21] In interacting primarily among themselves, subordinates avoid the high costs of interacting with leaders; and, although social approval from their peers is not a particularly valuable reward, it can be extracted with comparatively few costs—thus allowing for a sufficient profit. Conversely, leaders can

---

[19]Ibid., p. 208.

[20]Ibid., p. 221.

[21]As will be recalled from Chapter 15, these processes were insightfully described by George C. Homans in *The Human Group* (New York: Harcourt, Brace & World, 1950).

avoid the high costs (in terms of time and energy) of constantly competing and negotiating with inferiors regarding when and how compliance and esteem are to be bestowed upon them. Instead, by having relatively limited and well-defined contact with subordinates, they can derive the high rewards that come from compliance and esteem without incurring excessive costs in interacting with subordinates—thereby allowing for a profit.

Strains toward opposition   Thus far, I see Blau's exchange perspective as decidedly functional. Social exchange processes—attraction, competition, differentiation, and integration—are analyzed in terms of how they contribute to creating a legitimated set of normatively regulated relations. In a manner similar to Parsons' discussion of institutionalization, Blau emphasizes the importance of common values as a significant force in creating patterns of social organization. However, Blau is keenly aware that social organization is always rife with conflict and opposition, creating an inevitable dialectic between integration and opposition in social structures. Recognition of this fact leads Blau to assert:

> The functional approach reinforces the overemphasis on integrative social forces . . . whereas the dialectical perspective counteracts it by requiring explicit concern with disruptive tendencies in social structures. The pursuit of systematic analysis and the adoption of a dialectical perspective create a dilemma for the sociologist, who must rivet his attention on consistent social patterns for the sake of the former and on inconsistencies in accordance with the latter. This dilemma, like others, is likely to give rise to alternating developments, making him veer in one direction at one time and in the opposite at another.[22]

What I think is especially important about Blau's perspective is that, in adopting dialectical assumptions, he does not reject the useful tenets of functionalism. Blau recognizes that patterns of social organization are created and maintained *as well as* changed and broken down, leading him to seek the principles that can explain this spectrum of events. Thus, unlike Dahrendorf's conflict model, in which the organization of authority relations in imperatively coordinated associations (ICAs) and the opposition of quasi groups were merely taken as givens, Blau seeks to address the question of how, and through what processes, authority structures such as ICAs are created. In so doing, Blau is in a much better analytical position than Dahrendorf and other dialectical theorists to document how the creation of social structure can also set in motion forces for conflict and change. As I emphasized in my earlier discussion of conflict theory, to assert that conflict is endemic to authority relations in social structure and then to analyze how conflict changes structures is to define away the interesting theoretical question: under what conditions, in what types of structures, revealing what types of authority that have arisen through what

---

[22]Peter M. Blau, "Dialectical Sociology: Comments," *Sociological Inquiry* 42 (Spring 1972), p. 185. This article was written in reply to an attempt to document Blau's shift from a functional to a dialectical perspective; see Michael A. Weinstein and Deena Weinstein, "Blau's Dialectical Sociology," ibid., pp. 173–82.

processes, is what type of conflict likely to emerge? I feel that Blau's discussion of strains for integration represents an attempt to answer this question and provide a more balanced theoretical framework for discussing opposition and conflict in social systems.

Blau's exchange principles, which I summarized in Table 16–1, allow for the conceptualization of these strains for opposition and conflict. As Principle II–B on reciprocity documents, the failure to receive expected rewards in return for various activities leads actors to attempt to apply negative sanctions that, when ineffective, can drive people to violent retaliation against those who have denied them an expected reward. Such retaliation is intensified by the dynamics summarized in Principle III–B on justice and fair exchange, since when those in power violate such norms, they inflict excessive costs on subordinates, creating a situation that, at a minimum, leads to attempts to sanction negatively and, at most, to retaliation. Finally, Principle V on the inevitable imbalances emerging from multiple exchange relations emphasizes that to balance relations in one exchange context by meeting reciprocal obligations and conforming to norms of fairness is to put into imbalance other relations. Thus, the imbalances potentially encourage a cyclical process in which actors seek to balance previously unbalanced relations and thereby throw into imbalance currently balanced exchanges. In turn, exchange relations that are thrown into imbalance violate the norms of reciprocity and fair exchange, thus causing attempts at negative sanctioning and, under some conditions, retaliation. For Blau, then, built into all exchange relationships are sources of imbalance. When severely violating norms of reciprocity and fair exchange, these imbalances can lead to open conflict among individuals in group contexts.

I see these suggestive ideas as simply stating what can occur, without specifying the conditions under which the forces that they denote will actually be set into motion. Unfortunately, Blau provides few specific propositions that indicate when opposition will be activated. And when Blau does undertake a limited discussion of the conditions leading to increasingly intense forms of opposition, I see his propositions as resembling Dahrendorf's discussion of the technical, political, and social conditions of conflict-group organization.

In Table 16–3, I summarize Blau's ideas in propositional form.[23] To appreciate the degree to which these propositions resemble those in conflict theory, I suggest that Table 16–3 be compared with Tables 9–1 and 10–1. For, in the end, I believe that dialectical conflict theory and exchange theory are converging perspectives. In fact, I would go so far as to argue that conflict theory is a derivative of exchange theory, although I am sure that many would argue just the reverse. In either case, the perspectives converge.

From the discursive context in which the propositions in Table 16–3 are imbedded comes a conceptualization of opposition. Blau hypothesizes that, the more imbalanced exchange relations are experienced collectively, the greater is the sense of deprivation and the greater is the potential for opposition.

---

[23]Blau, *Exchange and Power*, pp. 224–52.

TABLE 16–3 Blau's Principles of Exchange Conflict

I. The more exchange relations between super- and subordinates become imbalanced, the greater is the probability of opposition to those with power.
   A. The more norms of reciprocity are violated by the superordinates, the greater is the imbalance.
   B. The more norms of fair exchange are violated by superordinates, the greater is the imbalance.
II. The more individuals experience collectively relations of imbalance with superordinates, the greater is their sense of deprivation and the greater is the probability of opposition to those with power.
   A. The less is the spatial dispersion of subordinates, the more likely are they to experience collectively relations of imbalance with superordinates.
   B. The more subordinates can communicate with one another, the more likely are they to experience collectively relations of imbalance with superordinates.
III. The more subordinates can experience collectively deprivations in exchange relations with superordinates, the more likely are they to codify ideologically their deprivations and the greater their opposition to those with power.
IV. The more deprivations of subordinates are ideologically codified, the greater is their sense of solidarity and the greater is the probability of opposition.
V. The greater is the sense of solidarity among subordinates, the more they can define their opposition as a noble and worthy cause and the greater is the probability of their opposition to those with power.
VI. The greater is the sense of ideological solidarity, the more likely are subordinates to view opposition as an end in itself and the greater is the probability of opposition to those with power.

Although he does not explicitly state the case, he appears to hold that increasing ideological codification of deprivations, the formation of group solidarity, and the emergence of conflict as a way of life will increase the intensity of the opposition—that is, members' emotional involvement in and commitment to opposition to those with power. I think that these propositions offer a suggestive lead for conceptualizing inherent processes of opposition in exchange relations. Moreover, unlike Dahrendorf's dialectical model, Blau's scheme presents theoretical insight into how the creation of relations of authority can also cause opposition. Indeed, I see Blau's conceptualization of the processes of institutionalization *and* conflict in terms of the same abstract exchange principles as representing a significant improvement over Dahrendorf's model, which failed to specify how latent conflicts first emerge in authority systems. I also believe that Blau's presentation represents an improvement over Homans' analysis of institutionalization and of the inherent conflict between the institutional and subinstitutional. In Blau's scheme there is a more adequate conceptualization of the process of institutionalization of those power relations from which opposition, innovation, and deviance ultimately spring.

Despite these strengths, however, I think that Blau's model can be improved by: (*a*) more precise formulation of the conditions under which exchange imbalances are likely for various types of social units; and then (*b*) specification of the conditions leading to various levels of intensity, violence, and duration in relations of opposition among various types of social units.

In looking back on Blau's discussion of microexchange processes, it is clear that he visualizes a series of basic exchange processes in human groupings: attraction, competition, differentiation, integration, and opposition. Of particular interest are the processes of differentiation in terms of power and how this pattern of differentiation creates strains for both integration and opposition—thus giving social reality a dialectical character.

Also noteworthy in the perspective is the attempt to utilize concepts developed in the analysis of elementary exchange processes in order to examine more complex exchange processes among the macrosocial units of social systems. Of great significance is Blau's recognition of the necessity for reformulating and supplementing elementary exchange concepts when analyzing more complex social processes. But, as I indicated earlier, Blau is now convinced that this strategy was unsuccessful. For he had hoped that exchange theory:

> could serve as a microsociological foundation for building a macrosociological theory of social structure. This is what I attempted . . . but I was more successful in the microsociological analysis of exchange processes than in employing the micro principles as the groundwork for building a vigorous macrostructural theory.[24]

And so, as a result, he shifted his theoretical approach to macrostructure, as I will discuss in Chapter 28. But I am not so convinced that Blau was correct in abandoning this exchange approach for macroprocesses; therefore I still think it useful to summarize the key elements of his macroexchange perspective.

Exchange Systems and Macrostructure

Although the general processes of attraction, competition, differentiation, integration, and opposition are evident in the exchanges among macrostructures, Blau sees several fundamental differences between these exchanges and those among microstructures.

1. In complex exchanges among macrostructures, the significance of "shared values" increases, for it is through such values that indirect exchanges among macrostructures are mediated.
2. Exchange networks among macrostructures are typically institutionalized. Although spontaneous exchange is a ubiquitous feature of social life, there are usually well-established historical arrangements that circumscribe the operation of the basic exchange processes of attraction, competition, differentiation, integration, and even opposition among collective units.
3. Since macrostructures are themselves the product of more elementary exchange processes, the analysis of macrostructures requires the analysis of more than one level of social organization.[25]

---

[24]Blau, "Contrasting Theoretical Perspectives."
[25]Ibid., pp. 253–311.

Mediating values   For Blau the "interpersonal attraction" of elementary exchange among individuals is replaced by shared values at the macro level. These values can be conceptualized as "media of social transactions" in that they provide a common set of standards for conducting the complex chains of indirect exchanges among social structures and their individual members. Such values are viewed by Blau as providing effective mediation of complex exchanges by virtue of the fact that the individual members of social structures have usually been socialized into a set of common values, leading them to accept them as appropriate. Furthermore, when coupled with codification into laws and enforcement procedures by those groups and organizations with power, shared values provide a means for mediating the complex and indirect exchanges among the macrostructures of large-scale systems. In mediating indirect exchanges among groups and organizations, shared values provide standards for the calculation of: (*a*) expected rewards, (*b*) reciprocity, and (*c*) fair exchange.

Thus, since individuals are not the units of complex exchanges, Blau emphasizes that, in order for complex patterns of social organization to emerge and persist, it is necessary for a "functional equivalent" of direct interpersonal attraction to exist. Values assume this function and assure that exchange can proceed in accordance with the principles presented in Table 16-1. And even when complex exchanges do involve people, their interactions are frequently so protracted and indirect that one individual's rewards are contingent on others who are far removed, requiring that common values guide and regulate the exchanges.

I think that there is considerable similarity between Blau's concern with "mediating values" and Parsons' analysis of "generalized media of exchange" (see Chapter 3).[26] Although the respective conceptualizations of the general classes and types of media differ, each is concerned with how social relationships utilize in varying contexts distinctive symbols, not only to establish the respective values of actions among exchange units but also to specify just how the exchange should be conducted. For, without shared values, exchange is tied to the direct interpersonal interactions of individuals. Since virtually all known social systems involve indirect exchange relations among various types of social units—from individuals and groups to organizations and communities—it is necessary to conceptualize just how this can occur. For Blau, mediating values are a critical condition for the development of complex exchange systems. Without them, social organization beyond face-to-face interaction would not be possible.

---

[26]See discussion in Chapter 3, as well as Talcott Parsons, "On the Concept of Political Power," *Proceedings of the American Philosophical Society* 107 (June 1963), pp. 232–62; Talcott Parsons, "On the Concept of Influence," *Public Opinion Quarterly* 7 (Spring 1963), pp. 37–67; and Talcott Parsons, "Some Problems of General Theory," in *Theoretical Sociology: Perspectives and Developments*, eds. J. C. McKinney and E. A. Tiryakian (New York: Appleton-Century-Crofts, 1970), pp. 28–68. See also T. S. Turner, "Parsons' Concept of Generalized Media of Social Interaction and Its Relevance for Social Anthropology," *Sociological Inquiry* 38 (Spring 1968), pp. 121–34.

Institutionalization Whereas values facilitate processes of indirect exchange among diverse types of social units, institutionalization denotes those processes that regularize and stabilize complex exchange processes.[27] As people and various forms of collective organization become dependent upon particular networks of indirect exchange for expected rewards, pressures for formalizing exchange networks through explicit norms increase. This formalization and regularization of complex exchange systems can be effective under three minimal conditions: (1) The formalized exchange networks must have profitable payoffs for most parties to the exchange. (2) Most individuals organized into collective units must have internalized through prior socialization the mediating values used to build exchange networks. And (3) those units with power in the exchange system must receive a level of rewards that moves them to seek actively the formalization of rules governing exchange relations.

Institutions are historical products whose norms and underlying mediating values are handed down from one generation to another, thereby limiting and circumscribing the kinds of indirect exchange networks that can emerge. Institutions exert a kind of external constraint on individuals and various types of collective units, bending exchange processes to fit their prescriptions and proscriptions. Institutions thus represent sets of relatively stable and general norms regularizing different patterns of indirect and complex exchange relations among diverse social units.

I see this conception of institutionalization as similar to the somewhat divergent formulations of both Parsons and Homans. Although institutions represent for both of them the regularization through norms of interaction patterns, Parsons visualizes institutions as normative structures, infused with values that allow for the patterning of interaction among diversely oriented and goal-seeking actors, whereas Homans considers institutions as the formalization through norms and generalized reinforcers of exchange relations that ultimately have payoffs for each individual involved. Despite their respective points of emphasis, however, both are concerned with the basic process through which norms emerge to facilitate the attainment of goals and rewards by social units. Both view the formalization of such institutional norms as allowing for expanded networks of interaction or exchange among various social units. Blau draws from both these perspectives by emphasizing, in a vein similar to Homans, that institutionalized patterns of interaction must have payoffs for the reward-seeking individuals involved and, in a way reminiscent of Parsons, that shared values must exist prior to effective institutionalization of indirect exchange relations. In this way, Blau apparently has sought to weld exchange-theoretical principles to the functionalist's concern with how values and norms account for the emergence and persistence of complex social systems.

In doing so, Blau, I sense, recognizes Homans' failure to develop concepts that describe the various types and classes of institutionalized exchange systems. In an effort to correct for this oversight, Blau develops a typology of

---

[27]Blau, *Exchange and Power*, pp. 273–80.

institutions embracing both the substance and style of the Parsonian formulation. Just as Parsons employed the pattern variables (see Chapter 3) to describe the values guiding institutionalized patterns, Blau attempts to classify institutions in terms of the values they appear to embody in their normative structure. He posits three generic types of institutions: (1) Integrative institutions "perpetuate particularistic values, maintain social solidarity, and preserve the distinctive character and identity of the social structure."[28] (2) Distributive institutions embody universalistic values and operate to "preserve the social arrangements that have been developed for the production and distribution of needed social facilities, contributions, and rewards of various kinds."[29] And (3) organizational institutions utilize values legitimating authority and serve "to perpetuate the authority and organization necessary to mobilize resources and coordinate collective effort in the pursuit of social objectives."[30]

However, Blau also recognizes that, in this form of typologizing, the potential is great for connoting an image of society as static and equilibrium maintaining. Thus, drawing from Homans' recognition that institutions are accepted only as long as they have payoffs for humans' primary needs and from Dahrendorf's concern with the inherent sources of conflict and change in all relations of authority, Blau stresses that all institutionalized exchange systems reveal a counterinstitutional component "consisting of those basic values and ideals that have not been realized and have not found expression in explicit institutional forms, and which are the ultimate source of social change."[31] To the extent that these values remain unrealized in institutionalized exchange relations, individuals who have internalized them will derive little payoff from existing institutional arrangements and will therefore feel deprived, seeking alternatives to dominant institutions. These unrealized values, even when codified into an opposition ideology advocating open revolution, usually contain at least some of the ideals and ultimate objectives legitimated by the prevailing culture. This indicates that institutional arrangements "contain the seeds of their potential destruction" by failing to meet all of the expectations of reward raised by institutionalized values.

Blau does not enumerate extensively the conditions for mobilization of individuals into conflict groups, but his scheme explicitly denotes the source of conflict and change: counterinstitutional values whose failure of realization by dominant institutional arrangements creates deprivations that, under unspecified conditions, can lead to conflict and change in social systems. In this way Blau attempts to avoid the predictable charges leveled against almost any form of functional analysis for failing to account for the sources of conflict,

---

[28]Ibid., p. 278.

[29]Ibid.

[30]Ibid., p. 279. It is of interest to note that Blau implicitly defines institutions in terms of their functions for the social whole. Although these functions are not made explicit, they are similar to Parsons' requisites. For example, integrative institutions appear to meet needs for latency; distributive institutions, for adaptation; and organizational institutions, for integration and goal attainment.

[31]Blau, *Exchange and Power*, p. 279.

deviance, and change in social systems. Unlike the Dahrendorf model of dialectical conflict, however, Blau's scheme does not just assert the pervasiveness of conflict and change in social systems but also attempts to document how opposition forces, culminating in conflict and change, are created by the very processes that lead to the institutionalization of power in complex exchange systems.

Such tendencies for complex exchange systems to generate opposition are explicable in terms of the basic principles of exchange. When certain mediating values are not institutionalized in a social system, exchange relations will not be viewed as reciprocated by those who have internalized these values. Thus, in accordance with Blau's principles on reciprocity (see Table 16-1), these segments of a collectivity are more likely to feel deprived and to seek ways of retaliating against the dominant institutional arrangements, which, from the perspective dictated by their values, have failed to reciprocate. For those who have internalized values that are not institutionalized, it is also likely that perceptions of fair exchange have been violated, leading them, in accordance with the principles of justice, to attempt to sanction negatively those arrangements that violate alternative norms of fair exchange. Finally, in institutionalized exchange networks, the balancing of exchange relations with some segments of a collectivity inevitably creates imbalances in relations with other segments (the imbalance principle in Table 16-1), thereby violating norms of reciprocity and fairness and setting into motion forces of opposition.

Unlike direct interpersonal exchanges, however, opposition in complex exchange systems is between large collective units of organization, which, in their internal dynamics, reveal their own propensities for integration and opposition. This fact requires that the analysis of integration and opposition in complex exchange networks be attuned to various levels of social organization. Such analysis needs to show, in particular, how exchange processes among macrostructures, whether for integration or for opposition, are partly influenced by the exchange processes occurring among their constituent substructures.

Levels of social organization   To a great extent, Blau believes, the "dynamics of macrostructures rests on the manifold interdependence between the social forces within and among their substructures."[32] The patterns of interdependence among the substructures of distinguishable macrostructures are various, including: (*a*) joint membership by some members of macrostructures in constituent substructures, (*b*) mobility of members among various substructures of macrostructures, and (*c*) direct exchange relations among the substructures of different macrostructure.

To discern these dynamics of substructures and how they influence exchanges among macrostructures, Blau first questions what generic types of substructures exist, resulting in the isolation of four classes of substructures:

---

[32]Ibid., p. 284.

"categories," "communities," "organized collectivities," and "social systems." Categories refer to an attribute, such as race, sex, and age, that "actually governs the relations among people and their orientations to each other."[33] Communities are "collectivities organized in given territories, which typically have their own government and geographical boundaries that preclude their being overlapping, though every community includes smaller and is part of larger territorial organizations."[34] Organized collectivities are "associations of people with a distinctive social organization, which may range from a small informal friendship clique to a large bureaucratized formal organization."[35] Social systems "consist not of the social relations in specific collectivities but of analytical principles or organization, such as the economy of a society or its political institutions."[36]

Values mediate the processes within these various types of substructures. However, discerning the complex relationships between the mediating values of substructures and those of macrostructures poses one of the most difficult problems of analysis. On the one hand, some values must cut across the substructures of a macrostructure if the latter is to remain minimally integrated; on the other hand, values of various substructures not only can segregate substructures from one another but can also generate conflict among them. Further, the relations among substructures involve the same basic exchange processes of attraction, competition, and differentiation in terms of the services they can provide for one another. It thus becomes evident that the analysis of exchange networks among macrostructures forces examination of the exchange processes of their substructures. Additionally, the relations among these substructures are complicated by the fact that they often have overlapping memberships or mobility of members between them, making the analysis of attraction, competition, differentiation, integration, and opposition increasingly difficult.

Blau simplifies the complex analytical task of examining the dynamics of substructures by positing that organized collectivities, especially formal organizations, are the most important substructures in the analysis of macrostructures. As explicit goal- (reward-) seeking structures that frequently cut across social categories and communities and that form the substructures of analytical social systems, collectivities are mainly responsible for the dynamics of macrostructures. Thus the theoretical analysis of complex exchange systems among macrostructures requires that primary attention be drawn to the relations of attraction, competition, differentiation, integration, and opposition among various types of complex organizations. In emphasizing the pivotal significance of complex organizations, Blau posits a particular image of society that should guide the ultimate construction of sociological theory.

---

[33]Ibid., p. 285.
[34]Ibid.
[35]Ibid.
[36]Ibid.

THE ORGANIZATIONAL BASIS OF SOCIETY

Organizations in a society must typically derive rewards from one another, thus creating a situation in which they are both attracted to, and in competition with, one another. Out of this competition, hierarchical differentiation between successful and less successful organizations operating in the same sphere emerges. Such differentiation usually creates strains toward specialization in different fields among less successful organizations as they seek to provide particular goods and services for dominant organizations and one another. If such differentiation and specialization among organizations are to provide effective means for integration, separate political organizations must also emerge to regulate their exchanges. Such political organizations possess power and are viewed as legitimate only as long as they are considered by individuals and organizations to follow the dictates of shared cultural values. Typically, political organizations are charged with several objectives: (1) regulating complex networks of indirect exchange by the enactment of laws; (2) controlling through law competition among dominant organizations, thereby assuring the latter of scarce resources; and (3) protecting existing exchange networks among organizations, especially those with power, from encroachment on these rewards by organizations opposing the current distribution of resources.

For Blau, then, it is out of the competition among organizations in a society that differentiation and specialization occur among macrostructures. Although mediating values allow differentiation and specialization among organizations to occur, it is also necessary for separate political organizations to exist and regularize, through laws and the use of force, existent patterns of exchange among other organizations. Such political organizations will be viewed as legitimate as long as they normatively regulate exchanges that reflect the tenets of mediating values and protect the payoffs for most organizations, especially the most powerful. However, the existence of political authority inevitably encourages opposition movements, for now opposition groups have a clear target—the political organizations—against which to address their grievances. As long as political authority remains diffuse, opposition organizations can only compete unsuccessfully against various dominant organizations. With the legitimation of clear-cut political organizations charged with preserving current patterns of organization, opposition movements can concentrate their energies against one organization, the political system.

In addition to providing deprived groups with a target for their aggressions, political organizations inevitably must aggravate the deprivations of various segments of a population, because political control involves exerting constraints and distributing resources unequally. Those segments of the population that must bear the brunt of such constraint and unequal distribution usually experience great deprivation in terms of the principles of reciprocity and fair exchange, which, under various conditions, creates a movement against the existing political authorities. To the extent that this organized opposition forces redistribution of rewards, other segments of the population are likely to feel constrained and deprived, leading them to organize into an opposition movement. These facts indicate that the organization of political authority assures that, in accordance with the principle of imbalance, attempts to balance one set of exchange relations among organizations throw into imbalance other exchange relations, causing the formation of opposition organizations. Thus, built into the structure of political authority in a society are inherent forces of opposition that give society a dialectical and dynamic character.

Echoing the assumptions of Dahrendorf and Coser, Blau conceptualizes opposition as representing "a regenerative force that interjects new vitality into a social structure and becomes the basis of social reorganization."[37] However, the extent to which opposition can result in dramatic social change is limited by several counterforces:[38] (1) The interdependence of the majority of organizations upon one another for rewards gives each a vested interest in the status quo, thus providing strong resistance to opposition organizations. (2) Dominant organizations that have considerable power to bestow rewards on other organizations independently of political organizations have a particularly strong vested interest in existing arrangements, thereby assuring their resistance to change-oriented organizations. (3) By virtue of controlling the distribution of valued resources, both dominant and political organizations are in a strategic position to make necessary concessions to opposition groups, thereby diffusing their effective organization. (4) Opposition movements must overcome the internalization of values by the majority; without control of the means of socialization, their ideological call for organization is likely to fall on unsympathetic ears. (5) Societies composed of exchange networks among complex organizations typically reveal high levels of social mobility up the organizational hierarchy, thus increasing the difficulties involved in organizing a stable constituency.[39]

In reviewing Blau's analysis of exchanges among organizations in a society, it is evident that he has attempted to cast many of the assumptions and propositions of Parsons, Dahrendorf, and Coser into an exchange perspective that extends Homans' insights beyond the analysis of individuals. In discussing the developments of exchange systems among organizations and of political authority, Blau focuses on the institutionalization of relations among what Parsons termed *social systems*. The fact that such institutionalization rests upon the internalization of shared values and that institutional patterns can be typologized in terms of the dominance of various clusters of values further underscores Blau's analytical debt to Parsons. In contrast with Parsons' less explicit analysis, Blau's discussion is concerned with mechanisms of social change. Embracing Marx's, Dahrendorf's, and Simmel's assumptions of the dialectical forces of opposition inherent in all micro and macro relations of

---

[37]Ibid., p. 301. Such a position is inevitable in light of Blau's explicitly stated belief that "our society is in need of fundamental reforms" (Blau, "Dialectical Sociology," p. 184).

[38]Blau, "Dialectical Sociology," p. 187; Blau, *Exchange and Power*, pp. 301–9.

[39]However, Blau recognizes that high rates of mobility in a society can also increase the sense of relative deprivation of those who are denied opportunities for advancement, thereby making them likely constituents of an opposition organization.

FIGURE 16–1   Blau's Image of Social Organization

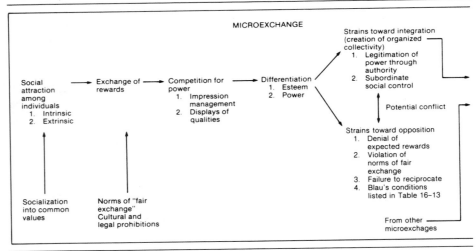

power and authority, Blau visualizes the source of conflict in the unbalanced exchange relations, violating norms of reciprocity and fairness, which are inevitable concomitants of some organizations having a disproportionate hold upon valued resources.

Blau does not enumerate the conditions leading to the organization of opposition as explicitly as did either Marx or Dahrendorf ("conflict groups" for Dahrendorf and "class" for Marx). But his debt to Marx's work (see Chapter 9) is evident in his analysis on how levels of deprivation are influenced by (1) the degree of ecological concentration and the capacity to communicate among the deprived, (2) the capacity to codify an opposition ideology, (3) the degree of social solidarity among the deprived, and (4) the degree to which opposition organization is politicized and directed against the political organizations (see Table 16-3). Furthermore, Blau's incorporation of Dahrendorf's key propositions is shown in his recognition that the capacity of the population's deprived segments to organize opposition is affected by such variables as the rate of social mobility, the capacity of dominant groups to make strategic concessions, and the number of cross-cutting conflicts resulting from multigroup affiliations. Finally, although Blau does not develop his argument extensively, he clearly has followed Simmel's and Coser's lead in emphasizing that conflict and opposition are regenerative forces in societies that "constitute countervailing forces against . . . institutional rigidities, rooted in vested powers as well as traditional values, and . . . [that] are essential for speeding social change."[40]

In sum, then, Blau has offered a most varied image of society. By incorporating—albeit in an unsystematic manner—the fruitful leads of sociology's

---

[40]Blau, *Exchange and Power*, p. 302.

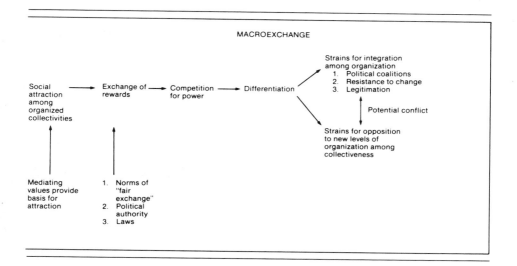

other dominant conceptual perspectives, he has indeed offered a suggestive theoretical prolegomenon that can serve as a guide to more explicit theoretical formulations. I have summarized both the micro- and macroprocesses of this prolegomenon in Figure 16-1. As is evident, Blau has synthesized several theoretical traditions, and in so doing he offers something for everyone. For the functionalist, Blau offers the concept of mediating values, types of institutions, and the counterpart of mechanisms of socialization and control that operate to maintain macrosocial wholes. For the conflict theorist, Blau presents a dialectical-conflict perspective emphasizing the inevitable forces of opposition in relations of power and authority. For those concerned with interactions among individuals, Blau's analysis of elementary exchange processes places considerable emphasis on the actions of people in interaction. And for the critic of Homans' reductionism, Blau provides an insightful portrayal of exchanges among emergent social structures, which leaves the integrity of sociological theorizing intact.

In offering a theoretical resting place for the major perspectives in sociology, Blau provides a useful example of how sociological theorizing should proceed: rather than becoming bogged down in controversies and debates among proponents of various theoretical positions, it is much wiser to borrow and integrate the useful concepts and assumptions of diverse perspectives. In offering this alternative to continued debate, Blau left a number of theoretical issues unresolved, but I think that the most important one is the question of how micro- and macroprocesses can be incorporated into one theoretical approach.[41]

---

[41]For examples of criticisms on Blau's work, see M. J. Mulkay, "A Conceptual Elaboration of Exchange Theory: Blau," in his *Functionalism, Exchange, Theoretical Strategy* (New York: Schocken

This issue is obscured by the insightfulness of his synthesis, and yet its resolution constitutes a critical next step in his theoretical strategy. And I suspect
that the failure to resolve it is what caused Blau to abandon this exchange
approach.

THE UNRESOLVED ISSUE: MICRO- VERSUS MACROANALYSIS

As I emphasized in Chapter 1, one of the issues that haunts sociological theory
is: To what extent are structures and processes at micro and macro levels of
social organization subject to analysis by the same concepts and laws? At what
levels of sociological organization do emergent properties require use of additional concepts and description in terms of their unique modes of analysis? In
what ways are groups, organizations, communities, or social systems similar
and different? These questions are extremely troublesome for sociological theorizing and constitute one of its most enduring problems. In his early exchange
approach, Blau tried to resolve this problem in several ways: (1) by assuming
that the basic exchange processes of attraction, competition, differentiation,
integration, and opposition occur at all levels of social organization; (2) by
explicating general exchange principles and incorporating abstract exchange
concepts that can account for the unfolding of these processes at all levels of
organization; (3) by enumerating additional concepts, such as mediating values
and institutionalization, to account for emergent phenomena at increasingly
macro levels of social organization; and (4) by classifying the generic types of
organization—categories, communities, organized collectivities, and social systems—that denote different levels of organization, requiring somewhat different
concepts for explication of their operation.

I see this as an interesting effort to bridge the micro/macro analytical gap
in sociological theorizing. A number of problems remain, however. First, Blau
defines organized collectivities so broadly that they include phenomena ranging
from small groups to complex organizations. It is likely that the concepts and
theoretical generalizations appropriate to the small primary group, the secondary group, a crowd, a social movement, a small organization, and a large
corporate bureaucracy will be somewhat different. Surely there are emergent
properties of social organization in a spectrum ranging from a small group to
a complex organization. In fact, aside from the study of community, most
subfields in sociology fall within Blau's category of organized collectivity. Thus
Blau has not resolved the problem of emergent properties; rather, I think that
it has been defined away with an excessively broad category that subsumes
most of the emergent properties of interest to sociologists.

Second, the addition of concepts to account for differences in levels of
organization only highlights the micro/macro gap without providing a sense

Books, 1971), especially pp. 206–12; Percy S. Cohen, *Modern Sociological Theory* (New York: Basic
Books, 1968), pp. 123–27; and Anthony Heath, "Economic Theory and Sociology: A Critique of
P. M. Blau's 'Exchange and Power in Social Life,'" *Sociology* 2 (September 1968), pp. 273–92.

of what concepts are needed to understand increasingly macro phenomena. What Blau does is to assert that there are elementary exchange processes that occur at macro levels of organization and that require the addition of the concept of mediating values if these emergent levels of organization are to be understood. But such an analysis begs the key question: when and at what levels of organization do such concepts become critical? Among dyads? Triads? Primary groups? Secondary groups? Small organizations? Large organizations? To phrase the issue differently, at what point do what kinds of values, operating in accordance with what laws, become analytically significant? Blau simply says that at some point mediating values become critical, and he thereby avoids answering the theoretically interesting question.

Third, Blau's presentation of exchange concepts and their incorporation into exchange principles is vague. My analysis and reformulation of Blau's ideas into tables and figures represent an attempt to make more explicit the implicit exchange principles employed in his analysis. Without explicit statements of the exchange laws that cut across levels of organization, Blau fails to address an issue that he claims to be of great significance in the opening pages of his major work: "The problem is to derive the social processes that govern the complex structures of communities and societies from the simpler processes that pervade the daily intercourse among individuals and their interpersonal relations."[42]

To make such derivations, it is necessary to formulate explicitly the laws from which derivations from simpler to more complex structures are to be made. Too often Blau hides behind the fact that, to use his words, his "intent is not to present a systematic theory of social structure; it is more modest than that."

Yet, even in his modesty, I believe that Blau synthesized diverse theoretical traditions into a suggestive exchange perspective. But he abandoned it, nonetheless. He simply stopped trying to view macroprocesses as fundamentally connected to microexchanges. I think that this shift in strategy was mistaken, but it does signal a dissatisfaction with his own exchange solution to the micro/macro issue.[43] My sense is that exchange theory is one of the few approaches that can bridge this gap with common principles. And thus my belief is that it should be utilized, not abandoned, in an effort to integrate interpersonal and structural theories.

---

[42]Blau, *Exchange and Power*, p. 2.

[43]Patrick Spread, "Blau's Exchange Theory, Support, and the Macro Structure," *The British Journal of Sociology* 35 (June 1984), pp. 157–73.

CHAPTER 17

♦

Rational Choice Theory

Michael Hechter

THE SOCIOLOGICAL REVIVAL OF UTILITARIANISM

As I indicated in Chapter 14, the roots of exchange theory reside not only in behaviorism but also in utilitarianism. Yet, for many years, explicitly utilitarian theories were recessive—indeed, almost "underground." The reason for this situation, I suspect, is the long tradition in sociology of debunking theories in which actors are viewed as "rational" and as seeking to "maximize their utilities." Émile Durkheim's unfair attacks on Spencer,[1] Karl Marx's effort to reform Adam Smith and change capitalist society,[2] Vilfredo Pareto's disillusionment with economics,[3] and Talcott Parsons' early critique of utilitarianism[4] all created a bias against thinking in the terms of classical economics. Hence the basic ideas of utilitarianism tended to come into sociological theory via behaviorism, as we have seen for George Homans'[5] work and as we shall see for Richard Emerson's/Karen Cook's[6] exchange network approach. Of course, utilitarian thinking has not been boycotted, as evidenced by the fact that Blau's[7] theory contains elements of classical economics. Moreover, well-known scholars like James Coleman[8] and Gary Becker[9] have "profitably" (no pun intended)

---

[1]Émile Durkheim, *The Division of Labor in Society* (New York: Macmillan, 1933; originally published in 1893).

[2]Karl Marx, *Grundrisse* (New York: Vintage Books, 1973; originally written in the 1850s).

[3]Vilfredo Pareto, *The Mind and Society* (New York: Harcourt Brace, 1935; originally published in 1916 under the title "Treatise in General Sociology").

[4]Talcott Parsons, *The Structure of Social Action* (New York: McGraw-Hill, 1937).

[5]See Chapter 15.

[6]See Chapter 27.

[7]See Chapter 16.

[8]For example, see: James S. Coleman, *Individual Interests and Collective Action: Studies of Rationality and Social Change* (New York: Cambridge University Press, 1986); "Social Theory,

argued for a theory of rational actors. Yet there has been a sustained antagonism against theories that seem too utilitarian, and so it is interesting to note that these kinds of theories have made a dramatic comeback in the last two decades.[10]

What do these theories assume, and why has sociology been antagonistic toward them? Utilitarian theories assume that human action is purposive and intentional, that it is guided by a well-ordered hierarchy of preferences, and that it is "rational" in several senses: (1) actors make calculations of utilities or preferences in selecting a course of action; (2) actors also calculate the costs (foregoing utility that can be gained from engaging in alternative courses of action) for each line of conduct; and (3) actors attempt to maximize their utilities in pursuing a particular option.

It is this maximization portion of utilitarianism that is perhaps the most controversial and has been at the core of most critiques of this approach. But this assumption alone cannot explain the hostility. What really upsets many sociologists is the methodological sidekick of much utilitarianism: individualism. Here the assumption is made that social outcomes and social structures are to be understood in terms of a theory about individual action. This is the same point of emphasis that aroused hostility toward Homans' behavioristic approach because it seems "reductionistic."[11] That is, it appears to relegate macro sociology to the theoretical dust pile, and, on the surface, it seems to underestimate the importance of emergent properties.

There is no doubt that rational choice theories start with principles about individuals' efforts to maximize utilities and are, to this extent, individualistic and reductionist. But their goal is to explain emergent properties, particularly social structures, collective behavior and movements among masses of individuals, and collective decisions or emergent social outcomes that result from individuals each making a particular kind of rational choice.[12] Moreover, current rational choice theories recognize that calculations of rational actors occur within an existing social structure that influences (1) the distribution of resources available to actors, (2) the distribution of opportunities to pursue various lines of conduct, and (3) the "rules of the game," or norms and sanctions of social structures.

Social Research, and a Theory of Action," *American Journal of Sociology* 91 (6), pp. 984–1014. See also *Foundations of Social Theory* (Cambridge, MA: Belknap Press, 1990).

[9]Gary S. Becker, *The Economic Approach to Human Behavior* (Chicago: University of Chicago Press, 1976) and *A Treatise on the Family* (Cambridge, MA: Harvard University Press, 1981).

[10]For examples, see: Jon Elster, ed., *Rational Choice* (New York: New York University Press, 1986); Anthony Heath, *Rational Choice and Social Exchange* (Cambridge, England: Cambridge University Press, 1976); Brian Barry and Russell Hardin, eds., *Rational Man and Irrational Society* (Beverly Hills, CA: Sage, 1982); Mancur Olson, *The Logic of Collective Action* (Cambridge, MA: Harvard University Press, 1965); Andrew Schotter, *The Economic Theory of Social Institutions* (Cambridge, England: Cambridge University Press, 1981); Herbert Simon, *Reason in Human Affairs* (Stanford, CA: Stanford University Press, 1983).

[11]See pages 319–320 in Chapter 15.

[12]Debra Friedman and Michael Hechter, "The Contribution of Rational Choice Theory to Macrosociological Research," *Sociological Theory* 6 (Fall 1988), pp. 201–18.

TABLE 17-1   Assumptions of Rational Choice Theory

I. Humans are purposive and goal oriented.

II. Humans have sets of hierarchically ordered preferences, or utilities.

III. In choosing lines of behavior, humans make rational calculations with respect to:
(a) the utility of alternative lines of conduct with reference to the preference hierarchy;
(b) the costs of each alternative in terms of utilities foregone; (c) the best way to maximize utility.

IV. Emergent social phenomena—social structures, collective decisions, and collective behavior—are ultimately the result of rational choices made by utility-maximizing individuals.

V. Emergent social phenomena that arise from rational choices constitute a set of parameters for subsequent rational choices of individuals in the sense that they determine: (a) the distribution of resources among individuals; (b) the distribution of opportunities for various lines of behavior; (c) the distribution and nature of norms and obligations in a situation.

Thus the rational choices of individuals are constrained by emergent social forces. The goal of rational choice theories is to explain these structures and other collective outcomes in terms of principles about individual action. The very constraints on individual actions are, themselves, the product of previous rational choices; hence they must be explained in terms of individual preferences, calculations, and maximization of utility. I doubt if this point of emphasis will silence the critics, but it is a conceptual tack that is worth pursuing.

The basic assumptions of rational choice theory are outlined in Table 17-1 and typify most sociological theories working in this utilitarian tradition. In this chapter I will outline one particularly interesting version of rational choice theory—Michael Hechter's theory of group solidarity[13]—that falls within the perspective enumerated in Table 17-1.

HECHTER'S RATIONAL CHOICE THEORY OF GROUP SOLIDARITY

Hechter's approach is intriguing because it addresses a topic of central concern to sociological theory: solidarity.[14] How do individuals create social solidarity and, in the process, resolve the Hobbesian problem of order? Group solidarity is seen by Hechter as the critical element of social order, and so, if he can explain solidarity in terms of rational choice principles, then he will have gone

---

[13]Michael Hechter, *Principles of Group Solidarity* (Berkeley: University of California Press, 1987); Michael Hechter, ed., *The Microfoundations of Macrosociology* (Philadelphia: Temple University Press, 1983); "Rational Choice Foundations of Social Order," in J. H. Turner, ed., *Theory Building in Sociology* (Newbury Park, CA: Sage, 1988).

[14]Solidarity is defined in rational choice terms as: "the greater the average proportion of each member's private resources contributed to collective ends, the greater the solidarity of the group" (Hechter, *Principles of Group Solidarity*, p. 18). For the present, however, I prefer to use the concept in a more connotative way since the above definition is unclear without an introduction to crucial concepts.

a long way toward demonstrating the viability of the utilitarian approach and silencing the critics (the former is, I think, easier than the latter).

The Critique of "Emergent Property" Theories

Hechter begins by criticizing normative, functional, and structural theories,[15] which are all presumed to be superior to utilitarian theories because they take account of emergent cultural and structural properties. But in Hechter's view they do no such thing. Normative theories rely too heavily on the internalization of norms, cultural prohibitions, and value commitments to explain solidarity; as a result, they cannot explain why actors would produce norms and values in the first place, why they would pay attention to them, and why they would internalize them and develop commitments. Functional theories depend upon a vague selection mechanism in which group survival or integration creates "selection pressures" for group solidarity; as a consequence, these theories cannot specify how and through what specific processes "survival needs" produce and reproduce "solidarity." Structural theories argue that those who share "common interests" as a result of ecological or structural parameters (like geographical location, organizational position, or social class) will develop solidarity, but these theories do not specify the mechanisms by which common interests translate into solidarity.

In Hechter's view, then, the traditional approaches do not explain how group obligations evolve and why group members honor them. The reason for this failing is that these "emergent property" theories do not adequately take account of individuals. What these theories fail to understand is a simple truth: normative obligations develop in groups because, under certain conditions, it is rational for individuals to construct them; and people conform to these obligations because, under certain conditions, it is the most rational thing to do. Of course, stated in this way, the very conditions making it rational for actors to construct obligations and then conform to them are not specified, making Adam Smith's "invisible hand of order" as vague as it was in 1789.[16] Thus the key to Hechter's rational choice approach is specifying these conditions.

Basic Concepts in Rational Choice Theories

In rational choice theories, individuals are seen as bearers of preferences, ordered hierarchically. The world is conceptualized as revealing scarcity; hence, people must make choices about which of their preferences they can fulfill. In making such choices, individuals are conceived to maximize their utilities in the sense of engaging in behaviors that are the most likely or probable to meet

---

[15]Ibid., pp. 1–39.

[16]Adam Smith, *The Wealth of Nations*, 2 vols. (London: Davis, 1805; originally published in 1789).

one's highest preferences.[17] In many contexts, especially those of interest to sociologists, individuals are dependent upon others in a group context for those resources, or "goods," that will maximize utilities. That is, they cannot produce the good for themselves and, hence, must rely on others to produce it for them or join others in its production. For example, if companionship and affection are high preferences, this "good" can be attained only in interaction with others, usually in groups; if money is a preference, then this good can usually be attained in modern settings through work in an organizational context. Thus the "goods" that meet individual preferences can often be secured only in a group; in fact, groups are conceptualized in rational choice theory as existing to provide or produce "goods" for their members.

Those goods that are produced by the activities of group members can be viewed as *joint goods*, because they are produced jointly in the coordinated activities of group members. Such joint goods vary along a critical dimension: their degree of "publicness." A highly public good is available not only to the members of the group but to others outside the group as well. Furthermore, once produced, its use by one person does not diminish its supply for another. For instance, radio waves, navigational aids, and roads are public goods because they can be used by those who did not produce them and because their use by one person does not (at least up to a point) preclude use by another. In contrast to public goods are *private goods*, which are produced for consumption by their producers. Moreover, consumption by one person decreases the capacity of others to consume the good. Private goods are thus kept out of the reach of others in order to assure that only a person, or persons, can consume them.

It is around the question of public goods that the basic problem of order for rational choice theorists revolves. This problem is phrased in terms of the *free-rider* dilemma. People are supposed to produce jointly public goods. It is rational to consume public goods without thought of contributing to their production. To avoid the costs of contributing to production is free-riding. If everybody free-rides, then the joint good will never be produced. How, then, is this dilemma avoided?

An answer to this question has been controversial in the larger literature in economics,[18] but the basic thrust of the argument is that, if a good is highly public, people can be coerced (through taxes, for example) to contribute to its production (say, national defense) or they can be induced to contribute by being rewarded (salaries, praise) for their contribution. Another way to prevent free-riding is to exclude those who do not contribute to production from consumption, thereby decreasing the degree of "publicness" of the good. This exclusion can result in a group that "throws out" noncontributing members or does not allow them to join in the first place. A final way to control free-riders is to impose user fees or prices for goods that are consumed.

Thus, for rational choice theory, the basis of social order revolves around creating group structures to produce goods that are consumed in ways that

---

[17]Recall Homans' "rationality principle" on page 313 in Chapter 15.

[18]See Mancur Olson, *The Logic of Collective Action*.

limit free-riding—that is, consumption without contributing in some way, directly or indirectly, to production. The sociologically central problem of social solidarity thus becomes one of understanding how rational egoists go about establishing groups that create normative obligations on their members to contribute and, then, enforcing their conformity—thereby diminishing the problem of free-riding. Solidarity is thus seen as a problem of social control.

The Basis of Social Control:
Dependence, Monitoring, and Sanctioning

In rational choice theory, groups exist to provide joint goods. The more dependent is an individual on a group for resources or goods that rank high in his or her preference hierarchy, the greater the potential power of the group over that individual. When people are dependent on a group for a valued good, it is rational for them to create rules and obligations that will assure access to this joint good. Such is particularly likely to be the case when the valued joint good is not readily available elsewhere, when individuals lack information about alternatives, when the costs of exiting the group are high, when moving or transfer costs are high, and when personal ties, as unredeemable sunk investments, are strong.

This basic idea is, of course, similar to Blau's proposition on the differentiation of power[19] and, as we will see in Chapter 27, Richard Emerson's notion of power-dependence relations.[20] But there is, I think, an important twist in Hechter's conceptualization: dependence is the incentive behind efforts to create normative obligations in order to assure that actors will get a joint good. The "power" in Hechter's conceptualization is that of "the group," although specific individuals are obviously involved. Moreover, Hechter adds an additional proposition: *the extensiveness of normative obligations in a group is related to the degree of dependence.* Thus dependence creates incentives not just for norms but for extensive norms that guide and regulate to a high degree.

Yet Hechter is quick to emphasize that "the extensiveness of a group alone . . . has no necessary implications for group solidarity."[21] What is crucial is that group members will comply with these norms. Compliance is related to a group's *control capacity*, which in turn is a function of (1) *monitoring* and (2) *sanctioning*. Monitoring is the process of detecting nonconformity to group norms and obligations, whereas sanctioning is the use of rewards and punishments to induce conformity. When the monitoring capacity of a group is low, then it becomes difficult to assure compliance to norms because conformity is a cost that rational individuals will seek to avoid, if they can. And, without monitoring, sanctioning cannot effectively serve as an inducement to conformity.

---

[19]See Table 16–2 on page 334.
[20]See pages 560–563.
[21]Hechter, *Principles of Group Solidarity*, p. 49.

For Hechter, then, solidarity is the product of dependence, monitoring, and sanctioning. But it is also related to the nature of the group, leading Hechter to distinguish types of groups in terms of the nature of joint goods produced.

Types of Groups

Hechter views control capacity—that is, monitoring and sanctioning—as operating differently in two basic types of groups. If a grouping produces a joint good for a market and does not itself consume the good, then control capacity can be potentially reduced, because the profits from the sale of the good can be used to "buy" conformity. Conformity can be bought, for example, because members are compensated for their labor; if they are highly dependent on a group for this compensation, then it is rational to conform to norms. But, since the same compensation can be achieved in other groups, it is less likely that dependence on the group is high, which thereby reduces the extensiveness of norms. The result is that monitoring and sanctioning must be high in such groups, for it would be rational for the individual to free-ride and take compensation without a corresponding effort to produce the marketable good. Yet, if monitoring and sanctioning are too intrusive and impose costs on individuals, then it is rational to leave the group and seek compensation elsewhere. Moreover, as we will see, extensive monitoring and sanctioning are costly and cut into profits—hence limiting the social control capacity of the group. The control capacity of these *compensatory groups*, as Hechter calls them, is thus problematic and assures that solidarity will be considerably lower than in *obligatory groups*.

Obligatory groups are those that produce a joint good for their members' own consumption. Under these conditions it is rational to create obligations for contributions from members; if dependence on the joint good is high, then there is considerable incentive for conformity because there is no easy alternative to the joint good (unlike the case in groups in which a generalized medium like money is employed as compensation). Moreover, monitoring and sanctioning can usually be more efficient since monitoring typically occurs as a byproduct of joint production of a good that the members consume and since the ultimate sanction—expulsion from the group—is very costly to members who value this good. In fact, as Hechter notes:[22]

> ... due to greater dependence, obligatory groups have lower sanctioning and monitoring costs. Since every group has one relatively costless sanction at its disposal—the threat of expulsion—then the greater the dependence of group members, the more weight this sanctioning causes. [Moreover,] ... monitoring and sanctioning are to some extent substitutable. If the value of the joint good is relatively large, the threat of expulsion can partly compensate for inadequate monitoring. The more one has to lose by noncompliance, the less likely one is to risk it. ...

---

[22]Ibid., p. 126.

There is an implicit variable in Hechter's analysis: group size. In general terms, compensatory groups organize larger numbers of individuals to produce marketable goods, whereas obligatory groups are smaller and provide goods for their members that cannot be obtained (or, only at great cost) in a market. Thus, not only will dependence be higher in obligatory groups, but monitoring and sanctioning will be considerably easier, thereby increasing solidarity, which Hechter defines as the extent to which members' private resources are contributed to a collective end. High contributions of private resources can occur only with extensive norms and high conformity—two conditions unlikely to prevail in compensatory groups. In Hechter's terms, then, high solidarity can be achieved only in obligatory groups in which dependence on a jointly produced and consumed good is high, in which monitoring is comparatively easy due to small size and to the fact that members can observe one another's production and consumption of the good, and in which sanctioning is built into the very nature of the good (that is, receiving a good is a positive sanction, whereas expulsion or not receiving the good is a very costly negative sanction). Under these conditions, people will commit their private resources—time, energy, and self—to the production of the joint good and, in the process, promote high solidarity.

For Hechter, then, high degrees of solidarity are possible only in obligatory groups, in which dependence, monitoring, and sanctioning are high. Figure 17–1 represents his argument, as I understand it. I have, however, modified it in several respects. First, I have added group size as a crucial variable, for, as obligatory groups get large, their monitoring and sanctioning capacity decreases. Second, I have added on the far left a variable that is perhaps more typical in many human groupings: the ratio of consumption to compensation. That is, many groupings involve a mixture of extrinsic compensation for the production of goods consumed by others *and* goods consumed by members. For example, groups composed of members working for a salary in an organization often develop solidarity because they also produce joint goods—friendship, approval, assistance, and the like. Indeed, at times solidarity develops around obligations that run counter to the official work norms of the organization. We need, I think, to conceptualize groups not so much as two polar types but as mixtures of external compensation for goods jointly produced but consumed by others and internal consumption of goods jointly produced and consumed by members. The greater the proportion of compensation to internal consumption, the less likely are the processes depicted in Figure 17–1 to operate; conversely, the less the ratio of compensation to internal consumption, the more likely are these processes to be activated in ways that produce solidarity. I do not think that this violates the intent of Hechter's typology—which I see as the polar extremes of these processes—because in fact he argues that control capacity of compensatory groups is increased if they also produce a joint good for their own consumption.

A larger-scale social system, such as a community or society, is a configuration of obligatory and compensatory groups. Solidarity will be confined mostly to obligatory groups, whereas the problems of free-riding will be most

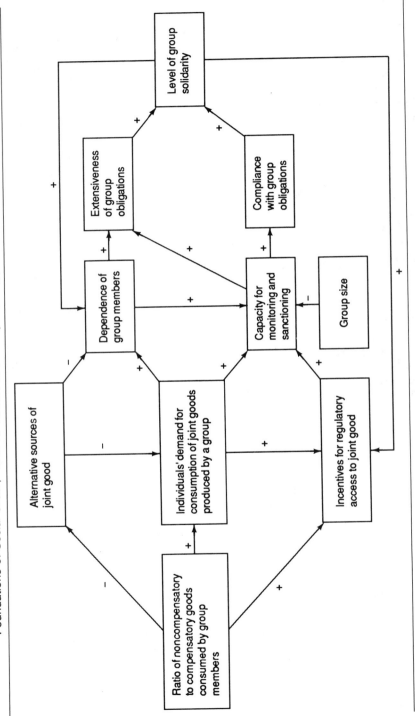

FIGURE 17–1 The Determinants of Group Solidarity. (Adapted, but extensively revised, from Michael Hechter, "Rational Choice Foundations of Social Order," in J. H. Turner, ed., *Theory Building*.)

evident in compensatory groups, unless they also develop joint goods that are highly valued and consumed by group members. Hechter has thus turned back to the basic distinctions that dominated early sociological theory—*gemeinschaft* vs. *geselleschaft*, primary vs. secondary groups, mechanical vs. organic solidarity, traditional vs. rational authority, folk vs. urban—and has sought to explain these distinctions in terms of the production of joint goods and the nature of the control process that stems from whether a joint good is consumed by members or produced for a market in exchange for extrinsic compensation. For Hechter, the nature of the joint good determines the level of dependence of individuals on the group and the control capacity of the group. High dependence and control are most likely when joint goods are consumed, and hence solidarity is high under these conditions. A society with only compensatory groups will, therefore, reveal low solidarity—as all of the early theorists recognized and lamented (except, perhaps, Georg Simmel). What is distinctive about Hechter's conceptualization is that it is tied to a utilitarian theory in which both high and low levels of solidarity follow from rational choices of individuals.

Patterns of Control in Compensatory and Obligatory Groups

In a vein similar to classical sociological theory, Hechter examines the process of "formalization"—if I can put a word in his mouth.[23] As groups get larger, informal controls become inadequate, even in obligatory groups. Of course, if compensatory groups proliferate and/or grow in size, the process of creating formal controls escalates to an even greater degree. There is, however, a basic dilemma in this process: formal monitoring and sanctioning are costly because they involve creating special agents, offices, procedures, and roles, thereby cutting into the production of goods. Obligatory groups can put off the process of formal controls because of high dependence and control that comes from joint consumption of a good that is valued; but when obligatory groups get too large, more formal controls become necessary. Compensatory groups can try to keep formal controls to a minimum, especially if they can create consumption of joint goods that reinforce the production norms for those goods that will be externally consumed; however, if they get too large, then they must also increase formal monitoring and sanctioning. In all these cases, it is rational for actors in groups to resist imposing formal controls because they are costly and cut into profits or joint consumption. But if free-riding becomes too widespread and cuts into production, then it is rational to begin to impose them.

To some extent the implementation of formal controls can be delayed, or mitigated, by several forces. One is common socialization, and groups often seek members who share similar outlooks and commitments in order to reduce the risks of free-riding and cut the costs of monitoring. Another is selection for altruism, especially in obligatory groups, in which unselfish members are

---

[23]Ibid., pp. 59–77, 104–24.

recruited. Yet there clearly are limits on how effective these forces can be in maintaining social control, especially in compensatory groups but also in obligatory groups as they get larger.

The result of this basic dilemma between the costs of free-riding, on the one side, and formalization of control, on the other, is for groups to seek "economizing" measures for monitoring and sanctioning.[24] These are particularly visible in compensatory groups in which formal social control is more essential to production, but elements of these "economizing tactics" can be seen in obligatory groups as they get larger or as they produce a joint good that creates problems of free-riding.

Hechter lists a number of monitoring and sanctioning economies. Let me first examine monitoring economies and then turn to sanctioning economies. One way to decrease monitoring costs is to increase the visibility of individuals in the group through a variety of techniques: designing the architecture of the group so that people are physically visible to one another, requiring members to engage in public rituals to reaffirm their commitments to the group, encouraging group decision making so that individuals' preferences are exposed to others, and administering public sanctions for behavior that exemplifies group norms. Another set of techniques for reducing monitoring costs includes those that have members share the monitoring burden, as is the case when rewards are given to groups rather than individuals (under the assumption that, if your rewards depend upon others, you will monitor their activity), when privacy is limited, when informants are rewarded, and when gossip is encouraged. A final economizing technique, which follows from socialization processes, is to minimize errors of interpretation of behavior through recruitment and training of members into a homogeneous culture.

Turning to economizing techniques for sanctioning, one is symbolic sanctioning through the creation of a prestige hierarchy and the differential rewarding of prestige to group members who personify group norms. Another technique is public sanctioning of deviance from group norms. A final sanctioning technique is to increase the exit costs of group members through geographical isolation from other groups, imposition of nonrefundable investments on entry to the group, and limitation of extragroup affiliations.

Yet there are limits to these monitoring and sanctioning economies, especially in compensatory groups. The result is that, at some point, a group must create formal agents and offices to monitor and control. Thus a formal organization always reveals not only some of the economizing processes listed above but also agents charged with monitoring and control—for example, foremen, comptrollers, supervisors, personnel offices, quality-control agents, and the like. Such monitoring and sanctioning are extremely costly, and so it is not surprising that organizations seek to economize here also.

In addition to the more general techniques listed above, a variety of mechanisms for reducing agency costs, or increasing productive efficiency and hence

---

[24]Ibid., pp. 126–46.

profitability, can be employed. For example, inside and outside contractors are often used to perform work (and to incur their own monitoring and sanctioning costs) at a set price for an organization; standardization of tools, work flow, and other features of work is another way to reduce the need for monitoring; assessment of only outputs (and ignoring how these are generated) is yet another technique for economizing on at least some phases of monitoring; setting production goals for each stage in production is still another technique; and, perhaps most effective, the creation of an obligatory group within the larger compensatory organization is the most powerful economizing technique, as long as the norms of the obligatory group correspond to those of the more inclusive compensatory organization (sometimes, however, just the opposite is the case, which thereby increases monitoring costs to even higher levels).

The Theory Summarized

Such are the key elements in Hechter's theory. In Table 17–2, I have tried to summarize the theory in somewhat more formal and abstract terms than in Hechter's work. When stated in this way, the theory has applicability to a wide range of empirical processes—social class formation, ethnic solidarity, complex organization, communities, and other social units. The propositions in Table 17–2 must be seen as building upon the assumptions delineated in Table 17–1; when this is done, it is clear that Hechter has tried to explain, as he phrases the matter, "the micro foundations of the macro social order."

Macrostructural Implications

Hechter sees the basic ideas of rational choice theory as useful in understanding more macrostructural processes among large populations of individuals.[25] For example, the basic theory as outlined in Figure 17–1 and Tables 17–1 and 17–2 can be used to explain processes within nation/states. A state or government imposes relatively extensive obligations on its citizens—pay your taxes, be loyal, be willing to die in war, and so on. The reason for this capacity is that citizens are often highly dependent on the state for public goods and cannot leave the society easily (because they like where they live, cannot incur the exit and transfer costs, enjoy many benefits from the joint goods produced by the citizenry, etc.). Compliance with the demands of the state involves more than dependence and extensive obligations; compliance also hinges on the state's control capacity. But how is it possible for the state to monitor and sanction all of its citizens who are organized into diverse configurations of obligatory and compensatory groups?

The answer to this question, Hechter argues, lies in economies of control. Such economies can be generated *within* and *between* groups. The key process is to get the citizens themselves monitoring and sanctioning one another within

---

[25]Ibid., pp. 168–86; Hechter, "Rational Choice Foundations of Social Order"; "Introduction" in *The Microfoundations of Macrosociology*.

TABLE 17–2 Hechter's Implicit Principles of Social Structure

I. The more members of a group jointly produce goods for consumption outside the group, the more their productive efforts will depend upon increases in the ratio of extrinsic compensation to intrinsic compensation.

II. The more members of a group jointly produce goods for their own consumption, the more their efforts will depend upon the development of normative obligations.

III. The more members of a group are dependent upon the group for a good, or for compensation, the greater is the power of the group to determine the decisions and behaviors of members.

   A. The less available are alternative sources of a good, or compensation, the greater is the dependence of members.

   B. The less available is information about alternative sources of a good, or compensation, the greater is the dependence of members.

   C. The greater the costs of exiting the group, the greater the dependence of members.

   D. The greater the moving or transfer costs from one group to another, the greater the dependence of members.

   E. The greater the intensity of personal ties among members of a group, the greater the dependence of members.

IV. The more a group produces a joint good for its own consumption and develops normative obligations to regulate productive activity, the more likely are conditions III-A, III-B, III-C, III-D, and III-E to be met; conversely, the more a group produces a joint good for the consumption of others outside the group, the less likely are these conditions to be met.

V. The more a group produces a joint good for its own consumption and develops extensive normative obligations to regulate productive activity, the more likely is social control through monitoring and sanctioning to be informal and implicit and, hence, less costly.

VI. The more a group produces a good for external consumption and must rely upon a high ratio of extrinsic over intrinsic compensation for members, the more likely is social control through monitoring and sanctioning to be formal and explicit and, hence, more costly.

VII. The larger the size of a group, the more likely is social control to be formal and explicit and the greater will be its cost.

VIII. The greater the cost of social control through monitoring and sanctioning, the more likely are economizing procedures to be employed in a group.

---

groups and, then, to link these groups together in ways that maximize dependence and control capacity. For example, Hechter sees Émile Durkheim's vision of an integrated society of cohesive "occupational groups," linked together through democratic political institutions and solidarity-promoting school systems, as essentially a proposal that follows from the very utilitarian principles that Durkheim rejected.[26]

Hechter further illustrates the potential usefulness of the rational choice approach by asking: why is Japan the most integrated of the industrial powers? The answer is dependence and control capacity. Japanese citizens are among the most highly dependent in the world, with their high standard of living,

---

[26]See, for example, Émile Durkheim, "Preface," *The Division of Labor in Society*, 2nd ed. (New York: Free Press, 1947; preface originally published in 1902).

unique language, lifetime employment, and strong social ties. Control is possible because individuals are attached to highly cohesive groups that are arrayed in patterns of hierarchical relations to more inclusive groups, which in turn are connected hierarchically to other groupings, and so on. The result is that the state monitors and sanctions a small number of megacorporations that, successively, monitor and sanction the dependent corporations and groups to which employees belong.

Thus we might want to add a macrostructural principle to the bottom of Table 17-2. It might go something like this: the greater are the size and number of groupings to be integrated in a population, the more economizing procedures will revolve around (1) dependence of the population and its constituent grouping on a central authority that supplies valued joint goods, (2) capacity of a central authority to monitor selectively key groupings to which individuals belong, (3) patterns of dependence and control within these key groupings, and (4) patterns of dependence and control between key groupings and even larger numbers of subgroupings.

CONCLUSION

What Hechter's theory demonstrates is that assumptions and principles about human behavior (see Tables 17-1 and 17-2) can be used to provide a provocative interpretation of emergent social processes. Moreover, this interpretation is in the mainstream of the sociological tradition with its focus on solidarity in groups and social control processes. Thus it may have been premature to write off the utilitarian theory—as I once did.[27]

Yet I should not leave Hechter's theory—and, by implication, other rational choice theories—without a few words of criticism. First, let us ask: what if we simply left off the notions of rational choice and utility maximization from Hechter's descriptions? Would we lose any insight? In other words, Hechter's argument can be made without reference to the rationality assumptions in Table 17-1. Indeed, they could be seen as superfluous to the emergent processes that he describes. Is this a fair criticism? It is probably too extreme, but it should not be ignored. Second, much of the discussion of rational choices is post hoc. If process *x* occurs, then it was the result of a rational decision; but if its opposite should occur, then this would be seen as rational too. What is being added here to our understanding? Is the after-the-fact insertion of caveats about rational actors adding to our understanding? Again, I am overstating the case, but it is worth considering. Third, much of Hechter's discussion is subject to a criticism similar to one that he made for functional and structural theories. Does Hechter give us a sense for the processes or mechanisms of structural events that is any less vague than "selection processes" or "interests"? Is not "rational choices maximizing utilities" just another gloss over the actual process among actors in groups and organizations? Perhaps I am

---

[27]Jonathan H. Turner, "Social Exchange Theory: Future Directions," in K. S. Cook, ed., *Social Exchange Theory* (Newbury Park, CA: Sage, 1987), pp. 223–38.

being too harsh, because all micro-to-macro explanations will suffer on this issue, but we should consider whether or not Hechter has provided us with a superior sense of the processes and mechanisms by which emergent structures operate.

Yet, having raised these questions, I think that Hechter and others have gone a long way in demonstrating that early sociologists—from Durkheim and Pareto to the young Talcott Parsons—were too harsh in their assessment of utilitarian theory in economics. There is something there that we should reconsider and incorporate into our theories.

PART 4

Interactionist Theorizing

CHAPTER 18

Early Interactionism
and Phenomenology

I think that some of the most intriguing questions in social theory concern the relationships between society and the individual. In what ways does one mirror the other? How does society shape individuals, and how do individuals create and maintain and change society? How are society and the personality of individuals interrelated and yet separate emergent phenomena? Indeed, as I have already emphasized for virtually all theorists in the previous chapters, the issue of micro versus macro sociological analysis boils down to a question of the relationship between the properties of individuals and interaction, on the one side, and those of social structure, on the other.

It was near the close of the 19th century that the grand analytical schemes of Marx, Durkheim, Spencer, and other Europeans were supplemented by a concern for the specific processes that link individuals to one another and to society. Instead of focusing on macrostructures and processes, such as evolution, class conflict, and the nature of the body social, attention shifted to the study of processes of social interaction and their consequences for the individual and the society. And out of this shift in concern came a variety of interactionist theories. For the most part, such theory is American, but we will need to be aware of important European contributions to the emergence of interactionism.

EARLY AMERICAN INSIGHTS

Modern interactionism draws its inspiration from a number of prominent thinkers in America and Europe, all of whom wrote between 1880 and 1935. Yet I believe that interactionism is indebted to the genius of one in particular, George Herbert Mead. Much like Darwin's great synthesis of the theory of evolution from his own studies and the speculations of others, so interactionist ideas were codified by George Herbert Mead. Mead borrowed ideas from others and combined them with his own insights to produce a synthesis that still stands as the conceptual core of modern interactionism. And so I will begin with Mead's synthesis, and then later I will turn to European thought and indicate some of the ways that it supplemented Mead's great work.

To appreciate Mead's feat, I think it best to review the contributions of three thinkers who most influenced him: William James, John Dewey, and Charles Horton Cooley. Each of these scholars provided a key concept that was to form the basis of Mead's synthesis.

William James and the Concept of "Self"

The Harvard psychologist William James (1842–1910) was perhaps the first social scientist to develop a clear concept of self. James recognized that humans have the capacity to view themselves as objects and to develop self-feelings and attitudes toward themselves. Just as humans can (a) denote symbolically other people and aspects of the world around them, (b) develop attitudes and feelings toward these objects, and (c) construct typical responses toward objects, so they can denote themselves, develop self-feelings and attitudes, and construct responses toward themselves. James called these capacities *self* and recognized their importance in shaping the way people respond in the world.

James developed a typology of selves: the "material self," which includes those physical objects that humans view as part of their being and as crucial to their identity; the "social self," which involves the self-feelings that individuals derive from associations with other people; and the "spiritual self," which embraces the general cognitive style and capacities typifying an individual.[1] This typology was never adopted by subsequent interactionists, but James's notion of the social self was to become a part of all interactionists' formulations.

James's concept of the social self recognizes that people's feelings about themselves arise out of interaction with others. As he noted, "a man has as many social selves as there are individuals who recognize him."[2] Yet James did not carry this initial insight very far. He was, after all, a psychologist who was more concerned with internal psychological functioning of individuals than with the social processes out of which the capacities of individuals arise.

Self and Social Process: Charles Horton Cooley

I believe that Charles Horton Cooley made two significant breakthroughs in the study of self.[3] First, he refined the concept of self, viewing it as the process in which individuals see themselves as objects, along with other objects, in their social environment. Second, he recognized that self emerges out of communication with others. As individuals interact with each other, they interpret

---

[1]William James, *The Principles of Psychology* (New York: Henry Holt, 1890), pp. 292–99 of vol. 1.

[2]Ibid., p. 294.

[3]Charles Horton Cooley, *Human Nature and the Social Order* (New York: Scribner's, 1902) and *Social Organization: A Study of the Larger Mind* (New York: Scribner's, 1916).

each other's gestures and thereby see themselves from the viewpoint of others. They imagine how others evaluate them, and they derive images of themselves or self-feelings and attitudes. Cooley termed this process *the looking glass self*: the gestures of others serve as mirrors in which people see and evaluate themselves, just as they see and evaluate other objects in their social environment.

Cooley also recognized that self arises out of interaction in group contexts. He developed the concept of "primary group" to emphasize that participation in front of the looking glass in some groups is more important in the genesis and maintenance of self than participation in other groups. Those small groups in which personal and intimate ties exist are the most important in shaping people's self-feelings and attitudes.

Cooley thus refined and narrowed James's notion of self and forced the recognition that it arises out of symbolic communication with others in group contexts. These insights were to influence profoundly the thought of George Herbert Mead.

Pragmatism and Thinking: The Contribution of John Dewey

John Dewey (1859–1952) was, for a brief period, a colleague of Cooley's at the University of Michigan. But more important was Dewey's enduring association with George Herbert Mead, whom he brought to the University of Chicago. As the chief exponent of a school of thought known as *pragmatism*, Dewey stressed the process of human adjustment to the world, in which humans constantly seek to master the conditions of their environment. And thus the unique characteristics of humans arise out of the *process* of adjusting to their life conditions.

What is unique to humans, Dewey argued, is their capacity for thinking. Mind is not a structure but a process that emerges out of humans' efforts to adjust to their environment. Moreover, mind is the unique capacity that allows humans to deal with conditions around them.

Dewey thus devoted considerable effort to understanding human consciousness. His basic questions were: How does mind work? And how does it facilitate adaptation to the environment? Mind for Dewey is the process of denoting objects in the environment, ascertaining potential lines of conduct, imagining the consequences of pursuing each line, inhibiting inappropriate responses, and, then, selecting a line of conduct that will facilitate adjustment. Mind is thus the process of thinking, which involves deliberation:

> Deliberation is a dramatic rehearsal (in imagination) of various competing possible lines of action. . . . Deliberation is an experiment in finding out what the various lines of possible action are really like. It is an experiment in making various combinations of selected elements . . . to see what the resultant action would be like if it were entered upon.[4]

---

[4]John Dewey, *Human Nature and Human Conduct* (New York: Henry Holt, 1922), p. 190. For an earlier statement of these ideas, see John Dewey, *Psychology* (New York: Harper & Row, 1886).

Dewey's conception of mind as a process of adjustment, rather than as a thing or entity, was to be critical in shaping Mead's thought. For, much as Cooley had done for the concept of self, Dewey had demonstrated that mind emerges and is sustained through interactions in the social world. This line of thought from both Cooley and Dewey was to prove decisive as Mead's great synthesis unfolded.

Main Currents of Thought in America: Pragmatism, Darwinism, and Behaviorism

At the time that Mead began to formulate his synthesis, I think the convergence of several intellectual traditions was crucial since it appears to have influenced the direction of his thought. Mead considered himself a behaviorist, but not of the mechanical stimulus/response type. In fact, many of his ideas were intended as a refutation of such prominent behaviorists as John B. Watson. Mead accepted the basic premise of behaviorism—that is, the view that reinforcement guides and directs action. He was, however, to use this principle in a novel way. Moreover, he rejected as untenable the methodological presumption of early behaviorism that it was inappropriate to study the internal dynamics of the human mind. James's, Cooley's, and Dewey's influence assured that Mead would rework the principle of reinforcement in ways that allowed for the consideration of mind and self.

Another strain of thought that shaped Mead's synthesis is pragmatism, as it was acquired through exposure with Dewey. As I noted, pragmatism sees organisms as practical creatures that come to terms with the actual conditions of the world. Coupled with behaviorism, pragmatism offered a new way of viewing human life: human beings seek to cope with their actual conditions, and they learn those behavioral patterns that provide gratification. The most important type of gratification is adjustment to social contexts.

This line of argument was buttressed in Mead's synthesis by yet another related intellectual tradition, Darwinism. Mead recognized that humans are organisms seeking to find a niche in which they can adapt. Historically this was true of humans as an evolving species; more important, it is true of humans as they discover a niche in the social world. Mead's commitment to behaviorism and pragmatism thus allowed him to apply the basic principle of Darwinian theory to each human: that which facilitates survival or adaptation of the organism will be retained.

In this way, behaviorist, pragmatist, and Darwinian principles blended into an image of humans as attempting to adjust to the world around them and as retaining those characteristics—particularly mind and self—that enable them to adapt to their surroundings. Mind, self, and other unique features of humans evolve out of efforts to survive in the social environment. They are thus capacities that arise from the processes of coping, adjusting, adapting, and achieving the ultimate gratification or reinforcement: survival. For this reason, Mead's analysis emphasizes the processes by which the infant organism acquires mind and self as an adaptation to society. But Mead did much more; he showed how society could survive only from the capacities for mind and self among individuals. From Mead's perspective, then, the capacities for mind, self, and society are intimately connected.

George Herbert Mead's Synthesis

The names of William James, Charles Horton Cooley, and John Dewey figure prominently in the development of interactionism, but George Herbert Mead brought their related concepts together into a coherent theoretical perspective that linked the emergence of the human mind, the social self, and the structure of society to the process of social interaction.[5] As I have emphasized, Mead appears to have begun his synthesis with two basic assumptions: (1) the biological frailty of human organisms forces their cooperation with one another in group contexts in order to survive; and (2) those actions within and among human organisms that facilitate their cooperation, and hence their survival or adjustment, will be retained. Starting from these assumptions, Mead was able to reorganize the concepts of others so that they denoted how mind, the social self, and society arise and are sustained through interaction.

Mind   Following Dewey's lead, Mead recognized that the unique feature of the human mind is its capacity to (1) use symbols to designate objects in the environment, (2) rehearse covertly alternative lines of action toward these objects, and (3) inhibit inappropriate lines of action and select a proper course of overt action. Mead termed this process of using symbols or language covertly *imaginative rehearsal,* revealing his conception of mind as a *process* rather than a structure. Further, as I will develop more fully later, the existence and persistence of society, or cooperation in organized groups, are viewed by Mead as dependent upon this capacity of humans to imaginatively rehearse lines of action toward one another and thereby select those behaviors that facilitate cooperation.

Much of Mead's analysis focuses not so much on the mind of mature organisms as on how this capacity first develops in individuals. Unless mind emerges in infants, neither society nor self can exist. In accordance with principles of behaviorism, Darwinism, and pragmatism, Mead stressed that mind arises out of a selective process in which an infant's initially wide repertoire of random gestures are narrowed as some gestures bring favorable reactions

---

[5]Mead's most important sociological ideas can be found in the published lecture notes of his students. His most important exposition of interactionism is found in his *Mind, Self, and Society,* ed. C. W. Morris (Chicago: University of Chicago Press, 1934). Other useful sources include George Herbert Mead, *Selected Writings* (Indianapolis: Bobbs-Merrill, 1964); and Anselm Strauss, ed., *George Herbert Mead on Social Psychology* (Chicago: University of Chicago Press, 1964). For excellent secondary sources on the thought of Mead, see Tamotsu Shibutani, *Society and Personality: An Interactionist Approach* (Englewood Cliffs, NJ: Prentice-Hall, 1962); Anselm Strauss, *Mirrors and Masks: The Search for Identity* (Glencoe, IL: Free Press, 1959); Bernard N. Meltzer, "Mead's Social Psychology," in *The Social Psychology of George Herbert Mead* (Ann Arbor, MI: Center for Sociological Research, 1964), pp. 10–31; Jonathan H. Turner, "Returning to Social Physics: Illustrations from George Herbert Mead," *Perspectives in Social Theory* 2 (1981) and *A Theory of Social Interaction* (Stanford, CA: Stanford University Press, 1988); also see Jonathan H. Turner and Leonard Beeghley, *The Emergence of Sociological Theory* (Belmont, CA: Wadsworth, 1981). For a more global overview of Mead's ideas, see John D. Baldwin, *George Herbert Mead: A Unifying Theory for Sociology* (Newbury Park, CA: Sage, 1986).

from those upon whom the infant is dependent for survival. Such selection of gestures facilitating adjustment can occur either through trial and error or through conscious coaching by those with whom the infant must cooperate. Eventually, through either of these processes, gestures come to have common meanings for both the infant and those in its environment. With this development, gestures now denote the same objects and carry similar dispositions for all the parties to an interaction. Gestures that have such common meanings are termed by Mead *conventional gestures.* These conventional gestures have increased efficiency for interaction among individuals because they allow for more precise communication of desires and wants as well as intended courses of action—thereby increasing the capacity of organisms to adjust to one another.

The ability to use and to interpret conventional gestures with common meanings represents a significant step in the development of mind, self, and society. By perceiving and interpreting gestures, humans can now assume the perspective (dispositions, needs, wants, and propensities to act) of those with whom they must cooperate for survival. By reading and then interpreting covertly conventional gestures, individuals are able to imaginatively rehearse alternative lines of action that will facilitate adjustment to others. Thus, by being able to put oneself in another's place, or to "take the role of the other," to use Mead's concept, the covert rehearsal of action can take on a new level of efficiency, since actors can better gauge the consequences of their actions for others and thereby increase the probability of cooperative interaction.

Thus, when an organism develops the capacity (1) to understand conventional gestures, (2) to employ these gestures to take the role of others, and (3) to imaginatively rehearse alternative lines of action, Mead believed that such an organism possesses "mind."

Self   Drawing from James and Cooley, Mead stressed that, just as humans can designate symbolically other actors in the environment, so they can symbolically represent themselves as an object. The interpretation of gestures, then, can not only facilitate human cooperation but also serve as the basis for self-assessment and evaluation. This capacity to derive images of oneself as an object of evaluation in interaction is dependent upon the processes of mind. What Mead saw as significant about this process is that, as organisms mature, the transitory "self-images" derived from specific others in each interactive situation eventually become crystallized into a more or less stabilized "self-conception" of oneself as a certain type of object. With these self-conceptions, individuals' actions take on consistency, since they are now mediated through a coherent and stable set of attitudes, dispositions, or meanings about oneself as a certain type of person.

Mead chose to highlight three stages in the development of self, each stage marking not only a change in the kinds of transitory self-images an individual can derive from role taking but also an increasing crystallization of a more stabilized self-conception. The initial stage of role taking in which self-images can be derived is termed *play.* In play, infant organisms are capable of assuming the perspective of only a limited number of others, at first only one or two.

Later, by virtue of biological maturation and practice at role taking, the maturing organism becomes capable of taking the role of several others engaged in organized activity. Mead termed this stage the *game* because it designates the capacity to derive multiple self-images from, and to cooperate with, a group of individuals engaged in some coordinated activity. (Mead typically illustrated this stage by giving the example of a baseball game in which all individuals must symbolically assume the role of all others on the team in order to participate effectively.) The final stage in the development of self occurs when an individual can take the role of the "generalized other" or "community of attitudes" evident in a society. At this stage individuals are seen as capable of assuming the overall perspective of a community, or general beliefs, values, and norms. This means that humans can both (1) increase the appropriateness of their responses to others with whom they must interact and (2) expand their evaluative self-images from the expectations of specific others to the standards and perspective of the broader community. Thus it is this ever-increasing capacity to take roles with an ever-expanding body of others that marks the stages in the development of self.

Society For Mead, society or institutions represent the organized and patterned interactions among diverse individuals.[6] Such organization of interactions is dependent upon mind. Without the capacities of mind to take roles and imaginatively rehearse alternative lines of activity, individuals could not coordinate their activities. Mead emphasized:

> The immediate effect of such role-taking lies in the control which the individual is able to exercise over his own response. The control of the action of the individual in a co-operative process can take place in the conduct of the individual himself if he can take the role of the other. It is this control of the response of the individual himself through taking the role of the other that leads to the value of this type of communication from the point of view of the organization of the conduct in the group.[7]

Society is also dependent upon the capacities of self, especially the process of evaluating oneself from the perspective of the generalized other. Without the ability to see and evaluate oneself as an object from this community of attitudes, social control would rest solely on self-evaluations derived from role taking with specific and immediately present others—thus making coordination of diverse activities among larger groups extremely difficult.[8]

Although Mead was vitally concerned with how society and its institutions are maintained and perpetuated by the capacities of mind and self, these concepts also allowed him to view society as constantly in flux and rife with potential change. The fact that role taking and imaginative rehearsal are on-

---

[6]For a more detailed analysis, see Jonathan H. Turner, "A Note on G. H. Mead's Behavioristic Theory of Social Structure," *Journal for the Theory of Social Behavior* 12 (July 1982), pp. 213–22.

[7]Mead, *Mind, Self, and Society*, p. 254.

[8]Ibid., pp. 256–57.

going processes among the participants in any interaction situation reveals the potential these processes give individuals for adjusting and readjusting their responses. Furthermore, the insertion of self as an object into the interactive process underscores the fact that the outcome of interaction will be affected by the ways in which self-conceptions alter the initial reading of gestures and the subsequent rehearsal of alternative lines of behavior. Such a perspective thus emphasizes that social organization is both perpetuated and altered through the adjustive capacities of mind and the mediating impact of self:

> Thus the institutions of society are organized forms of group or social activity—forms so organized that the individual members of society can act adequately and socially by taking the attitudes of others toward these activities. . . . [But] there is no necessary or inevitable reason why social institutions should be oppressive or rigidly conservative, or why they should not rather be, as many are, flexible and progressive, fostering individuality rather than discouraging it.[9]

In this passage, I think, is a clue to Mead's abiding distaste for rigid and oppressive patterns of social organization. He viewed society as a *constructed* phenomenon that arises out of the adjustive interactions among individuals. As such, society can be altered or reconstructed through the processes denoted by the concepts of mind and self. However, I believe that Mead went one step further and stressed that change is frequently unpredictable, even by those emitting the change-inducing behavior. To account for this indeterminacy of action, Mead used two concepts first developed by William James, the "I" and the "me."[10] For Mead the "I" points to the impulsive tendencies of individuals, and the "me" represents the self-image of behavior after it has been emitted. With these concepts Mead emphasized that the "I," or impulsive behavior, cannot be predicted because the individual can only "know in experience" (the "me") what has actually transpired and what the consequences of the "I" have been.

In sum, then, society for Mead represents those constructed patterns of coordinated activity that are maintained by, and changed through, symbolic interaction among and within actors. Both the maintenance and the change of society, therefore, occur through the processes of mind and self. Although many of the interactions causing both stability and change in groups are viewed by Mead as predictable, the possibility for spontaneous and unpredictable actions that alter existing patterns of interaction is also likely.

This conceptual legacy had a profound impact on a generation of American sociologists prior to the posthumous publication of Mead's lectures in 1934. Yet, despite the suggestiveness of Mead's concepts, I think that they fail to address some important theoretical issues.

The most important of these issues concerns the vagueness of his concepts in denoting the nature of social organization or society and the precise points

---

[9]Ibid., pp. 261–62.

[10]See James, *Principles of Psychology*, pp. 135–76.

of articulation between society and the individual. Mead viewed society as organized activity, regulated by the generalized other, in which individuals make adjustments and cooperate with one another. Such adjustments and the cooperation are seen as possible by virtue of the capacities of mind and self. Whereas mind and self emerged out of existent patterns of social organization, the maintenance or change of such organization is viewed by Mead as a reflection of the processes of mind and self. Although these and related concepts of the Meadian scheme point to the mutual interaction of society and the individual and although the concepts of mind and self denote crucial processes through which this dependency is maintained, they do not allow for the analysis of variations in patterns of social organization and in the various ways individuals are implicated in these patterns. To note that society is coordinated activity and is maintained or changed through the role-taking and self-assessment processes of individuals offers only a broad conceptual portrait of the linkages between the individual and society. Indeed, Mead's picture of society does not indicate how variable types of social organization reciprocally interact with variable properties of self and mind. Thus, in the end, Mead's concepts appeared to emphasize that society shapes mind and self, whereas mind and self affect society—a simple but profound observation for the times but one that needed supplementation.

The difficult task of filling in the details of this broad portrait began only five decades ago, as researchers and theorists began to encounter the vague and circular nature of Mead's conceptual perspective. The initial efforts at documenting more precisely the points of articulation between society and the individual led to attempts at formulating a series of concepts that could expose the basic units from which society is constructed. In this way the linkages between society and the individual could be more adequately conceptualized.

Conceptualizing Structure and Role

And so, although Mead's synthesis provided the initial conceptual breakthrough, it did not satisfactorily resolve the problem of how participation in the *structure* of society shaped individual conduct, and vice versa. In an effort to resolve this vagueness, sociological inquiry began to focus on the concept of role. Individuals were seen as playing roles associated with positions in larger networks of positions. With this vision, efforts to understand more about social structures and how individuals are implicated in them intensified during the 1920s and 1930s. This line of inquiry was to become known as *role theory*.

Robert Park and Role Theory

Robert Park, who came to the University of Chicago near the end of Mead's career, was one of the first to extend Mead's ideas through an emphasis on roles. As he observed, "everybody is always and everywhere, more or less consciously, playing a role."[11] But Park stressed that roles are linked to structural positions in society and that self is intimately linked to playing roles within the confines of the positions of social structure:

> The conceptions which men form of themselves seem to depend upon their vocations, and in general upon the role they seek to play in communities and social groups in which they live, as well as upon the recognition and status which society accords them in these roles. It is status, i.e., recognition by the community, that confers upon the individual the character of a person, since a person is an individual who has status, not necessarily legal, but social.[12]

Park's analysis stresses the fact that self emerges from the multiple roles that people play.[13] In turn, roles are connected to positions in social structures. I see this kind of analysis as shifting attention to the nature of society and how its structure influences the processes outlined in Mead's synthesis.

Jacob Moreno and Role Theory

Inspired in part by Mead's concept of role taking and by his own earlier studies in Europe, Jacob Moreno was one of the first to develop the concept of role playing. In *Who Shall Survive* and in many publications in the journals that he founded in America, Moreno began to view social organization as a network of roles that constrained and channeled behavior.[14] In his early works, Moreno distinguished different types of roles: (*a*) "psychosomatic roles," in which behavior is related to basic biological needs, as conditioned by culture, and in which role enactment is typically unconscious; (*b*) "psychodramatic roles," in which individuals behave in accordance with the specific expectations of a particular social context; and (*c*) "social roles," in which individuals conform to the more general expectations of various conventional social categories (for example, worker, Christian, mother, and father).

Despite the suggestiveness of these distinctions, I see their importance as coming not so much from their substantive content as from their intent: to conceptualize social structures as organized networks of expectations that require varying types of role enactments by individuals. In this way, analysis can move beyond the vague Meadian conceptualization of society as coordinated activity regulated by the generalized other to a conceptualization of social

---

[11]Robert E. Park, "Behind Our Masks," *Survey Graphic* 56 (May 1926), p. 135. For a convenient summary of the thrust of early research efforts in role theory, see Ralph H. Turner, "Social Roles: Sociological Aspects," *International Encyclopedia of the Social Sciences* (New York: Macmillan, 1968).

[12]Robert E. Park, *Society* (New York: Free Press, 1955), pp. 285–86.

[13]Indeed, Park studied briefly with Simmel in Berlin and apparently acquired insight into Simmel's study of the individual and the web of group affiliations (see later discussion). Coupled with his exposure to William James at Harvard, who also stressed the multiple sources of self, it is clear that Mead's legacy was supplemented by Simmel and James through the work of Robert Park.

[14]Jacob Moreno, *Who Shall Survive* (Washington, DC, 1934); rev. ed. (New York: Beacon House, 1953).

organization as various *types* of interrelated role enactments regulated by varying *types* of expectations.

Ralph Linton and Role Theory

Shortly after Moreno's publication of *Who Shall Survive*, the anthropologist Ralph Linton further conceptualized the nature of social organization, and the individual's embeddedness in it, by distinguishing among the concepts of role, status, and individuals:

> A status, as distinct from the individual who may occupy it, is simply a collection of rights and duties. . . . A *role* represents the dynamic aspect of status. The individual is socially assigned to a status and occupies it with relation to other statuses. When he puts the rights and duties which constitute the status into effect, he is performing a role.[15]

In this passage are a number of important conceptual distinctions. Social structure reveals several distinct elements: (*a*) a network of positions, (*b*) a corresponding system of expectations, and (*c*) patterns of behavior that are enacted with regard to the expectations of particular networks of interrelated positions. In retrospect, these distinctions may appear self-evident and trivial, but I feel that they made possible the subsequent elaboration of many interactionist concepts:

1. Linton's distinctions allow for the conceptualization of society in terms of clear-cut variables: the nature and kinds of interrelations among positions and the types of expectations attending these positions.
2. The variables Mead denoted by the concepts of mind and self can be analytically distinguished from both social structure (positions and expectations) and behavior (role enactment).
3. By conceptually separating the processes of role taking and imaginative rehearsal from both social structure and behavior, the points of articulation between society and the individual can be more clearly marked, since role taking pertains to covert interpretations of the expectations attending networks of statuses and role denotes the enactment of these expectations as mediated by self.

Thus, by offering more conceptual insight into the nature of social organization, Park, Moreno, and Linton provided a needed supplement to Mead's suggestive concepts. For now it would be possible to understand more precisely the interrelations among mind, self, and society.

EARLY EUROPEAN INSIGHTS

At the same time that American theory turned to the study of interaction and roles, European thinking was making a similar shift in emphasis. The more

---

[15]Ralph Linton, *The Study of Man* (New York: Appleton-Century-Crofts, 1936), p. 28.

macro-oriented work of Max Weber and Émile Durkheim became increasingly concerned with meaning and how society "gets inside" the individual. But neither Weber nor Durkheim was as interactionistic in approach as Georg Simmel. In the end, however, the main theoretical contribution to interactionism from Europe came from philosophical phenomenology. And so I will briefly summarize Weber's, Durkheim's, and Simmel's interactionist ideas, and then I will turn to the early phenomenologists Edmund Husserl and Alfred Schutz.

Georg Simmel and "Sociation"

Georg Simmel was perhaps the first European sociologist to begin a serious exploration of interaction, or "sociability" as he called it. In so doing, he elevated the study of interaction from the taken-for-granted.[16] For Simmel, as for the first generation of American sociologists in Chicago, the macrostructures and processes studied by functional and some conflict theories—class, the state, family, religion, evolution—are ultimately reflections of the specific interactions among people. These interactions result in emergent social phenomena, but considerable insight into the latter can be attained by understanding the basic interactive processes that first give and then sustain their existence.

In Chapter 9 I discussed Simmel's analysis of the forms of conflict, and in Chapter 14 I summarized his exchange theory. But Simmel's study of interaction extends beyond just the analysis of conflict and exchange, for he was concerned with understanding the forms and consequences of many diverse types of interactions. Some of his most important insights, which influenced American interactionists, concern the relationship between the individual and society. In his famous essay on "the web of group affiliations," for example, Simmel emphasized that human personality emerges from, and is shaped by, the particular configuration of a person's group affiliations.[17] What people are— that is, how they think of themselves and are prepared to act—is circumscribed by their group memberships. As he emphasized, "the genesis of the personality [is] the point of intersection for innumerable social influences, as the end-product of heritages derived from the most diverse groups and periods of adjustment."[18]

Although Simmel did not analyze in great detail the emergence of human personality, his "formal sociology" did break away from the macro concerns of early German, French, and British sociologists. He began in Europe a mode of analysis that, as I have just summarized, became the prime concern of the

---

[16]Georg Simmel, "Sociability," in *The Sociology of Georg Simmel*, ed. K. H. Wolff (New York: Free Press, 1950), pp. 40–57. For an excellent secondary account of Simmel's significance for interactionism, see Randall Collins and Michael Makowsky, *The Discovery of Society* (New York: Random House, 1972), pp. 138–42.

[17]*Conflict and the Web of Group Affiliations*, trans. R. Bendix (Glencoe, IL: Free Press, 1955; originally published in 1922).

[18]Ibid., p. 141.

first generation of American sociologists. Simmel thus could be considered one of the first European interactionists.

Émile Durkheim's Metamorphosis

In his *Division of Labor in Society*, Émile Durkheim portrayed social reality as an emergent phenomenon, *sui generis*, and as not reducible to the psychic states of individuals. And yet, in his later works, such as *The Elementary Forms of the Religious Life*, Durkheim began to ask: How does society rule the individual? How is it that society "gets inside" individuals and guides them from within? Why do people share common orientations and perspectives?[19] Durkheim never answered these questions effectively, for his earlier emphasis on social structures prevented him from seeing the micro reality of interactions among individuals implicated in macro social structures. But I think it significant that the most forceful advocate of the sociologistic position became intrigued with the relationship between the individual and society. I see two critical lines of interactionist thought in *Elementary Forms*: (1) the analysis of ritual and (2) the concern with categories of thought.

> 1. The implicit theory of ritual argues that people's sense of solidarity is reinforced by ritual activity. Durkheim emphasized that religious rituals are, in essence, the worship of society. But there are more general implications to his ideas: people's sense of attachment to the social order and to one another is very much dependent upon ritual performances. That is, interpersonal rituals are vital to the maintenance of the macro social order.[20]
>
> 2. The concern with thought in *Elementary Forms* also influenced social theory. Durkheim emphasized that "the collective conscience" is not "entirely outside us" and that people's definitions of, and orientations to, situations are related to the organization of subjective consciousness. The categories of this consciousness, however, reflect the structural arrangements of society. Hence, varying macro structures generate different forms of thought and perception of the world. Such forms feed back and reinforce social structures.

Number (1) above was to exert considerable influence in interactionist thinking, whereas number (2) was to form the core idea behind much "structuralist" social theory (see Chapter 25 on structuralism). Although Durkheim's ideas on ritual and categories of thought were never as rigorously or systematically developed as his earlier work on macro processes (see Chapter 2 on the emergence of functionalism), these were perhaps his most original ideas, since

---

[19]Émile Durkheim, *The Elementary Forms of the Religious Life* (New York: Free Press, 1954; originally published in 1912).

[20]Randall Collins develops this idea in *Conflict Sociology* (New York: Academic Press, 1975) and other works. See Chapter 12. But perhaps the most significant adoption of Durkheim's emphasis on ritual was by the late Erving Goffman, the subject of Chapter 22.

much of his early macrostructural emphasis was borrowed from Herbert Spencer (see Chapter 2). But the desire to discover the underlying interactive basis of social structure led Durkheim to significant insights, which, as we will see in Chapter 22 on dramaturgy, constituted the conceptual basis for a distinctive line of interactionist theorizing.

Max Weber and "Social Action"

Max Weber was also becoming increasingly concerned with the micro social world, although I think that his most important insights are in the area of macro and historical sociology. Yet Weber's definition of sociology was highly compatible with the flourishing American school of interactionism. For Weber, sociology is "that science which aims at the interpretative understanding of social behavior in order to gain an explanation of its causes, its course, and its effects."[21] Moreover, the behavior to be studied by sociology is seen by Weber as social action that includes:

> all human behavior when and insofar as the acting individual attaches a subjective meaning to it. Action in this sense may be overt, purely inward, or subjective; it may consist of positive intervention in a situation, of deliberately refraining from such intervention, or passively acquiescing in the situation. Action is social insofar as by virtue of the subjective meaning attached to it by the acting individual (or individuals), it takes account of the behavior of others and is thereby oriented in its course.[22]

Thus Weber recognized that the reality behind the macrostructures of society—classes, the state, institutions, and nations—are the meaningful and symbolic interactions among people. Moreover, Weber's methodology stresses the need for understanding macrostructures and processes "at the level of meaning." For, in the real world, actors interpret and give meaning to the reality around them and act in terms of these meanings. And yet, despite this key insight, Weber's actual analysis of social structures—class, status, party, change, religion, bureaucracy, and the like—rarely follows his own methodological prescriptions. As with other European thinkers, he tended to focus on social and cultural *structures* and the impact of these structures upon one another. The interacting and interpreting person is often lost amid Weber's elaborate taxonomies of structures and analyses of historical events. And it is just this failing that was to attract the attention of Alfred Schutz, who, more than any other European thinker, was to translate phenomenology into a perspective that could be incorporated into interactionist theory.

European Phenomenology

Phenomenology began as the project of the German philosopher Edmund Husserl (1859–1938).[23] In his hands, I think, this project showed few signs of being

---

[21]Max Weber, *Basic Concepts in Sociology* (New York: Citadel Press, 1964), p. 29.

[22]Max Weber, *The Theory of Social and Economic Organization* (New York: Free Press, 1947; originally published after Weber's death), p. 88.

[23]For some readable, general references on phenomenology, see George Psathas, ed., *Phenom-*

anything more than an orgy of subjectivism. Yet, as I will emphasize, the German social thinker Alfred Schutz was to take Husserl's concepts and transform them into an interactionist analysis that has exerted considerable influence on modern-day interactionism. I am sure that Schutz's migration to the United States in 1939 facilitated this translation, especially as he came into contact with American interactionism, but my sense is that all of his most important ideas were formulated before he came. His subsequent work while in America involved an elaboration of basic ideas originally developed in Europe. I will first analyze Husserl's original phenomenology and then turn to Schutz's critique of Max Weber and his transformation of Husserl's ideas into an interactionist perspective.

Edmund Husserl's project   As the father of phenomenology, there can be little doubt that Edmund Husserl's thought has profoundly influenced contemporary social science. Yet, as I have suggested, his ideas have been transformed, especially by Alfred Schutz. Indeed, Husserl would be upset by what is now attributed to his genius. As Z. Bauman notes: "It took guile, utterly illegitimate as viewed from the Husserlian perspective, to devise a social science which would claim to be the brain-child, or logical consequence of the phenomenological project."[24]

Husserl's ideas, then, have been selectively borrowed and used in ways that he would not condone to develop modern phenomenology and various forms of interactionist thought. In reviewing Husserl's contribution, therefore, I think it best to focus more on what was borrowed than on the details of his complete philosophical scheme. I will therefore highlight several features of his work: (1) the basic philosophical dilemma, (2) the properties of consciousness, (3) the critique of naturalistic empiricism, and (4) the philosophical alternative to social science.[25]

enological Sociology (New York: John Wiley, 1973); Richard M. Zaner, The Way of Phenomenology: Criticism as a Philosophical Discipline (New York: Pegasus, 1970); Peter L. Berger and Thomas Luckman, The Social Construction of Reality (Garden City, NY: Doubleday, 1966); Herbert Spiegelberg, The Phenomenological Movement, vols. 1 and 2, 2nd ed. (The Hague: Martinus Nijhoff, 1969); Hans P. Neisser, "The Phenomenological Approach in Social Science," Philosophy and Phenomenological Research 20 (1959), pp. 198-212; Stephen Strasser, Phenomenology and the Human Sciences (Pittsburgh: Duquesne University Press, 1963); Maurice Natanson, ed., Phenomenology and the Social Sciences (Evanston, IL: Northwestern University Press, 1973); and Quentin Lauer, Phenomenology: Its Genesis and Prospect (New York: Harper Torchbooks, 1965).

[24]Bauman, "On the Philosophical Status of Ethnomethodology," The Sociological Review 21 (February 1973), p. 6.

[25]Husserl's basic ideas are contained in the following: Phenomenology and the Crisis of Western Philosophy (New York: Harper & Row, 1965; originally published in 1936); Ideas: General Introduction to Pure Phenomenology (London: Collier-Macmillan, 1969; originally published in 1913); and "Phenomenology," in The Encyclopedia Britannica, 14th ed., vol. 17, col. 699-702, 1929. For excellent secondary analyses, see Helmut R. Wagner, "The Scope of Phenomenological Sociology," in Phenomenological Sociology, ed. G. Psathas, pp. 61-86, and "Husserl and Historicism," Social Research 39 (Winter 1972), pp. 696-719; Aron Gurwitsch, "The Common-Sense World as Social Reality," Social Research 29 (Spring 1962), pp. 50-72; Robert J. Antonio, "Phenomenological Sociology," in Sociology: A Multiple Paradigm Science, ed. G. Ritzer (Boston: Allyn & Bacon, 1975), pp. 109-12; Robert Welsh Jordan, "Husserl's Phenomenology as an 'Historical Science,'" Social Research 35 (Summer 1968), pp. 245-59.

1. Basic questions confronting all inquiry are: What is real? What actually exists in the world? How is it possible to know what exists? For the philosopher Husserl, these were central questions that required attention. Husserl reasoned that humans know about the world only through experience. All notions of an external world, "out there," are mediated through the senses and can be known only through mental consciousness. The existence of other people, values, norms, and physical objects is always mediated by experiences as these register on people's conscious awareness. One does not directly have contact with reality; contact is always indirect and mediated through the processes of the human mind.

Since the process of consciousness is so important and central to knowledge, philosophic inquiry must first attempt to understand how this process operates and how it influences human affairs. It is this concern with the process of consciousness—of how experience creates a sense of an external reality—that was to become the central concern of phenomenology.

2. Husserl initially made reference to the "world of the natural attitude." Later he was to use the phrase *lifeworld*. In either case, with these concepts he emphasized that humans operate in a taken-for-granted world that permeates their mental life. It is the world that humans sense to exist. It is composed of the objects, people, places, ideas, and other things that people see and perceive as setting the parameters for their existence, for their activities, and for their pursuits.

This lifeworld or world of the natural attitude *is* reality for humans. I see two features of Husserl's conception of natural attitude as influencing modern interactionist thought: (*a*) The lifeworld is taken for granted. It is rarely the topic of reflective thought, and yet it structures and shapes the way people act and think. (*b*) Humans operate on the presumption that they experience the same world. Since people experience only their own consciousness, they have little capacity to directly determine if this presumption is correct. Yet people act *as if* they experienced a common world.

Human activity, then, is conducted in a lifeworld that is taken for granted and that is presumed to be experienced collectively. This fact brought Husserl back to his original problem: how do humans break out of their lifeworld and ascertain what is real? If people's lifeworld structures their consciousness and their actions, how is an objective science of human behavior and organization possible? These questions led Husserl to criticize what he termed *naturalistic science,* or the position that I advocated in Chapter 1.

3. As I stressed in Chapter 1, science assumes that a factual world exists out there, independent of, and external to, human senses and consciousness. Through the scientific method this factual world can be directly known. With successive efforts at its measurement, increasing understanding of its properties can be ascertained. But Husserl challenged this vision of science: if one can know only through consciousness and if consciousness is structured by an implicit lifeworld, then how can objective measurement of some external and real world be possible? How is science able to measure objectively an external

world when the only world that individuals experience is the lifeworld of their consciousness?

4. Husserl's solution to this problem is a philosophical one. He advocated what he termed the search for the *essence of consciousness*. To understand social events, the basic process through which these events are mediated—that is, consciousness—must be comprehended. The substantive *content* of consciousness, or the lifeworld, is not what is important; the abstract processes of consciousness, per se, are to be the topic of philosophic inquiry.

Husserl advocated what he termed the *radical abstraction of the individual* from interpersonal experience. Investigators must suspend their natural attitude and seek to understand the fundamental processes of consciousness, per se. One must discover, in Husserl's words, "Pure Mind." To do this, it is necessary to perform "epoch"—that is, to see if the substance of one's lifeworld can be suspended. Only when divorced from the substance of the lifeworld can the fundamental and abstract properties of consciousness be exposed and understood. And with understanding of these properties, real insight into the nature of reality would be possible. For, if all that humans know is presented through consciousness, it is necessary to understand the nature of consciousness in abstraction from the specific substance or content of the lifeworld.

I should caution here that Husserl was not advocating Max Weber's method of *verstehen,* or sympathetic introspection into an investigator's own mind. Nor was he suggesting the unstructured and intuitive search for people's definitions of situations. These methods would, he argued, only produce data on the substance of the lifeworld and would be no different than the structured measuring instruments of positivism. Rather, Husserl's goal was to create an abstract theory of consciousness that bracketed out, or suspended, any presumption of "an external social world out there."

Not surprisingly, I think, Husserl's philosophical doctrine failed. He never succeeded in developing an abstract theory of consciousness, radically abstracted from the lifeworld. But his ideas set into motion a new line of thought that was to become the basis for modern phenomenology and for its elaboration into ethnomethodology and other forms of theory. Indeed, recall Jurgen Habermas' critical project in which the lifeworld is a prominent concept (Chapter 13). Or, as I will emphasize for ethnomethodology in Chapter 23 and in the work of various interactionists, Husserl's ideas have endured, although in a dramatically transformed way. This transformation was, I feel, the major accomplishment of Alfred Schutz's early work. Indeed, he converted Husserl's phenomenological project into a type of interactionism.

The phenomenological interactionism of Alfred Schutz   Alfred Schutz migrated to the United States in 1939 from Austria, after spending a year in Paris. With his interaction in American intellectual circles and the translation of his early works into English over the last decades, Schutz's contribution to sociological theorizing is becoming increasingly recognized.[26]

---

[26]For the basic ideas of Alfred Schutz, see his *The Phenomenology of the Social World* (Ev-

As I have emphasized, his contribution resides in his ability to blend Husserl's radical phenomenology with Max Weber's action theory and American interactionism. This blend was, in turn, to stimulate the further development of phenomenology, the emergence of ethnomethodology, and the refinement of other theoretical perspectives.

Schutz's work begins with a critique of his compatriot Max Weber, who employed the concept of social action in his many and varied inquiries.[27] Social action occurs when actors are consciously aware of each other and attribute meanings to their common situation. For Weber, then, a science of society must seek to understand social reality "at the level of meaning." Sociological inquiry must penetrate people's consciousness and discover how they view, define, and see the world. Weber advocated the method of *verstehen*, or sympathetic introspection. Investigators must become sufficiently involved in situations to be able to get inside the subjective world of actors. Causal and statistical analysis of complex social structures would be incomplete and inaccurate without such *verstehen* analysis.

Schutz's first major work addressed Weber's conception of action. Schutz's analysis is critical and detailed, and I need not summarize it here, except to note that the basic critique turns on Weber's failure to use his *verstehen* method and to explore *why*, and through what processes, actors come to share common meanings. In Schutz's eye, Weber simply assumes that actors share subjective meanings, leading Schutz to ask: Why and how do actors come to acquire common subjective states in a situation? How do they create a common view of the world? This is the problem of "intersubjectivity," and it is central to Schutz's intellectual scheme. As Richard D. Zaner summarizes:

> How is it possible that although I cannot live in your seeing of things, cannot feel your love and hatred, cannot have an immediate and direct perception of your mental life as it is for you—how is it that I can nevertheless share your thoughts, feelings, and attitudes? For Schutz the "problem" of intersubjectivity is here encountered in its full force.[28]

I should, of course, stress how close this concern with intersubjectivity is to G. H. Mead's analysis of role taking, which was the process by which actors anticipate one another's dispositions to act.

Schutz was more influenced in his early years by Husserl's phenomenology than by anything Mead had written. And yet, as Mead would have directed, Schutz departs immediately from Husserl's strategy of holding the individual in radical abstraction and of searching for Pure Mind or the abstract laws of consciousness. He accepts Husserl's notion that humans hold a natural attitude

anston, IL: Northwestern University Press, 1967; originally published in 1932); *Collected Papers*, vols. 1, 2, 3 (The Hague: Martinus Nijhoff, 1964, 1970, and 1971, respectively). For excellent secondary analyses, see Maurice Natanson, "Alfred Schutz on Social Reality and Social Science," *Social Research* 35 (Summer 1968), pp. 217–44.

[27]Schutz, *The Phenomenology of the Social World*, 1921.

[28]Richard M. Zaner, "Theory of Intersubjectivity: Alfred Schutz," *Social Research* 28 (Spring 1961), p. 76.

and lifeworld that is taken for granted and that shapes who they are and what they will do. He also accepts Husserl's notion that people perceive that they share the same lifeworld and act *as if* they lived in a common world of experiences and sensations. Moreover, Schutz acknowledges the power of Husserl's argument that social scientists cannot know about an external social world out there independently of their own lifeworld.

Having accepted these lines of thought from Husserl, however, Schutz advocates Weber's strategy of sympathetic introspection into people's consciousness. Only by observing people in interaction, rather than in radical abstraction, can the processes whereby actors come to share the same world be discovered. Social science cannot come to understand how and why actors create a common subjective world independently of watching them do so. This abandonment of Husserl's phenomenological project liberated phenomenology from philosophy and allowed sociologists to study empirically what Schutz considered the most important social reality: the creation and maintenance of intersubjectivity—that is, a common subjective world among pluralities of interacting individuals.

With his immigration to the United States, Schutz's phenomenology came under the influence of early symbolic interactionists, particularly G. H. Mead and W. I. Thomas. But I am not sure that Mead and other Chicago School interactionists exerted great influence on Schutz, who did devote some attention to symbolic interactionism.[29] Indeed, their influence may have been subtle, if not subliminal. The early symbolic interactionist's concern with the process of constructing shared meanings was, of course, similar to Schutz's desire to understand intersubjectivity. Schutz thus found immediate affinity with W. I. Thomas' "definition of the situation," since this concept emphasizes that actors construct orientations to, and dispositions to act in, situations. Moreover, Thomas' recognition that definitions of situations are learned from past experiences while being altered in present interactions appears to have influenced Schutz's conceptualization of the process of intersubjectivity. Mead's recognition that mind is a social process, arising out of interaction and yet facilitating interaction, probably had considerable appeal for Schutz. But, if there was any direct influence from Chicago, I think it was Mead's concept of role taking. For Schutz became vitally concerned with the process whereby actors come to know one another's role and to typify one another as likely to behave in certain ways. Additionally, Mead's concept of the generalized other may have influenced Schutz in that actors are seen as sharing a "community of attitudes"—or, in other words, common subjective states. Yet, although there is considerable affinity and perhaps some cross-fertilization between early symbolic interactionists' conceptualizations and Schutz's phenomenology, Schutz was to inspire a line of sociological inquiry that often challenges the interactionism initiated by Mead.

Having reviewed some of the intellectual influences on Schutz, let me now summarize his scheme and indicate some of its implications for sociological

---

[29]See Schutz, *Collected Papers*, for references to interactionists.

theorizing. Unfortunately, Schutz died just as he was beginning a systematic synthesis of his ideas; as a result, only a somewhat fragmented but suggestive framework is evident in his collective work. But his early analysis of Weber, Husserl, and interactionism led to a concern with a number of key issues: (1) How do actors create a common subjective world? (2) What implications does this creation have for how social order is maintained?

All humans, Schutz asserted, carry in their minds rules, social recipes, conceptions of appropriate conduct, and other information that allows them to act in their social world. Extending Husserl's concept of lifeworld, Schutz views the sum of these rules, recipes, conceptions, and information as the individual's "stock knowledge at hand." Such stock knowledge gives people a frame of reference or orientation with which they can interpret events as they pragmatically act on the world around them.

Several features of this stock knowledge at hand are given particular emphasis by Schutz:

1. People's reality *is* their stock knowledge. For the members of a society, stock knowledge constitutes a "paramount reality"—a sense of an absolute reality that shapes and guides all social events. Actors use this stock knowledge and sense of reality as they pragmatically seek to deal with others in their environment.
2. The existence of stock knowledge that bestows a sense of reality on events gives the social world, as Schutz agreed with Husserl, a taken-for-granted character. The stock knowledge is rarely the object of conscious reflection but rather an implicit set of assumptions and procedures that are silently used by individuals as they interact.
3. Stock knowledge is learned. It is acquired through socialization within a common social and cultural world, but it becomes *the* reality for actors in this world.
4. People operate under a number of assumptions that allow them to create a sense of a "reciprocity of perspectives." That is, others with whom an actor must deal are considered to share an actor's stock knowledge at hand. And, although these others may have unique components in their stock knowledge because of their particular biographies, these can be ignored by actors.
5. The existence of stock knowledge, its acquisition through socialization, and its capacity to promote reciprocity of perspectives all operate to give actors in a situation a *sense* or *presumption* that the world is the same for all and that it reveals identical properties for all. What often holds society together is this presumption of a common world.
6. The presumption of a common world allows actors to engage in the process of typification. Action in most situations, except the most personal and intimate, can proceed through mutual typification as actors use their stock knowledge to categorize one another and to adjust their responses to these typifications.[30] With typification, actors can effec-

---

[30]Ralph H. Turner's emphasis on role differentiation and accretion is an example of how these ideas have been extended by role theorists. See Chapter 21.

tively deal with their world, since every nuance and characteristic of their situation do not have to be examined. Moreover, typification facilitates entrance into the social world; it simplifies adjustment because it allows for humans to treat each other as categories, or as "typical" objects of a particular kind.

These points of emphasis in Schutz's thought represent, as I noted earlier, a blending of ideas from European phenomenology and American interactionism. The emphasis on stock knowledge is clearly borrowed from Husserl, but it is highly compatible with Mead's notion of the generalized other. The concern with the taken-for-granted character of the world as it is shaped by stock knowledge is also borrowed from Husserl but is similar to early interactionists' discussions of habit and routine behaviors. The emphasis on the acquired nature of stock knowledge coincides with early interactionists' discussions of the socialization process. The concern with the reciprocity of perspectives and with the process of typification owes much to Husserl and Weber but is very compatible with Mead's notion of role taking, by which actors read one another's role and perspective.

But the major departure from interactionism should also be emphasized: actors operate on an unverified *presumption* that they share a common world, and this *sense* of a common world and the practices that produce this sense are crucial in maintaining social order. In other words, social organization may be possible not so much by the substance and content of stock knowledge nor by the actual fact of reciprocity of perspectives nor by successful typification as that by the presumption actors share intersubjective states. Schutz did not carry this line of inquiry far, but it was to inspire new avenues of phenomenological inquiry.

In sum, then, I think that Schutz is primarily responsible for liberating Husserl's concern with the basic properties and processes of consciousness from radical abstraction. Schutz brought Husserl's vision of a lifeworld back into the process of interaction. In so doing, he began to ask how actors come to share, or presume that they share, intersubjective states. He made Husserl's ideas more compatible with interactionists' concern with socialization and role taking as well as with their emphasis on pragmatic actors seeking to cope with their world. But Schutz gave these concerns a new twist: humans act *as if* they see the world in similar ways, and they deal with one another *as if* others could be typified and categorized. This shift in emphasis to the interpersonal processes by which actors create a sense of a shared world was to inspire challenges to traditional interactionism and role theory and, in the process, to correct for some deficiencies in Meadian interactionism.

MODERN INTERACTIONISM: A REVIEW

The Meadian legacy has directly inspired a theoretical perspective that can best be termed *symbolic interactionism*. This perspective, which I will discuss in the next chapter, focuses on how the symbolic processes of role taking, imaginative rehearsal, and self-evaluation by individuals attempting to adjust to one another are the basis for social organization. While accepting the analytical importance of these symbolic processes, a more recent theoretical tradition has placed conceptual emphasis on the vision of social structure connoted by Park's, Moreno's, and Linton's concepts. Although not as clearly codified or as unified as the symbolic interactionist position, this theoretical perspective—the subject of Chapter 20—can be labeled *structural role theory,* because it focuses primarily analytical attention on the *structure of status networks and attendant expectations* as they circumscribe the internal symbolic processes of individuals and the eventual enactment of roles. As a reaction to this emphasis on structure, more process versions of role theory have emerged, which are explored in Chapter 21.

In some respects, the distinction between symbolic interactionism and the two role theoretic approaches may initially appear arbitrary, since each perspective relies heavily on the thought of George Herbert Mead and since both are concerned with the relationship between the individual and society. Yet, despite the fact that each represents a variant of interactionism, I think that there are great differences in emphasis between symbolic interactionism and all role theories. However, Ralph H. Turner's process approach comes close to uniting symbolic interactionism and role theory into a unified interactionist perspective (see Chapter 21).

In Chapter 22, Erving Goffman's work is examined. Here the late Durkheimian emphasis on ritual is highlighted and blended with more phenomenological concerns on the taken-for-granted character of much interaction. In so doing, Goffman either ignores or criticizes much work within the Meadian tradition. Finally, I will examine, in Chapter 23, the major interactionist perspective inspired by phenomenology: ethnomethodology. This perspective challenges much interactionist theory, but, as I will emphasize, this challenge has been rather shrill and, at times, silly. It is much wiser, I believe, for ethnomethodology to be seen as an important supplement to traditional interactionist theory.

CHAPTER 19

Symbolic Interactionism

Herbert Blumer and Manford Kuhn

All interactionist theory draws upon the theoretical synthesis provided by G. H. Mead. Yet various contemporary thinkers have selectively borrowed Mead's ideas; as a result, interactionist theorizing reveals considerable diversity. Even within a particular interactionist school of thought, there is conceptual controversy. Nowhere is this kind of controversy more evident than within *symbolic interactionism,* the theoretical perspective most closely tied to Mead's ideas.

The poles around which this controversy rages are termed the Iowa and Chicago Schools of symbolic interactionism.[1] These labels are, I feel, somewhat archaic and inappropriate, for several reasons. First, the principal exponent of the Chicago School, Herbert Blumer, left Chicago more than 30 years ago; it is true that he took over Mead's social psychology course at Chicago upon the latter's death and that he has defined himself as the principal interpreter of Mead's thought, but the association with Chicago is now so remote as to make the label seem rather contrived. Second, although many who follow Blumer's general approach were trained at the University of Chicago, an equal number who are less sympathetic to Blumer's advocacy were also trained there. Moreover, many of these Chicago-trained interactionists are more in tune with the principal figure of the Iowa School, Manford Kuhn. Third, Kuhn's use of quantitative measures of interactionist ideas while at the State University at Iowa was, in fact, pioneered by members of the Chicago department in the 1930s.[2]

---

[1] Bernard N. Meltzer and Jerome W. Petras, "The Chicago and Iowa Schools of Symbolic Interactionism," in *Human Nature and Collective Behavior*, ed. T. Shibutani (Englewood Cliffs, NJ: Prentice-Hall, 1970).

[2] Jonathan H. Turner, "The Rise of Scientific Sociology," *Science* 227 (March 15, 1985); Jonathan H. Turner and Stephen Park Turner, *American Sociology* (Warsaw: Polish Scientific Publishers, 1992) and *The Impossible Science* (Newberry Park, CA: Sage, 1990). See also: Martin Bulmer, *The Chicago School of Sociology* (Chicago: University of Chicago Press, 1984), and Lester R. Kurtz, *Evaluating Chicago Sociology* (Chicago: University of Chicago Press, 1984).

And, fourth, symbolic interactionists are located all over the country and employ such a diversity of approaches that simple labels are much less relevant than they were a few decades ago.[3]

Yet, having said this, I will use the distinction between the Iowa and Chicago Schools in this chapter. For I think that Kuhn's positivistic approach and Blumer's critique of deductive theory do represent differences. Symbolic interactionists still gravitate toward Kuhn's or Blumer's advocacy. And so, if I am to represent the scope and diversity of views within symbolic interactionism, a detailed review of Blumer's and Kuhn's respective approaches is a reasonable place to begin.

SYMBOLIC INTERACTIONISM: POINTS OF CONVERGENCE

Before turning to the points of divergence between the Iowa and Chicago Schools, let me review the common legacy of Mead's assumptions that all symbolic interactionists employ. These points of convergence are, I believe, what make symbolic interactionism a distinctive theoretical perspective.

Humans as Symbol Users

Symbolic interactionists, as their name implies, place enormous emphasis on the capacity of humans to create and use symbols. In contrast to other animals, whose symbolic capacities are limited or nonexistent, the very essence of humans and the world that they create flows from their ability to symbolically represent one another, objects, ideas, and virtually any phase of their experience. Without the capacity to create symbols and to use them in human affairs, patterns of social organization among humans could not be created, maintained, or changed. Humans have become, to a very great degree, liberated from instinctual and biological programming and thus must rely on their symbol-using powers to adapt and survive in the world.

---

[3]For examples, consult Larry T. Reynolds, *Interactionism: Exposition and Critique*, 2nd ed. (Dix Hills, NY: General Hall, 1990); Jerome G. Manis and Bernard N. Meltzer, eds., *Symbolic Interaction: A Reader in Social Psychology* (Boston: Allyn & Bacon, 1972); J. Cardwell, *Social Psychology: A Symbolic Interactionist Approach* (Philadelphia: F. A. Davis, 1971); Alfred Lindesmith and Anselm Strauss, *Social Psychology* (New York: Holt, Rinehart & Winston, 1968); Arnold Rose, ed., *Human Behavior and Social Process* (Boston: Houghton Mifflin, 1962); Tamotsu Shibutani, *Society and Personality* (Englewood Cliffs, NJ: Prentice-Hall, 1961); C. K. Warriner, *The Emergence of Society* (Homewood, IL: Dorsey Press, 1970); Gregory Stone and H. Farberman, eds., *Symbolic Interaction: A Reader in Social Psychology* (Waltham, MA: Xerox Learning Systems, 1970); John P. Hewitt, *Self and Society: A Symbolic Interactionist Social Psychology* (Boston: Allyn & Bacon, 1976); Robert H. Laver and Warren H. Handel, *Social Psychology: The Theory and Application of Symbolic Interactionism* (Boston: Houghton Mifflin, 1977); Sheldon Stryker, *Symbolic Interactionism* (Menlo Park, CA: Benjamin/Cummings, 1980); Clark McPhail, "The Problems and Prospects of Behavioral Perspectives," *The American Sociologist* 16 (1981), pp. 172–74.

Symbolic Communication

Humans use symbols to communicate with one another. By virtue of their capacity to agree upon the meaning of vocal and bodily gestures, humans can effectively communicate. Symbolic communication is, of course, extremely complex, since people use more than word or language symbols in communication. They also use facial gestures, voice tones, body countenance, and other symbolic gestures in which there is common meaning and understanding.

Interaction and Role Taking

By reading and interpreting the gestures of others, humans communicate and interact. They become able to mutually read each other, to anticipate each other's responses, and to adjust to each other. Mead termed this basic capacity "taking the role of the other," or role taking—the ability to see the other's attitudes and dispositions to act. Interactionists still emphasize the process of role taking as the basic mechanism by which interaction occurs. For example, the late Arnold Rose, who was one of the leaders of contemporary interactionism, indicated that role taking "means that the individual communicator imagines—evokes within himself—how the recipient understands that communication."[4] Or, as another modern interactionist, Sheldon Stryker, emphasizes, role taking is "anticipating the responses of others with one in some social act."[5] And, as Alfred Lindesmith and Anselm Strauss emphasize, role taking is "imaginatively assuming the position or point of view of another person."[6]

Without the ability to read gestures and to use these gestures as a basis for putting oneself in the position of others, interaction could not occur. And, without interaction, social organization could not exist.

Interaction, Humans, and Society

Just as Mead emphasized that mind, self, and society are intimately connected, so contemporary interactionists analyze the relation between the genesis of "humanness" and patterns of interaction. What makes humans unique as a species and enables each individual to possess distinctive characteristics is the result of interaction in society. Conversely, what makes society possible are the capacities that humans acquire as they grow and mature in society.

I find that current symbolic interactionists tend to emphasize the same human capacities as Mead: the genesis of mind and self. Mind is the capacity to think—to symbolically denote, weigh, assess, anticipate, map, and construct

---

[4]Rose, *Human Behavior*, p. 8.

[5]Sheldon Stryker, "Symbolic Interaction as an Approach to Family Research," *Marriage and Family Living* 2 (May 1959), pp. 111–19. See also his "Role-Taking Accuracy and Adjustment," *Sociometry* 20 (December 1957), pp. 286–96.

[6]Lindesmith and Strauss, *Social Psychology*, p. 282.

courses of action. Although Mead's term, *mind,* is rarely used today, the processes that this term denotes are given great emphasis. As Rose indicates: "Thinking is the process by which possible symbolic solutions and other future courses of action are examined, assessed for their relative advantages and disadvantages in terms of the values of the individual, and one of them chosen for action."[7]

Moreover, the concept of mind has been reformulated to embrace what W. I. Thomas termed the *definition of the situation.*[8] With the capacities of mind, actors can name, categorize, and orient themselves to constellations of objects—including themselves as objects—in all situations. In this way they can assess, weigh, and sort out appropriate lines of conduct.[9]

As the concept of the definition of the situation underscores, self is still a key concept in the interactionist literature. Present emphasis in the interactionist orientation is on (*a*) the emergence of self-conceptions—relatively stable and enduring conceptions that people have about themselves—and (*b*) the ability to derive self-images—pictures of oneself as an object in social situations. Self is thus a major object that people inject into their definitions of situations. It shapes much of what they see, feel, and do in the world around them.

Society, or relatively stable patterns of interaction, is seen by interactionists as possible only by virtue of people's capacities to define situations and, most particularly, to view themselves as objects in situations. Society can exist by virtue of human capacities for thinking and defining as well as for self-reflection and evaluation.

In sum, these points of emphasis constitute the core of the interactionist approach. Humans create and use symbols. They communicate with symbols. They interact through role taking, which involves the reading of symbols emitted by others. What makes them unique as a species—the existence of mind and self—arises out of interaction. Conversely, the emergence of these capacities allows for the interactions that form the basis of society.

AREAS OF DISAGREEMENT AND CONTROVERSY

From this initial starting point, Blumer and Kuhn often diverge, as do more recent advocates of their respective positions.[10] I see the major areas of dis-

---

[7]Rose, *Human Behavior,* p. 12.

[8]W. I. Thomas, "The Definition of the Situation," in *Symbolic Interaction,* eds. J. Manis and B. Meltzer, pp. 331–36.

[9]For clear statements on the concept of "definition of the situation" as it is currently used in interactionist theory, see Lindesmith and Strauss, *Social Psychology,* pp. 280–83.

[10]The following comparison draws heavily from Meltzer's and Petras' "The Chicago and Iowa Schools" but extends their analysis by drawing from the following works of Blumer and Kuhn. For Blumer, *Symbolic Interactionism: Perspective and Method* (Englewood Cliffs, NJ: Prentice-Hall, 1969); "Comment on 'Parsons as a Symbolic Interactionist,' " *Sociological Inquiry* 45 (Winter 1975), pp. 59–62. For Kuhn, "Major Trends in Symbolic Interaction Theory in the Past Twenty-Five Years," *Sociological Quarterly* 5 (Winter 1964), pp. 61–84; "The Reference Group Reconsidered," *Sociological Quarterly* 5 (Winter 1964), pp. 6–21; "Factors in Personality: Socio-Cultural Determinants as Seen Through the Amish," in *Aspects of Culture and Personality,* ed. F. L. Kittsu

agreement as revolving around the following issues: (1) What is the nature of the individual? (2) What is the nature of interaction? (3) What is the nature of social organization? (4) What is the most appropriate method for studying humans and society? And (5) what is the best form of sociological theorizing? Let me now examine each of these five controversial questions in more detail.

The Nature of the Individual

Both Blumer and Kuhn have emphasized the ability of humans to use symbols and to develop capacities for thinking, defining, and self-reflecting. However, there is considerable disagreement over the degree of structure and stability in human personality. Blumer emphasized that humans have the capacity to view themselves as objects and to insert any object into an interaction situation. Therefore, human actors are not pushed and pulled around by social and psychological forces but are *active creators* of the world to which they respond. Interaction and emergent patterns of social organization can be understood only by focusing on these capacities of individuals to create symbolically the world of objects to which they respond. For there is always the potential for spontaneity and indeterminacy in human behavior. If humans can invoke any object into a situation, they can radically alter their definitions of that situation and, hence, their behaviors. Self is but one of many objects to be seen in a situation; other objects from the past, present, or anticipated future can also be evoked and can provide a basis for action.

In contrast, Kuhn emphasized the importance of people's "core self" as an object. Through socialization, humans acquire a relatively stable set of meanings and attitudes toward themselves. The core self will shape and constrain the way people will define situations by circumscribing the cues that will be seen and the objects that will be injected into social situations. Human personality is thus structured and comparatively stable, giving people's actions a continuity and predictability. And, if it is possible to know the expectations of those groups that have shaped a person's core self and that provide a basis for its validation, then human behavior could, in principle, be highly predictable.[11] As Kuhn and Hickman noted:

> As self theory views the individual, he derives his plans of action from the roles he plays and the statuses he occupies in the groups with which he feels identified—his reference groups. His attitudes toward himself as an object are the best indexes to these plans of action, and hence to the action itself, in that

---

(New York: Abelard-Schuman, 1954); "Self-attitudes by Age, Sex, and Professional Training," in *Symbolic Interaction*, eds. G. Stone and H. Farberman, pp. 424–36; T. S. McPartland, "An Empirical Investigation of Self-Attitude," *American Sociological Review* 19 (February 1954), pp. 68–76; "Family Impact on Personality," in *Problems in Social Psychology*, eds. J. E. Hulett and R. Stagner (Urbana: University of Illinois Press, 1953); C. Addison Hickman and Manford Kuhn, *Individuals, Groups, and Economic Behavior* (New York: Dryden Press, 1956).

[11]For an interesting methodological critique of Kuhn's self-theory, see Charles W. Tucker, "Some Methodological Problems of Kuhn's Self Theory," *Sociological Quarterly* 7 (Winter 1966), pp. 345–58.

they are the anchoring points from which self-evaluations and other-evalua-
tions are made.[12]

The Nature of Interaction

Both Blumer and Kuhn have stressed the process of role taking, in which
humans mutually emit and interpret each other's gestures. From the information gained through this interpretation of gestures, actors are able to rehearse covertly various lines of activity and then emit those behaviors that can
allow cooperative and organized activity. As might be expected, however, Blumer and Kuhn disagree over the degree to which interactions are actively
constructed. Blumer's scheme, and that of most of the Chicago School, emphasizes the following points:

1. In addition to viewing each other as objects in an interaction situation, actors select and designate symbolically additional objects in any
interaction. (*a*) One of the most important of these objects is the self. On
the one hand, self can represent the transitory images that an actor derives
from interpreting the gestures of others; on the other, self can denote the
more enduring conceptions of one as an object that an actor brings to and
interjects into the interaction. (*b*) Another important class of objects are
the varying types of expectation structures—for example, norms and values—that may exist to guide interaction. (*c*) Finally, because of the human
organism's capacity to manipulate symbols, almost any other object—
whether another person, a set of standards, or a dimension of self—may be
inserted into the interaction.

2. It is because of the objects in interaction situations that actors have
various dispositions to act. Thus, in order to understand the potentials for
action among groups of individuals, it is necessary to understand the world
of objects that they have symbolically designated.

3. In terms of the particular cluster of objects and of the dispositions
to act that they imply, each actor arrives at a definition of the situation.
Such a definition serves as a general frame of reference within which the
consequences of specific lines of conduct are assessed. This process is
termed *mapping.*

4. The selection of a particular line of behavior involves complex symbolic processes. At a minimum, actors typically evaluate: (*a*) the demands
of others immediately present; (*b*) the self-images they derive from role
taking, not only with others in the situation but also with those not actually
present; (*c*) the normative expectations they perceive to exist in the situation; and (*d*) the dispositions to act toward any additional objects they
may inject symbolically into the interaction.

5. Once behavior is emitted, redefinition of the situation and perhaps
remapping of action may occur as the reactions of others are interpreted

---

[12]Hickman and Kuhn, *Individuals, Groups, and Economic Behavior,* pp. 224–25.

and as new objects are injected into, and old ones discarded from, the interaction.

Thus, by emphasizing the interpreting, evaluating, defining, and mapping processes, Blumer stressed the creative, constructed, and changeable nature of interaction. Rather than constituting the mere vehicle through which preexisting psychological, social, and cultural structures inexorably shape behavior, the symbolic nature of interaction assures that social, cultural, and psychological structures will be altered and changed through shifts in the definitions and behaviors of humans.

In contrast to Blumer's scheme, Kuhn stressed the power of the core self and the group context to constrain interaction. Much interaction is released rather than constructed, as interacting individuals follow the dictates of the self-attitudes and the expectations of their respective roles. Although Kuhn would certainly not have denied the potential for constructing and reconstructing interactions, he tended to view individuals as highly constrained in their behaviors by virtue of their core self and the requirements of their mutual situation.

The Nature of Social Organization

Symbolic interactionism tends to concentrate on the interactive *processes* by which humans form social relationships rather than the *end products* of interactions. Moreover, both Blumer and Kuhn, as well as other interactionists, have tended to emphasize the micro processes among individuals within small-group contexts. Blumer has consistently advocated a view of social organization as temporary and constantly changing, whereas Kuhn typically focused on the more structured aspects of social situations. Additionally, Blumer argued for a view of social structure as merely one of many objects that actors employ in their definition of a situation.

As Blumer has emphasized:

1. Since behavior is a reflection of the interpretive, evaluational, definitional, and mapping processes of individuals in various interaction contexts, social organization represents an active fitting together of action by those in interaction. Social organization must therefore be viewed as more of a process than a structure.

2. Social structure is an emergent phenomenon that is not reducible to the constituent actions of individuals, but it is difficult to understand patterns of social organization without recognizing that they represent an interlacing of the separate behaviors among individuals.

3. Although much interaction is repetitive and structured by clear-cut expectations and common definitions of the situation, its symbolic nature reveals the potential for new objects to be inserted or old ones altered and abandoned in a situation. The result is that reinterpretation, reevaluation, redefinition, and remapping of behaviors can always occur. Social structure must therefore be viewed as rife with potential for alteration and change.

4. Thus, patterns of social organization represent emergent phenomena that can serve as objects that define situations for actors. However, the very symbolic processes that give rise to and sustain these patterns can also operate to change and alter them.

In contrast to this emphasis, Kuhn usually sought to isolate the more structured features of situations. In conformity with what I term *structural role theory* in the next chapter, Kuhn saw social situations as constituting relatively stable networks of positions with attendant expectations or norms. Interactions often create such networks, but, once created, people conform to the expectations of those positions in which they have anchored their self-attitudes.

From this review of different assumptions of Blumer and Kuhn about the nature of individuals, interaction, and social organization, it is clear that Chicago School interactionists view individuals as potentially spontaneous, interaction as constantly in the process of change, and social organization as fluid and tenuous.[13] Iowa School interactionists are more prone to see individual personality and social organization as structured, with interactions being constrained by these structures.[14] These differences in assumptions have resulted in, or perhaps have been a reflection of, varying conceptions of how to investigate the social world and how to build theory.

The Nature of Methods

E. L. Quarantelli and Joseph Cooper have observed that Mead's ideas provide for contemporary interactionists a "frame of reference within which an observer can look at behavior rather than a specific set of hypotheses to be tested."[15] There can be little doubt that this statement is true. Yet there is much literature attempting to test some of the implications of Mead's ideas, especially those about self, with standard research protocols. This diversity in methodological approaches is underscored by the contrasting methodologies of Blumer and Kuhn. Indeed, the many students and students-of-students of these two figures tend to use Mead's ideas either as a sensitizing framework or as an inspiration for narrow research hypotheses.

---

[13]Prominent thinkers leaning toward Blumer's position include Anselm Strauss, Alfred Lindesmith, Tamotsu Shibutani, and Ralph Turner. For a critique of Blumer's use of Mead, see Clark McPhail and Cynthia Rexroat, "Mead vs. Blumer," *American Sociological Review* 44 (1979), pp. 449–67.

[14]Prominent Iowa School interactionists include Frank Miyamoto, Sanford Dornbusch, Simon Dinitz, Harry Dick, Sheldon Stryker, and Theodore Sarbin. For a list of studies by Kuhn's students, see Harold A. Mulford and Winfield W. Salisbury II, "Self-Conceptions in a General Population," *Sociological Quarterly* 5 (Winter 1964), pp. 35–46. Again, these individuals and those in note 12 would resist this classification, since none advocates as extreme a position as Kuhn or Blumer. Yet the tendency to follow either Kuhn's or Blumer's assumptions is evident in their work and in that of many others.

[15]E. L. Quarantelli and Joseph Cooper, "Self-conceptions and Others: A Further Test of Meadian Hypotheses," *Sociological Quarterly* 7 (Summer 1966), pp. 281–97.

Methodological approaches to studying the social world follow from thinkers' assumptions about what they can, or will, discover. The divergence of Blumer's and Kuhn's methodologies thus reflects their varying assumptions about the operation of symbolic processes. Ultimately, I think that their differences boil down to the question of causality. Whether or not events are viewed as the result of deterministic causes will influence the methodologies that are employed. Thus, to appreciate why Blumer and Kuhn diverge in their methodological approaches, it is necessary to see how their assumptions about individual interaction and social organization have become translated into a set of causal images that dictate different methodological approaches.

Diverging assumptions about causality For Blumer, the Meadian legacy challenges the utility of theoretical perspectives that underemphasize the internal symbolic processes of actors attempting to fit together their respective behaviors into an organized pattern.[16] Rather than being the result of system forces, societal needs, and structural mechanisms, social organization is the result of the mutual interpretations, evaluations, definitions, and mappings of individual actors. Thus the symbolic processes of individuals cannot be viewed as a neutral medium through which social forces operate, but instead these processes must be viewed as shaping the ways social patterns are formed, sustained, and changed.

Similarly, as I emphasized earlier, Blumer's approach challenges theoretical perspectives that view behavior as the mere releasing of propensities built into a structured personality. Just as patterns of social organization must be conceptualized as in a continual state of potential flux through the processes of interpretation, evaluation, definition, and mapping, so the human personality must also be viewed as a constantly unfolding process rather than a rigid structure from which behavior is mechanically released. By virtue of the fact that humans can make varying and changing symbolic indications to themselves, they are capable of altering and shifting behavior. Behavior is not so much released as it is constructed by actors making successive indications to themselves.

For Blumer, social structures and normative expectations are objects that must be interpreted and then used to define a situation and to map out the prospective behaviors that ultimately create and sustain social structures. From this perspective, overt behavior at one point generates self-images that serve as objects for individuals to symbolically map subsequent actions at another point in time. At the same time, existing personality traits, such as self-conceptions, self-esteem, and internalized needs, mediate each successive phase of interpretation of gestures, evaluation of self-images, definition of the situation, and mapping of diverse behaviors.

---

[16]See particularly Blumer's "Society as Symbolic Interaction," in *Human Behavior and Social Process*, ed. A. Rose, pp. 179–92 (reprinted in Blumer's *Symbolic Interactionism*); "Sociological Implications of the Thought of George Herbert Mead," in *Symbolic Interactionism*; and "The Methodological Position of Symbolic Interactionism," in *Symbolic Interactionism*.

In such a scheme, causality is difficult to discern. Social structures do not cause behavior, since they are only one class of objects inserted into an actor's seemingly unpredictable symbolic thinking. Similarly, self is only another object inserted into the definitional process. Action is thus created out of the potentially large number of objects that actors can insert into situations. In this vision of the social world, then, behavior does not reveal clear causes. Indeed, the variables influencing an individual's definition of the situation and action are of the actor's own choosing and apparently not subject to clear causal analysis.

In contrast to this seemingly *in*deterministic view of causality, Kuhn argued that the social world is deterministic. The apparent spontaneity and indeterminacy of human behavior are simply the result of insufficient knowledge about the variables influencing people's definitions and actions. For, if the social experiences of individuals can be discerned, it is possible to know what caused the emergence of their core self. With knowledge of the core self, of the expectations that have become internalized as a result of people's experiences, and of the particular expectations of a given situation, it is possible to understand and predict people's definitions of situations and their conduct. Naturally, this level of knowledge is impossible with current methodological techniques, but insight into deterministic causes of behavior and emergent patterns of social organization is possible *in principle*. For Kuhn, then, methodological strategies should therefore be directed at seeking the causes of behaviors.

Diverging methodological protocols   These differing assumptions about causality have shaped divergent methodological approaches within symbolic interactionism. The extremes of these diverging approaches are, I feel, best illustrated by Blumer's and Kuhn's respective approaches.

Blumer mounted a consistent and persistent line of attack on sociological theory and research.[17] His criticism questions the utility of current research procedures for unearthing the symbolic processes from which social structures and personality are built and sustained. Rather than letting the nature of the empirical world dictate the kinds of research strategies used in its study, Blumer and others have argued that present practices allow research strategies to determine what is to be studied:

> Instead of going to the empirical social world in the first and last instances, resort is made instead to a priori theoretical schemes, to sets of unverified concepts, and to canonized protocols of research procedure. These come to be the governing agents in dealing with the empirical social world, forcing research to serve their character and bending the empirical world to their premises.[18]

For Blumer, then, the fads of research protocol blind investigators and theorists to the real character of the social world. Such research and theoretical protocols force analysis away from the direct examination of the empirical

---

[17]See, in particular, Blumer's "Methodological Position."
[18]Ibid., p. 33.

world in favor of preconceived notions of what is true and how these truths should be studied. In contrast, the processes of symbolic interaction dictate that research methodologies should respect the character of empirical reality and adopt methodological procedures that encourage its direct and unbiased examination.[19]

To achieve this end, the research act itself must be viewed as a process of symbolic interaction in which researchers take the role of those individuals whom they are studying. To do such role taking effectively, researchers must study interaction with a set of concepts that, rather than prematurely structuring the social world for investigators, sensitize them to interactive processes. This approach would enable investigators to maintain the distinction between the concepts of science and those of the interacting individuals under study. In this way the interpretive and definitional processes of actors adjusting to one another in concrete situations can guide the refinement and eventual incorporation of scientific concepts into theoretical statements on the interactive processes that make up society.

Blumer has advocated a twofold process of research. First, it involves "exploration" in which researchers prepare to observe concrete situations and then revise their observations as new impressions of the situation arise. Second, exploration must be followed by a process of "inspection," whereby researchers use their observations to dictate how scientific concepts are to be refined and incorporated into more abstract and generic statements of relationships among concepts. In this dual research process, investigators must understand each actor's definition of the situation, the relationship of this definition to the objects actors perceive in the situation, and the relationship of objects to specific others, groups, and expectations in both the actor's immediate and remote social worlds.[20] In this way the research used to build the abstract concepts and propositions of sociological theory is connected to the empirical world of actors interpreting, evaluating, defining, and mapping the behaviors that sustain patterns of social organization.

A major area of controversy over Blumer's methodological position concerns the issue of operationalization of concepts. How is it possible to operationalize concepts, such as self and definition of the situation, so that different investigators at different times and in different contexts can study the same phenomena? I see Blumer as defining away the question of how the interpretative, evaluational, definitional, and mapping processes are to be studied in terms of clear-cut operational definitions. For him, however, such questions "show a profound misunderstanding of both scientific inquiry and symbolic

---

[19]For an interesting discussion of the contrasts between these research strategies, see Llewellyn Gross, "Theory Construction in Sociology: A Methodological Inquiry," in *Symposium on Sociological Theory*, ed. L. Gross (New York: Harper & Row, 1959), pp. 531–63.

[20]For an eloquent and reasoned argument in support of Blumer's position, see Norman K. Denzin, *The Research Act: A Theoretical Introduction to Sociological Methods* (Chicago: Aldine, 1970), pp. 185–218; and Norman K. Denzin, "Symbolic Interactionism and Ethnomethodology," in *Understanding Everyday Life: Toward a Reconstruction of Sociological Knowledge*, ed. J. D. Douglas (Chicago: Aldine, 1970), pp. 259–84.

interactionism."[21] The concepts and propositions of symbolic interactionism allow for the direct examination of the empirical world; therefore, "their value and their validity are to be determined in that examination and not in seeing how they fare when subjected to the alien criteria of an irrelevant methodology." According to Blumer, these "alien criteria" embrace a false set of assumptions about just how concepts should be attached to events in the empirical world. In general, these "false" assumptions posit that, for each abstract concept, a set of operational definitions should guide researchers, who then examine the empirical cases denoted by the operational definition.

Blumer has consistently emphasized current deficiencies in the attachment of sociological concepts to actual events in the empirical world:

> This ambiguous nature of concepts is the basic deficiency in social theory. It hinders us in coming to close grips with our empirical world, for we are not sure of what to grip. Our uncertainty as to what we are referring obstructs us from asking pertinent questions and setting relevant problems for research.[22]

Blumer argues that it is only through the methodological processes of exploration and inspection that concepts can be attached to the empirical. Rather than seeking a false sense of scientific security through rigid operational definitions, sociological theory must accept the fact that the attachment of abstract concepts to the empirical world must be an *ongoing process* of investigators exploring and inspecting events in the empirical world.

In sum, then, Blumer's presentation of the methodological position of symbolic interactionism questions the current research protocols. As an alternative, he advocates (*a*) the more frequent use of the exploration/inspection process, whereby researchers seek to understand the symbolic processes that shape interaction, and (*b*) the recognition that only through ongoing research activities can concepts remain attached to the fluid interaction processes of the empirical world. In turn, this methodological position has profound implications for the construction of theory in sociology, as we will see shortly.

In contrast to Blumer's position, Kuhn's vision of a deterministic world led him to emphasize the commonality of methods in all the sciences. The key task of methodology is to provide operational definitions of concepts so that their implications can be tested against the actual facts of social life. Most of Kuhn's career was thus devoted to taking the suggestive but vague concepts of Mead's framework and developing measures of them. He sought to find replicable measures of such concepts as self, social act, social object, and reference group.[23] His most famous measuring instrument—the Twenty Statements Test (TST)—can serve to illustrate his strategy. With the TST he sought to measure core self—the more enduring and basic attitudes that people have about themselves—by assessing people's answers to the question "What kind

---

[21]Blumer, "Methodological Position," p. 49.

[22]Herbert Blumer, "What Is Wrong with Social Theory?" *American Sociological Review* 19 (August 1954), pp. 146–58.

[23]Meltzer and Petras, "The Chicago and Iowa Schools."

of person are you?"[24] For example, the most common variant of the TST reads as follows:

> In the spaces below, please give 20 different answers to the question, "Who Am I?" Give these as if you were giving them to yourself, not to somebody else. Write fairly rapidly, for the time is limited.

Answers to such questions can be coded and scaled so that variations in people's self-conceptions can be linked to either prior social experiences or behaviors. Thus Kuhn sought to find empirical indicators of key concepts. These indicators would allow for recorded variations in one concept, such as self, to be linked to variations in other measurable concepts. In this way, Mead's legacy could be tested and used to build a theory of symbolic interactionism.

The Nature and Possibilities of Sociological Theory

Blumer's and Kuhn's assumptions about the nature of the individual, inter-action, social organization, causality, and methodology are reflected in their different visions of what theory is, should be, and can be. Again, Blumer and Kuhn stand at opposite poles, with most interactionists standing between these extremes and yet leaning toward one or the other.

Blumer's theory-building strategy   Blumer's assumptions, image of causal processes, and methodological position have all come to dictate a par-ticular conception of sociological theory. The recognition that sociological con-cepts do not come to grips with the empirical world is seen by Blumer as the result not only of inattention to actual events in the empirical world but also of the kind of world it is. The use of more definitive concepts referring to classes of precisely defined events is perhaps desirable in theory building, but it may be impossible, given the nature of the empirical world. Since this world is composed of constantly shifting processes of symbolic interaction among actors in various contextual situations, the use of concepts that rip only some of the actual ongoing events from this context will fail to capture the contextual nature of the social world. More important, the fact that social reality is ul-timately "constructed" from the symbolic processes among individuals assures that the actual instances denoted by concepts will shift and vary, thereby defying easy classification through rigid operational definitions.

These facts, Blumer argued, require the use of "sensitizing concepts," which, although lacking the precise specification of attributes and events of definitive concepts, do provide clues and suggestions about where to look for certain classes of phenomena. As such, sensitizing concepts offer a general sense of what is relevant and thereby allow investigators to approach flexibly a shift-ing empirical world and feel out and pick one's way in an unknown terrain. The use of this kind of concept does not necessarily reflect a lack of rigor in sociological theory but rather a recognition that, if "our empirical world pre-

---

[24]For an important critique of this methodology, see Tucker, "Some Methodological Problems."

sents itself in the form of distinctive and unique happenings or situations and if we seek through the direct study of this world to establish classes of objects and relations between classes, we are . . . forced to work with sensitizing concepts."[25]

The nature of the empirical world may preclude the development of definitive concepts, but sensitizing concepts can be improved and refined by flexibly approaching empirical situations denoted by sensitizing concepts and then by assessing how actual events stack up against the concepts. Although the lack of fixed benchmarks and definitions makes this task more difficult for sensitizing concepts than for definitive ones, the progressive refinement of sensitizing concepts is possible through "careful and imaginative study of the stubborn world to which such concepts are addressed."[26] Furthermore, sensitizing concepts formulated in this way can be communicated and used to build sociological theory; and, although formal definitions and rigid classifications are not appropriate, sensitizing concepts can be explicitly communicated through descriptions and illustrations of the events to which they pertain.

In sum, the ongoing refinement, formulation, and communication of sensitizing concepts must inevitably be the building blocks of sociological theory. With careful formulation they can be incorporated into provisional theoretical statements that specify the conditions under which various types of interaction are likely to occur. In this way the concepts of theory will recognize the shifting nature of the social world and thereby provide a more accurate set of statements about social organization.

The nature of the social world and the type of theory it dictates have profound implications for just how such theoretical statements are to be constructed and organized into theoretical formats. Blumer's emphasis on the constructed nature of reality and on the types of concepts that this fact necessitates has led him to emphasize inductive theory construction. In inductive theory, generic propositions are abstracted from observations of concrete interaction situations. This emphasis on induction is considered desirable, since current attempts at deductive theorizing in sociology usually do not involve rigorous derivations of propositions from each other or a scrupulous search for the negative empirical cases that would refute propositions.[27] These failings, Blumer contends, assure that deductive sociological theory will remain unconnected to the events of the empirical world and, hence, unable to correct for errors in its theoretical statements. Coupled with the tendency for fads of research protocol to dictate research problems and methods used to investigate them, it appears unlikely that deductive theory and the research it inspires can unearth those processes that would confirm or refute its generic statements. In the wake of this theoretical impasse, then, it is crucial that sociological theorizing refamiliarize itself with the actual events of the empirical world. As

---

[25]Blumer, "What Is Wrong with Social Theory?", p. 150.

[26]Ibid.

[27]Blumer, "Methodological Position."

Blumer argued: "No theorizing, however ingenious, and no observance of scientific protocol, however meticulous, are substitutes for developing familiarity with what is actually going on in the sphere of life under study."[28] Without such inductive familiarity, sociological theory will remain a self-fulfilling set of theoretical prophecies bearing little relationship to the phenomena it is supposed to explain.

Kuhn's theory-building strategy   Kuhn advocated a more deductive format for sociological theorizing than Blumer. Although his own work does not reveal great deductive rigor, he held that subsumption of lower-order propositions under more general principles is the most appropriate way to build theory. He visualized his "self-theory" as one step in the building of theory. By developing general statements on how self-attitudes emerge and shape social action, it would be possible to understand and predict human behavior. Moreover, less general propositions about aspects of self could be subsumed under general propositions about processes of symbolic interaction.

Although Kuhn recognized that symbolic interaction theory had become partitioned into many suborientations, including his own self-theory, he held a vision of a unified body of theoretical principles. As he concluded in an assessment of interactionist theory: "I would see in the next 25 years of symbolic interaction theory an accelerated development of research techniques, on the one hand, and a coalescing of most of the separate subtheories."[29]

Thus the ultimate goal of theory is consolidation or "coalescing" of testable lower-order theories under a general set of symbolic interactionist principles. Self-theory, as developed by Kuhn, would be but one set of derivations from a more general system of interactionist principles. In contrast to Blumer, then, theory for Kuhn was ultimately to form a unified system from which specific propositions about different aspects or phases of symbolic interactionism could be derived.

The Chicago and Iowa Schools: An Overview

In Table 19-1, I have summarized the points of convergence and divergence of the Chicago and Iowa Schools, or of Blumer and Kuhn. The two right columns explore a number of issues over which there is disagreement among interactionists. But I think it important to place these disagreements within the context of the points where symbolic interactionists agree, as I have done in the left column.

I should caution again that distinctions between a Chicago and Iowa school are hazardous when examining the work of a particular symbolic interactionist. The distinction denotes only *tendencies* to view humans, interaction, social organization, methods, and theory in a particular way. Few symbolic inter-

---

[28]Ibid., p. 39.
[29]Kuhn, "Major Trends," p. 80.

TABLE 19–1   Convergence and Divergence in the Chicago and Iowa Schools of Symbolic Interactionism

Theoretical Issues	Convergence of Schools
The nature of humans	Humans create and use symbols to denote aspects of the world around them.
	What makes humans unique are their symbolic capacities.
	Humans are capable of symbolically denoting and involving objects, which can then serve to shape their definitions of social situations and, hence, their actions.
	Humans are capable of self-reflection and evaluation. They see themselves as objects in most social situations.
The nature of interaction	Interaction is dependent upon people's capacities to emit and interpret gestures.
	Role taking is the key mechanism of interaction, for it enables actors to view the other's perspective, as well as that of others and groups not physically present.
	Role taking and mind operate together by allowing actors to use the perspectives of others and groups as a basis for their deliberations, or definitions of situations, before acting. In this way, people can adjust their responses to each other and to social situations.
The nature of social organization	Social structure is created, maintained, and changed by processes of symbolic interaction.
	It is not possible to understand patterns of social organization—even the most elaborate—without knowledge of the symbolic processes among individuals who ultimately make up this pattern.
The nature of sociological methods	Sociological methods must focus on the processes by which people define situations and select courses of action.
	Methods must focus on individual persons.
The nature of sociological theory	Theory must be about processes of interaction and seek to isolate out the conditions under which general types of behaviors and interactions are likely to occur.

TABLE 19-1 (*continued*)

Chicago School	Iowa School
Humans with minds can introject any object into a situation.	Humans with minds can define situations, but there tends to be consistency in terms of the objects that they introject into situations.
Although self is an important object, it is not the only object.	Self is the most important object in the definition of a situation.
Humans weigh, assess, and map courses of action before action, but humans can potentially alter their definitions and actions.	Humans weigh, assess, and map courses of action, but they do so through the prism of their core self and the groups in which this self is anchored.
Interaction is a constant process of role taking with others and groups.	Interaction is dependent upon the process of role taking.
Others and groups thus become objects that are involved in people's definitions of situations.	The expectations of others and norms of the situation are important considerations in arriving at definitions of situations.
Self is another important object that enters into people's definitions.	People's core self is the most important consideration and constraint on interaction.
People's definitions of situations involve weighing and assessing objects and then mapping courses of action.	
Interaction involves constantly shifting definitions and changing patterns of action and interaction.	Interaction most often involves actions that conform to situational expectations as mitigated by the requirements of the core self.
Social structure is constructed by actors adjusting their responses to each other.	Social structures are composed of networks of positions with attendant expectations or norms.
Social structure is one of many objects that actors introject into their definitions of situations.	Although symbolic interactions create and change structures, once these structures are created they operate to constrain interaction.
Social structure is subject to constant realignments as actors' definitions and behaviors change, forcing new adjustments from others.	Social structures are thus relatively stable, especially when people's core self is invested in particular networks of positions.
Sociological methods must seek to penetrate the actors' mental world and see how they construct courses of action.	Sociological methods must seek to measure with reliable instruments actors' symbolic processes.
Researchers must be attuned to the multiple, varied, ever-shifting, and often indeterminate influences on definitions of situations and actions.	Research should be directed toward defining and measuring those variables that causally influence behaviors.
Research must therefore use observational, biographical, and unstructured interview techniques if it is to penetrate people's definitional processes and take account of changes in these processes.	Research must therefore use structured measuring instruments, such as questionnaires, to get reliable and valid measures of key variables.
Only sensitizing concepts are possible in sociology.	Sociology can develop precisely defined concepts with clear empirical measures.
Deductive theory is thus not possible in sociology.	Theory can thus be deductive, with a limited number of general propositions subsuming lower-order propositions and empirical generalizations on specific phases of symbolic interaction.
At best, theory can offer general and tentative descriptions and interpretations of behaviors and patterns of interaction.	Theory can offer abstract explanations that can allow for predictions of behavior and interaction.

actionists follow literally either Blumer's or Kuhn's positions. These positions merely represent the boundaries within which symbolic interactionists work.

SYMBOLIC INTERACTIONISM: A CONCLUDING COMMENT

Blumer's vision of symbolic interactionism advocates a clear-cut strategy for building sociological theory. The emphasis on the interpretive, evaluative, definitional, and mapping processes of actors has come to dictate that it is only through induction from these processes that sociological theory can be built. Further, the ever-shifting nature of these symbolic processes necessitates that the concepts of sociological theory be sensitizing rather than definitive, with the result that deductive theorizing should be replaced by an inductive approach. Thus, whether as a preferred strategy or as a logical necessity, the Blumer interactionist strategy is to induce generic statements, employing sensitizing concepts, from the ongoing symbolic processes of individuals in concrete interaction situations.

Such a strategy is likely to keep theorizing attuned to the processual nature of the social world. However, I feel that this approach (and, to a lesser extent, Kuhn's) has not been able to link conceptually the processes of symbolic interaction to the formation of different patterns of social organization. Furthermore, the utility of induction from the symbolic exchanges among individuals for the analysis of interaction among more macro, collective social units has yet to be demonstrated. Unless these problems can be resolved, I do not think it wise to follow exclusively the strategy of Blumer and others of his persuasion. Until symbolic interactionists of all persuasions demonstrate in a more compelling manner than is currently the case that the inductive approach utilizing sensitizing concepts can account for more complex forms of social organization, pursuit of its strategy will preclude theorizing about much of the social world.

Yet I feel that both Blumer's and Kuhn's strategies call attention to some important substantive and theoretical issues that are often ignored in social theory. First, it is necessary that sociological theorizing be more willing to undertake the difficult task of linking conceptually structural categories to classes of social processes that underlie these categories. For this task, symbolic interactionism has provided a wealth of suggestive concepts. Second, macrosociological theorizing has traditionally remained detached from the processes of the social world it attempts to describe. Much of the detachment stems from a failure to define concepts clearly or provide operational clues about what processes in the empirical world they denote. To the extent that symbolic interactionist concepts can supplement such theorizing, they will potentially provide a bridge to actual empirical processes and thereby help attach sociological theory to the events it purports to explain.

Symbolic interactionism has great potential for correcting the past inadequacies of sociological theory, but it has yet to demonstrate exactly *how* this corrective influence is to be exerted. At present, symbolic interactionism seems capable of analyzing micro social patterns and their impact on personality, particularly the self. Although interactionism has provided important insights in the study of socialization, deviance, and micro social processes, it has yet to demonstrate any great potential for analyzing complex, macro social patterns. At best, it can provide in its present form a supplement to macroanalysis by giving researchers a framework and measuring instruments to analyze micro processes within macro social events.

I find that much of symbolic interaction, especially Blumer's advocacy, consists of gallant assertions that "society is symbolic interaction" but without indicating what types of emergent structures are created, sustained, and changed by what types of interaction in what types of contexts. Much like the critics' allegations about Parsons' social system, Dahrendorf's imperatively coordinated association, Homans' institutional piles, or Blau's organized collectivities, social structural phenomena emerge somewhat mysteriously and are then sustained or changed by vague references to interactive processes. The vagueness of the links between the interaction process and its social structural products leaves symbolic interactionism with a legacy of assertions but little in the way of carefully documented statements about how, when, where, and with what probability interaction processes operate to create, sustain, and change varying patterns of social organization.[30] It is to this goal that interactionist theory must redirect its efforts.

---

[30]Probably the most significant effort to overcome this problem is by Tamotsu Shibutani, *Social Processes: An Introduction to Sociology* (Berkeley: University of California Press, 1986).

Structural Role Theory

One of the most ambiguous concepts in sociology is "role." For example, is a role simply overt behavior? Is it a conception of appropriate behavior? Is it normatively expected behavior? Is it behavior enacted by virtue of incumbency in a status position? Or is it all of these? The lack of a definitive answer to these questions has led some to advocate that sociology abandon the concept.[1] I think that this solution is far too extreme. Instead, we must try to clarify our conceptualization of roles, and so, in this and the next chapter, I will present the range of approaches in the analysis of roles.

My sense is that there is a structural approach to roles at one extreme and a more processual strategy at the other.[2] This diversity mirrors the range of approaches within interactionism in general. We might view this range as a continuum and use the analogy of a play and a game to illustrate its dimensions.[3] At one pole of this continuum, individuals are seen as players in the theater, whereas at the other end, players are considered to be participants in a pick-up game. When human action is seen as occurring in a theater, interaction is likely to be viewed as highly structured by the script, director, other actors, and the audience. When conceptualized as a game, interaction is more likely to be seen as less structured and as influenced by the wide range of tactics available to participants.

In this chapter I will examine the more structural approach to role theory.[4] I will not examine any one scholar's work because, in truth, there is no one

---

[1]See, for example, Anthony Giddens, *The Constitution of Society* (Berkeley: University of California Press, 1984).

[2]For two recent efforts to play down this distinction, see Warren Handel, "Normative Expectations and the Emergence of Meaning as Solutions to Problems: Convergence of Structural and Interactionist Views," *American Journal of Sociology* 84 (1979), pp. 855–81; and Jerold Heiss, "Social Roles," in *Social Psychology: Sociological Perspectives*, eds. M. Rosenberg and R. H. Turner (New York: Basic Books, 1981), pp. 94–132.

[3]Bernard Farber, "A Research Model: Family Crisis and Games Strategy," in *Kinship and Family Organization*, ed. B. Farber (New York: John Wiley, 1966), pp. 430–34. Walter Wallace (*Sociological Theory* [Chicago: Aldine, 1969], pp. 34–35) has more recently chosen this same analogy to describe symbolic interactionism.

[4]For the first early analytical statements, see Jacob Moreno, *Who Shall Survive*, rev. ed. (New

theoretical approach. Rather, there is a large number of research-oriented scholars who have contributed collectively to a structural conceptualization of roles. Indeed, one of the major problems of the structural approach is the lack of theoretical synthesis. As a result, my efforts in this chapter will revolve around framing the issues that need to be theoretically synthesized.

THE CONCEPTUAL THRUST OF STRUCTURAL ROLE THEORY

The thrust of the structural role perspective, as it flowed from a mixture of Park's, Simmel's, Moreno's, Linton's, and Mead's insights, has often been captured by quoting a famous passage from Shakespeare's *As You Like It*:

> All the world's a stage
> and all the men and women merely players:
> They have their exits and their entrances;
> And one man in his time plays many parts. (Act II, scene vii)

The analogy is then drawn between the players on the stage and the actors of society.[5] Just as players have a clearly defined part to play, so actors in society occupy clear positions; just as players must follow a written script, so actors in society must follow norms; just as players must obey the orders of a director, so actors in society must conform to the dictates of those with power or those of importance; just as players must react to each other's performance on the stage, so members of society must mutually adjust their responses to one another; just as players respond to the audience, so actors in society take the role of various audiences or "generalized others"; and just as players with varying abilities and capacities bring to each role their unique interpretations, so actors with varying self-conceptions and role-playing skills have their own styles of interaction.

Despite its simplicity, I think that the analogy is appropriate. As we will come to appreciate, the role theoretic perspective supports the thrust of Shakespeare's passage. At the outset, however, I should caution again that, despite its pervasiveness in sociology, role analysis is far from being a well-articulated perspective.

IMAGES OF SOCIETY AND THE INDIVIDUAL

Shakespeare's passage provides, I feel, the general outline of what role theorists assume about the social world. In the concept of *stage* are assumptions about

---

York: Beacon House, 1953; originally published in 1934); and Ralph Linton, *The Study of Man* (New York: Appleton-Century-Crofts, 1936).

[5]For an example of this form of analogizing, see Bruce J. Biddle and Edwin Thomas, *Role Theory: Concepts and Research* (New York: John Wiley, 1966), pp. 3–4. For the best-known dramaturgical model, see Erving Goffman, *The Presentation of Self in Everyday Life* (Garden City, NY: Doubleday, 1959). In fact, I will describe Goffman's scheme as "dramaturgy" in Chapter 22.

the nature of social organization; in the concept of *players* are implicit assumptions about the nature of the individual; and in the vision of men and women as "merely players" who have "their exits and their entrances" are a series of assumptions about the relationship of individuals to patterns of social organization.

The Nature of Social Organization

For structural role theorists, the social world is viewed as a network of variously interrelated *positions*, or statuses, within which individuals enact roles.[6] For each position, as well as for groups and classes of positions, various kinds of expectations about how incumbents are to behave can be discerned. Thus, social organization is ultimately composed of various networks of statuses and expectations.[7]

Statuses are typically analyzed in terms of how they are interrelated to one another to form various types of social units. In terms of variables such as size, degree of differentiation, and complexity of interrelatedness, status networks are classified into forms ranging from various types of groups to larger forms of collective organization. There has been some analysis of their formal properties, but status networks are rarely analyzed independently of the types of expectations attendant upon them. Part of the reason for this close relation between form and content is that the types of expectations that typify particular networks of positions represent one of their defining characteristics. It is usually assumed that the behavior emitted by incumbents is not an exclusive function of the structure of positions, per se, but also of the kinds of expectations that inhere in these positions.

The range of expectations denoted by role theoretic concepts is diverse. Pursuing the dramaturgical analogy to a play, I see three general classes of expectations as typifying structural role theory's vision of the world: (*a*) expectations from the "script," (*b*) expectations from other "players," and (*c*) expectations from the "audience."

Expectations from the script   Much of social reality can be considered to read like a script in that, for many positions, there are norms specifying just how an individual ought to behave. The degree to which activity is regulated by norms varies under different conditions; thus one of the questions to be resolved by role theory concerns the conditions under which norms vary in

---

[6]For a recent effort to bring together structurally oriented conceptions of role, see Bruce Biddle, *Role Theory: Expectations, Identities, and Behaviors* (New York: Academic Press, 1979).

[7]See, for example, Jacob L. Moreno, "Contributions of Sociometry to Research Methodology in Sociology," *American Sociological Review* 12 (June 1947), pp. 287–92; Jacob L. Moreno, ed., *The Sociometry Reader* (Glencoe, IL: Free Press, 1960); Oscar A. Oeser and Frank Harary, "Role Structures: A Description in Terms of Graph Theory," *Human Relations* 15 (May 1962), pp. 89–109; and Darwin Cartwright and Frank Harary, "Structural Balance: A Generalization of Heider's Theory," *Psychological Review* 63 (September 1956), pp. 277–93.

terms of such variables as scope, power, efficacy, specificity, clarity, and degree of conflict with one another.

Expectations from other players   In addition to the normative structuring of behavior and social relations, role theory also focuses on the demands emitted by the other players in an interaction situation. Such demands, interpreted through role taking of others' gestures, constitute one of the most important forces shaping human conduct.

Expectations from the audience   A final source of expectations comes from the audiences of individuals occupying statuses. These audiences can be real or imagined, constitute an actual group or a social category, and involve membership or simply a desire to be a member. It is necessary only that the expectations imputed by individuals to such variously conceived audiences be used to guide conduct. As such, the audiences comprise a frame of reference, or reference group, that circumscribes the behavior of actors in various statuses.[8]

In sum, then, structural role theory views the social world as organized in terms of expectations from a variety of sources, whether the script, other players, or various audiences. Just which types of expectations are attendant upon a given status, or network of positions, is one of the important empirical questions that follows from this assumption.

Although structural role theory implicitly assumes that virtually the entire social spectrum is structured in terms of statuses and expectations, rarely is this whole spectrum studied. In fact, I find that role analysis usually concentrates on restricted status networks, such as the types of expectations evident in more micro social units like small groups. Such an emphasis can be seen as representing a strategy for analytically coping with the incredible complexity of the entire status network and attendant expectations of a society or of some of its larger units. In this delimitation of inquiry, however, I believe that there is an implicit view of the social order as structured only by certain basic kinds of micro groups and organizations. Larger social phenomena, such as social classes or nation/states and relations among them, are less relevant because there is a presumption that these phenomena can be understood in terms of their constituent groups and organizations.

This emphasis on the microstructures of society is perhaps inevitable in light of the fact that role theory ultimately attempts to account for types of

---

[8]Mead's concept of the generalized other anticipated this analytical concern with reference groups. For some of the important conceptual distinctions in the theory of reference-group behavior, see Robert K. Merton, "Continuities in the Theory of Reference Groups and Social Structure," *Social Theory and Social Structure* (New York: Free Press, 1968) pp. 225–80; Tamotsu Shibutani, "Reference Groups as Perspectives," *American Journal of Sociology* 60 (May 1955), pp. 562–69; Harold H. Kelley, "Two Functions of Reference Groups," in *Readings in Social Psychology*, ed. G. E. Swanson et al. (New York: Henry Holt, 1958); Ralph H. Turner, "Role-Taking, Role Standpoint, and Reference Group Behavior," *American Journal of Sociology* 61 (January 1956), pp. 316–28; and Herbert Hyman and Eleanor Singer, eds., *Readings in Reference Group Behavior* (New York: Free Press, 1968).

role performances by individuals. Although macro patterns of social organization are viewed as providing much of the "structure" for these performances, society cannot be conceptualized independently of its individual incumbents and their performances.

The Nature of the Individual

Structural role theorists generally conceptualize the individual in terms of two basic attributes: (a) self-related characteristics and (b) role-playing skills and capacities. The self-related concepts of role theory are diverse, but I see them as concerned with the impact of self-conceptions on people's interpretation of the expectations guiding conduct in a particular status. Role-playing skills denote those capacities of individuals to perceive various types of expectations and then, with varying degrees of competence and with different role-playing styles, to follow a selected set of expectations. These two attributes—self and role-playing skills—are obviously interrelated, since self-conceptions will mediate the perception of expectations and the way roles are enacted, while role-playing skills will determine the kinds of self-images derived from an interaction situation and those involved in the construction of a stable self-conception.

This view of the individual parallels Mead's portrayal of mind and self. For Mead and contemporary role theorists, the capacity to take roles and mediate self-images through a stable self-conception is what distinguishes the human organism. Although this conceptualization of self and role-playing capacities offers the potential for visualizing unique interpretations of expectations and for analyzing spontaneous forms of role playing, the opposite set of assumptions is more often connoted in structural role theory. That is, emphasis is on the ways that individuals conform to what is expected of them by virtue of occupying a particular status. The degree and form of conformity are usually seen as the result of a variety of internal processes operating on individuals. Depending on the interactive situation, these internal processes are conceptualized in terms of variables such as (1) the degree to which expectations have been internalized as part of an individual's need structure,[9] (2) the extent to which negative or positive sanctions are perceived by the individual to accompany a particular set of expectations,[10] (3) the degree to which expectations are used as a yardstick for self-evaluation,[11] and (4) the extent to which expectations represent either interpretations of others' actual responses or merely

---

[9]For example, Talcott Parsons, *The Social System* (New York: Free Press, 1951), pp. 1–94; and William J. Goode, "Norm Commitment and Conformity to Role-Status Obligations," *American Journal of Sociology* 66 (November 1960), pp. 246–58.

[10]For example, B. F. Skinner, *Science and Human Behavior* (New York: Macmillan, 1953), pp. 313–55, 403–19; Biddle and Thomas, *Role Theory*, pp. 27–28; Marvin E. Shaw and Philip R. Costanzo, *Theories of Social Psychology* (New York: McGraw-Hill, 1970), pp. 332–33.

[11]Harold Kelley, "Two Functions of Reference Groups"; Ralph H. Turner, "Role-Taking, Role Standpoint" and "Self and Other in Moral Judgement," *American Sociological Review* 19 (June 1954), pp. 254–63.

anticipations of their potential responses.[12] Just which combination of these internal processes operates in a particular interaction situation depends upon the nature of the statuses and attendant expectations. Although this complex interactive process has yet to be codified into an inventory of theoretical statements, it remains one of the principal goals of role theory.

From this conceptualization, the individual is assumed to be not so much a creative role entrepreneur who tries to change and alter social structure through varied and unique responses as a pragmatic performer who attempts to cope with and adjust to the variety of expectations inhering in social structure. These implicit assumptions about the nature of the individual are consistent with Mead's concern with the adaptation and adjustment of the human organism to society, but they clearly underemphasize the creative consequences of mind and self for the construction and reconstruction of society. Thus, structural role theory has tended to expand conceptually upon only part of the Meadian legacy. This tendency assures some degree of assumptive one-sidedness, which is understandable in light of the role theorists' concern for sorting out only certain types of dynamic interrelationships between society and the individual.

The Articulation between the Individual and Society

The point of articulation between society and the individual is denoted by the concept of role and involves individuals who are incumbent in statuses and who employ self and role-playing capacities to adjust to various types of expectations. Despite agreement over these general features of role, current conceptualizations differ.[13] Depending upon which component of role is emphasized, I see three alternative conceptualizations.[14]

Prescribed roles   When conceptual emphasis is placed upon the expectations on individuals in status positions, then the social world is assumed to be composed of relatively clear-cut prescriptions. The individual's self and role-playing skills are seen as operating to meet such prescriptions, with the result that analytical emphasis is drawn to the degree of conformity to the demands of a particular status.

Subjective roles   Since all expectations are mediated through the prism of self, they are subject to interpretations by individuals in statuses. When

---

[12]R. Turner, "Role-Taking, Role Standpoint."

[13]For summaries of the various uses of the concept, see Lionel J. Neiman and James W. Hughes, "The Problem of the Concept of Role—A Re-survey of the Literature," *Social Forces* 60 (December 1981), pp. 141–49; Ragnar Rommetveit, *Social Norms and Roles: Explorations in the Psychology of Enduring Social Pressures* (Minneapolis: University of Minnesota Press, 1955); Biddle and Thomas, *Role Theory*; Shaw and Costanzo, *Theories of Social Psychology*, pp. 334–38; and Morton Deutsch and Robert M. Krauss, *Theories in Social Psychology* (New York: Basic Books), pp. 173–77.

[14]Deutsch and Krauss, *Theories in Social Psychology*, p. 175; Daniel J. Levinson, "Role, Personality, and Social Structure in the Organizational Setting," *Journal of Abnormal and Social Psychology* 58 (March 1959), pp. 170–80.

conceptual emphasis falls upon the perceptions and interpretations of expectations, the social world is seen as structured in terms of individuals' subjective assessments of the interaction situation. As a consequence, the interpersonal style of individuals who interpret and then adjust to expectations is given primary consideration.

Enacted roles   Ultimately, expectations and the subjective assessment by individuals of these expectations are manifested in behavior. When conceptual priority is given to overt behavior, the social world is viewed as a network of interrelated behaviors. The more stress that is placed upon overt role enactment, the less evident will be analyses of either expectations or individual interpretations of them.

When viewed separately from one another, I think it is clear that these three conceptual notions are inadequate. For overt human behavior obviously involves a subjective assessment of various types of expectations. In fact, a review of the research and theoretical literature on role theory demonstrates that the prescriptive, subjective, or enacted components of roles receive *varying degrees* of emphasis and that theoretical efforts usually deal with the complex causal relationships among these components.

Perhaps more than any conceptual perspective, role theory portrays images of causality rather than an explicit set of causal linkages. Part of the reason for this vagueness stems from the fact that the label "role theory" embraces a wide number of specific perspectives in a variety of substantive areas. Despite these qualifications, I think that role theorists have tended to develop concepts that denote specific interaction processes without revealing the precise ways these concepts are interrelated. To the extent that I can bring into focus role theory's causal images, they appear to emphasize the deterministic consequences of social structure on interaction. However, rarely are larger, more inclusive units of culture and structure included in this causal analysis. Rather, concern tends to be with the impact of specific norms, others, and reference groups associated with particular clusters of status positions on (*a*) self-interpretations and evaluations, (*b*) role-playing capacities, or (*c*) overt role behavior. Although there is considerable variability in the role theoretic literature, my sense is that self-interpretations and evaluations are usually viewed as having a deterministic impact on role-playing capacities, which then circumscribe overt role behavior. I have depicted these causal images in Figure 20-1.

As I have portrayed in the middle section of Figure 20-1, the specific causal images evident in the literature are more complex than the general portrayal at the top of the figure. Expectations are still viewed as determinative, but role theorists have frequently emphasized the reciprocal nature of causal processes. That is, certain stages in the causal sequence are seen as feeding back and affecting the subsequent causal relations among analytical units, which are portrayed in the middle section of Figure 20-1. Although there are numerous potential interconnections among these units, I find that structural role theory has emphasized only a few of these causal linkages, as is designated by the

FIGURE 20–1 The Causal Imagery of Role Theory

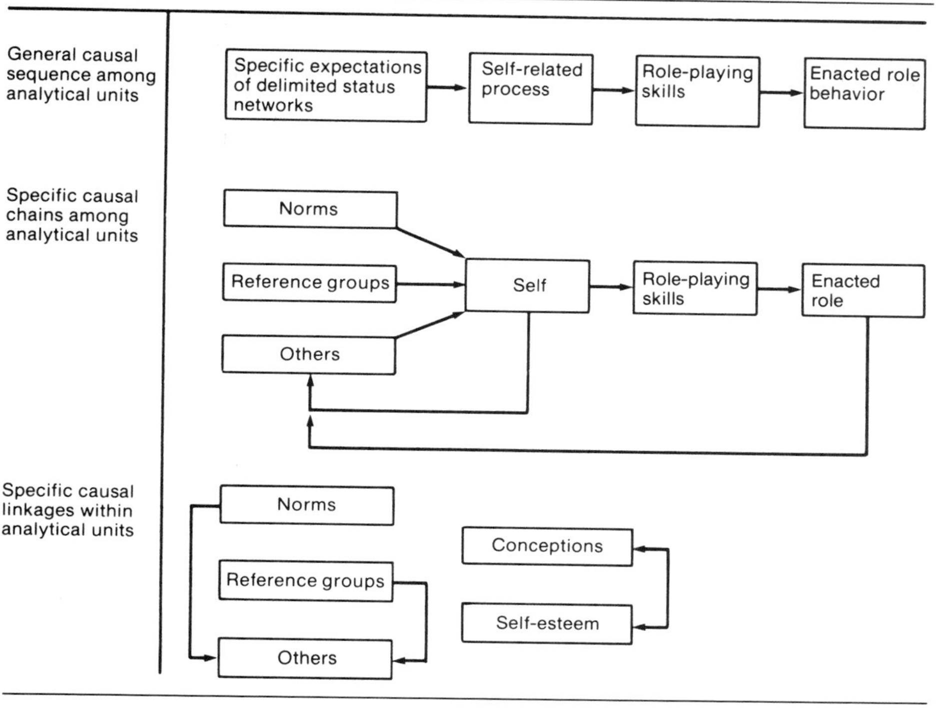

arrows in the figure. Furthermore, to the extent that specific units of analysis are conceptualized for expectations, self-variables, role-playing skills, and overt behavior, only some of the causal interconnections among these units are explored, as I have indicated at the bottom of Figure 20–1.

With respect to the interrelations among expectations, self, role-playing skills, and overt behavior, I see structural role theory as being primarily concerned with conceptualizing how different types of expectations are mediated by self-interpretations and evaluations and then circumscribed by role-playing skills in a way that a given style of role performance is evident. This style is then typically analyzed in terms of its degree of conformity to expectations.[15] However, at each stage in this sequence, certain feedback processes are also emphasized so that the degree of significance of norms, others, or reference groups for the maintenance of individuals' self-conceptions is considered critical in influencing which expectations are most likely to receive the most attention. For this causal nexus, emphasis is on the degree of embeddedness of

---

[15]The study of deviance, socialization, and role playing in complex organizations and small groups has profited from this form of analysis.

self in certain groups,[16] the degree of intimacy with specific others,[17] and the degree of commitment to, or internalization of, certain norms.[18] Another prominent feedback process that has received considerable attention is the impact of overt behavior at one point in time on the expectations of others as they shape the individual's self-conception and subsequent role behavior at another point in time. In this context the childhood and adult socialization of the individual and the emergence of self have been extensively studied,[19] as has the emergence of deviant behavior.[20]

With respect to interrelations within the analytical units of the overall causal sequence, I think that the arrows at the bottom of Figure 20-1 portray the current theoretical emphasis of the literature. In regard to the interrelations among types of expectations, analytical attention appears to be on how specific others personify group norms or the standards of reference groups. In turn, these significant others are often viewed as deterministically linking the self-interpretations and evaluations of an individual to either the norms of a group or the standards of a reference group.[21] With respect to the relations among the components of self, analysis appears to have followed the lead of William James[22] by focusing on the connections between the self-esteem and the self-conception of an individual.[23] In turn, the interaction between self-esteem and other components of self is viewed as the result of reactions from various others who affect the self-images of the individual. Finally, all these components interact in complex ways to shape the individual's overt behavior.

Thus, looking back at Figure 20-1, it is evident that only some processes have been extensively studied. Relatively little theoretical attention has been

---

[16]For example, see Norman Denzin, "Symbolic Interactionism and Ethnomethodology," in *Understanding Everyday Life: Toward a Reconstruction of Sociological Knowledge*, ed. J. D. Douglas (Chicago: Aldine, 1970), pp. 259–84. For the most thorough set of such studies in this area, see Sarbin's examinations of the intensity of self-involvement and role-playing behavior: Theodore R. Sarbin, "Role Theory," in *Handbook of Social Psychology*, ed. G. Lindzey, vol. 1 (Reading, MA: Addison-Wesley, 1954), pp. 223–58; Theodore R. Sarbin and Norman L. Farberow, "Contributions to Role-Taking Theory: A Clinical Study of Self and Role," *Journal of Abnormal and Social Psychology* 47 (January 1952), pp. 117–25; and Theodore R. Sarbin and Bernard G. Rosenberg, "Contributions to Role-Taking Theory," *Journal of Social Psychology* 42 (August 1955), pp. 71–81.

[17]See, for example, Tamotsu Shibutani, *Society and Personality* (Englewood Cliffs, NJ: Prentice-Hall, 1961), pp. 367–403.

[18]John Finley Scott, *Internalization of Norms: A Sociological Theory of Moral Commitment* (Englewood Cliffs, NJ: Prentice-Hall, 1971), pp. 127–215; William Goode, "Norm Commitment and Conformity"; B. F. Skinner, *Science and Human Behavior*.

[19]For example, see Anselm Strauss, *Mirrors and Masks* (Glencoe, IL: Free Press, 1959), pp. 100–18; Shibutani, *Society and Personality*, pp. 471–596; Orville G. Brim and Stanton Wheeler, *Socialization after Childhood* (New York: John Wiley, 1966).

[20]See Edwin Lemert, *Social Pathology* (New York: McGraw-Hill, 1951); Thomas Scheff, "The Role of the Mentally Ill and the Dynamics of Mental Disorder: A Research Framework," *Sociometry* 26 (December 1963), pp. 436–53; Howard Becker, *Outsiders: Studies in the Sociology of Deviance* (New York: Free Press, 1963).

[21]R. Turner, "Role-Taking, Role Standpoint"; Shibutani, *Society and Personality*, pp. 249–80.

[22]William James, *Principles of Psychology*, 2 vols. (New York: Henry Holt, 1980).

[23]See Shibutani, *Society and Personality*, pp. 433–46.

drawn, I feel, to the following potential connections indicated in the figure: (*a*) broader social and cultural structure and specific patterns of interaction, (*b*) enacted role behaviors and their effect on role-playing capacities, (*c*) these role-playing capacities and self, and (*d*) enacted roles and the self-assessments that occur *independently* of role taking with specific others or groups.[24] Rather, concern has been focused on the relations between self and expectations as they affect, and are affected by, enacted roles.

PROBLEMS AND ISSUES IN BUILDING ROLE THEORY

Constructing Propositions

To this point, I see role theoretic concepts as providing only a means for categorizing and classifying expectations, self, role-playing capacities, role enactment, and relationships among these analytical units. The use of concepts is confined primarily to classification of different phenomena, whether attention is drawn to the forms of status networks,[25] types and sources of expectations,[26] relations of self to expectations,[27] or the enactment of roles.[28] In the future, as they begin the difficult task of building interrelated inventories of propositions, structural role theorists, I believe, will confront several theoretical problems.

First, they must fill in the gaps of their causal imagery. To continue emphasizing only some causal links while ignoring others will encourage skewed sets of theoretical statements. Particularly crucial in this context will be the development of propositions that specify the linkages between concepts de-

---

[24]There is much experimental literature on the impact of various types of "contrived" role playing on attitudes and other psychological attributes of individuals; as yet the findings of these studies have not been incorporated into the role theoretic framework. For some examples of these studies, see Bert T. King and Irving L. Janis, "Comparison of the Effectiveness of Improvised versus Non-Improvised Role-Playing in Producing Opinion Changes," *Human Relations* 9 (May 1956), pp. 177–86; Irving L. Janis and Bert T. King, "The Influence of Role Playing on Opinion Change," *Journal of Abnormal and Social Psychology* 49 (April 1954), pp. 211–18; Paul E. Breer and Edwin A. Locke, *Task Experience as a Source of Attitudes* (Homewood, IL: Dorsey Press, 1955); Theodore Sarbin and V. L. Allen, "Role Enactment, Audience Feedback, and Attitude Change," *Sociometry* 27 (June 1964), pp. 183–94.

[25]For example, see Davis, *Human Society*; Merton, *Social Theory and Social Structure*; Benoit, "Status, Status Types"; and Biddle and Thomas, *Role Theory*, pp. 23–41.

[26]Davis, *Human Society*; Richard T. Morris, "A Typology of Norms," *American Sociological Review* 21 (October 1956), pp. 610–13; Alan R. Anderson and Omar K. Moore, "The Formal Analysis of Normative Concepts," *American Sociological Review* 22 (February 1957), pp. 9–16; Williams, *American Society*.

[27]R. Turner, "Role-Taking, Role Standpoint"; and Sarbin and Allen, "Role Enactment, Audience Feedback."

[28]Goffman, *Presentation of Self*; Sarbin, "Role Theory"; John H. Mann and Carola H. Mann, "The Effect of Role-Playing Experience on Role-Playing Ability," *Sociometry* 22 (March 1959), pp. 69–74; and William J. Goode, "A Theory of Role Strain," *American Sociological Review* 25 (August 1960), pp. 483–96.

noting more inclusive social and cultural variables, on the one hand, and concepts pointing to specific interaction variables, on the other.

Another problem is that the current propositions that do exist in the literature will have to be reformulated so that they specify when certain role processes are likely to occur. For example, in the theory of reference-group behavior, propositions assert that the use of a particular group as a frame of reference is likely to occur when (a) contact with members of a reference group is likely, (b) dissatisfaction with alternative group memberships exists, (c) perception of potential rewards from a group is likely, (d) the perception of that group's standards is possible, and (e) perception of the availability of significant others in the group is possible.[29] Although these propositions are suggestive, they offer few clues as to what forms of contact, what levels of dissatisfaction, what types of rewards and costs, which group standards, and what type of significant others serve as conditions for use as a frame of reference by an individual. Furthermore, many relevant variables are not included in these propositions. For instance, in order to improve the theory of reference-group behavior, I think it necessary to incorporate theoretical statements on the intensity of self-involvement, the capacity to assume roles in a group, the nature of group standards, and their compatibility with various facets of an individual's self-conception.

These are the problems that a theoretical perspective attempting to link social/structural and individual personality variables will inevitably encounter. When psychological variables, such as self-concept, self-esteem, and role-playing capacities, are seen as interacting with cultural and structural variables, such as status, norm, reference group, and others, the resulting inventory of theoretical statements will become complex. Such an inventory must not only delve into the internal states of individuals but must also cut across several levels of emergent phenomena—at a minimum, the individual, the immediate interaction situation, and the more inclusive structural and cultural contexts within which the interaction occurs. Yet, despite the difficulties involved, I think that structural role theory is crucial to resolving some of the theoretical gaps in sociology, especially those that concern the micro versus macro debate (see Chapter 1).

Methodological Implications

The potential utility of role theory derives from its concern with the complex interrelations among the expectations of social structure, the mediation of these expectations through self-conceptions and role-playing capacities of actors in statuses, and the resulting enactment of role behaviors. Measurement of role enactment does not pose a major methodological obstacle, since it is the most observable of the phenomena studied by role theorists. However, to the extent that such overt behavior reflects the impact of expectations and self-related

---

[29]Example drawn from summary in Alvin Boskoff, *The Mosaic of Sociological Theory* (New York: Thomas Y. Crowell, 1972), pp. 49–51.

variables, several methodological problems become evident. The complexity of the interrelations between role behavior, on the one hand, and self and expectations, on the other, as well as the difficulty of finding indicators of these interrelations, pose a series of methodological problems that continue to make it difficult to construct bodies of theoretical statements on the relation between society and the individual.

Since one of the assumed links between society and the individual revolves around the expectations that confront individuals, it is theoretically crucial that various types of expectations and the ways they affect individuals be measurable. One method is to infer expectations from observed behavior. The most obvious problem with this approach is that expectations can be known only after the emission of the behavior they are supposed to circumscribe. Therefore the concept of expectations as inferred from behavior has little theoretical utility, since it cannot be measured independently of behavior. Hence role behavior cannot be predicted from the content of expectations and their relationship to self. An alternative method involves (a) the accumulation of verbal accounts of individuals prior to a particular interaction sequence,[30] (b) the inference of what types of expectations are guiding conduct, and (c) the prediction of how role behavior will unfold in terms of these expectations. This method has the advantage of making predictions about the impact of expectations, but it suffers from the fact that, much like inferences drawn from role enactment, expectations are not measurable independently of the individual who is to be guided by them.[31] The end result of these methodological dilemmas is for expectations to represent inferences that are difficult to discern independently of the behavior—whether verbal accounts or role enactment—that they are supposed to guide.

A final alternative to this methodological problem is for researchers to become active participants in social settings and from this participation to derive some "intuitive sense" of the kinds of expectations that are operating on actors. From this intuitive sense it is then presumed that more formal conceptual representation of different types of expectations and of their varying impact on selves and behavior can be made. The principal drawback to such an approach is that one researcher's intuitive sense is not another's, and the result may be that the expectation structure observed by different researchers in the same situation is a "negotiated" product as researchers attempt to achieve consensus as to exactly what this structure is to be. However, to the extent that such a negotiated conceptualization has predictive value, it would represent an indicator of the expectation structure that is, in part at least, derived in-

---

[30]There are many ways to accumulate such accounts, ranging from informal observations and unstructured interviews with subjects to highly structured interviews and questionnaires. All of these have been employed by role theorists, whether in natural settings or the small-group laboratory.

[31]This dilemma anticipates the discussion of ethnomethodology to be undertaken in a later chapter, for, as the ethnomethodologist would argue, these verbal accounts are the "reality" that guides conduct. From this perspective, the assumption of a "really real" world, independent of an actor's mental construction of it, is considered to be unfounded.

dependently of the verbal statements and behavior of those whose conduct it is supposed to guide.

In sum, then, I believe that studying expectations is a difficult enterprise. Since most bodies of sociological theory assume the existence of an expectation structure, it is crucial that these methodological problems be exposed because they have profound implications for theory building. The most important of these implications concerns the possibility of building theory with concepts that are not measurable, even in principle. Examining either role behavior or verbal statements and then inferring the existence of expectations tend to make specific propositions tautologous, since variation in behavior is explained by variation in phenomena inferred from such behavior. The use of participant or observational techniques overcomes this problem, but it presents the equally perplexing question of how different researchers are to replicate, and hence potentially refute, one another's findings. If conceptualization of the expectation structure is a negotiated product, the subsequent investigation of similar phenomena by different investigators would require renegotiation. When the nature of phenomena that are incorporated into theoretical statements is not explicitly defined and classified in terms of independently verifiable, clear-cut, and agreed-upon standards but is instead the product of negotiation, then the statements are not refutable, even in principle. As products of negotiation, they have little utility for building a scientific body of knowledge. Ultimately, I imagine that the severity of these problems is a matter of subjective assessment. Some of these problems would appear to be fundamental, whereas others stem from the inadequacies of current research techniques.

Whether problems with conceptualizing expectations are seen as either fundamental or technical, they are compounded by the methodological problems of measuring self-related variables. How is it possible to derive operational indicators of self-conceptions, self-esteem, and intrapsychic assessments of the situation? As the discussion of Blumer and Kuhn in the last chapter stressed, verbal accounts and observational techniques have typically been used to tap these dimensions of interaction. Although they are technically inadequate in that their accuracy can be questioned, verbal and observational accounts do not raise the same fundamental questions as do expectations, since they do seem measurable in principle. The problems arise only when attempts are made to link these self-related variables to expectations that are presumed to have an independent existence that guides the creation and subsequent operation of self-related processes. Although the independent existence of norms, others, reference groups, and the like is intuitively pleasing, just how these phenomena are to be conceptualized and measured separately from the self-related processes they are assumed to circumscribe remains a central problem of interactionism in general.

The substantive criticisms of structural role theory revolve around the overly circumscribed vision of human behavior, and presumably of social organization, that it connotes. Although it can be argued that current role theory is too diverse to be vulnerable to this line of criticism, I think that both theory and research in this tradition portray a highly structured view of social reality.

Structural role theory assumes the social world to be organized in terms of status networks and corresponding clusterings of expectations. Whereas these expectations are viewed as mediated by self and role-playing capacities (subjective role), the main thrust is on how individuals adjust and adapt to the demands of the "script," other "actors," and the "audiences" of the "play." Undoubtedly, much social action is structured in this way, but my sense is that analysis is loaded in the direction of assuming too much structure and order in the social world.

The conspicuous conceptualization of role conflicts[32] (conflicts among expectations), role strain[33] (the impossibility of meeting all expectations), and anomie[34] (the lack of clear-cut expectations) in role theory would seemingly balance this overly structured conception of reality. Yet, frequently, strain, conflict, and anomie are viewed as deviant situations that represent exceptions to the structure of the normal social order. What is critical, then, is that these concepts be elaborated upon and inserted into current theoretical statements. In this way they can serve to specify the conditions under which the social world is less circumscribed by social structure.

The causal imagery of role theory also contributes to an overly structured vision of social reality. Figure 20-1 emphasizes that the causal thrust of role theory is on the way expectations, as mediated by selves and role-playing capacities, circumscribe role enactment. Although attention is drawn to the feedback consequences of role enactment for expectations, this analysis usually concerns how behaviors of individuals alter the reactions of others in such a way that self-conceptions are reinforced or changed.

The determinative consequences of role enactments for changes and alterations in social structure are, I believe, ignored in this mode of structural analysis. In focusing primarily on how changes of behavior affect self-conceptions, role theory has underemphasized the fact that behavior can also force changes in the organization of status networks, norms, reference groups, the responses of others, and other features of social structure. Until structural role theory stresses the consequences of role enactment, not only for self-related variables but also for the properties of structure, it will continue to conceptualize the social world as excessively ordered.

Some of the logical problems of role theoretic analysis further contribute to this conception of the social world. The vagueness of just how and under what conditions social structure affects self and role enactment leaves much of role analysis with the empty assertion that society shapes individual conduct. If role theoretic assumptions are to have theoretical significance, it is essential to specify just when, where, how, and through what processes this circum-

---

[32]For example, see John W. Getzels and E. C. Guba, "Role, Role Conflict, and Effectiveness," *American Sociological Review* 19 (February 1954), pp. 164–75; Talcott Parsons, *Social System*, pp. 280–93; Merton, *Social Theory and Social Structure*, pp. 369–79; and Robert L. Kahn et al., *Organizational Stress: Studies in Role Conflict and Ambiguity* (New York: John Wiley, 1974).

[33]Goode, "Theory of Role Strain."

[34]Merton, *Social Theory and Social Structure*, pp. 121–51.

scription of role behavior occurs. In fact, in the absence of theoretical specificity, I think that a subtle form of functionalism is connoted: the needs of social structure and the individual require that behavior be circumscribed. This functionalism is further sustained by the classificatory nature of role theoretic concepts. In denoting the types of interrelations among society, self, and behavior without indicating the conditions under which these relationships are likely to exist, these concepts appear to denote what processes must occur without indicating when, where, and how they are to occur.[35]

Finally, the methodological problems of measuring expectations separately from the very processes that they are supposed to circumscribe make even more mysterious just how and in what ways social structure affects individual conduct. Again, the inability to measure this crucial causal nexus leaves the role theorist with the uninteresting assertion that society shapes and guides individual conduct.

These problems have led others to propose a more processual version of role theory. Yet it would be foolish, I think, to ignore this more structural orientation. Social structures do exist; they do reveal normative expectations; they do order people's access to reference groups; and they do circumscribe people's options and self-evaluations. It has become very "trendy" in social theory to ignore these facts, especially after the effort to kill off Parsonian functionalism. I think that the more process-oriented theorists had a legitimate criticism, but they have often overreacted and posited a social universe in which people are interpersonal entrepreneurs and gadflies. Nonetheless, let us now turn to the more processual approach of Ralph H. Turner, keeping in mind that roles are often enacted in social structures and that theoretical sociology must confront this fact. For although structural role theory has very clear problems, it does attempt to analyze an important dynamic of social reality.

---

[35]Although most structural role theorists would vehemently deny it, they have followed Parsons' strategy for building theory, except on a more micro level. In developing concepts to classify and order role-related phenomena, without also developing clear-cut propositions, they have created a conceptual order without indicating when and how the order denoted by concepts is maintained (or broken down or reconstructed).

CHAPTER 21

Process Role Theory

Ralph H. Turner

Within interactionist theorizing there is considerable diversity. Most of this diversity, however, revolves around one central question: to what extent do the interactions of individuals represent an enactment of expectations inhering in social structure, or, conversely, to what degree are lines of interaction fluid and negotiated anew in each encounter among individuals? Among symbolic interactionists the Blumer versus Kuhn debate addresses this issue, and among role theorists this is the central question.[1] In the last chapter I presented a summary and analysis of more structural versions of role theory, and so in this chapter I will concentrate on the more process-oriented role theory of Ralph H. Turner.[2]

THE CRITIQUE OF STRUCTURAL THEORY

Over the course of several decades, Ralph H. Turner has mounted a consistent line of criticism against role theory.[3] This criticism incorporates several lines of attack: (1) Role theory presents an overly structured vision of the social world, with its emphasis upon norms, status positions, and the enactment of normative expectations. (2) Role theory tends to concentrate an inordinate amount of research and theory-building effort on "abnormal" social processes, such as role conflict and role strain, thereby ignoring the normal processes of human interaction. (3) Role theory is not theory but rather, a series of disjointed and unconnected propositions. (4) Role theory has not utilized to the degree required Mead's concept of role taking as its central dynamic.

---

[1] See Chapter 19.

[2] I am often asked if I am related to Ralph H. Turner. To avoid charges of familial favoritism, let me emphasize that we are not related to each other.

[3] See, for example, Ralph H. Turner, "Role-Taking: Processes versus Conformity," in *Human Behavior and Social Processes*, ed. A. Rose (Boston: Houghton Mifflin, 1962), pp. 20-40.

It is criticism on the overly structural view of roles that has, I believe, been the main impetus to Turner's alternative approach. Indeed, he has recently emphasized the overly structural bias in role theory, even among those who have tried to reconcile structural and processual accounts of roles.[4] He attacks structural approaches with a series of questions. First, does the structural conception of role add anything that is not covered by the concept of normative expectation? For, if roles are merely enactments of expectations, why even introduce the concept of role? Second, can the structural analysis of roles deal with roles that are not lodged in some organizational structure? Most roles, Turner contends, are not attached to a clear structure; if conceptualization of roles requires that we examine roles with structural positions and expectations, then analysis is dramatically limited. Third, can structural orientations to roles deal with situations where new roles are created or old ones changed by efforts of people to realize certain values or beliefs? For, if roles are part of structure, how is it possible to analyze the emergence of new roles, often in defiance of structural dictates?

These questions led Turner to posit a series of unresolved issues in role theory, which, he claims, structural role theory cannot handle. One issue is whether or not roles are acts of conformity to norms or creative constructions of actors. Turner believes that roles involving normative conformity are, in reality, exceptional cases that occur when a repressive structure limits opportunities, when people receive few rewards from their roles, and when people are insecure about their capabilities. But these are relatively rare situations, and so, in most social contexts, people negotiate their respective roles.

Another issue is whether or not roles are inventories of specific behaviors or more general gestalts and configurations of meaning about lines of conduct. It is rare, Turner argues, for actors to be able to list precisely the behaviors of a role; more typically, they can communicate only general attitudes, styles, and loosely defined behavioral options associated with a role. Only in rigid and pathological structures are roles so clearly defined as to constitute a list of appropriate behaviors.

Yet another issue is: are roles fixed and predetermined, or do individuals negotiate which roles they are going to play in a situation? Even in highly structural situations, roles are negotiated. And in most social encounters, individuals must actively make roles for themselves and negotiate with others for their right to play a given role.

What emerges from Turner's questions and presentation of unresolved issues in role theory is a view of roles as general configurations of responses that people negotiate as they form social relations. They are not mere enactments of expectations, and they are not always tied to positions in structures. In light

---

[4]Ralph H. Turner, "Unanswered Questions in the Convergence between Structuralist and Interactionist Role Theories," in *Perspectives on Sociological Theory*, eds. S. N. Eisenstadt and H. J. Helle (London: Sage Publications, 1985). In particular, he addresses Warren Handel, "Normative Expectations and the Emergence of Meaning as Solutions to Problems: Convergence of Structural and Interactionist Views," *American Journal of Sociology* 84 (1979), pp. 855–81.

of these considerations, Turner offers a conceptualization of roles that emphasizes the process of interaction over the dictates of social structures.

INTERACTION AND ROLES

The Role-Making Process

Turner utilizes and extends Mead's concept of role taking to describe the nature of social action. Turner assumes that "it is the tendency to shape the phenomenal world into roles which is the key to the role-taking as the core process in interaction."[5] Like Mead and Blumer, Turner stresses the fact that actors emit gestures or cues—words, bodily countenance, voice inflections, dress, facial expressions, and other gestures—as they interact. Actors use these gestures to "put themselves in the other's role" and to adjust their lines of conduct in ways that can facilitate cooperation. This is essentially Mead's definition of taking the role of the other, or role taking.

Turner then extends Mead's concept. He first argues that cultural definitions of roles are often vague and even contradictory. At best, they provide a general framework within which actors must construct a line of conduct. Thus actors *make* their roles and communicate to others *what* role they are playing. Turner then argues that humans act *as if* all others in their environment are playing *identifiable roles*.[6] Humans assume others to be playing a role, and this assumption is what gives interaction a common basis. Operating with this folk assumption, people then read gestures and cues in an effort to determine what role others are playing.[7] This effort is facilitated by others creating and asserting their roles, with the result that they actively emit cues as to what roles they are attempting to play.

For Turner, then, role taking is also "role making." Humans make roles in three senses: (1) They are often faced with only a loose cultural framework in which they must make a role to play. (2) They assume others are playing a role and thus make an effort to discover the underlying role behind a person's acts. (3) Humans seek to make a role for themselves in all social situations by emitting cues to others that give them claim on a particular role. This role-taking process as it becomes transformed into a role-making process is the underlying basis for all human interaction. It is what ultimately allows people to interact and cooperate with one another.

The "Folk Norm of Consistency"

As people interact with one another, Turner argues, they assess behavior not in terms of its conformity to imputed norms or positions in a social structure

---

[5]Turner, "Role-Taking: Processes versus Conformity."

[6]Ibid., "Social Roles: Sociological Aspects," *International Encyclopedia of the Social Sciences* (New York: Macmillan, 1968).

[7]"The Normative Coherence of Folk Concepts," *Research Studies of the State College of Washington* 25 (June 1957).

but rather, in regard to its consistency. Humans seek to group one another's behavior into coherent wholes or gestalts, and, by doing so, they can make sense of one another's actions, anticipate one another's behavior, and adjust to one another's responses. If another's responses are inconsistent and not seen as part of an underlying role, then interaction will prove difficult. Thus there is an implicit "norm of consistency" in people's interactions with one another. Humans attempt to assess the consistency of others' actions in order to discern the underlying role that is being played.

With the concepts of role making and the norm of consistency, Turner shifts the analysis of roles toward a position that symbolic interactionists such as Blumer might appreciate. This shift is underscored by a third major assumption with which Turner approaches the analysis of roles: interaction is always a tentative process.

The Tentative Nature of Interaction

Turner echoes Blumer's position when he states that "interaction is always a tentative process, a process of continuously testing the conception one has of the role of the other."[8] Humans are constantly interpreting additional cues emitted by others and using these new cues to see if they are consistent with those previously emitted and with the imputed roles of others. If they are consistent, then the actor will continue to adjust responses in accordance with the imputed role of the other. But as soon as inconsistent cues are emitted, the identification of the other's role will undergo revision. Thus a given imputation of a particular role to another person will persist only as long as it provides a stable framework for interaction. The tentative nature of the role-making process points to another facet of roles: the process of role verification.

The Process of Role Verification

Actors seek to verify that behaviors and other cues emitted by people in a situation do indeed constitute a role. Turner argues that such efforts at verification or validation are achieved by the application of external and internal criteria. The most often used internal criterion is the degree to which an actor perceives a role to facilitate interaction. External criteria can vary, but in general they involve assessment of a role by important others, relevant groups, or commonly agreed-upon standards. When an imputed role is validated or verified in this way, then it can serve as a stable basis for continued interaction among actors.

Self-Conceptions and Role

All humans reveal self-conceptions of themselves as certain kinds of objects. Humans develop self-attitudes and feelings out of their interactions with others,

---

[8]Turner, "Role-Taking: Processes versus Conformity," p. 23.

but, as Turner and all role theorists emphasize, actors attempt to present themselves in ways that will reinforce their self-conceptions.[9] Since others will always seek to determine an actor's role, it becomes necessary for an actor to inform others, through cues and gestures, about the degree to which self is anchored in a role. Thus actors will signal one another about their self-identity and the extent to which their role is consistent with their self-conception. For example, roles not consistent with a person's self-conception will likely be played with considerable distance and disdain, whereas those that an individual considers central to self-definitions will be played much differently.[10]

Let me now sum up Turner's basic assumptions. An emphasis on the behavioral aspect of role is retained, since it is through behavioral cues that actors impute roles to one another. The notion that roles are conceptions of expected behaviors is also preserved, for the assignment of a role to a person invokes an expectation that a certain type and range of responses will ensue. The view that roles are the norms attendant on status positions is given less emphasis but not ignored, since norms and positions can be the basis for assigning and verifying roles.[11] And the conception of roles as parts that people learn to play is preserved, for people are able to denote one another's roles by virtue of their prior socialization into a common role repertoire.

Not only do these assumptions embrace the major points of emphasis in prominent definitions of role, but they also help reconcile the differences between symbolic interactionism and more structural versions of role theory. Turner's assumptions employ the key concepts of Mead's synthesis while taking cognizance of Blumer's emphasis on the processes that underlie patterns of joint action. These assumptions also point to the normal processes of interaction but are sufficiently general to embrace the possibility of conflictual and stressful interactions. And these assumptions about the role-making process do not preclude the analysis of structured interaction, since formal norms and status positions are often the major cues for ascertaining the roles of people as well as a source of verification for imputed roles.

Assumptions are, however, only as good as the theory that they can generate. I think it particularly significant that Turner delineates an explicit strategy for translating these assumptions into theoretical propositions. My sense is that the execution of this strategy will help interactionism bridge the gap between suggestive assumptions and concepts, on the one hand, and the current

---

[9]Ralph H. Turner, "The Role and the Person," *American Journal of Sociology* 84 (1978), pp. 1–23.

[10]Turner, "Social Roles: Sociological Aspects." Turner has extensively analyzed this process of self-anchorage in roles. See, for example, Ralph H. Turner, "The Real Self: From Institution to Impulse," *American Journal of Sociology* 81 (1970), pp. 989–1016; Ralph H. Turner and Victoria Billings, "Social Context of Self-Feeling," in *The Self-Society Interface: Cognition, Emotion, and Action*, J. Howard and P. Callero, eds. (Cambridge, England: Cambridge University Press, 1990); Ralph H. Turner and Steven Gordon, "The Boundaries of the Self: The Relationship of Authenticity to Inauthenticity in the Self-Conception," in *Self-Concept: Advances in Theory and Research*, ed. M. D. Lynch et al. (Cambridge, MA: Ballinger, 1981).

[11]Ralph H. Turner, "Rule Learning as Role Learning," *International Journal of Critical Sociology* 1 (September 1974).

plethora of narrow empirical propositions, on the other. Let me now outline the basic elements of this strategy.

THE STRATEGY FOR BUILDING ROLE THEORY

Although Turner accepts the process orientation of Blumer, he is committed to developing interactionism into "something akin to axiomatic theory."[12] He recognizes that, in its present state, role theory is segmented into a series of narrow propositions and hypotheses and that role theorists have been reluctant "to find unifying themes to link various role processes."

Turner's strategy is to use propositions from the large number of research studies to build more formal and abstract theoretical statements. The goal is to maintain a productive dialogue between specific empirical propositions and more abstract theoretical statements.

The Concepts of Role Theory

Turner argues against rigid definitions of concepts when beginning the theory-building process. It is more useful, he contends, to begin with loosely defined concepts such as "actor," "role," "other," and "situation." The concepts will take on greater clarity as propositions incorporating them are developed. Moreover, early emphasis on concept formation "turns our attention from empirical to definitional concerns, from dynamic to static questions, and many a theory-building enterprise becomes hopelessly diverted into creating an elegant system that neither suggests nor generates new empirical propositions."[13]

Turner adopts Blumer's position that theorists must begin with sensitizing concepts. As I will document shortly, however, he uses these concepts in ways that will allow for the formulation of more precise definitions and propositions.

Sorting Out Tendencies

Without definitive concepts and with a large body of segmented propositions, an alternative way of linking sensitizing concepts to observe empirical regularities is necessary. Turner advocates the use of what he terms *main tendency* propositions to link concepts to empirical regularities and to consolidate the thrust of these regularities.[14] What Turner seeks is a series of statements that highlight what tends to occur in the normal operation of systems of interaction. These statements are not true propositions because they are not of the form: under $C_1, C_2, C_3, \ldots, C_n$, x varies with y. Rather, they are statements of the form: in most normal situations, event x tends to occur. These are not state-

---

[12]Ralph H. Turner, "Strategy for Developing an Integrated Role Theory," *Humboldt Journal of Social Relations* 7 (Winter 1980), pp. 123–39, and "Role Theory as Theory," unpublished manuscript.

[13]Turner, "Strategy," pp. 123–24.

[14]Turner, "Social Roles: Sociological Aspects."

ments of covariance but statements of what is presumed to typically transpire in the course of interaction.

Turner provides a long list of main tendency propositions with respect to a number of issues: (*a*) the emergence and character of roles, (*b*) role as an interactive framework, (*c*) roles in relation to actors, (*d*) role in organizational setting, (*e*) role in societal setting, and (*f*) role and the person. Let me summarize some of these tendency propositions.

Emergence and character of roles   In these propositions, Turner presents some observations about the nature of the social world as a series of empirical tendencies. I should emphasize, however, that such tendencies are observed through the heavy prism of Turner's assumptions.

1. In any interactive situation, behavior, sentiments, and motives tend to be differentiated into units that can be termed roles; once differentiated, elements of behavior, sentiment, and motives that appear in the same situation tend to be assigned to existing roles. (Tendencies for role differentiation and accretion.)
2. In any interactive situation, the meaning of individual actions for ego (the actor) and for any alter is assigned on the basis of the imputed role. (Tendencies for meaningfulness.)
3. In connection with every role, there is a tendency for certain attributes of actors, aspects of behavior, and features of situations to become salient cues for the identification of roles. (Tendencies for role cues.)
4. The character of a role—that is, its definition—will tend to change if there are persistent changes in either the behaviors of those presumed to be playing the role or the contexts in which the role is played. (Tendencies for behavioral correspondence.)
5. Every role tends to acquire an evaluation in terms of rank and social desirability. (Tendencies for evaluation.)

These propositions both reassert Turner's assumptions about the social world and provide several points of elaboration. People are seen as viewing the world in terms of roles; they are visualized as employing a folk norm to seek consistency of behaviors and to assign behavioral elements to an imputed role (role differentiation and accretion). Actors are viewed as interpreting situations by virtue of imputing roles to one another (meaningfulness tendency). Humans are observed to use cues of other actors' attributes and behaviors, as well as the situation, to identify roles (role cues). When role behaviors or situations are permanently altered, the definition of role will also undergo change (behavioral correspondence). And humans tend to evaluate roles by ranking them in terms of power, prestige, and esteem, while assessing them with regard to their degree of social desirability and worth (tendency for evaluation).

Role as an interactive framework   In these tendency propositions, Turner elaborates his assumption that interaction cannot proceed without the identification and assignment of roles. These propositions specify the ways in which roles provide a means for interaction to occur.

6. The establishment and persistence of interaction tend to depend upon the emergence and identification of ego and alter roles. (Tendency for interaction in terms of roles.)
7. Each role tends to form as a comprehensive way of coping with one or more relevant alter roles. (Tendency for role complementarity.)
8. There is a tendency for stabilized roles to be assigned the character of legitimate expectations and to be seen as the appropriate way to behave in a situation. (Tendency for legitimate expectations.)

In these propositions, interaction is seen as dependent upon the identification of roles. Moreover, roles tend to be complements of others—as is the case with wife/husband, parent/child, boss/employee roles—and thus operate to regularize interaction among complementary roles. Finally, roles that prove useful and that allow for stable and fruitful interaction are translated into expectations that future transactions will and should occur as in the past.

Role in relation to actor   These propositions concern the relationship between actors and the roles that provide the framework for interaction.

9. Once stabilized, the role structure tends to persist, regardless of changes in actors. (Tendency for role persistence.)
10. There is a tendency to identify a given individual with a given role and a complementary tendency for an individual to adopt a given role for the duration of the interaction. (Tendency in role allocation.)
11. To the extent that ego's role is an adaptation to alter's role, it incorporates some conception of alter's role. (Tendency for role taking.)
12. Role behavior tends to be judged as adequate or inadequate by comparison with a conception of the role in question. (Tendency to assess role adequacy.)
13. The degree of adequacy in role performance of an actor determines the extent to which others will respond and reciprocate an actor's role performance. (Tendency for role reciprocity.)

Thus, once actors identify and assign one another to roles, the roles persist, and new actors will tend to be assigned to those roles that already exist in a situation. Humans also tend to adopt roles for the duration of an interaction, while having knowledge of the roles that others are playing. Additionally, actors carry with them general conceptions of what a role entails and what constitutes adequate performance. Finally, the adequacy of a person's role performance greatly influences the extent to which the role, and the rights, privileges, and complementary behaviors that it deserves, will be acknowledged.

Role in organizational settings   Turner rejects an overly structural view of roles, but he still recognizes that many roles are enacted in structured contexts, necessitating a listing of additional tendencies in the role-making processes.

14. To the extent that roles are incorporated into an organizational setting, organizational goals tend to become crucial criteria for role differentiation, evaluation, complementarity, legitimacy or expectation, consensus, allocation, and judgments of adequacy. (Tendency for organization goal dominance.)

15. To the extent that roles are incorporated into an organizational setting, the right to define the legitimate character of roles, to set the evaluations on roles, to allocate roles, and to judge role adequacy tends to be lodged in particular roles. (Tendency for legitimate role definers.)

16. To the extent that roles are incorporated into an organizational setting, differentiation tends to link roles to statuses in the organization. (Tendency for status.)

17. To the extent that roles are incorporated into an organizational setting, each role tends to develop as a pattern of adaptation to multiple alter roles. (Tendency for role-sets.)

18. To the extent that roles are incorporated into an organizational setting, the persistence of roles is intensified through tradition and formalization. (Tendency for formalization.)

When roles are lodged in an organization, then, its goals and key personnel become important in the role-making process. Moreover, it is primarily within organizations that status and role become merged. In this way, Turner incorporates Linton's insight that status and role *can* become highly related, but he does not abandon Mead's and Blumer's emphasis that much interaction occurs in contexts in which roles are not circumscribed by networks of clearly defined status positions. Turner also recognizes that roles in structured situations develop as ways of adapting to a number of other roles that are typically assigned by role definers or required by organizational goals. Finally, roles within organizations tend to become formalized in that written agreements and tradition come to have the power to maintain a given role system and to shape normative expectations.

Role in societal setting   Many roles are identified, assumed, and imputed in relation to a broader societal context. And so, in the tendencies listed below, Turner first argues that people tend to group behaviors in different social contexts into as few unifying roles as is possible. Thus people will identify a role as a way of making sense of disparate behaviors in different contexts. At the societal level, values are the equivalent of goals in organizational settings for identifying, differentiating, allocating, evaluating, and legitimating roles. Finally, all people tend to assume multiple roles in society, but they tend to assume roles that are consistent with one another. Thus:

19. Similar roles in different contexts tend to become merged, so as to be identified as a single role recurring in different relationships. (Tendency for economy of roles.)

20. To the extent that roles refer to more general social contexts and situations, differentiation tends to link roles to social values. (Tendency for value anchorage.)
21. The individual in society tends to be assigned and to assume roles consistent with one another. (Tendency for allocation consistency.)

Role and the person   The "person" is a concept employed by Turner to denote "the distinctive repertoire of roles" that an individual enacts in relevant social settings. The concept of person is his means for summarizing the way in which individuals cope with their roles. In the generalizations below, Turner observes that people seek to resolve tensions among roles and to avoid contradictions between self-conceptions and roles. These propositions are, to a very great extent, elaborations of Turner's assumptions about the relationship between self-conceptions and role.

22. Actors tend to act so as to alleviate role strain arising out of role contradiction, role conflict, and role inadequacy and to heighten the gratifications of high role adequacy. (Tendency to resolve strain.)
23. Individuals in society tend to adopt as a framework for their own behavior and as a perspective for interpretation of the behavior of others a repertoire of role relationships. (Tendency to be socialized into common culture.)
24. Individuals tend to form self-conceptions by selective identification of certain roles from their repertoires as more characteristically "themselves" than other roles. (Tendency to anchor self-conception.)
25. The self-conception tends to stress those roles that supply the basis for effective adaptation to relevant alters. (Adaptation of self-conception tendency.)
26. To the extent that roles must be played in situations that contradict the self-conception, those roles will be assigned role distance, and mechanisms for demonstrating lack of personal involvement will be employed. (Tendency for role distance.)

These 26 generalizations represent only the first step in Turner's theoretical strategy. These empirical tendencies incorporate the loosely defined concepts of interactionism, but they link these concepts to actual events presumed to occur in the social world. Naturally, more tendencies might be discerned and recorded. The list of 26 is sufficient, I think, to illustrate what Turner views as the next step in developing a more integrated role theory.

Generating and Organizing Empirical Propositions

As I have emphasized, the tendency propositions are not true propositions. They do not reveal relations of covariance among variables. However, Turner believes that the tendency propositions can help generate true empirical propositions of the form: x varies with y. This is done by attempting to determine the empirical conditions that shape the degree or rate of variation in a tendency proposition. For example, Tendency Proposition 22 becomes the dependent variable in a search for independent variables that can specify the conditions under which actors "tend to act so as to alleviate role strain." The tendency proposition thus provides an initial set of guidelines for developing true propositions about relationships among variables.

Furthermore, since the tendency propositions are grouped together, as was done in the sections above, the true empirical propositions will be organized around related tendencies. As such, the propositions are less scattered and disparate than would be the case if the search for propositions had not begun with the delineation of certain normal tendencies. Yet I still find the transition from the tendency propositions to the true empirical propositions rather vague. Nonetheless, Turner's propositions about the person and role can illustrate the potential of his theory-building strategy (the last section in the tendency propositions above).[15]

The person is viewed as the repertoire of roles that he or she plays. And although "people are normally quite different actors in different roles, and even have varying senses of 'who they are,'" humans also use roles as a means for self-identification and self-validation.[16] Some roles are more important to individuals and resist compartmentalization, or separation, from a person's self-concept. Roles that arouse strong self-feelings, which people appear to play across situations, refuse to abandon, and embellish with associated attitudes, are likely to involve considerable merger of the individual's self with the role.

Thus, to generate an empirical proposition from these tendencies, we need to know the conditions under which individuals become identified with roles, using the role for purposes of self-identification and validation. Tendency Propositions 24 and 26 can be dependent variables organizing a search for the conditions under which individuals tend "to form self-conceptions by selective identification of certain roles from their repertoires as more characteristically themselves than other roles" and to show "role distance" and "a lack of personal involvement" in situations that "contradict the self-conception." For all the other tendency propositions listed, a similar search could be initiated in an effort to discover the empirical conditions that influence their rate, degree, and extent of occurrence.[17]

Turner divides his discussion of person and role into two general lines of analysis: (1) how do others in the situation discover a merger of role and person, and (2) how do the individuals come to lodge their selves in certain roles? Each of these questions organizes the propositions that I have listed, respectively, in Tables 21-1 and 21-2.

In Table 21-1, Turner examines the simple situation in which an actor's interaction with others is confined to one situation. When so confined, others

---

[15]Turner, "Role and the Person." See also note 10.

[16]Ibid., p. 2.

[17]In his most recent effort, "Strategy for Developing an Integrated Role Theory," Turner has focused on "role allocation" and "role differentiation" because he views these as the two most critical tendencies. My example of "the person and role" results in the same explanatory principles. Thus this illustration provides additional examples of Turner's strategy.

TABLE 21–1   Others and Role Merger in One Situation

I. The more inflexible the allocation of actors to a role, the greater the tendency for members of the social circle to identify the role with the person, and the stronger the tendency for the actors to accept that identification for themselves.

II. The more comprehensively and strictly differentiated the role, the greater the tendency for members of the social circle to identify the role with the person, and the stronger the tendency for actors to accept that identification for themselves.

III. The more conflictual the relationship between roles, the greater the tendency for members of the social circle to identify the role with the person, and the stronger the tendency for the actors to accept that identification for themselves.

IV. The higher and more consistent the judgments of role adequacy, the greater the tendency for members of the social circle to identify the role with the person, and the stronger the tendency for the actors to accept that identification for themselves.

V. The more difficult the role is thought to be, the greater the tendency for members of the social circle to identify the role with the person, and the stronger the tendency for actors to accept that identification for themselves.

VI. The more polar the evaluation of a role as favorable or unfavorable, the greater the tendency for members of the social circle to identify the role with the person, and the stronger the tendency for the actors to accept that identification for themselves.

VII. The more polar the social rank of a role as high or low, the greater the tendency for members of the social circle to identify the role with the person, and the stronger the tendency for the actors to accept that identification for themselves.

VIII. The greater the potential power vested in a role, the greater the tendency for members of the social circle to identify the role with the person, and the stronger the tendency for the actors to accept that identification for themselves.

IX. The greater the discretion vested in a role, the greater the tendency for members of the social circle to identify the role with the person, and the stronger the tendency for the actors to accept that identification for themselves.

X. The greater the extent to which members of a social circle are bonded to role incumbents by ties of identification, the greater the tendency for them to identify the role with the person, and the stronger the tendency for the actors to accept that identification for themselves.

XI. The more intimate the role relationship between an actor's social circle and alter roles, the greater the tendency for them to identify the role with the person, and the stronger the tendency for the actor to accept that identification.

in that situation will identify the role as integral to the person's self-identification, and the person will accept this identification by others under the conditions listed in each of the 11 propositions. For example, if there is little flexibility in assuming a role (Proposition I), then merger of person and role is increased. In Proposition II, the more clearly defined and distinguishable the role, the greater the tendency for person/role merger. In Proposition III, the more that roles among person and others are in conflict, the greater the role/person merger—and so on for the remaining propositions. What is critical in these and other propositions is that they reveal covariance between variables and thus constitute true propositions.

The social settings of actors often overlap. People see one another in different settings, but, as they interact, they seek to discover which role is part of a person's self-identification. The five propositions in Table 21–2 are ex-

TABLE 21–2 Others, Person, and Role in Multiple Settings

I. The broader the setting in which a role is lodged, the greater the tendency for others to identify the role with the person, and the stronger the tendency for the actors to accept that identification for themselves.

II. The more a role in one setting determines allocation and performance of roles in other settings, the greater the tendency for others to identify the role with the person, and the stronger the tendency for the actors to accept that identification for themselves.

III. The more conspicuous and widely recognizable the role cues, the greater the tendency for others to identify the role with the person, and the stronger the tendency for the actors to accept that identification for themselves.

IV. The more a role exemplifies the goals and nature of the group or organization in which it is lodged, the greater the tendency for others to identify the role with the person, and the stronger the tendency for the actors to accept that identification for themselves.

V. The more that allocation to a role is understood to be temporary and the role discontinuous in content with respect to preceding and succeeding roles, the greater the tendency for community members not to identify the role with the person, and the stronger the tendency for the actors not to identify role with self.

amples of some of the considerations that Turner views as operating in multiple social settings to influence role/person merger. People in multiple settings will be likely to identify the role with the person, and the person will likely accept this identification when the role cuts across many social settings (I), the role influences the other roles that a person can play and/or how they are played (II), the role is highly conspicuous (III), or the role exemplifies, or personifies, the nature of the social unit in which it is lodged (IV). Proposition V is an example of a condition under which roles will not merge with the person. When a role is short-lived and inconsistent with other roles played by a person, then the individual will not identify with the role.

The propositions in Tables 21–1 and 21–2 address the question of how others in situations discover a person/role merger. However, people are rarely passive, accepting the labels of others. People actively seek to dictate the merger of their self with a role. This fact is stressed with the concept of role making, thus requiring still more propositions on self and role merger.

Turner expands upon Mead's insight that the emergence of self-conceptions in individuals facilitates interaction and the functioning of society. Self provides individuals with a way to discriminate among roles and to partition them in terms of their importance and significance. If individuals could not do this, they would emotionally exhaust themselves. They could disrupt the flow of society by trying to play all roles equally well and with the same degree of intensity. Additionally, self allows actors to maintain an identity across roles and to resist sanctions that would force them to act in contradictory ways. And, finally, self gives action consistency and coherence across roles, which allows others to anticipate that an actor will behave in a given way, thereby enabling others to adjust their response to the actor.

The propositions in Table 21–3 offer some insight into these processes. People are likely to locate their self in roles that are highly evaluated, that

TABLE 21–3 Individual Efforts at Role Merger

I. The more highly evaluated a role, the greater the tendency to locate self in that role.

II. The more adequately a role can be performed, the greater the tendency to locate self in that role.

III. The higher the evaluation of roles among a repertoire of roles that can be played adequately, the greater the tendency to locate self in the roles of highest evaluation.

IV. The more visible and readily appraisable the role performance, the more the tendency to locate self in highly evaluated roles will be modified by the tendency to locate self in roles that can be played with high degrees of adequacy.

V. The more the scope of an individual's social world exceeds the boundaries of the social circle of a given role, the greater the tendency to use evaluations of the larger community rather than those of a specific social circle to locate self in a role.

VI. The more intrinsic (as opposed to extrinsic) the benefits derived from enacting a role, the greater the tendency to locate self in that role.

VII. The greater the investment of time and effort in gaining or maintaining the opportunity to claim a role or in learning to play a role, the greater the tendency to locate self in that role.

VIII. The greater the sacrifice made in gaining or maintaining the opportunity to claim a role or in learning to play a role, the greater the tendency to locate self in that role.

IX. The more publicly a role is played and the more actors must explain and justify a role, the greater the tendency to locate self in that role.

X. The more unresolved role strain encountered in a role has been prolonged, the greater the tendency to locate self in that role.

---

they can perform well, that are both highly evaluated *and* played well (actors will, by implication, avoid roles of high evaluation that they cannot play well), that are visible and more readily subject to evaluation by others, that are comprehensive and cut across social contexts, that provide personal and subjectively defined benefits, that involve the expenditure of time and effort, that involve sacrifice in reaching, that are publicly played and in need of public justification, and that involve prolonged role strain and effort to eliminate strain.

The propositions on the person and role listed in Tables 21–1, 21–2, and 21–3 are only tentative. Similar empirical propositions need to be developed for Turner's other tendency statements. Yet even clearly articulated empirical propositions represent only crude groupings of statements and tendencies. Such propositions are the result of speculation, and they can, no doubt, suggest additional propositions. But these empirical propositions are not organized *deductively*—that is, in a way that would allow them to be deduced from a small number of abstract propositions or "axioms." The next step in Turner's strategy, therefore, involves an effort to generate explanatory propositions.

Developing Explanatory Propositions

In recent years Turner, along with various collaborators, has sought to develop general explanatory propositions that can account for why empirical propositions cohere around a main tendency. That is, are there some underlying processes that can explain why the tendency should occur and why the empirical propositions with the tendency as the dependent variable should hold true?

In his most recent reworking of explanatory propositions, Turner and Paul Colomy[18] posit three sets of principles revolving around (1) functionality, (2) representation, and (3) tenability. In articulating these principles, Turner has come to emphasize, at least implicitly, the primacy of role differentiation and accretion (Tendency Proposition 1 dealing with the emergence and character of roles, on page 431). Thus, to a great extent, other role processes follow from the questions: How do the elements making up a role cluster together (accretion)? And how do they become differentiated from other elements? The answers to these questions resides in the interrelated dynamics of *functionality, representation,* and *tenability.*

Functionality   If roles are used to achieve goals or ends in an efficient manner, and are organized in a way that facilitates the achievement of goals, then considerations of functionality are highlighted. Concern is with creating and differentiating roles in an efficient division of labor that gets things done. Turner's intent here is to argue that functionality can explain patterns of role differentiation and accretion, as well as other empirical tendencies, as these tendencies are organized by various empirical propositions (see Tables 21–1, 21–2, 21–3). Turner intends propositions about functionality, and those about representation and tenability also, to stand at the top of a deductive system of explanation. Functionality is, therefore, one of Turner's "laws" of social organization, at least among "roles" as a basic property of all patterns of social organization.

In Table 21–4, I have reduced Turner's and Colomy's more extensive discussion[19] down to several key principles. Let me review each. Principle I is a tautology, because the defining characteristics of functionality are used to explain its effects on role differentiation and accretion (and other empirical tendencies as well). But it is a useful tautology in two senses. First, it is more than a circular definition, because the proposition highlights a master dynamic in social organization and states that, when activity among actors is oriented to achieving goals through efficient and effective means, then the way in which roles are created and differentiated will be skewed toward considerations of functionality over other basic processes—namely, considerations of representation and tenability (to be discussed shortly and as summarized in Tables 21–5 and 21–6). Second, and more important, a tautology like Principle I in Table 21–4 encourages us to look for additional conditions that increase (or decrease) the salience of functionality. Turner and Colomy discuss two such conditions, and I have listed them as Propositions I-A and I-B in Table 21–4. Proposition I-A argues that, when there is potential incompatibility of goals and means among actors or roles, and when there are latent or even overt conflicts of

---

[18]Ralph H. Turner and Paul Colomy, "Role Differentiation: Orienting Principles," *Advances in Group Processes* 5 (1987), pp. 1–47.
[19]Ibid.

TABLE 21–4   Explanatory Principles on Functionality

I. The more activity among individuals involves efforts to realize goals in an efficient and effective manner, the greater are considerations of functionality in the differentiation and accretion of the roles organizing their activity.

  A. The more those collaborative and goal-directed roles organizing individuals' activities reveal potential incompatibility among their goals and preferred means, and/or the more they evidence conflicts of interests, the more likely are incompatible and conflictual roles to be differentiated and separated in ways that encourage the achievement of goals and, hence, the greater will be considerations of functionality.

  B. The more collaborative and goal-directed roles organizing individuals' activities involve recruitment from pools of actors who are differentiated by their abilities and dispositions, the more likely are roles to be formed and differentiated in ways that match dispositions and abilities to the achievement of goals and, hence, the greater are considerations of functionality.

II. The more considerations of functionality guide the differentiation and accretion of roles, the greater will be considerations of tenability over representation in modifying these roles, to the extent that they can be modified.

III. The more functionally differentiated roles recruit members from homogeneous pools of individuals, and the longer these actors play their roles, the more likely are these roles to develop representational elements, and the greater will be their resistance to change.

---

interests among actors who are collaborating to achieve a goal, efforts to cluster role elements and differentiate roles will employ the criterion of functionality—that is, how to create a division of labor among roles so as to get things done and avoid points of incompatibility and conflict. Proposition I-B argues that, when collaboration among actors involves recruiting role incumbents who range in their abilities and dispositions, then once again considerations of functionality will prevail in accretion, differentiation, and other role processes.

Propositions II and III in Table 21–4 seek to delineate some of the relations of functionality to the other two basic role organizing processes (representation and tenability). Proposition II states that, when considerations of functionality are guiding the organization of roles, tenability will be more significant than representation in modifying functional roles. Turner recognizes that, although one process may dominate the organization of roles, the other two processes can impinge upon the way in which this dominance process operates; so, it is necessary to state which other processes will intercede as well as the conditions that increase, or decrease, their effects on the dominant process. Since we have not discussed these other processes—tenability and representation—it is somewhat difficult to explain propositions in detail, but let me at least lay the argument out in its broad contours.

Tenability concerns how well roles "fit" the individual and provide a source of gratification (and cost), whereas representation denotes the process of imbuing roles with values and other cultural overlays. Thus Proposition II argues that functionality tends to keep considerations of value relevance minimal, but it does allow for actors to maneuver to make roles tenable in terms of their self-concept, expectations, power, and other resources that they, as persons, bring to a role. Proposition III emphasizes, however, that values and other evaluative cultural elements can impinge upon functionally organized roles when the incumbents come from similar or homogeneous backgrounds and when they have played their functional roles vis-à-vis one another for a prolonged period of time. Under these conditions, roles become imbued with values—a sense of their rightness and wrongness—and as a consequence they become difficult to change because to do so would violate incumbents' moral sensibilities.

Of course, I am sure that Turner would not see the "laws" presented in Table 21-4 (and 21-5 and 21-6 also) as the last word on functionality. Rather, these are tentative and incomplete statements, and our goal should be to fine-tune these and develop additional propositions on the conditions increasing functionality and/or modifying functionality through increases in the salience of representation and tenability. But, as stated, these principles can provide a means for deductively organizing the tendency and empirical propositions articulated by Turner, and by others as well.

Representation   This role organizing process is the most recent addition to Turner's evolving scheme. Turner's long-standing affinity to Durkheimian sociology—especially the notions of "collective representations" and "collective conscience"—was, no doubt, the stimulus for this reconceptualization, and his collaboration with the functional theorist Paul Colomy would seem quite natural when exploring representational processes.[20] As might be expected from a functional-oriented sociology, representational processes revolve around the separation and combination of the elements making up roles so as to accentuate value differences and similarities among roles and role incumbents. This point of emphasis was, of course, anticipated in Turner's early tendency propositions (see, for example, number 20 on page 434), but now the infusion of roles with value-laden content is seen as ever more fundamental to role-related dynamics. As with the propositions on functionality, Table 21-5 arrays what I see as the critical processes in Turner's and Colomy's discussion.

Like its counterpart on functionality, Principle I is a tautology, but it is a useful one nonetheless. It simply states that a fundamental dynamic of role differentiation, accretion, and presumably other role processes is their infusion with cultural values; once this occurs, role processes are dramatically transformed in ways that make the representation of values a paramount consideration for the roles themselves, and for their incumbents. It is propositions I-A, I-B, I-C, I-D, I-E, and I-F, however, that specify more precisely and nontautologously the conditions under which representational forces begin to influence the content and organization of roles. Proposition I-A emphasizes that, when roles are involved in conflict and competition, they tend to polarize around values. Conflict generally escalates the ideological content of activity;

---

[20]See, for example, Jeffrey C. Alexander and Paul Colomy, "Toward Neo-Functionalism," *Sociological Theory* 3 (1985), pp. 11–23.

TABLE 21-5 Explanatory Principles on Representation

I. The more activity among individuals involves the embodiment of cultural values, the greater are considerations of representation in the differentiation and accretion of roles.

   A. The more roles are implicated in intergroup conflict and competition, the more likely are these roles to polarize around value premises and, hence, the greater are considerations of representation.

   B. The more different roles involve behaviorally similar and potentially interchangeable elements, the more likely are differences among roles to be marked by values and, hence, the greater are considerations of representation.

   C. The more the elements of a role-set reveal contradictions and disjuncture, the more likely are points of incompatibility and disjuncture to be the focus of value-laden processes and, hence, the greater are considerations of representation.

      1. The more a role-set consists of value-laden elements that are incompatible with considerations of functionality, the more likely are these elements to be marked and reaffirmed by ceremonies and rituals.

      2. The more a role-set consists of incompatible value elements, the more likely are the discordant elements to be differentiated out into separate roles.

      3. The more a role-set consists of idealized elements that are difficult to realize in practice, the more likely are incumbents in these roles to emit cues marking their distance from those elements that do not realize the ideal.

   D. The more a role, or set of roles, recruits from pools of homogeneous incumbents, the more likely are points of similarities to be marked by values and, hence, the greater are considerations of representation.

   E. The more a role is interdependent with other roles revealing high levels of representational content, the more likely that this role will assume this content as one of its elements and, hence, the greater will be considerations of representation.

   F. The more a role, or set of roles, occurs in a context of value dissensus, conflict, incompatibility, and other problematic conditions, the more likely are efforts to resolve this problematic content to depend upon the reassertion, reapplication, or realignment of values to roles and, hence, the greater are considerations of representation.

II. The more considerations of representation guide the differentiation and accretion of roles, the greater will be considerations of tenability over functionality in modifying these roles, to the extent that they can be modified.

---

and when roles are in conflict, or organized to pursue conflict, they become heavily infused with the value premises that mobilize actors in conflict and legitimate its pursuit. Proposition I-B addresses the question of "interchangeability," as Turner calls it. Many behaviors are very similar in terms of the elements involved (that is, they would appear to be interchangeable, at least on the surface), but they can take on very different meanings depending upon the context. Proposition I-B argues that, when behaviors are similar, a critical basis for distinguishing them as elements of different roles is their infusion with values that provide a way to "see" these behaviors as distinctive and thereby provide a way to bestow different meanings on them. Proposition I-C addresses the related issues of conflict and discrepancy within and between roles. The basic idea is that, if a role-set contains inconsistent elements, if it clashes with other organizing processes, particularly functionality, or if it contains value elements that are highly idealized and impossible to realize in practice, then one mechanism for resolving these problems is to infuse the role(s) with value content. For example, a role-set containing some inconsistent elements might have the unifying elements of the underlying role emphasized by value content; or, clashes with functionality can be partially resolved by heightening the representational character of the role(s), which, reciprocally, would decrease the functionality of the role(s) in a way that reduces the conflict; or, the impossibility of realizing the idealized elements of a role in actual practice can be mitigated by accentuating the value and worthiness of those elements that can be put into actual practice. Propositions I-B-1, I-B-2, and I-B-3 add even greater specificity to Turner's argument. I-B-1 states that, when representational processes come into conflict with functionality considerations, the value elements are usually marked and reaffirmed through explicit rituals and ceremonies in a way that emphasizes the value content and, at the same time, allows instrumental activity to proceed. I-B-2 argues that, when role-sets reveal incompatible elements, especially those involving values, there are usually efforts to differentiate out new roles in order to absorb the discordant elements. And I-B-3 states that, if actors cannot behave in ways that realize all the value elements of a role, then they will emit cues and gestures expressing distance from those elements that do not correspond to the idealized version of the role.

Proposition I-D repeats Proposition III in Table 21–4 on functionality, but this time in the context of representational processes. When actors in roles come from similar backgrounds, the role's representational character will increase as they play their roles and reaffirm their common background. Proposition I-E argues that those roles that are interconnected with roles already having strong representational elements will themselves take on these value premises. Proposition I-F argues that, when roles are played in the context of value conflict, value inconsistency, value ambiguity, and other problematic moral situations, these roles will tend to take on representational elements as a way to resolve the conflict, mend the inconsistency, or remove the ambiguity.

Finally, Turner and Colomy offer only one clear proposition on the relation of representational roles to questions of functionality and tenability. Proposition II argues that, when roles are highly representational, modification of the role will revolve more around issues of tenability (the fit of the role to the personal cost/reward calculations of incumbents) than around functionality (the fit of the role to the realization of collective goals). This is why, I assume, Proposition III in Table 21–4 on functionality emphasizes that, when functional roles do become infused with representational content, they become highly resistant to change. But, as Proposition II in Table 21–5 implies, this situation is rare; if modification of a representational role is to occur, it will be in terms of tenability considerations.

Tenability   Turner contends that there is a tendency for roles to evolve so as to provide the kinds of resources incumbents find gratifying and to minimize the costs that they must endure. Thus roles form and differentiate in ways that allow individuals to gain some personal reward, and so the issue of tenability revolves around the "fit" between the individual and a role in terms of the individual's calculations of costs and rewards. There is always a pressure

TABLE 21-6 Explanatory Principles on Tenability

I. The more activity among individuals involves the calculation of personal costs and rewards, the greater are considerations of tenability in the differentiation and accretion of roles.

  A. The more roles involve costs and rewards that are consensually valued in the broader social context, the more likely are roles to become implicated in the distribution of power and, hence, the greater is the likelihood that considerations of tenability will be linked to the broader system of inequality.

   1. The more roles are made tenable by their fit with the distribution of power, the more likely are those high in power to be incumbent in roles high in rewards, and the more likely are these incumbents to use their power to resist efforts to reduce the rewards or increase the costs associated with their roles.

   2. The more roles are made tenable by their fit with the distribution of power, the more likely are those low in power to be incumbent in roles with low rewards and/or high costs, and the more likely are they to use expressive ritual and communication to increase the rewards and/or reduce the costs associated with their roles.

  B. The less roles involve costs and rewards that are consensually valued in the broader social context, the more likely are roles to become implicated in self-definitional processes and, hence, the greater is the likelihood that tenability will be linked to individuals' self-conceptions.

   1. The more roles are made tenable by their fit with incumbents' self-conceptions, the more likely are incumbents to merge self and role, and the greater will be problems of tenability posed by discrepancies between elements of a role and self-conceptions.

   2. The more roles are made tenable by their fit with incumbents' self-conceptions, the more likely are those with power to use their resources to mitigate against change in their roles and to make their roles predictable.

II. The more considerations of tenability guide the differentiation and accretion of roles, the greater will be considerations of representation over functionality in modifying these roles, to the extent that they can be modified.

III. The more considerations of tenability revolve around self-related processes in the differentiation and accretion of roles, the greater is the likelihood that these roles will develop representational elements congruent with self-conceptions, and the more these roles will become resistant to change.

---

for tenability in roles as individuals try to reduce their costs and increase their rewards. Table 21-6 summarizes Turner's propositions on tenability processes in a manner comparable to my review of functionality and representation.

Like its counterparts for functionality and representation, Proposition I is a tautology simply asserting that, as individuals begin to calculate personal costs and rewards in a role, its dynamics are altered toward tenability processes. Proposition I-A presents Turner's views on an important condition influencing tenability. When the costs and rewards associated with roles are consensually valued, incumbency in roles will reflect the distribution of power in the broader social context. That is, those with external bases of power in the wider social arena will use that power to secure those roles that have high rewards and low costs, whereas those without power will become incumbent in roles with fewer rewards and perhaps higher costs. Thus roles will reflect the more general system of inequality when there is consensus over the worth of resources received, and costs incurred, in a role.

With this basic insight, Propositions I-A-1 and I-A-2 specify in more detail some of the dynamics involved with power as the principal basis of individuals' efforts to make their roles tenable. Proposition I-A-1 states that those incumbents with power will resist efforts to reduce the rewards or increase the costs associated with a role, whereas Proposition I-A-2 argues that those low in power and incumbent in less gratifying and more costly roles will use expressive rituals and forms of communication as a way to increase their rewards and make costs more endurable.

Proposition I-B turns to the situation whereby the rewards and costs of a role are not consensually valued. Under this condition, tenability shifts from fitting external power to the appropriate level of rewards in a role to finding a fit between an individual's self-conception and the rewards and costs associated with a role. Proposition I-B-1 emphasizes that this process can create problems for the individual, since, once people view roles as having to be worthy of their self-conception, such merger of self and role escalates tenability problems when there are discrepancies between the role and self-conception. People who merge role and self thus become vulnerable when forced to engage in elements of a role that are at odds with their sense of self. Proposition I-B-2 reevokes the power variable by noting that, when those with power merge self with a role, they use their power and control of other resources to mitigate external influences that might change the role, and they seek to make their roles predictable.

Proposition II states the general relationship between tenability, on the one hand, and functionality and representation, on the other. If tenability is the basis for role differentiation and accretion, then representational considerations more than functionality will modify the role, if indeed any modifications occur. Proposition III simply extends this idea and notes that tenable roles tend to develop representational overlays that are congruent with people's self-conceptions, thereby making roles doubly resistant to change.

PROCESS ROLE THEORY: A BRIEF ASSESSMENT OF TURNER'S APPROACH

In assessing Turner's strategy, let me return to some of the issues of substance, method, and theory that have plagued interactionism in general. With regard to substantive issues, Turner's assumptions emphasize the fluid nature of interactive processes, but, in contrast to Blumer, he holds a deterministic view of causality. Moreover, unlike Blumer, he does not make exaggerated claims about the range and scope of his theoretical efforts, which are about individuals and the microprocesses by which they come to terms with one another in varying types of social contexts. There is no claim that all social events can be understood by his role theory. Rather, to the extent that attention to the interactive processes of individuals is considered important, Turner argues that his approach is the most appropriate.

With respect to method, Turner's empirical work on a variety of topics attests to his recognition that operationalization of concepts is critical.[21] Observational techniques are an important research tool, but survey and experimental techniques that utilize structured measuring instruments are also deemed appropriate.

Finally, more than most interactionists, Turner has been concerned with building theory. He recognizes that it is necessary to move beyond sensitizing frameworks and to explore strategies for developing deductive relations among propositions. Although I think that his functionality, representation, and tenability propositions will require revision and supplementation, he has at least developed a strategy that may yield more powerful propositions.

In sum, then, I feel Turner's theory is one of the best efforts to incorporate all varieties of symbolic interactionism and role theory into a conceptual framework and strategy that stresses theory building and theory testing. The general direction and thrust of his approach should, I feel, be emulated if interactionism is to remain a viable theoretical perspective.

---

[21]For representative examples of the variety of empirical research conducted by Turner, see "The Navy Disbursing Officer as a Bureaucrat," *American Sociological Review* 12 (June 1947), pp. 342–48; "Moral Judgment: A Study in Roles," *American Sociological Review* 17 (January 1952), pp. 70–77; "Occupational Patterns of Inequality," *American Journal of Sociology* 50 (March 1954), pp. 437–47; "Zoot-Suiters and Mexicans: Symbols in Crowd Behavior" (with S. J. Surace), *American Journal of Sociology* 62 (July 1956), pp. 14–20; "The Changing Ideology of Success: A Study of Aspirations of High School Men in Los Angeles," *Transactions of the Third World Congress of Sociology* 5 (1956), pp. 35–44; "An Experiment in Modification of the Role Conceptions," *Yearbook of the American Philosophical Society* (1959), pp. 329–32; "Some Family Determinants of Ambition," *Sociology and Social Research* 46 (July 1962), pp. 397–411; *The Social Context of Ambition* (San Francisco: Chandler & Sharp, 1964); "Ambiguity and Interchangeability in Role Attribution" (with Norma Shosid), *American Sociological Review* 41 (December 1976); and "The True Self Method for Studying Self-Conception," *Symbolic Interaction* 4 (1981), pp. 1–20.

CHAPTER 22

Dramaturgical Theory

Erving Goffman

THE INTERACTION ORDER

Erving Goffman was one of the most creative analysts of micro social processes. His domain of inquiry was, as he phrased the issue just before his death in 1982, the *interaction order*, or micro processes revolving around face-to-face behavior and interaction among individuals.[1] Although his works span many different topics, Goffman emphasized that people spend much of their waking life moving about in space, making fleeting as well as engrossing contact with others, going to meetings, attending performances (lectures, movies, plays, television, etc.), and celebrating occasions. These kinds of activities, Goffman argued, have not been given sufficient attention in sociological theory, despite the fact that they constitute such a large proportion of human daily experience.

Unlike many who advocate fine-grained microanalysis, however, Goffman did not proclaim that the interaction order is all that is real. Rather, he simply argued that this interaction order constitutes a distinctive realm of reality that reveals its own unique dynamics. For, ". . . to speak of the relative autonomous forms of life in the interaction order . . . is not to put forward these forms as somehow prior, fundamental, or constitutive of the shape of macroscopic phenomena."[2] Goffman recognized that, at best, there is a "loose coupling" of the micro and macro realms. Macro phenomena, such as commodities markets, urban land-use values, economic growth, and society-wide stratification, are not going to be explained by micro-level analysis.[3] Of course, one can supple-

---

[1]Erving Goffman, "The Interaction Order," *American Sociological Review* (February 1983), pp. 1–17. For a recent debate on Goffman's view of the interaction order, see: Anne W. Rawls, "The Interaction Order Sui Generis. Goffman's Contribution to Social Theory," *Sociological Theory* 5 (2, 1987), pp. 136–49; and Stephan Fuchs, "The Constitution of Emergent Interaction Orders, A Comment on Rawls," *Sociological Theory* 5 (1, 1988), pp. 122–24.

[2]Goffman, "The Interaction Order," p. 9.

[3]Ibid.

ment macro-level explanations by recording how individuals interact in various types of settings and encounters, but these analyses will not supplant macro-level explanations. Conversely, what transpires in the interaction order cannot be explained solely by macroprocesses. Rather, macro-level phenomena are always transformed in ways unique to the individuals involved in interaction.

To be sure, macro phenomena constrain and circumscribe interaction and, at times, guide the general form of the interaction, but the inherent dynamics of the interaction itself preclude a one-to-one relation to these structural parameters. Indeed, the form of interaction can often be at odds with macro-structures, operating smoothly in ways that contradict these structures without dramatically changing them. Thus the crude notion that interaction is constrained by macrostructures in ways that "reproduce" social structure does not recognize the autonomy of the interaction order. For interaction ". . . is not an expression *of* structural arrangements in any simple sense; at best it is an expression advanced *in regard to* these arrangements. Social structures don't 'determine' culturally standard displays (of interaction rituals), they merely help select from the available repertoire of them."[4] Thus there is a "loose coupling" of "interactional practices and social structures, a collapsing of strata and structures into broader categories, the categories themselves not corresponding one-to-one to anything in the structural world. . . ."[5] There is, then, "a set of transformation rules, or a membrane selecting how various externally relevant social distinctions will be managed within the interaction."[6]

These "transformations" are, however, far from insignificant phenomena. What people do as they make contact and interact creates for them a sense of order. Much of what gives the social world a sense of "being real" are the practices of individuals as they deal with one another in various types of situations. For it is out of the rules of interpersonal contact that "we owe our unshaking sense of realities."[7] Moreover, although a single gathering and episode of interaction may not have great social significance, it is ". . . through these comings together [that] much of our social life is organized."[8] The interaction order is thus a central topic of sociological theory.

Goffman's approach to this domain is unique.[9] For, although we must "keep faith with the spirit of natural science, and lurch along, seriously kidding ourselves that our rut has a forward direction,"[10] we must not become overenamored with the mature sciences. Social life is "ours to study naturalistically,"

---

[4]Ibid., p. 11. Italics in original.

[5]Ibid.

[6]Ibid.

[7]Erving Goffman, *Encounters: Two Studies in the Sociology of Interaction* (Indianapolis: Bobbs-Merrill, 1961), p. 81.

[8]Erving Goffman, *Behavior in Public Places: Notes on the Social Organization of Gatherings* (New York: Free Press, 1963), p. 234.

[9]For a recent set of essays on Goffman's analysis, see: Paul Drew and Anthony Wootton, eds., *Erving Goffman: Exploring the Interaction Order* (Cambridge, England: Polity Press, 1988).

[10]Goffman, "The Interaction Order," p. 2.

but our study should not be rigid.[11] Instead, ad hoc observation, cultivation of anecdotes, creative thinking, illustrations from literature, examination of books of etiquette, personal experiences, and many other sources of unsystematic data should guide inquiry into the micro order. Indeed, "human life is only a small irregular scab on the face of nature, not particularly amenable to deep systematic analysis."[12] Yet it can be studied in the spirit of scientific inquiry. And it is this spirit of science that will guide my review of the theoretical significance of Goffman's sociology. My concern will be not so much with the voluminous examples and wide range of topics in Goffman's work as with the core structure of his theory as it evolved over the years.

THE DRAMATURGICAL METAPHOR

In both his first and last major works—*The Presentation of Self in Everyday Life* and *Frame Analysis*, respectively—Goffman analogized to the stage and theater. Hence the designation of his work as "dramaturgical" has become commonplace.[13] This designation is, however, somewhat misleading because it creates the impression that there is a script, a stage, an audience, props, and actors playing roles. Such imagery is, I think, more in tune with standard role theory (discussed in Chapter 20). There is another sense in which dramaturgy is used to denote Goffman's work: individuals are actors who "put on" a performance, often cynical and deceptive, for one another and who manipulate the script, stage, props, and roles for their own purposes. This more cynical view of Goffman's dramaturgy is perhaps closer to the mark,[14] but it too is somewhat misleading. Commentators have often portrayed Goffman as presenting a kind of "con man"[15] view of human social interaction—a metaphor that captures some of the examples and topics of his approach. And yet this emphasis tends to obscure the more fundamental processes denoted by Goffman's theorizing.

I will use the metaphor of dramaturgy in a less extreme sense. In Goffman's work there is concern with a cultural script, or normative roles; there is a heavy emphasis on how individuals manage their impressions and play roles; there is a concern with stages and props (physical space and objects); there is an emphasis on stag*ing*, or the manipulation of gestures as well as spacing, props, and other physical aspects of a setting; there is a view of self as situational, determined more by the cultural script, stage, and audience than by enduring

---

[11]Ibid., p. 17.

[12]Ibid.

[13]Erving Goffman, *The Presentation of Self in Everyday Life* (Garden City, NY: Anchor Books, 1959) and *Frame Analysis: An Essay on the Organization of Experience* (Boston: Northeastern University Press, 1986; originally published in 1974 by Harper & Row).

[14]See, for example, Randall Collins, *Theoretical Sociology* (San Diego: Harcourt Brace Jovanovich, 1988), pp. 203–7, 291–98, and *Three Sociological Traditions* (New York: Oxford University Press, 1985).

[15]For example, see: R. P. Cuzzort and E. W. King, *Twentieth-Century Social Thought*, 4th ed. (Fort Worth, TX: Holt, Rinehart & Winston, 1989), Chap. 12.

and transituational configurations of self-attitudes and self-feelings; and there is a particular emphasis on how performances create a theatrical ambience—a mood, definition, and sense of reality.

The above metaphor provides only an orientation of Goffman's approach. We need to "fill in" this orientation with more details. To do so, I will review Goffman's most important works and try to pull the diverse vocabulary and concepts together into a more unified theoretical perspective—dramaturgical in its general contours but more than just a clever metaphor.[16]

THE PRESENTATION OF SELF

The Presentation of Self in Everyday Life[17] is Goffman's first major work and is largely responsible for the designation of Goffman as a dramaturgical theorist. The basic line of argument is that individuals deliberately "give" and inadvertently "give off" signs that provide others with information about how to respond. Out of such mutual use of "sign-vehicles," individuals develop a "definition of the situation," which is a "plan for cooperative activity" but which, at the same time, is ". . . not so much a real agreement as to what exists but rather a real agreement as to whose claims concerning what issues will be temporarily honored."[18] In constructing this overall definition of a situation, individuals engage in performances in which each orchestrates gestures to "present oneself" in a particular manner as a person having identifiable characteristics and deserving of treatment in a certain fashion. These performances revolve around several interrelated dynamics.

First, a performance involves the creation of a *front*. A front includes the physical "setting" and the use of the physical layout, its fixed equipment (decor, furniture, and the like), and other "stage props" to create a certain impression. A front also involves (*a*) "items of expressive equipment" (emotions, energy, and other capacities for expression); (*b*) "appearance," or those signs that tell others of an individual's social position and status as well as the "ritual state" of the individual with respect to social, work, or recreational activity;[19] and (*c*) "manner," or those signs that inform others about the role that an individual expects to play.[20] As a general rule, people expect consistency in these elements of their fronts—use of setting and its props, mobilization of expressive equipment, social status, expression of ritual readiness for various types of activity, and efforts to assume certain roles.[21] There is a relatively small number of

---

[16]See Stephan Fuchs, "Second Thoughts on Emergent Interaction Orders," *Sociological Theory* 7 (1, 1989), pp. 121–23.

[17]Goffman, *The Presentation of Self in Everyday Life* (see note 13 for full reference).

[18]Ibid., pp. 9–10.

[19]Note: it is in this discussion that Randall Collins (see Chapter 12) took his idea of situations as either work-practical, ceremonial, or social.

[20]Note the similarity of this idea to Ralph H. Turner's discussion (see Chapter 21) on "role making."

[21]Goffman, *The Presentation of Self*, pp. 24–25.

fronts, and people know them all. Moreover, fronts tend to be established, institutionalized, and stereotypical for various kinds of settings, with the result that "when an actor takes an established role, usually he finds that a particular front has already been established for it."[22]

Second, in addition to presenting a front, individuals use gestures in what Goffman termed *dramatic realization*, or the infusion into activity of signs that highlight commitment to a given definition of a situation. The more a situation creates problems in presenting a front, Goffman argued, the greater will be efforts at dramatic realization.[23]

Third, performances also involve *idealizations*, or efforts to present oneself in ways that "incorporate and exemplify the officially accredited values of society."[24] When individuals are mobile, moving into a new setting, efforts at idealization will be most pronounced. Idealization creates a problem for individuals, however, because, if the idealization is to be effective, individuals must suppress, conceal, and underplay those elements of themselves that might contradict more general values.

Fourth, such efforts at concealment are part of a more general process of *maintaining expressive control*. Because minor cues and signs are read by others and contribute to a definition of a situation, actors must regulate their muscular activity, their signals of involvement, their orchestration of front, and their ability to be fit for interaction. The most picayune discrepancy between behavior and the definition of a situation can unsettle the interaction, for ". . . the impression of reality fostered by a performance is a delicate, fragile thing that can be shattered by very minor mishaps."[25]

Fifth, individuals can also engage in *misrepresentation*. The eagerness of one's audience to read gestures and determine one's front makes that audience vulnerable to manipulation and duping.[26]

Sixth, individuals often attempt to engage in *mystification*, or the maintenance of distance from others as a way to keep them in awe and in conformity to a definition of a situation. Such mystification is, however, limited primarily to those of higher rank and status.

Seventh, individuals seek to make their performances seem *real* and to avoid communicating a sense of contrivance. Thus individuals must communicate, or at least appear to others, as sincere, natural, and spontaneous.

These procedures for bringing off a successful performance and thereby creating an overall definition of a situation are the core of Goffmanian sociology. They become elaborated and extended in subsequent works, but Goffman never

---

[22]Ibid., p. 27.

[23]Ibid., p. 32.

[24]Ibid., p. 35. Note how this idea parallels the notion of "representational" aspect of roles developed by Ralph H. Turner and Paul Colomy (see Chapter 21).

[25]Ibid., p. 56.

[26]This is a consistent theme in Goffman's work that, I think, he overemphasized. See, for example, Erving Goffman, *Strategic Interaction* (Philadelphia: University of Pennsylvania Press, 1969).

TABLE 22-1 Goffman's Propositions on Interaction and Performance

---

I. The more individuals in a setting make visual and verbal contact, the greater are their efforts to use gestures to orchestrate a performance.

  A. The more individuals can present a coherent front, the greater is their ability to orchestrate a performance.

    1. The more individuals can control physical space, props, and equipment in a setting, the greater is their ability to present a coherent front.

    2. The more individuals can control their expressive equipment in a setting, the greater is their ability to present a coherent front.

    3. The more individuals can control signals of their propensity for types of ritual activity, the greater is their ability to present a coherent front.

    4. The more individuals can control the signals of their status outside and inside the interaction, the greater is their ability to present a coherent front.

    5. The greater the ability of individuals to control those signals pertaining to identifiable roles, the greater is their ability to present a coherent front.

  B. The more individuals can accentuate those signals relevant to a situation, the greater is their ability to orchestrate a performance.

  C. The more individuals can incorporate and exemplify general cultural values, the greater is their ability to orchestrate a performance.

  D. The more individuals can maintain control of their expressive gestures, the greater is their ability to orchestrate a performance.

  E. The more individuals can imbue a situation with their own mystique, the greater is their ability to orchestrate a performance.

  F. The more individuals can signal their sincerity, the greater is their ability to orchestrate a performance.

II. The more individuals in a setting can orchestrate their performances and, at the same time, accept one another's performances, the more likely are they to develop a common definition of the situation.

III. The more individuals can develop a common definition of the situation, the greater will be the ease of their interaction.

---

abandoned the idea that fundamental to the interaction order are the efforts of individuals to orchestrate their performances, even in deceptive and manipulative ways, so as to maintain a particular definition of the situation. These ideas are presented propositionally in Table 22-1.

Although the propositions in Table 22-1 constitute only an opening chapter in *The Presentation of Self*, they are by far the most enduring portions of this first major work. The rest of the book is concerned with performances sustained by more than one individual. Goffman introduced the concept of *team* to denote performances that are presented by individuals who must cooperate to bring off a particular definition of the situation. Often two teams must present performances to each other, but more typically one team constitutes a performer and the other an audience. Team performers generally move between a *front region*, or frontstage, where they coordinate their performances before an audience, and a *back region*, or backstage, where team members can relax. Goffman also introduced the notion of *outside*, or the residual region beyond the frontstage and backstage. Frontstage behavior is polite, maintaining a decorum appropriate to a team performance (for example, selling cars, serving food, meeting students, etc.), whereas backstage behavior is more informal and is geared toward maintaining the solidarity and morale of team performers. When outsiders or members of the audience intrude upon performers in the backstage, a tension is created because team members are caught in their nonperforming roles.[27]

A basic problem of all team performances is maintaining a particular definition of the situation in front of the audience. This problem is accentuated when there are large rank or status differences among team members,[28] when the team has many members,[29] when the front- and backstages are not clearly partitioned, and when the team must hide information contrary to its image of itself. To counteract these kinds of problems, social control among the team's members is essential. When members are backstage, such control is achieved through morale-boosting activities, such as denigrating the audience, kidding one another, shifting to informal address, and engaging in stage talk (talk about performances on frontstage). When they are onstage, control is sustained by realigning actions revolving around subtle communications among team members that, hopefully, the audience will not understand.

Breaches of the performances occur when a team member acts in ways that call into question the definition of the situation created by the team's performances. Attempts to prevent such incidents involve further efforts at social control, especially a backstage emphasis on (a) playing one's part and not emitting unmeant gestures, (b) showing loyalty to the team and not the audience, and (c) exercising foresight and anticipating potential problems with the team or the audience. Team members are assisted in social control by members of the audience, which (a) tends to stay away from the backstage, (b) acts disinterested when exposed to backstage behavior, and (c) employs elaborate etiquette (exhibiting proper attention and interest, inhibiting their own potential performances, avoiding faux pas) in order to avoid a "scene" with the team.

What is true of teams and audiences is, Goffman implied, also true of individuals. Interaction involves a performance for others who constitute an audience. One seeks to sustain a performance when moving to the frontstage, while relaxing a front when moving to a backstage region. People try very hard to avoid mistakes and faux pas that could breach the definition of the situation, and they are assisted in this effort by others in their audience who exercise tact and etiquette in order to avoid a scene. Such are the themes of *The Presentation of Self*, and most of Goffman's work represents a conceptual elaboration of them. The notion of "teams" recedes, but a general model elaborating these themes into a theory of interaction among individuals emerges.

FOCUSED INTERACTION

Goffman generally employed the terms *unfocused* and *focused* to denote two basic types of interaction. *Unfocused interaction* "consists of interpersonal

---

[27]Goffman, *The Presentation of Self*, pp. 137–38.

[28]Ibid., p. 92.

[29]Ibid., p. 141.

communications that result solely by virtue of persons being in one another's presence, as when two strangers across the room from each other check up on each other's clothing, posture, and general manner, while each modifies his own demeanor because he himself is under observation."[30] As I will explore later, such unfocused interaction is, Goffman argued, an important part of the interaction order, for much of what people do is exchange glances and monitor each other in public places. *Focused interaction*, in contrast, "occurs when people effectively agree to sustain for a time a single focus of cognitive and visual attention, as in a conversation, a board game, or a joint task sustained by a close face-to-face circle of contributors."[31]

Encounters

Focused interaction occurs within what Goffman termed *encounters*, which constitute one of the core structural units of the interaction order. Goffman mentioned encounters in his first work, *The Presentation of Self in Everyday Life*, but their full dimensions are explored in his next book, *Encounters*.[32] There, an encounter is defined as focused interaction revealing the following characteristics:[33]

1. a single visual and cognitive focus of attention
2. a mutual and preferential openness to verbal communication
3. a heightened mutual relevance of acts
4. an eye-to-eye ecological huddle, maximizing mutual perception and monitoring
5. an emergent "we" feeling of solidarity and flow of feeling
6. a ritual and ceremonial punctuation of openings, closings, entrances, and exits
7. a set of procedures for corrective compensation for deviant acts

To sustain itself, an encounter develops a "membrane," or penetrable barrier to the larger social world in which the interaction is located. Goffman typically conceptualized the immediate setting of an encounter as a *gathering*, or the assembling in space of co-present individuals; in turn, gatherings are lodged within a more inclusive unit, the *social occasion*, or the larger undertaking sustained by fixed equipment, distinctive ethos and emotional structure, program and agenda, rules of proper and improper conduct, and preestablished sequencing of activities (beginning, phases, high point, and ending). Thus encounters emerge from episodes of focused interaction within gatherings that are lodged in social occasions.[34]

---

[30]Goffman, *Encounters*, p. 7 (see note 7 for full reference).

[31]Ibid.

[32]Ibid.; see note 7.

[33]Ibid., p. 18. Note the similarity of this definition to Randall Collins' portrayal of "interaction rituals" (see Chapter 12).

[34]Goffman, *Behavior in Public Places*, pp. 18–20.

The membrane of an encounter, as well as its distinctive characteristics listed above, are sustained by a set of *rules*. In *Encounters*, Goffman lists several; later, in what is probably his most significant work, *Interaction Ritual*, he lists several more.[35] Let me combine both discussions by listing the rules that guide focused interaction in encounters:

1. *Rules of irrelevance*, which "frame" a situation as excluding certain materials (attributes of participants, psychological states, cultural values and norms, etc.).[36]
2. *Rules of transformation*, which specify how materials moving through the membrane created by rules of irrelevance are to be altered so as to fit into the interaction.
3. *Rules of "realized resources,"* which circumscribe the schema for expression and interpretation that is to be constructed and used by participants.
4. *Rules of talk*, which are the procedures, conventions, and practices guiding the flow of verbalizations with respect to:[37]
   a. maintaining a single focus of attention.
   b. establishing "clearance cues" for determining when one speaker is done and another can begin.
   c. determining how long and how frequently any one person can hold the floor.
   d. regulating interruptions and lulls in the conversation.
   e. sanctioning participants whose attention wanders to matters outside the conversation.
   f. assuring that nearby people do not interfere with the conversation.
   g. guiding the use of politeness and tact, even in the face of disagreements.
5. *Rules of self respect*, which encourage participants to honor with tact and etiquette their respective efforts to present themselves in a certain light.

Interaction is thus guided by complex configurations of rules that individuals learn how to use and apply in different types of encounters, logged in varying types of gatherings and social occasions. The "reality" of the world is, to a very great extent, sustained by people's ability to invoke and use these rules to sustain encounters.[38] When these rules are operating effectively, individuals develop a "state of euphoria," or what Randall Collins (see Chapter 12) terms enhanced "emotional energy." However, encounters are vulnerable to "dysphoria" or tension when these rules do not exclude troublesome external ma-

---

[35]Goffman, *Encounters*, pp. 20–33; and Erving Goffman, *Interaction Ritual: Essays on Face-to-Face Behavior* (Garden City, NY: Anchor Books, 1967), p. 33.

[36]The concept of frame, which Goffman said that he took from Gregory Bateson, is to become central in Goffman's later work. See closing section in this chapter.

[37]These rules come from: Goffman, *Interaction Ritual*, p. 33.

[38]Here Goffman anticipates much of ethnomethodology, which is examined in Chapter 23.

terials or fail to regulate the flow of interaction. Such failures are seen by Goffman as *incidents* or *breaches*; when they can be effectively handled by tact and corrective procedures, they are then viewed as *integrations* because they are blended into the ongoing encounter. The key mechanism for avoiding dysphoria and maintaining the integration of the encounter is the use of ritual.

Ritual

In his *Interaction Ritual*,[39] Goffman's great contribution is the recognition that minor, seemingly trivial, and everyday rituals—such as "Hello, how are you?" "Good morning," "Please, after you," and so on—are crucial to the maintenance of social order. In his words, he "reformulated Durkheim's social psychology in a modern dress"[40] by recognizing that, when individuals gather and begin to interact, their behaviors are highly ritualized. That is, actors punctuate at each phase of interpersonal contact with stereotypical sequences of behavior that invoke the rules of the encounter and, at the same time, become the medium or vehicle by which the rules are followed. Rituals are thus essential for (*a*) mobilizing individuals to participate in interaction; (*b*) making them cognizant of the relevant rules of irrelevance, transformation, resource use, and talk; (*c*) guiding them during the course of the interaction; and (*d*) helping them correct for breaches and incidents.

Among the most significant are those rituals revolving around deference and demeanor. *Deference* pertains to interpersonal rituals that express individuals' respect for others, their willingness to interact, their affection and other emotions, and their engagement in the encounter. In Goffman's words, deference establishes "marks of devotion" by which an actor "celebrates and confirms his relationship to a recipient."[41] As a result, deference contains a "kind of promise, expressing in truncated form the actor's avowal and pledge to treat the recipient in a particular way in the on-coming activity."[42] Thus, seemingly innocuous gestures—"It's nice to see you again," "How are things?" "What are you doing?" "Goodbye," "See you later," and many other stereotypical phrases as well as bodily movements—are rituals that present a demeanor invoking relevant rules and guiding the opening, sequencing, and closing of the interaction.

Deference rituals, Goffman argued, can be of two types: (1) avoidance rituals and (2) presentational rituals. *Avoidance rituals* are those that an individual uses to keep distance from another and to avoid violating the "ideal sphere" that lies around the other. Such rituals are most typical among unequals. *Presentational rituals* "encompass acts through which the individual makes specific attestations to recipients concerning how he regards them and how he will treat

---

[39]Goffman, *Interaction Ritual* (see note 35 for full reference).

[40]Particularly Durkheim's later work, as it culminated in *Elementary Forms of the Religious Life* (New York: Free Press, 1947; originally published in 1912).

[41]Goffman, *Interaction Ritual*, pp. 56–57.

[42]Ibid., p. 60.

them in the on-coming interaction."[43] Goffman sees interaction as constantly involving a dialectic between avoidance and presentational rituals as individuals respect each other and maintain distance while trying to make contact and get things done.[44]

In contrast, *demeanor* is "that element of the individual's ceremonial behavior conveyed through deportment, dress, and bearing which serves to (inform) those in his immediate presence that he is a person of certain desirable or undesirable qualities."[45] Through demeanor rituals individuals present images of themselves to others and, at the same time, communicate that they are reliable, trustworthy, and tactful.

Thus it is through deference and demeanor rituals that individuals plug themselves into an encounter by invoking relevant rules and demonstrating their capacity to follow them, while simultaneously indicating their respect for others and presenting themselves as certain kinds of individuals. It is the enactment of such deference and demeanor rituals in concrete gatherings, especially encounters but also including unfocused situations, that provides a basis for the integration of society. For "throughout . . . ceremonial obligations and expectations, a constant flow of indulgences is spread through society, with others who are present constantly reminding the individual that he must keep himself together as a well demeaned person and affirm the sacred quality of these others."[46]

Roles

In presenting themselves to others, individuals also seek to play a particular role. Thus, as people present a front, invoke relevant rules, and emit rituals, they also try to orchestrate a role for themselves. For Goffman, then, a *role* is "a bundle of activities visibly performed before a set of others and visibly meshed into the activities these others perform."[47] In the terms of Ralph Turner's analysis (see Chapter 21), individuals attempt to "make a role" for themselves; if successful, this effort contributes to the overall definition of the situation.

In trying to establish a role, the individual "must see to it that the impressions of him that are conveyed in the situation are compatible with role-appropriate personal qualities effectively imputed to him."[48] Thus individuals in a situation are expected to try to make roles for themselves that are consistent with their demeanor, their self as performed before others, and their front (stage props, expressive equipment, appearance). And, if the inconsistency between the attempted role and these additional aspects of a performance becomes

---

[43]Ibid.

[44]Ibid., pp. 75–76.

[45]Ibid., p. 77.

[46]Ibid., p. 91.

[47]Goffman, *Encounters*, p. 96.

[48]Ibid., p. 87.

evident, then others in the situation are likely to sanction the individual through subtle cues and gestures. These others are driven to do so because discrepancy between another's role and other performance cues disrupts the definition of the situation and the underlying sense of reality that this definition promotes. Thus role is contingent on the responses and reactions of others, and, because their sense of reality is partially dependent on successful and appropriate role assumption, an individual will have difficulty changing a role in a situation once it is established.

Oftentimes, however, persons perceive a role to be incompatible with their image of themselves in a situation. Under these conditions they will display what Goffman termed *role distance*, whereby a "separation" of the person from a role is communicated. Such distancing, Goffman argued,[49] allows the individual to (a) release the tension associated with a role considered to be "beneath his (her) dignity," (b) present additional aspects of self that extend beyond the role, and (c) remove the burden of "complete compliance to the role," thereby making minor transgressions less dramatic and troublesome for others.

Role distance is but an extreme response to the more general process of *role embracement*. For any role, individuals will reveal varying degrees of attachment and involvement in the role. One extreme is role distance, whereas the other extreme is what Goffman termed *engrossment*, or complete involvement in a role. In general, Goffman argued,[50] those roles in which individuals can direct what is going on are likely to involve high degrees of embracement, whereas those roles in which the individual is subordinate will be played with considerable role distance.

As is evident, then, the assumption of a role is connected to the self-image that actors project in their performances. Although the self that one reveals in a situation is dependent upon the responses of others who can confirm or disconfirm that person's self in a situation, the organization of a performance onstage before others is still greatly circumscribed by self.

Self

Goffman's view of self is highly situational and contingent on the responses of others. Although one of the main activities of actors in a situation is to present themselves in a certain way, Goffman was highly skeptical about a "core" or "transituational" self-conception that is part of an individual's "personality." In almost all his works he took care to emphasize that individuals do not have an underlying "personality" or "identity" that is carried from situation to situation. For example, in his last major book, *Frame Analysis*,[51] he argued that people in interaction often presume that the presented self provides a glimpse at a more coherent and core self, but in reality this is simply a folk

---

[49]Ibid., p. 113.
[50]Ibid., p. 107.
[51]Goffman, *Frame Analysis* (see note 13 for full reference).

presumption because, in fact, there is "... no reason to think that all these gleanings about himself that an individual makes available, all these pointings from his current situation to the way he is in his other occasions, have anything very much in common."[52]

Yet, even though there is no transituational or core self, people's efforts to present images of themselves in a particular situation and others' reactions to this presentation are central dynamics in all encounters. For individuals constantly emit demeanor cues that project images of themselves as certain kinds of persons, or, in the vocabulary of *The Presentation of Self*, they engage in "performances." In *Interaction Ritual*, Goffman rephrases this argument somewhat, and, in so doing, he refines his views on self. In encounters, an individual acts out *a line*, which is "a pattern of verbal and nonverbal acts by which he expresses his view of the situation and, through this, his evaluation of the participants, especially himself."[53] In developing a line, an individual presents *a face*, which is "the positive social value a person effectively claims for himself by the line others assume he has during a particular contact."[54] Individuals seek to stay *in face* or to *maintain face* by presenting an image of themselves through their line that is supported by the responses of others and, if possible, sustained by impersonal agencies in a situation. Conversely, a person is *in wrong face* or *out of face* when the line emitted is inappropriate and unaccepted by others. Thus, although a person's social face "can be his most personal possession and the center of his security and pleasure, it is only on loan to him from society; it will be withdrawn unless he conducts himself in a way that is worthy of it...."[55]

As noted earlier, Goffman argued that a key norm in any encounter is "the rule of self respect," which requires individuals to maintain face and, through tact and etiquette, the face of others. Thus, by virtue of tact or the "language of hint..., innuendo, ambiguities, well-placed phrases, carefully worked jokes, and so on,"[56] individuals sustain each other's face; in so doing, they confirm the definition of the situation and promote a sense of a common reality. It is for this reason that, once established, a given line and face in an encounter are difficult to change, since to alter face (and the line by which it is presented) would require redefining the situation and recreating a sense of reality. And, since face is "on loan" to a person from the responses of others, the individual must incur high costs—such as embarrassment or a breach of the situation— to alter a line and face.

Face engagements are usually initiated with eye contact, and, once initiated, they involve ritual openings appropriate to the situation (as determined by length of last engagement, amount of time since previous engagement, level of inequality, etc.). During the course of the face engagement, each individual

---

[52]Ibid., p. 299.

[53]Goffman, *Interaction Ritual*, p. 5.

[54]Ibid.

[55]Ibid., p. 10.

[56]Ibid., p. 30.

uses tact to maintain, if possible, each other's face and to sanction, if necessary, each other into their appropriate line. In particular, participants seek to avoid "a scene" or breach in the situation, and so they use tact and etiquette to save their own face and that of others. Moreover, as deemed appropriate for the type of encounter (as well as for the larger gathering and more inclusive social occasion), individuals will attempt to maintain what Goffman sometimes termed *the territories of self*, revolving around such matters as physical props, ecological space, personal preserve (territory around one's body), and conversational rights (to talk and be heard), which are necessary for people to execute their line and maintain face.[57] In general, the higher the rank of individuals, the greater their territories of self in an encounter.[58] To violate such territories disrupts or breaches the situation, forcing remedial action on the part of participants to restore their respective lines, face, definitions of the situation, and sense of reality.

Talk

Throughout his work, but especially in later books such as *Frame Analysis*[59] and in numerous essays (see those collected in *Forms of Talk*[60]), Goffman emphasized the significance of verbalizations for focusing people's attention. For when "talk" is viewed interactionally, "it is an example of that arrangement by which individuals come together and sustain matters having a ratified, joint, current, and running claim upon attention, a claim which lodges them together in some sort of intersubjective, mental world."[61] Thus, in Goffman's view, "no resource is more effective as a basis for joint involvement than speaking" because it fetches "speaker and hearer into the same interpretation schema that applies to what is thus attended."[62]

Talk is thus a crucial mechanism for drawing individuals together, focusing their attention, and adjudicating an overall definition of the situation. Because talk is so central to focusing interaction, it is normatively regulated and ritualized. One significant norm is the prohibition against *self-talk*, except under a few conditions, because, when people talk to themselves, it "warns others that they might be wrong in assuming a jointly maintained base of ready mutual intelligibility. . . ."[63] Moreover, other kinds of quasi talk are also regulated and ritualized. For example, *response cues* or "exclamatory interjections which are not full-fledged words"—"Oops," "Wow," "Oh," and "Yikes"—are regulated as

---

[57]Erving Goffman, *Relations in Public: Micro Studies of the Public Order* (New York: Harper Colophon Books, 1972; originally published in 1971 by Basic Books), pp. 38–41.

[58]Ibid., pp. 40–41.

[59]See note 13.

[60]Erving Goffman, *Forms of Talk* (Philadelphia: University of Pennsylvania Press, 1981).

[61]Ibid., pp. 70–71.

[62]Ibid., p. 71.

[63]Ibid., p. 85.

to when they can be used and the way they are uttered.[64] Verbal fillers—"ah," "uh," "um," and the like—are also ritualized and are used to facilitate "conversational tracking." In essence, they indicate that "the speaker does not have, as of yet, the proper word but is working on the matter" and that he or she is still engaged in the conversation. Even seemingly emotional cues and tabooed expressions, such as all the "four-letter words," are not so much an expression of emotion as "self-other alignment" and assert that "our inner concerns should be theirs." Such outbursts are normative and ritualized because this "invitation into our interiors tends to be made only when it will be easy to other persons present to see where the voyage takes them."[65]

In creating a definition of the situation, Goffman argued, talk operates in extremely complex ways. When individuals talk, they create what Goffman termed a *footing*, or assumed foundation for the conversation and the interaction. Because verbal symbols are easily manipulated, people can readily change the footing or basic premises underlying the conversation.[66] Such shifts in footing are, however, highly ritualized and usually reveal clear markers. For example, when a person says something like "Let's not talk about that," the footing of the conversation is shifted, but in a ritualized way; similarly, when someone utters a phrase like "That's great, but what about . . .?" this person is also changing the footing through ritual.

Shifts in footing raise a question that was increasingly to dominate Goffman's later works: the issue of *embedding*. Goffman came to recognize that conversations are *layered* and, hence, embedded in different footings. There are often multiple footings for talk, as when someone "says one thing but means another" or when a person "hints" or "implies" something else. These "layerings" of conversations, which embed them in different contexts, are possible because speech is capable of generating subtle and complex meanings. For example, irony, sarcasm, puns, wit, double-entendres, inflections, shadings, and other manipulations of speech demonstrate the capacity of individuals to shift footings and contextual embeddings of a conversation (for example, think of a conversation in a work setting involving romantic flirtations; it will involve constant movement in footing and context). Yet, for encounters to proceed smoothly, these alterations in footing are, to some extent, normatively regulated and ritualized, enabling individuals to sustain a sense of common reality (of course, as Goffman often emphasized and, in my view, overemphasized, people can seek to manipulate others and change footings and embeddings without the others' full awareness).

Talk is thus a critical dimension of focused interaction. Without it the gestures and cues that people can emit are limited and lack the subtlety and complexity of language. And, as Goffman began to explore this complexity in

---

[64]Ibid., p. 120.

[65]Ibid., p. 121.

[66]Goffman termed this *reframing* (see later section) when it involved a shift in frame. However, footing and refooting can occur *within* an existing frame, and so a person can change the footing of a conversation without breaking or changing a frame.

TABLE 22–2 Goffman's General Propositions on Focused Interaction

I. The probability of an encounter is a positive and additive function of the degree to which:
    A. Social occasions create gatherings of proximate individuals.
    B. Gatherings allow for face-to-face contact revolving around talk.

II. The degree of viability of an encounter is a positive and multiplicative function of:
    A. The availability of relevant normative rules to guide participants with respect to such issues as:
        1. Irrelevance, or excluded matters.
        2. Transformation, or how external matters are to be incorporated.
        3. Resource use, or what local resources are to be drawn upon.
        4. Talk, or how verbalizations are to be ordered.
        5. Self-respect, or the maintenance of lines and face.
    B. The availability of ritual practices that can be used to:
        1. Regulate talk and conversation.
        2. Express appropriate deference and demeanor.
        3. Invoke and punctuate normative rules.
        4. Repair breaches to the interaction.
    C. The capacity of individuals to present acceptable performances with respect to:
        1. Lines, or directions of conduct.
        2. Roles, or specific clusters of rights and duties.
        3. Faces, or specific presentations of personal characteristics.
        4. Self, or particular images of oneself.

---

later works, earlier notions about "definitions of situations" seemed too crude, because people could construct multiple, as well as subtly layered, definitions of any situation—an issue that I will explore near the end of this chapter. But, for our purposes here, the critical point is that talk focuses attention and pulls actors together, forcing their interaction on a face-to-face basis. But, despite the complexity of how this focusing can be done, talk is still normatively and ritually regulated in ways that produce a sense of shared reality for individuals.[67]

To summarize the dynamics of focused interaction, I present the general line of argument as propositions in Table 22–2. Proposition II emphasizes that rules, rituals, and acceptable performances are *multiplicatively* related, which means that, when combined, the effects of rules, rituals, and performances accelerate one another's impact on the focused interaction in an encounter. That is, the effects of normative rules are dramatically increased when invoked and punctuated with rituals; the appropriateness of performances projecting roles, self, and lines is escalated when done with rituals and in accordance with normative rules; and so on for each combination of the variables in Propositions II-A, II-B, and II-C. However, these mutually reinforcing effects can be disrupted by problematic conditions. I have alluded to these as "breaches" and other events that cause "scenes," but let me close this section by exploring a

---

[67]Again, this line of emphasis is what gives Goffman's work an ethnomethodological flair. See next chapter.

point that Goffman continually emphasized: the fragility of the interaction order.

Disruption and Repair in Focused Interaction

Goffman stressed that disruption in encounters is never a trivial matter:[68]

> Social encounters differ a great deal in the importance that participants give to them but, whether crucial or picayune, all encounters represent occasions when the individual can become spontaneously involved in the proceedings and derive from this a firm sense of reality. And this feeling is not a trivial thing, regardless of the package in which it comes. When an incident occurs . . . then the reality is threatened. Unless the disturbance is checked, unless the interactants regain their proper involvement, the illusion of reality will be shattered. . . .

When a person emits gestures that contradict normative rules, present a contradictory front, fail to enact appropriate rituals, seek an inappropriate role, attempt a normatively or ritually incorrect line, or present a wrong face, there is potential for *a scene*. From the point of view of the person, there is a possibility of *embarrassment*, to use Goffman's favorite phrase; once embarrassed, an individual's responses can further degenerate in an escalating cycle of ever greater levels of embarrassment. From the perspective of others, a scene disrupts the definition of the situation and threatens the sense of reality so necessary for them to feel comfortable in a situation. For, implicitly, individuals assume that people are reliable and trustworthy, that they are what they appear to be, that they are competent, and that they can be relied upon; thus, when a scene occurs, these implicit assumptions are called into question and threaten the organization of the encounter (and, potentially, the larger gathering and social occasion).

It is for this reason that an individual will seek to repair a scene caused by the use of inappropriate gestures, and others will use tact to assist the individual in such repair efforts. The sense of order of a situation is thus sustained by a variety of corrective responses on the part of individuals and by the willingness of others to use tact in ignoring minor mistakes and, if this is not possible, to employ tact to facilitate an offending individual's corrective efforts. People thus "disattend" much potentially discrepant behavior, and, when this is no longer an option, they are prepared to accept apologies, accounts, new information, excuses, and other ritually and normatively appropriate efforts at repair.

Of course, this willingness to accept people as they are, to assume their competence, and to overlook minor interpersonal mistakes makes them vulnerable to manipulation and deceit. Goffman spent many pages in all his work dwelling on such matters. Indeed, he appeared fascinated by con jobs, dishonesty, fabrications, and other manipulative practices, and it is this fascination

---

[68]Goffman, *Interaction Ritual*, p. 135.

that too many commentators have emphasized. I will not pursue these discourses, because they obscure the more fundamental dynamics that make interpersonal manipulation possible. For underneath the obvious cynicism of Goffman's analysis is a view of a moral order, sustained by norms, rituals, roles, tolerance, and tact. It is in his analysis of these "moral processes" that Goffman's enduring contribution resides—not in his constant regresses into manipulative and cynical behaviors of seemingly amoral individuals.

UNFOCUSED INTERACTION

Goffman was one of the few sociologists to recognize that behavior and interaction in public places, or in unfocused settings, are important features of the interaction order and, by extension, of social organization in general. Such simple acts as walking down the street, standing in line, sitting in a waiting room or on a park bench, standing in an elevator, going to and from a public restroom, and many other activities represent a significant realm of social organization. These unfocused situations in which people are co-present but not involved in prolonged talk and "face encounters" represent a crucial topic of sociological inquiry—a topic that is often seen as trivial but that, in fact, embraces much of people's time and attention. In two works, *Relations in Public* and *Behavior in Public Places*, Goffman explored the dynamics of unfocused gatherings.[69]

Unfocused gatherings are like focused interactions in their general contours: they are normatively regulated; they call for performances by individuals; they include the presentation of a self; they involve the use of rituals; they have normatively and ritually appropriate procedures for repair; and they depend upon a considerable amount of etiquette, tact, and inattention. Let me explore each of these features in somewhat greater detail.

Much like a focused interaction, unfocused gatherings involve normative rules concerning spacing, movement, positioning, listening, talking, and self-presentation. But, unlike focused interaction, norms do not have to sustain a well-defined membrane. For there is no closure, intense focus of attention, or face-to-face obligations in unfocused encounters. Rather, rules pertain to how individuals are to comport themselves *without* becoming the focus of attention and involved in a face encounter. Rules are thus about how to move, talk, sit, stand, present self, apologize, and perform other acts necessary to sustain public order without creating a situation requiring the additional interpersonal "work" of focused interaction.

When in public, individuals still engage in performances, but, since the audience is not involved in a face engagement or prolonged tracks of talk, the presentation can be more muted and less animated. Goffman used a variety of terms to describe these presentations, two of the most frequent being *body*

---

[69]See notes 57 and 8, respectively, for full references.

idiom[70] and *body gloss*.[71] Both of these terms denote the overall configuration of gestures, or demeanor, that an individual makes available and gleanable to others. (Conversely, others are constantly *scanning* to determine the content of others' body idiom and body gloss.) Such demeanor denotes a person's direction, speed, resoluteness, purpose, and other aspects of a course of action. In his *Relations in Public*, Goffman enumerated three types of body gloss:[72] (1) *orientation gloss*, or gestures giving evidence to others confirming that a person is engaged in a recognizable and appropriate activity in the present time and place; (2) *circumspection gloss*, or gestures indicating to others that a person is not going to encroach on or threaten the activity of others; and (3) *overplay gloss*, or gestures signaling that a person is not constrained or under duress and is, therefore, fully in charge and control of his or her other movements and actions. Thus the public performance of an individual in unfocused interaction revolves around providing information that one is of "sound character and reasonable competency."[73]

In public and during unfocused interactions, the territories of self become an important consideration. Goffman listed various kinds of territorial considerations that can become salient during unfocused interaction, including:[74] (*a*) fixed geographical spaces attached to a particular person, (*b*) egocentric preserves of nonencroachment that surround individuals as they move in space, (*c*) personal spaces that others are not to violate under any circumstances, (*d*) stalls or bounded places that an individual can temporarily claim, (*e*) use spaces that can be claimed as an individual engages in some instrumental activity, (*f*) turns or the claimed order of doing or receiving something relative to others in a situation, (*g*) possessional territory or objects identified with self and arrayed around an individual's body, (*h*) informational preserve or the body of facts about a person that is controlled and regulated, and (*i*) conversational preserve or the right to control who can summon and talk to an individual. Depending upon the type of unfocused interaction, as well as on the number, age, sex, rank, position, and other characteristics of the participants, the territories of self will vary; but in all societies there are clearly understood norms about which configuration of the territories listed above is relevant, and to what degree it can be invoked.

These territories of self are made visible through what Goffman termed *markers*. Markers are signals and objects that denote the type of territorial claim, its extent and boundary, and its duration. Violation of these markers involves an encroachment upon a person's self and invites sanctioning, perhaps creating a breach or scene in the public order. Indeed, seemingly innocent acts— like inadvertently taking someone's place, butting in line, cutting someone off, and the like—can become a violation or befoulment of another's self and, as a

---

[70]Goffman, *Behavior in Public Places*, p. 35.

[71]Goffman, *Relations in Public*, p. 8.

[72]Ibid., pp. 129–38.

[73]Ibid., p. 162.

[74]Ibid., Chapter 2.

result, invite an extreme reaction. Thus, social organization in general depends upon the capacity of individuals to read those markers that establish their territories of self in public situations.

Violations of norms and territories create breaches and potential scenes, even when individuals are not engaged in focused interaction. These are usually repaired through ritual activity, such as (*a*) *accounts* explaining why a transgression has occurred (ignorance, unusual circumstances, temporary incompetence, unmindfulness, etc.), (*b*) *apologies* (some combination of expressed embarrassment or chagrin, clarification that the proper conduct is known and understood, disavowal and rejection of one's behavior, penance, volunteering of restitution, etc.), and (*c*) *requests*, or a preemptive asking for license to do something that might otherwise be considered a violation of a norm or a person's self.[75] The use of these ritualized forms of repair sustains the positioning, movement, and smooth flow of activity among people in unfocused situations; without these repair rituals, tempers would flair and other disruptive acts would overwhelm the public order.

The significance of ritualized responses for repair only highlights the significance of ritual in general for unfocused interaction. As individuals move about, stand, sit, and engage in other acts in public, these activities are punctuated with rituals, especially as people come close to contact with each other. Nods, smiles, hand gestures, bodily movements, and, if necessary, brief episodes of talk (especially during repairs) are all highly ritualized, involving stereotyped sequences of behavior that reinforce norms and signal individuals' willingness to get along with and accommodate each other.

In addition to ritual, much unfocused interaction involves tact and inattention. By simply ignoring or quietly tolerating small breaches of norms, self, and ritual practices, people are able to gather and move about without undue tension and acrimony. In this way, unfocused interactions are made to seem uneventful, enabling individuals to cultivate a sense of obdurate reality in the subtle glances, nods, momentary eye contact, shifting of direction, and other acts of public life.

In Table 22–3 the key propositions implied in Goffman's discussion are enumerated. The relationships among the variables listed in the propositions are multiplicative in that each one accelerates the effects of the other in maintaining order in public situations of unfocused gatherings. It is this interactive effect that allows order to be sustained without reliance upon focused talk and conversation.

FRAMES AND THE ORGANIZATION OF EXPERIENCE

Goffman's last major work, *Frame Analysis: An Essay on the Organization of Experience*,[76] is hardly an "essay" but rather, an 800-page treatise on phenom-

---

[75]Ibid., pp. 109–20.
[76]See note 13 for full reference.

TABLE 22–3 Goffman's General Propositions on Unfocused Interaction

I. The degree of order in unfocused interaction is a positive and multiplicative function of:
   A. The clarity of normative rules regulating behavior in ways to limit face encounters and talk.
   B. The capacity of individuals to provide demeanor cues with respect to:
      1. Orientation, or the appropriateness of activities at the present time and place.
      2. Circumspection, or the willingness to avoid encroachment on, and threat to, others.
      3. Overplays, or the capacity to control and regulate conduct without duress and constraint.
   C. The capacity of individuals to signal with clear markers those configurations of normatively appropriate territories of self with respect to:
      1. Fixed geographical spaces that can be claimed.
      2. Egocentric preserves of nonencroachment that can be claimed during movement in space.
      3. Personal spaces that can be claimed.
      4. Stalls of territory that can be temporarily used.
      5. Use spaces that can be occupied for instrumental purposes.
      6. Turns of performing or receiving goods that can be claimed.
      7. Informational preserves that can be used to regulate facts about individuals.
      8. Possessional territory and objects identified with, and arrayed around, self.
      9. Conversational preserves that can be invoked to control talk.
   D. The availability of configurations of normatively appropriate repair rituals revolving around:
      1. Accounts, or explanations for transgressions.
      2. Apologies, or expressions of embarrassment, regret, and penance for mistakes.
      3. Requests, or redemptive inquiries about making a potential transgression.
   E. The availability and clarity of rituals to reinforce norms and to order conduct by restricting face engagements among individuals.
   F. The availability of ritualized procedures for ignoring minor transgressions of norms and territories of self (tact and etiquette).

enology, or the subjective organization of experience in social situations. It is a dense and rambling work, but it nonetheless returns to a feature of interaction that guided Goffman's work from the very beginning: the construction of "definitions of situations." That is, how is it that people define the reality of situations?

What Is a Frame?

The concept of *frame* appears in Goffman's first major work, *The Presentation of Self,* and periodically thereafter. Surprisingly, he never offered a precise definition of this term, but the basic idea is that people "interpret" events or "strips of activity" in situations with a "schemata" that cognitively encircles or *frames* what is occurring.[77] The frame is much like a picture frame in that it marks off the boundary of the pictured events, encapsulating and distin-

---

[77]See Goffman, *Frame Analysis*, p. 10, for the vagueness of his portrayal.

guishing them from the surrounding environment. Goffman's early discussion of the "rules of irrelevance" in encounters—that is, considerations, characteristics, aspects, and events in the external world to be excluded during a focused interaction—represented an earlier way of communicating the dynamics of framing. Thus, as people look at the world, they impose a frame that defines what is to be pictured on the inside and what is to be excluded by what Goffman termed the *rim of the frame* on the outside. Human experience is organized in terms of frames, which provide an interpretive "framework" or "frame of reference" for designating events or "strips of activity."

Primary Frames

Goffman argued that, ultimately, interpretations of events are anchored in a *primary framework*, which is a frame that does not depend upon some prior interpretation of events.[78] Primary frameworks are thus anchored in the *real world*, at least from the point of view of the individual's organization of experience. People tend to distinguish, Goffman emphasized, between natural and social frameworks.[79] A *natural frame* is anchored in purely physical means for interpreting the world—body, ecology, terrain, objects, natural events, and the like. A *social frame* is lodged in the world created by acts of intelligence and social life. These two kinds of primary frameworks can vary enormously in their organization; some are clearly organized as "entities, postulates, and roles," whereas most others are not clearly shared, "providing only a lore of understanding, an approach, a perspective."[80] All social frameworks involve rules about what is to be excluded beyond the rim of the frame, what is to be pictured inside the frame, and what is to be done when acting within the frame. Yet, as humans look at, and act in, the world, they are likely to apply several frameworks, giving the organization of experience a complexity that Goffman only alluded to in his earlier works.

Although Goffman stressed that he was not analyzing social structure with the concept of frames, he was clearly developing a neo-Durkheimian argument, recasting Durkheim's social psychology—especially the notion of how the "collective conscience rules people from within"—into more complex and dynamic terms.[81] For, as he argued:[82]

> Taken all together, the primary frameworks of a particular social group constitute a central element of its culture, especially insofar as understandings emerge concerning principal classes of schemata, the relations of these classes to one another, and the sum total of forces and agents that these interpretive designs acknowledge to be loose in the world.

---

[78]Ibid., p. 21.

[79]Ibid., pp. 21–24.

[80]Ibid., p. 21.

[81]See Durkheim, *Elementary Forms of the Religious Life.*

[82]Goffman, *Frame Analysis*, p. 27.

Yet, for the most part, Goffman concentrated on the dynamics of framing within the realms of personal experiences and the interaction order, leaving Durkheim's concerns about macrocultural processes to others. Indeed, as was so often the case with Goffman, he became so intrigued with the interpersonal manipulation of frames for deceitful purposes and with the fluidity and complexity of framing in contingent situations that it is often difficult to discern whether or not the analysis is still sociological. I will thus exclude much that is said in *Frame Analysis* because it moves far into psychology and, perhaps, philosophy, linguistics, and extreme phenomenology (see next chapter).

Keys and Keying

What makes framing a complex process, Goffman argued, is that frames can be *transformed*. One basic way to transform a primary frame is to engage in *keying*, which is "a set of conventions by which a given activity, one already meaningful in terms of some primary framework, is transformed into something patterned on this activity but seen by the participants to be something quite else."[83] For example, a theatrical production of a family setting is a keying of "real families"; or a hobby, such as woodworking, is a keying of a more primary set of occupational activities; or a daydream about a love affair is a keying of real love; or a sporting event is a keying of some more primary activity (running, fleeing, fighting, etc.); or practicing and rehearsing are keyings of real performances; or joking about someone's "love life" is a keying of love affairs; and so on. Primary frameworks are seen by people as "real," whereas keyings are seen as "less real"; and the more one rekeys a primary framework—say, performing a keying of a keying of a keying (and so on)—the "less real" is the frame. The *rim* of the frame is still ultimately a primary framework, anchored in some natural or social reality, but humans have the capacity to continually rekey and *layer* or *laminate* their experiences. Thus terms like "definition of the situation" do not adequately capture this layering of experience through keying, nor do such terms adequately denote the multiplicity of frameworks that people can invoke because of their capacities for shifting primary frameworks and rekeying existing ones.

Fabrications

The second type of frame transformation—in addition to keying—is *fabrication*, which is "the intentional effort of one or more individuals to manage activity so that a party of one or more others will be induced to have a false belief about what it is that is going on."[84] Unlike a key, a fabrication is not a copy (or a copy of a copy) of some primary framework, but an effort to make others think that something else is going on. Hoaxes, con games, and strategic ma-

---

[83]Ibid., pp. 43–44.
[84]Ibid., p. 83.

nipulations all involve fabrications—getting others to frame a situation in one way while others manipulate them in terms of another, hidden, framework.

The Complexity of Experience

Thus, as people interact, they frame situations in terms of primary frameworks, but they can also key these primary frames and fabricate new ones for purposes of deception and manipulation. From Goffman's viewpoint, an interaction can involve many keyings and rekeyings (that is, layers and laminations of interpretation) as well as fabrications. In fact, once keying or fabrication occurs, further keying and fabrication are facilitated since they initiate an escalating movement away from a primary framework. Goffman never stated how much fabrication (and fabrication on fabrications) and keying (or keying on keyings) can occur in interaction, but he does see novels, dramatic theater, and cinema as providing the vehicles for the deepest layering of interaction (since each is an initial keying of some primary frame in the real world, which then opens the almost infinite possibilities for rekeying and fabrication).

Yet there are procedures in interaction—typically ritualized and normatively regulated—for bringing participants' experiences back to the original primary frame—that is, for wiping away layers of keyings and fabrications. For example, when people have become caught up in benign ridicule and joking about something, a person saying "seriously now . . ." is trying to wipe the slate clean of keyings (and rekeyings) and come back to the primary frame on which the mocking and joking were based. Such rituals (with the normative obligation to "try to be serious for a moment") seek to reanchor the interaction in "the real world."

In addition to keyings and deliberate fabrications, individuals can *misframe* events—whether from ignorance, ambiguity, error, or initial disputes over framing among participants. Such misframing can persist for a time, but eventually individuals seek to *clear the frame* by getting a correct reading of information so as to be able to reframe the situation correctly. Such efforts to clear the frame become particularly difficult and problematic, however, when fabrication has been at least part of the reason for the misframing.[85]

Thus, framing is a very complex process—one that, Goffman contended, sociologists have not been willing to address seriously. The world that people experience is not unitary, and it is subject to considerable manipulation, whether for benign or deceptive purposes. And, because of humans' capacity for symbol use (especially talk), the processes of reframing, keying, and fabrication can create highly layered and complex experiences. And yet, during interaction, people seek to maintain a common frame (a need that, Goffman was all too willing to assert, makes them vulnerable to manipulation through fabrication). For without a common frame—even one that has been keyed or fabricated—the interaction cannot proceed smoothly. Unfortunately, Goffman

---

[85]See: ibid., p. 449, for a list of the conditions that make individuals inadvertently vulnerable to misframing. And see p. 463 for how misframing can be the result of deception and manipulation.

concentrated on framing in terms of "organizing experience," per se, rather than on framing as it organizes experience during focused and unfocused interaction. Hence the analysis of framing is provocative and suggestive, but too often it wanders away from the sociologically relevant topic—"the interaction order."

CONCLUSION

Erving Goffman's works represent a truly seminal breakthrough in the analysis of social interaction. The emphasis on self-presentations, norms, rituals, and frames presents sociological theory with many important conceptual leads[86] that have been adopted by such diverse theoretical traditions as role theory (Chapters 20 and 21), phenomenology and ethnomethodology (Chapter 23), interaction ritual theory (Chapter 12), critical theory (Chapter 13), structuralism (Chapter 25), and structuration theory (Chapter 26). Moreover, Goffman's recognition that the "interaction order" is a distinct realm that cannot explain more macrostructural phenomena and that, at best, can only be "loosely coupled" with these macro processes represents a tempered and reasoned position on the micro/macro question—as I will explore in Chapters 29, 30, 31, and 32. Thus I have devoted considerable space to Goffman's work because it is so central to much theorizing in sociology.

There are, however, some questionable points of emphasis that should, in closing, be mentioned. First, Goffman's somewhat cynical and manipulative view of humans and interaction—which I have consistently downplayed in this chapter—often takes analysis in directions that are not as fundamental to human organization as he implied. Second, Goffman's rather extreme situational view of self as only a projected image and perhaps a mirage in each and every situation is probably overdrawn. The denial of a core self or permanent identity certainly runs against the mainstream of interactionist theorizing and, I suspect, reality itself. And, third, Goffman's work tends, at times, to wander into a rather extreme subjectivism and interpersonal nihilism where experience is too layered and fickle, while interaction is too fluid and changeable by the slightest shift in frame and ritual. Yet, even with these points of criticism, Goffman's sociology represents a monumental achievement—certainly equal to that of George Herbert Mead, Alfred Schutz, Émile Durkheim, and any of the contemporary theorists who have undertaken to understand micro social processes.

---

[86]For my use of these ideas from Goffman, see: Jonathan H. Turner, *A Theory of Social Interaction* (Stanford, CA: Stanford University Press, 1988).

CHAPTER 23

The Ethnomethodological Challenge

In recent years a more phenomenological form of interactionism has emerged. This alternative interactionism goes by a number of labels, but I see *ethnomethodology* as the most descriptive. As this label underscores, ethnomethodology examines the "folk" (ethno) "methods" that people use in dealing with one another. I should emphasize at the outset, however, that ethnomethodologists would resent being discussed as interactionists.[1] Indeed, they see themselves as proposing a radical new paradigm that challenges existing conceptions of social reality, from functionalism to symbolic interactionism. Yet, as I will argue, this challenge is more bluster than substance. In fact, I do not see ethnomethodology as a challenge; rather, it represents an important supplement to current interactionist theorizing.

But why do ethnomethodologists so often see themselves as paradigmatic messiahs? The answer resides in ethnomethodologists' contention that they have discovered an alternative reality that cannot be conceptualized by existing theoretical perspectives. And they raise a question that owes its inspiration to Husserl and Schutz: how do sociologists and other groups of humans create and sustain for each other the *presumption* that the social world has a real character? A "more real" phenomenon for those who propose this question revolves around the complex ways people (laypersons and sociologists alike) go about consciously and unconsciously constructing, maintaining, and altering their "sense" of an external social reality. In fact, the cement that holds society together may not be the values, norms, common definitions, exchange payoffs, role bargains, interest coalitions, and the like of current social theory, but people's explicit and implicit "methods" for creating the presumption of a social order.

---

[1]See, for example, Thomas P. Wilson, "Normative and Interpretative Paradigms in Sociology," in *Understanding Everyday Life*, ed. J. Douglas (London: Routledge & Kegan Paul, 1970).

Such is the general profile of their challenge. In viewing it, I think that ethnomethodologists have uncovered some important social processes. But they have also been a bit pretentious, if not preposterous, in their claims about the "new reality" that they have uncovered. But let me not judge ethnomethodology too severely before I have discussed it in more detail. I will begin with a brief overview of its origins and then turn to a sampling of its major figures and other claims.

THE ORIGINS OF ETHNOMETHODOLOGY

Ethnomethodology borrows and extends ideas from phenomenology and, despite disclaimers to the contrary, from Meadian-inspired symbolic interactionism. In extending the ideas of these schools of thought, however, ethnomethodology claims to posit a different view of the world. And so, to appreciate just how ethnomethodology differs from more traditional forms of sociological theory, I will assess it in relation to those perspectives from which it tries to dissociate itself, but from which it still has drawn considerable inspiration.

Blumer's Interactionism and Ethnomethodology

As I emphasized in Chapter 19, Herbert Blumer's interactionism emphasizes the constructed and fluid nature of interaction. Because actors possess extensive symbolic capacities, they are capable of: (*a*) introjecting new objects into situations, (*b*) redefining situations, and (*c*) realigning their joint actions. As Blumer emphasizes, "it would be wise to recognize that any given [act] is mediated by acting units interpreting the situations with which they are confronted."[2] Whereas these situations consist of norms, values, roles, beliefs, and social structures, these are merely types of many "objects" that can be symbolically introjected and reshuffled to produce new definitions of situations.

These ideas advocate a concern with how meanings, or definitions, are created by actors interacting in situations. The emphasis is on the *process* of interaction and on how actors create common meanings in dealing with one another. This line of inquiry is also pursued by ethnomethodologists. The ethnomethodologist also focuses on interaction and on the creation of meanings in situations. But there is an important shift in emphasis that corresponds to Alfred Schutz's analysis of the lifeworld: In what ways do people create a *sense* that they share a common view of the world? And how do people arrive at the *presumption* that there is an objective, external world? Blumer's interactionism stresses the process of creating meaning, but it acknowledges the existence of an external social order. Ethnomethodology suspends, or "brackets," in Edmund Husserl's terms, the issue of whether or not there is an external world of norms, roles, values, and beliefs. Instead, it concentrates on how interaction creates among actors a *sense* of a factual world "out there."

---

[2]Herbert Blumer, "Society as Symbolic Interaction," in *Human Behavior and Social Process,* ed. A. Rose (Boston: Houghton Mifflin, 1962).

Goffman's Dramaturgical Analysis and Ethnomethodology

To the dismay of many ethnomethodologists, Paul Attewell has argued that the work of Erving Goffman represents a significant source of inspiration for ethnomethodology.[3] As was emphasized in Chapter 22, Goffman's work has often been termed the *dramaturgical school* of interactionism because it focuses on the ways that actors manipulate gestures to create an impression in a particular social scene. Goffman tends to emphasize the process of impression management, per se, and not the purposes or goals toward which action is directed. Much of Goffman's analysis thus concentrates on the form of interaction itself rather than on the structures it creates, sustains, or changes.[4] For example, Goffman has insightfully analyzed how actors validate self-presentations, how they justify their actions through gestures, how they demonstrate their membership in groups, how they display social distance, how they adjust to physical stigmas, and how they interpersonally manipulate many other situations.

This concern with the management of social scenes is also prominent in ethnomethodological analysis. Ethnomethodologists share Goffman's concern with the techniques by which actors create impressions in social situations, but their interest is not with individuals' impression management, per se, but with how actors create a *sense of a common reality*. This too is an important aspect of Goffman's work. For Goffman, much of what individuals do in interaction revolves around efforts to sustain a common view and sense of reality. Like Schutz, Goffman argued that humans not only actively try to manipulate impressions in order to create and sustain a view of reality but also try to ignore discrepant information in an effort to keep a situation intact. Moreover, humans take for granted much information and work to keep from questioning the situation, lest their common view of reality come apart. All of these themes are a part of ethnomethodology; although Goffman appears to have been highly critical of this perspective, he nonetheless contributed to its development.

Alfred Schutz, Phenomenology, and Ethnomethodology

Schutz's phenomenology, as I noted in Chapter 18, liberated phenomenology from Husserl's philosophical project.[5] Schutz asserted the importance of study-

---

[3]Paul Attewell, "Ethnomethodology Since Garfinkel," *Theory and Society* 1 (1974), pp. 179–210.

[4]This is not always true, especially in some of his more institutional works, such as *Asylums* (Garden City, NY: Anchor Books, 1961). Other representative works by Goffman include *The Presentation of Self in Everyday Life* (Garden City, NY: Doubleday, 1959); *Interaction Ritual* (Garden City, NY: Anchor Books, 1967); *Encounters* (Indianapolis: Bobbs-Merrill, 1961); *Stigma* (Englewood Cliffs, NJ: Prentice-Hall, 1963).

[5]See, in particular, Alfred Schutz, *Collected Papers I: The Problem of Social Reality*, ed. Maurice Natanson (The Hague: Martinus Nijhoff, 1962); Alfred Schutz, *Collected Papers II: Studies in Social Theory*, ed. Arvid Broderson (The Hague: Martinus Nijhoff, 1964); and Alfred Schutz, *Collected Papers III: Studies in Phenomenological Philosophy* (The Hague: Martinus Nijhoff,

ing how interaction creates and maintains a "paramount reality." He was also concerned with how actors achieve reciprocity of perspectives and how they construct a taken-for-granted world that gives order to social life.

This emphasis on the taken-for-granted nature of the world and the importance of this lifeworld for maintaining an actor's sense of reality becomes a prime concern of ethnomethodologists. Indeed, many ethnomethodological concepts are borrowed or adapted from Husserl's and Schutz's phenomenology. Yet ethnomethodologists adapt phenomenological analysis to the issue of how social order is maintained by the practices that actors use to create a sense that they share the same lifeworld.

Let me now summarize my argument thus far. Ethnomethodology draws from and extends the concerns of interactionists such as Blumer and Goffman and the phenomenological projects of Husserl and Schutz. It emphasizes the process of interaction, the use of interpersonal techniques to create situational impressions, and the importance of perceptions of consensus among actors. In extending interactionism and phenomenology, ethnomethodologists often think that they posit a different vision of the social world and an alternative orientation for understanding the question of how social organization is created, maintained, and changed. As Mehan and Wood note, ethnomethodologists "have chosen to ask not how order is possible, but rather to ask *how a sense of order is possible*." (I have added the emphasis here.)[6]

THE NATURE OF ETHNOMETHODOLOGY

Metaphysics or Methodology?

Ethnomethodology has often been misunderstood. Part of the reason for this misunderstanding stems from the vagueness of the prose of some ethnomethodologists.[7] One form of such misinterpretation asserts that ethnomethodology

---

1966). For adaptations of these ideas to interactionism and ethnomethodology, see Alfred Schutz and Thomas Luckmann, *The Structure of the Lifeworld* (Evanston, IL: Northwestern University Press, 1973), as well as Thomas Luckmann, ed., *Phenomenology and Sociology* (New York: Penguin Books, 1978). See also Robert C. Freeman, "Phenomenological Sociology and Ethnomethodology," in *Introduction to the Sociologies of Everyday Life*, eds. J. Douglas et al. (Boston: Allyn & Bacon, 1980).

[6]Hugh Mehan and Houston Wood, *The Reality of Ethnomethodology* (New York: John Wiley, 1975), p. 190. This is an excellent statement of the ethnomethodological perspective.

[7]See, for example, Harold Garfinkel, *Studies in Ethnomethodology* (Englewood Cliffs, NJ: Prentice-Hall, 1967). However, recent portrayals of the ethnomethodological position have done much to clarify this initial vagueness. See, for example, Mehan and Wood, *The Reality of Ethnomethodology*; D. Lawrence Wieder, *Language and Social Reality* (The Hague: Mouton, 1973); Don H. Zimmerman and Melvin Pollner, "The Everyday World as Phenomenon," in *Understanding Everyday Life*, ed. J. D. Douglas (Chicago: Aldine, 1970), pp. 80–103; Don H. Zimmerman and D. Lawrence Wieder, "Ethnomethodology and the Problem of Order: Comment on Denzin," in *Understanding Everyday Life*, pp. 285–95; Randall Collins and Michael Makowsky, *The Discovery of Society* (New York: Random House, 1972), pp. 209–13; George Psathas, "Ethnomethods and Phenomenology," *Social Research* 35 (September 1968), pp. 500–20; Roy Turner, ed., *Ethnomethodology* (Baltimore: Penguin Books, 1974); Thomas P. Wilson and Don H. Zimmerman, "Ethnomethodology, Sociology and Theory," *Humboldt Journal of Social Relations* 7 (Fall-Winter

represents a "corrective" to current sociological theorizing because it points to sources of bias among scientific investigators. From this position it is assumed that ethnomethodology can serve to check the reliability and validity of investigators' observations by exposing not only their biases but also those of the scientific community accepting their observations. Although ethnomethodology might be used for this purpose, if one were so inclined, those who advocate this use have failed, I think, to grasp the main thrust of the ethnomethodological position. For the ethnomethodologist, emphasis is not on questions about the reliability and validity of investigators' observations, but on the methods used by scientific investigators and laypersons alike to construct, maintain, and perhaps alter what each considers and believes to be a valid and reliable set of statements about order in the world. The methodology in the ethnomethodological perspective does not address questions about the proper, unbiased, or truly scientific search for knowledge; rather, ethnomethodology is concerned with the common methods people employ—whether scientists, homemakers, insurance salespersons, or laborers—to create a sense of order about the situations in which they interact. I think that the best clue to this conceptual emphasis can be found in the word *ethnomethodology* itself—*ology*, "study of"; *method*, "the methods [used by]"; and *ethno*, "folk or people."

Another related source of misunderstanding in commentaries on ethnomethodology comes from those who assume that this perspective simply uses "soft" research methods, such as participant observation, to uncover some of the taken-for-granted rules, assumptions, and rituals of members in groups.[8] This interpretation would appear to transform ethnomethodology into a research-oriented variant of the symbolic interactionist perspective.[9] Such a variant of ethnomethods would now represent a more conscientious effort to get at actors' interpretative processes and the resulting definitions of the situation. By employing various techniques for observation of, and participation in, the symbolic world of those interacting individuals under study, a more accurate reading of how situations are defined, how norms emerge, and how social action is controlled could be achieved. Although ethnomethodologists do employ observation and participant methods to study interacting individuals, their concerns are not the same as those of symbolic interactionists. Like all dominant forms of sociological theorizing, interactionists posit that common definitions, values, and norms emerge from interaction and serve to regulate how people perceive the world and interact with one another. For the interactionist, concern is with the conditions under which various types of explicit and implicit definitions, norms, and values emerge and thereby resolve the problem of how social organization is possible. In contrast, ethnomethodologists are interested

1979–80), pp. 52–88; Warren Handel, *Ethnomethodology: How People Make Sense* (Englewood Cliffs, NJ: Prentice-Hall, 1982); and George Psathas, ed., *Everyday Language: Studies in Ethnomethodology* (New York: Irvington, 1979).

[8]For example, see Norman K. Denzin, "Symbolic Interactionism and Ethnomethodology," *American Sociological Review* 34 (December 1969), pp. 922–34.

[9]For another example of this interpretation, see Walter L. Wallace, *Sociological Theory* (Chicago: Aldine, 1969), pp. 34–36.

in *how* members come to agree upon an *impression* that there are such things as rules, definitions, and values. Just what types of rules and definitions emerge is not a central concern of the ethnomethodologist, since there are more fundamental questions: Through *what types of methods* do people go about seeing, describing, and asserting that rules and definitions exist? How do people use their beliefs that definitions and rules exist to describe for one another the social order?

Again, the methods of ethnomethodology do not refer to a new and improved technique on the part of scientific sociology for deriving a more accurate picture of people's definitions of the situation and of the norms of social structure (as is the case with interactionists). For the ethnomethodologist, emphasis is on the *methods employed by those under study* in creating, maintaining, and altering their presumption that a social order actually exists out there in the real world.

Concepts and Principles of Ethnomethodology

Alfred Schutz postulated one basic reality—the paramount—in which people's conduct of their everyday affairs occurs. Most contemporary ethnomethodologists, however, are less interested in whether or not there is one or multiple realities, lifeworlds, or natural attitudes. Far more important in ethnomethodological analysis is the development of concepts and principles that can help explain how people's sense of reality is constructed, maintained, and changed. Although ethnomethodology has yet to develop a unified body of concepts or propositions, I see a conceptual core in the ethnomethodological perspective. Let me review some of its key elements.

Reflexive action and interaction[10]   Much interaction operates to sustain a particular vision of reality. For example, ritual activity directed toward the gods sustains the belief that gods influence everyday affairs. Such ritual activity is an example of reflexive action; it operates to maintain a certain vision of reality. Even when the facts would seem to contradict a belief, human interaction upholds the contradicted belief. For instance, should intense prayer and ritual activity not bring forth the desired intervention from the gods, the devout, rather than reject beliefs, proclaim that they did not pray hard enough, that their cause was not just, or that the gods in their wisdom have a greater plan. Such behavior is reflexive; it upholds or reinforces a belief, even in the face of evidence that the belief may be incorrect.

Much human interaction is reflexive. Humans interpret cues, gestures, words, and other information from one another in a way that sustains a particular vision of reality. Even contradictory evidence is reflexively interpreted

---

[10]For an early discussion of this phenomenon, see Garfinkel, *Studies in Ethnomethodology*. A more readable discussion can be found in Mehan and Wood, *The Reality of Ethnomethodology*, pp. 137–78.

to maintain a body of belief and knowledge. The concept of reflexivity thus focuses attention on how people in interaction go about maintaining the presumption that they are guided by a particular reality. Much of ethnomethodological inquiry addresses this question of how reflexive interaction occurs. That is, what concepts and principles can be developed to explain the conditions under which different reflexive actions among interacting parties are likely to occur?

The indexicality of meaning   The gestures, cues, words, and other information sent and received by interacting parties have meaning in a *particular context*. Without some knowledge of the context—the biographies of the interacting parties, their avowed purpose, their past interactive experiences, and so forth—it would easily be possible to misinterpret the symbolic communication among interacting individuals. This fact of interactive life is denoted by the concept of *indexicality*.[11] To say that an expression is indexical is to emphasize that the meaning of that expression is tied to a particular context.

This phenomenon of indexicality draws attention to the problem of how actors in a context construct a vision of reality in that context. They develop expressions that invoke their common vision about what is real in their situation. The concept of indexicality thus directs an investigator's attention to actual interactive contexts in order to see how actors go about creating indexical expressions—words, facial and body gestures, and other cues—to create and sustain the presumption that a particular reality governs their affairs.

With these two key concepts, reflexivity and indexicality, I think that the interactionists' concern with the process of symbolic communication is retained, while much of the phenomenological legacy of Schutz is rejuvenated. Concern is with how actors use gestures to create and sustain a lifeworld, body of knowledge, or natural attitude about what is real. The emphasis is not on the content of the lifeworld, but on the methods or techniques that actors use to create, maintain, or even alter a vision of reality. As Mehan and Wood note, "the ethnomethodological theory of the reality constructor is about the *procedures* that accomplish reality. It is not about any specific reality."[12] This emphasis has led ethnomethodologists to isolate the general types of methods employed by interacting actors.

Some general interactive methods   When analytical attention focuses on the methods that people use to construct a sense of reality, the task of the theorist is to isolate the general types of interpersonal techniques that people employ in interaction. Aaron Cicourel, for example, has summarized a number of such techniques or methods isolated by ethnomethodologists: (1) searching

---

[11]Garfinkel, *Studies in Ethnomethodology*; Garfinkel and Sacks, "The Formal Properties of Practical Actions," in *Theoretical Sociology*, eds. J. C. McKinney and E. A. Tiryakian (New York: Appleton-Century-Crofts, 1970).

[12]Mehan and Wood, *The Reality of Ethnomethodology*, p. 114.

for the normal form, (2) doing reciprocity of perspectives, and (3) using the et cetera principle.[13]

Searching for the normal form   If interacting parties sense that ambiguity exists over what is real and that their interaction is strained, they will emit gestures to tell each other to return to what is "normal" in their contextual situation. Actors are presumed to hold a vision of a normal form for situations or to be motivated to create one; hence much of their action is designed to reach this form.

Doing a reciprocity of perspectives   Borrowing from Schutz's formulation, ethnomethodologists have emphasized that actors operate under the presumption, and actively seek to communicate the fact, that they would have the same experiences were they to switch places. Furthermore, until they are so informed by specific gestures, actors can ignore differences in perspectives that might arise from their unique biographies. Thus much interaction will be consumed with gestures that seek to assure others that a reciprocity of perspectives does indeed exist.

Using the et cetera principle   In examining an actual interaction, much is left unsaid. Actors must constantly "fill in" or "wait for" information necessary to "make sense" of another's words or deeds. When actors do so, they are using the et cetera principle. They are agreeing not to disrupt the interaction by asking for the needed information; they are willing to wait or to fill in. For example, the common phrase "you know," which usually appears after an utterance, is often an assertion by one actor to another invoking the et cetera principle. The other is thus informed not to disrupt the interaction or the sense of reality in the situation with a counter utterance, such as "No, I do not know."

These three general types of folk methods are but examples of what ethnomethodologists seek to discover. There are certainly more folk methods. For some ethnomethodologists the ultimate goal of theory is to determine the conditions under which these and other interpersonal techniques will be used to construct, maintain, or change a sense of reality. Yet I have found few such propositions in the ethnomethodological literature. But let me try to imagine the nature of propositions, should they ever be developed.

Two general ethnomethodological propositions   I see ethnomethodological propositions as following several assumptions: (1) Social order is maintained by the use of techniques that give actors a sense that they share a common reality. (2) The substance of the common reality is less important in maintaining social order than the actors' acceptance of a common set of

---

[13]Aaron V. Cicourel, *Cognitive Sociology* (London: Macmillan, 1973), pp. 85–88. It should be noted that these principles are implicit in Garfinkel's *Studies in Ethnomethodology.*

techniques. With these assumptions, I offer two examples of what ethnomethodological propositions might look like if ethnomethodologists ever got around to generating them.

1. The more actors fail to agree on the use of interactive techniques, such as the et cetera principle, the search for the normal form, and the reciprocity of perspectives, the more likely is interaction to be disrupted and, hence, the less likely is social order to be maintained.
2. The more interaction proceeds on the basis of different, taken-for-granted visions of reality, the more likely is interaction to be disrupted and, hence, the less likely is social order to be maintained.

These propositions can perhaps be visualized as general laws from which more specific propositions on how actors go about constructing, maintaining, or changing their sense of reality could be developed. What is needed in ethnomethodology is to discover the *specific conditions* under which particular folk techniques are likely to be used to create a sense of a common world among interacting individuals. Indeed, I think that ethnomethodologists should worry less about whether or not they have discovered a radically new reality and instead direct their efforts to developing some general propositions about the conditions under which actors will use various "folk methods."

Varieties of Ethnomethodological Inquiry

Garfinkel's pioneering inquiries   Harold Garfinkel's *Studies in Ethnomethodology* firmly established ethnomethodology as a distinctive theoretical perspective.[14] Although the book is not a formal theoretical statement, the studies and the commentary in it established the domain of ethnomethodological inquiry. Subsequent ethnomethodological research and theory begin with Garfinkel's insights and take them in a variety of directions.

Garfinkel's work established ethnomethodology as a field of inquiry that seeks to understand the methods people employ to make sense out of their world. He places considerable emphasis on language as the vehicle by which this reality construction is done. Indeed, for Garfinkel, interacting individuals' efforts to account for their actions—that is, to represent them verbally to others—are the primary method by which the world is constructed. In Garfinkel's terms, *to do* interaction is *to tell* interaction; or, in other words, the primary folk technique used by actors is verbal description. In this way people use their accounts to construct a sense of reality.

Garfinkel places enormous emphasis on indexicality—that is, on the fact that members' accounts are tied to particular contexts and situations. An utterance, Garfinkel notes, indexes much more than it actually says; it also evokes connotations that can be understood only in the context of a situation. Garfinkel's work was thus the first to stress the indexical nature of interpersonal

---

[14]Garfinkel, *Studies in Ethnomethodology.*

cues and to emphasize that individuals seek to use accounts to create a sense of reality.

In addition to laying much of the groundwork for current ethnomethodology, Garfinkel and his associates conducted a number of interesting empirical studies in an effort to validate their assumptions about what is real. One line of empirical inquiry is known as the *breeching experiment,* in which the normal course of interaction is deliberately interrupted. For example, Garfinkel reports a series of conversations in which student experimenters challenged every statement of selected subjects. The end result was a series of conversations revealing the following pattern:

Subject: I had a flat tire.

Experimenter: What do you mean, you had a flat tire?

Subject: (appears momentarily stunned and then replies in a hostile manner): What do you mean, "What do you mean?" A flat tire is a flat tire. That is what I meant. Nothing special. What a crazy question![15]

In this situation the experimenter was apparently violating an implicit rule for this type of interaction and thereby aroused not only the hostility of the subject but also a negative sanction, "What a crazy question!" Seemingly, in any interaction there are certain background features that everyone should understand and that should not be questioned in order for all parties to be able to "conduct their common conversational affairs without interference."[16] Such implicit methods appear to guide a considerable number of everyday affairs and are critical for the construction of at least the perception among interacting humans that an external social order exists. In this conversation, for example, the et cetera principle and the search for the normal form are invoked by the subject. Through breeching, Garfinkel hoped to discover implicit ethnomethods by forcing actors to *actively* engage in the process of reality reconstruction after the situation had been disrupted.

Other research strategies also yielded insights into the methods parties use in an interaction for constructing a sense of reality. Garfinkel and his associates summarized the "decision rules" jurors employed in reaching a verdict.[17] By examining a group such as a jury, which must by the nature of its task develop an interpretation of what really happened, the ethnomethodologist might achieve some insight into the generic properties of the processes of constructing a *sense of social reality.* From the investigators' observations of jurors, it appeared that "a person is 95 percent juror before [coming] near the court," indicating that, through their participation in other social settings and through instructions from the court, they had accepted the "official" rules for reaching a verdict. However, these rules were altered somewhat as participants came together in

---

[15]Ibid., p. 42.

[16]Ibid.

[17]Ibid., pp. 104–15.

an actual jury setting and began the "work of assembling the 'corpus' which serves as grounds for inferring the correctness of a verdict."[18] Because the inevitable ambiguities of the cases before them made it difficult for strict conformity to the official rules of jury deliberation, new decision rules were invoked in order to allow jurors to achieve a "correct" view of "what actually happened." But, in their retrospective reporting to interviewers of how they reached the verdicts, jurors typically invoked the "official line" to justify the correctness of their decisions. When interviewers drew attention to discrepancies between the jurors' ideal accounts and their actual practices, jurors became anxious, indicating that somewhat different rules had been used to construct the corpus of what really happened.

In sum, I think that these two examples of Garfinkel's research strategy illustrate the general intent of much ethnomethodological inquiry: to penetrate natural social settings or create social settings in which the investigator can observe humans attempting to assert, create, maintain, or change the rules for constructing the appearance of consensus over the structure of the real world. By focusing on the process or methods for constructing a reality rather than on the substance or content of the reality itself, research from the ethnomethodological point of view can potentially provide a more interesting and relevant answer to the question of "how and why society is possible." Garfinkel's studies have stimulated a variety of research and theoretical strategies. Several of the most prominent strategies are briefly discussed below.

Harvey Sacks's linguistic analysis   Until his untimely death in 1976, Harvey Sacks exerted considerable influence within ethnomethodology. Although his work is not well known outside ethnomethodological circles, it represents an attempt to extend Garfinkel's concern with verbal accounts while eliminating some of the problems posed by indexicality.

Sacks was one of the first ethnomethodologists to articulate the phenomenological critique of sociology and to use this critique to build what he thought was an alternative form of theorizing.[19] The basic thrust of Sacks's critique can be stated as follows: Sociologists assume that language is a resource used in generating concepts and theories of the social world. In point of fact, however, sociologists are confusing resource and topic. In using language, sociologists are creating a reality; their words are not a neutral vehicle but *the* topic of inquiry for true sociological analysis.[20]

Sacks's solution to this problem in sociology is typical of phenomenologists. If the pure properties of language can be understood, then it would be possible to have an objective social science without confusing resource with subject matter. Sacks's research tended to concentrate on the formal properties of language-in-use. Typically, Sacks would take verbatim transcripts of actors in

---

[18]Ibid., p. 110.

[19]Harvey Sacks, "Sociological Description," *Berkeley Journal of Sociology* 8 (1963), pp. 1–17.

[20]Harvey Sacks, "An Initial Investigation of the Usability of Conversational Data for Doing Sociology," in *Studies in Interaction*, ed. D. Sudnow (New York: Free Press, 1972).

interaction and seek to understand the formal properties of the conversation while ignoring its substance. Such a tactic resolved the problem of indexicality, since Sacks simply ignored the substance and context of conversation and focused on its form. For example, "sequences of talk" among actors might occupy his attention.[21]

Sacks thus began to take ethnomethodology into formal linguistics, a trend that has continued and, in fact, seems to dominate current ethnomethodology. More important, he sought to discover universal forms of interaction—that is, abstracted patterns of talk—that might apply to all conversations. In this way he began to search for the laws of reality construction among interacting individuals.

Aaron Cicourel's cognitive approach   In his *Method and Measurement in Sociology*, Aaron Cicourel launched a line of attack on sociology similar to Sacks's.[22] The use of mathematics, he argues, will not remove the problems associated with language because mathematics is a language that does not necessarily correspond to the phenomena that it describes. Instead, it "distorts and obliterates, acts as a filter or grid for that which will pass as knowledge in a given era."[23] The use of statistics similarly distorts: events cannot be counted, averaged, and otherwise manipulated. Such statistical manipulations soon make sociological descriptions inaccurate and mold sociological analysis to the dictates of statistical logic.

In a less severe tone, Cicourel also questions Garfinkel's assertion that interaction and verbal accounts are the same process.[24] Cicourel notes that humans see, sense, and feel much that they cannot communicate with words. Humans use "multiple modalities" for communicating in situations. Verbal accounts represent crude and incomplete translations of what is actually communicated in interaction. This recognition has led Cicourel to rename his brand of ethnomethodology: *cognitive sociology.*

The details of his analysis are, I think, less important than the general intent of his effort to transform sociological research and theory. Basically, he has sought to uncover the universal "interpretive procedures" by which humans organize their cognitions and give meaning to situations.[25] It is through these interpretive procedures that people develop a sense of social structure and are able to organize their actions. These interpretive procedures are universal and invariant in humans, and their discovery would allow for understanding of how humans create a sense of social structure in the world around them.

---

[21]His best-known study, for example, is the coauthored article with Emmanuel Schegloff and Gail Jefferson, "A Simplest Systematics for the Analysis of Turn Taking in Conversation," *Language* 50 (1974), pp. 696–735.

[22]Aaron V. Cicourel, *Method and Measurement in Sociology* (New York: Free Press, 1964).

[23]Ibid., p. 35.

[24]Aaron V. Cicourel, "Cross Modal Communication," in *Linguistics and Language Science, Monograph 25*, ed. R. Shuy (Washington, DC: Georgetown University Press, 1973).

[25]See, for example, his *Cognitive Sociology* and "Basic Normative Rules in the Negotiation of Status and Role," in *Recent Sociology No. 2*, ed. H. P. Dreitzel (New York: Macmillan, 1970).

Zimmerman's, Pollner's, and Wieder's situational approach Sacks and Cicourel have focused on the universal properties, respectively, of language use and cognitive perception/representation. This concern with invariance, or universal folk methods, has become increasingly prominent in ethnomethodological inquiry. In a number of essays, for example, Don Zimmerman, D. Lawrence Wieder, and Melvin Pollner have developed an approach that seeks to uncover the universal procedures people employ to construct a sense of reality.[26] Their position is perhaps the most clearly stated of all ethnomethodologies, drawing inspiration from Garfinkel but extending his ideas. Let me summarize their argument:

1. In all interaction situations, humans attempt to construct the appearance of consensus over relevant features of the interaction setting.
2. These setting features can include attitudes, opinions, beliefs, and other cognitions about the nature of the social setting in which they interact.
3. Humans engage in a variety of explicit and implicit interpersonal practices and methods to construct, maintain, and perhaps alter *the appearance* of consensus over these setting features.
4. Such interpersonal practices and methods result in the assembling and disassembling of what can be termed an *occasioned corpus*—that is, the *perception* by interacting humans that the current setting has an orderly and understandable structure.
5. This appearance of consensus is not only the result of agreement on the substance and content of the occasioned corpus but also a reflection of each participant's compliance with the rules and procedures for assemblage and disassemblage of this consensus. In communicating, in however subtle a manner, that parties accept the implicit rules for constructing an occasioned corpus, they go a long way in establishing consensus over what is out there in the interaction setting.
6. In each interaction situation, the rules for constructing the occasioned corpus will be unique in some respects and hence not completely generalizable to other settings—thus requiring that humans in each and every interaction situation use interpersonal methods in search for agreement on the implicit rules for the assemblage of an occasioned corpus.
7. Thus, by constructing, reaffirming, or altering the rules for constructing an occasioned corpus, members in a setting are able to offer one another the appearance of an orderly and connected world out there, which compels certain perceptions and actions on their part.

It is from these kinds of assumptions about human interaction that Zimmerman's, Pollner's, and Wieder's ethnomethodology takes its subject matter. Rather than focusing on the actual content and substance of the occasioned corpus and on the ways members believe it to force certain perceptions and

---

[26]See, for example, Zimmerman and Pollner, "The World as a Phenomenon"; Zimmerman and Wieder, "Ethnomethodology and the Problem of Order"; Wieder, *Language and Social Reality*.

actions, attention is drawn primarily to the *methods humans use* to construct, maintain, and change the *appearance* of an orderly and connected social world. These methods are directly observable and constitute a major portion of people's actions in everyday life. In contrast, the actual substance and content of the occasioned corpus are not directly observable and can only be inferred. Furthermore, in concentrating on the *process* of creating, sustaining, and changing the occasioned corpus, it can be asked: Is not the process of creating for one another the appearance of a stable social order more critical to understanding how society is possible than the actual substance and content of the occasioned corpus? Is there anything more to society than members' beliefs that it is out there forcing them to do and see certain things? If this fact is true, order is the result not of the particular structure of the corpus but of the human capacity to continually assemble and disassemble the corpus in each and every interaction situation. This perspective suggests to ethnomethodologists like Zimmerman, Pollner, Wieder, and many others that theoretical attention should therefore be placed on the ongoing process of assembling and disassembling the appearance of social order and on the particular methods people employ in doing so.

NEW PARADIGM OR IMPORTANT SUPPLEMENT?

I think that ethnomethodology has uncovered a series of interpersonal processes that traditional interactionists have failed to conceptualize. The implicit methods that people use to communicate a sense of social order are, I believe, a very crucial dimension of social interaction and organization. The theoretical goal is to specify the generic conditions under which various folk methods are used by individuals.

The rather extreme polemics about the "new reality" exposed by ethnomethodologists are, I feel, best forgotten.[27] The few findings of ethnomethodologists do not sustain their metaphysical vision of the world any more than they support interactionist formulations. Indeed, the findings of ethnomethodologists blend very nicely with those of symbolic interactionists, role theorists, and other theoretical traditions. Ethnomethodologists have, therefore, rather overstated their case. I do not doubt that people's sense of sharing a common world is an important property of interaction and organization, but it is not the only interactive dynamic. And to the degree that ethnomethodologists assert that their domain of inquiry is *the only* reality, they make themselves look foolish.[28]

It is time, I think, for ethnomethodologists to get off their high horse and begin to integrate their ideas into social theory. For too much effort has been

---

[27]For an interesting discussion of this assumption that challenges the ethnomethodological perspective, see Bill Harrell, "Symbols, Perception, and Meaning," in *Sociological Theory: Inquiries and Paradigms*, ed. L. Gross (New York: Harper & Row, 1967), pp. 104-27.

[28]I agree with Lewis Coser on this issue. See his "Two Methods in Search of a Substance," *American Sociological Review* 40 (1975), pp. 691-700. I also agree with him about the "other method" discussed in this article.

spent challenging and attacking "normal sociology"; the goals of theoretical cumulation would be better served if there were a conscious and concerted attempt by ethnomethodologists to integrate their ideas with those of mainstream interactionism.[29] For a recent conciliatory overview of ethnomethodology by an ethnomethodologist, see: Deidre Boden, "The World as It Happens: Ethnomethodology and Conversation Analysis" in *Frontiers of Social Theory*, ed. George Titzer (New York: Columbia University Press, 1990), pp. 185–213. See also: John Heritage, *Garfinkel and Ethnomethodology* (Cambridge: Polity Press, 1984). See also the extensive bibliographies of these works for further references.

---

[29]Actually, many nonethnomethodologists, such as Jurgen Habermas (see Chapter 13), Randall Collins (see Chapter 12), and Anthony Giddens (see Chapter 26), have used the ideas of ethnomethodologists to build interesting theory—far more interesting than that produced by the ethnomethodologists themselves.

PART 5

Structural Theorizing

CHAPTER 24

---◆---

The Origins of Structural Theorizing

THE IDEA OF SOCIAL STRUCTURE

Despite sociologists' frequent use of the concept *social structure*, its meaning remains unclear. My sense is that notions of structure are employed as metaphors for describing social interaction and relations that endure over time, but, beyond this, structure is not conceptualized precisely. Throughout previous chapters we have come across labels to denote structure—labels like "social system," "imperatively coordinated associations," "corporate unit," "institutionalization," "strains toward integration," "organizational system," and so on. All of these and other concepts emphasize that human social relations become patterned and comparatively stable over time, and most theoretical perspectives in sociology try to understand how and why this is so. But curiously, even as theorists make these efforts, their conceptualization of social structure remains vague. In the group of chapters in Part V, I will examine several representative approaches that make the concept of structure a central issue. And, as we will see, even when an entire theoretical program is built around the idea of structure, the conceptualization of structure is still vague. Moreover, proponents of various "structural sociologies" reveal somewhat different views of the universe—making the chapters in this section perhaps the most diverse examined thus far.

Each of these approaches is eclectic, borrowing from those perspectives that I have already summarized, and each owes its inspiration to a number of early thinkers who framed the general problems of structural analysis and who provided many of the critical concepts. These early thinkers are by now familiar, because I have examined them in other contexts. But in this chapter I want to focus more precisely on how they each conceptualized social structure and how their respective views have stimulated more modern structural theorizing.[1] As I do so, it will be evident that many of the basic issues that trouble

---

[1]For one recent effort to clarify the "structuralisms" that pervade social theory, see Bruce H. Mayhew, "Structrualism versus Individualism: Part I, Shadowboxing in the Dark" and "Struc-

sociological theorizing become highly visible in structural theories. For example, can sociology be a science, or is there something different and unique about human structures? Can social structure be conceptualized independently of the mental micro processes from which it is constituted, or must it be analyzed in terms of these micro processes? And can the basic dimensions—both macro and micro—of structure be isolated and analyzed by theory, or is structure merely a reification?[2] As is evident, structural theorizing forces us to deal with sociology's more difficult questions. Let me now trace the origins of these questions in the structural thought of sociology's early masters.

KARL MARX AND STRUCTURAL ANALYSIS

There have been explicit efforts to reinterpret Marx by French structuralists, who, I should emphasize, are not discussed in the following chapters.[3] Despite the failure of such reinterpretations of Marx in structuralist terms, Marx's ideas have nonetheless influenced the development of less mystical analysis of social structure.[4] This influence revolves around Marx's conceptions of (1) system reproduction and (2) system contradiction.[5] I have imposed more modern terminology on Marx here, but the ideas are his. Let me briefly review each.

1. Marx was concerned with how patterns of inequality "reproduce" themselves. That is, how is inequality in power and wealth sustained? And how are social relations structured to sustain these inequalities? This metaphor of social reproduction has been picked up by many structural theorists who like to view social structures as "reproduced" by the repetitive social encounters among individuals.

From this metaphorical use of Marx's ideas comes a vision of social structure as the distribution of resources among actors who use their respective resources in social encounters and, in the process, reproduce social structure and its attendant distribution of resources. Thus structure *is* the symbolic, material, and political resources that actors have in their en-

---

turalism versus Individualism: Part II, Ideological and Other Obfuscations," *Social Forces* 59 (2 and 3, 1981), pp. 335–75 and 627–48, respectively. See also a volume edited by Ino Rossi, *Structural Sociology* (New York: Columbia University Press, 1982), for representative analyses.

[2]For example, see Douglas W. Maynard and Thomas P. Wilson, "On the Reification of Social Structure," *Current Perspectives in Social Theory* 1 (1980), pp. 287–322.

[3]Such is the case because I do not think that they have generated any theory, even if we include meta-theory as part of sociological theorizing. Structuralism is, of course, a broad intellectual movement, and its most Marxist exponent was Louis Althusser, *For Marx* (New York: Penguin Books, 1969) and *Politics and History* (London: NLB, 1977). For an interesting secondary review, see also Alison Assiter, "Althusser and Structuralism," *The British Journal of Sociology* 35 (2, 1984), pp. 272–94.

[4]See section on structuralism, pp. 500–507.

[5]See Karl Marx, *A Contribution to the Critique of Political Economy* (New York: International Publishers, 1970); *The Economic and Philosophic Manuscripts of 1844* (New York: International Publishers, 1964); and *Capital: A Critical Analysis of Capitalist Production* (New York: International Publishers, 1967).

counters; and, as they employ these resources to their advantage, they reproduce the structure of their social relations because they sustain their respective shares of resources. For example, those who can control where others are physically located, who have access to coercion, who control communication channels, and who can manipulate the flow of information can "structure" successive encounters with others who have fewer resources in ways that reproduce inequalities. In so doing, they reproduce structure, but structure is tied to the interactive encounters among individuals using their resources. It is not something "out there" beyond actors and their interactions in concrete situations.

2. Marx was, of course, interested in changing capitalist society, and so he needed to introduce a concept that could break the vicious cycle of system reproduction. His analysis of Hegel's dialectics gave him the critical idea of "contradiction." Material social arrangements contain, he argued, patterns of social relations that are self-transforming. For example, capitalists must concentrate large pools of labor around machines in urban areas, which allow workers to communicate their grievances and to organize politically to change the very nature of capitalism. Thus, private expropriation of profits by capitalists stands in contradiction to the socially organized production process; over time this underlying contradiction will produce conflictual social relations—i.e., a revolution by workers—that change and transform the nature of social relations and, hence, social structure.[6]

Again, my sense is that this notion of contradiction is borrowed in a metaphorical way. It is used to introduce conflict and change into structural analysis. It posits, in essence, that the distribution of resources that sustains or reproduces social relations is inherently subject to redistribution, under certain conditions. The main condition is inequality. That is, structures reproduced through unequal distribution of symbolic, material, and political resources will exhibit underlying contradictions that will, in turn, generate pressures for interactions that *do not* reproduce the structure but instead transform it in some way through redistribution of resources.

THE STRUCTURALISM OF ÉMILE DURKHEIM

Émile Durkheim's work has directly influenced structural analysis in contemporary social theory. His emphasis on the "collective conscience" and its relation to the ordering of social relations was to become a central theme in structuralist analysis.[7] In Durkheim's later works he argued that a "certain number of essential ideas . . . of space, class, number, cause, substance . . . [that] correspond to the most universal properties of things . . . are like the solid

---

[6]Karl Marx and Friedrich Engels, *The Communist Manifesto* (New York: International Publishers, 1971).

[7]Émile Durkheim, *The Division of Labor in Society* (New York: Free Press, 1947).

frame which encloses all thought."[8] But what determines the content of these "essential ideas"? Durkheim's answer was "the structure of society." That is, the categories by which we perceive the world are determined by the structure and pattern of our social relations, and it is in this way that human categories of thought reinforce the structure of society.

This line of argument was to become the main ingredient of structuralism, but early modern structuralists tended to turn "Durkheim on his head" in their insistence that "deep structures" of thought, sometimes seen to be lodged in human biology, determine the structure of social relations. Durkheim argued just the opposite: structures of thought reflect social structural arrangements, and, in so doing, they reinforce and reproduce those arrangements.[9] Some of the particulars of Durkheim's analysis are not, I feel, very illuminating,[10] but the general approach can be found in much contemporary structural analysis. But this modern analysis takes a slightly different tack: social structure *is*, to a very great extent, the mental categories that actors have about their social relations. People use their ideas about how to approach and organize social relations, and, in so doing, they reproduce these relations and give them continuity across time.

Thus ideas are not floating around freely detached from people; cognitions about social structures are part of individuals' stocks of knowledge. Similarly, social structure is not some ex cathedra entity wholly external to individuals that pushes them around; it is embodied in actors' cognitions and then used as an interactive resource in social relations to produce, reproduce, and potentially change patterns of social relations.

This line of argument, which strikes me as a bit vague but which is asserted with sincerity and seriousness by many modern thinkers, shifts attention away from macrostructure toward microstructural analysis of what individuals in specific encounters actually *think and do* with their stocks of knowledge about social relations. As analysis shifts to the micro foundations of the macro order, Durkheim's later ideas on ritual often serve as a conceptual springboard.

Durkheim made a simple observation about religious rituals: society is personified in the sacred forces of religions, and so the worship of these forces is, in fact, the worship of society.[11] Rituals addressed toward gods and supernatural forces are thus an affirmation of the power of society; as people enact these rituals, they legitimate existing social relations. In his more discursive statements about the Arunta aborigines in *The Elementary Forms of Religious Life*, Durkheim expanded these ideas somewhat.[12] Rituals among individuals

[8]Émile Durkheim, *The Elementary Forms of Religious Life* (New York: Free Press, 1947; originally published in 1912), p. 9.

[9]So argued some of the early linguists who borrowed from Durkheim. But modern linguistics has turned Durkheim, and those early French thinkers who borrowed from him, upside down.

[10]Indeed, at times they are embarrassingly bad. For example, see Émile Durkheim and Marcel Mauss, *Primitive Classification* (Chicago: University of Chicago Press, 1963; originally published in 1903).

[11]Durkheim, *The Elementary Forms.*

[12]Ibid.

who are gathered together promote a sense of solidarity, a convergence of outlooks, and an increase in social attachments. Using rather exaggerated accounts of ritual among Arunta aborigines, Durkheim believed that data on the periodic gatherings among traditional peoples offered a window on the past; they were, in a sense, "living fossils" about the nature of the first human societies. As a result, Durkheim concluded that society is built from religious rituals that promote social attachments and solidarity as well as a convergence of outlooks.

Contemporary structural theories often extend these ideas of Durkheim. It is argued that the basis of social structure revolves not just around formal ceremonies but also around the daily interaction rituals that people perform in all social encounters. These rituals are essential, it is asserted, for maintaining continuity across social encounters and for giving people a sense of psychological security. Social structure is thus built from interaction rituals that connect successive encounters together and give individuals psychological security as well as a sense of attachment to a larger social order.

THE STRUCTURALISM OF CLAUDE LÉVI-STRAUSS

In Chapter 14 on the emergence of exchange theory, I briefly examined Claude Lévi-Strauss's analysis of bridal exchanges. In *The Elementary Structures of Kinship*,[13] Lévi-Strauss only hints at the more philosophical view of the world that his analysis of kinship implies.[14] Most of his book examines the varying levels of social solidarity that emerge from direct and indirect bridal exchanges among kin groups. Yet in many ways *The Elementary Structures of Kinship* is a transitional work because it begins to depart from the earlier foundations provided by Émile Durkheim and Marcel Mauss.[15] Indeed, these departures from Durkheim and Mauss signal that Lévi-Strauss was about to "turn Durkheim on his head" in much the same way that Marx was to revise Hegel. For, as Durkheim and Mauss had argued in *Primitive Classification*, human cognitive categories reflect the structure of society.[16] In contrast, Lévi-Strauss's structuralism came to the opposite conclusion: the structure of society is but a surface manifestation of fundamental mental processes.

Lévi-Strauss came to this position under the influence of structural linguistics, as initially chartered by Ferdinand de Saussure[17] and Roman Jakobson.

---

[13]Claude Lévi-Strauss, *The Elementary Structures of Kinship* (Paris: University of France, 1949).

[14]Actually, an earlier work, "The Analysis of Structure in Linguistics and in Anthropology," *Word* 1 (1945), pp. 1–21, provided a better clue as to the form of Lévi-Strauss's structuralism.

[15]Marcel Mauss, *The Gift* (New York: Free Press, 1954; originally published in 1924), is given particular credit. It must be remembered, of course, that Mauss was Durkheim's student and son-in-law.

[16]Émile Durkheim and Marcel Mauss, *Primitive Classification*. Originally published in 1903, this is a rather extreme and unsuccessful effort to show how mental categories directly reflect the spatial and structural organization of a population. It is a horribly flawed work, but it is the most extreme statement of Durkheim's sociologistic position.

[17]Ferdinand de Saussure, *Course in General Linguistics* (New York: McGraw-Hill, 1966); originally compiled posthumously by his students from their lecture notes in 1915.

Ferdinand de Saussure is typically considered the father both of structural linguistics[18] and of Lévi-Strauss's structuralism. In reality, as a contemporary of Durkheim, de Saussure was far more committed to Durkheim's vision of reality than is typically acknowledged. Yet he made a critical breakthrough in linguistic analysis of the 19th century: speech is but a surface manifestation of more fundamental mental processes.[19] Language is not speech or the written word; rather, it is a particular way of thinking, which, in true Durkheimian fashion, de Saussure viewed as a product of the general patterns of social and cultural organization among people. This distinction of speech as a mere surface manifestation of underlying mental processes was to increasingly be used as a metaphor for Lévi-Strauss's structuralism. Of course, this metaphor is as old as Plato's view of reality as a mere reflection of universal essences and as recent as Marx's dictum that cultural values and beliefs, as well as institutional arrangements, are reflections of an underlying substructure of class relations.

Lévi-Strauss also borrows from the early-20th-century linguist Roman Jakobson the notion that the mental thought underlying language occurs in terms of *binary contrasts*, such as good/bad, male/female, yes/no, black/white, and human/nonhuman. Moreover, drawing from Jakobson and others, Lévi-Strauss views the underlying mental reality of binary opposites as organized, or mediated, by a series of "innate codes" or rules that can be used to generate many different social forms: language, art, music, social structure, myths, values, beliefs, and so on.[20] Thus, over the decades, Lévi-Strauss's structuralism became concerned with understanding cultural and social patterns in terms of the universal mental processes that are rooted in the biochemistry of the human brain.[21] It is in this sense that Lévi-Strauss's structuralism is mentalistic and reductionistic. And, for this reason, I do not think that it is sociologically very useful.[22] Indeed, it is more philosophy than science. Yet, since it now generates so much intellectual interest, even in sociology, let me at least summarize its basic line of argument.[23]

---

[18]I suspect that this is a retrospectively bestowed title by those looking for their intellectual roots.

[19]Such analysis had focused on the written languages of Europe, seeking their evolutionary origins.

[20]Actually, Jakobson simply argued that children's phonological development occurs as a system in which contrasts are critical—for example, "papa versus mama" or the contrasts that children learn between vowels and consonants. Lévi-Strauss appears to have added the jargon of information theory and computer technology.

[21]See, for example, Claude Lévi-Strauss, "Social Structure," in *Anthropology Today*, ed. A. Kroeber (Chicago: University of Chicago Press, 1953), pp. 524–53; *Structural Anthropology* (Paris: Plon, 1958; trans. 1963 by Basic Books); and *Mythologiques: le cru et le cuit* (Paris: Plon, 1964).

[22]For relevant critiques, see Marvin Harris, *The Rise of Anthropological Theory* (New York: Thomas Y. Crowell, 1968), pp. 464–513; and Eugene A. Hammel, "The Myth of Structural Analysis" (Addison-Wesley Module, no. 25, 1972).

[23]See, for example, Mirian Glucksmann, *Structuralist Analysis in Contemporary Social Thought* (London: Routledge & Kegan Paul, 1974). A more sympathetic review of Lévi-Strauss, as well as a more general review of structuralist thought, can be found in Tom Bottomore and Robert Nisbet, "Structrualism," in their *A History of Sociological Analysis* (New York: Basic Books, 1978).

Without enumerating the full complexity (and vagueness) of his approach, there are several critical steps of Lévi-Strauss's theoretical (philosophical) strategy:

1. The empirically observable must be viewed as a *system* of relationships among components—whether these components be elements of myths and folk tales or positions in a kinship system.
2. It is appropriate to construct "statistical models" of these observable systems to summarize the empirically observable relationships among components.
3. Such models, however, are only a surface manifestation of more fundamental forms of reality. These forms are the result of using various codes or rules to organize different binary opposites. Such forms can be visualized through the construction of "mechanical models," which articulate the logical results of using various rules to organize different binary oppositions.
4. The tendencies of statistical models will reflect, imperfectly, the properties of the mechanical model. But it is the latter that is "more real."
5. The mechanical model is built from rules and binary oppositions that are innate to humans and rooted in the biochemistry and neurology of the brain.

Steps 1 and 2 are about as far as Lévi-Strauss went in the first publication of *The Elementary Structures of Kinship*. Subsequent work on kinship and on myths has invoked at least the rhetoric of Steps 3, 4, and 5. What makes structuralism distinctive, therefore, is the commitment to the assumptions and strategy implied in these last steps. The major problem with this strategy is that it is untestable, for, if mechanical models are never perfectly reflected in the empirical world, how is it possible to confirm or disconfirm the application of rules to binary opposites? As Marshall Sahlins sarcastically remarked, "What is apparent is false and what is hidden from perception and contradicts it is true."[24] Yet, despite such criticisms, the imagery communicated by Lévi-Strauss, especially in Steps 1 and 3, has influenced a great deal of structural theorizing. Although the extremes of Lévi-Strauss's approach have not been adopted by many, the idea of structure as involving "grammars" and "codes" that guide actors in their actions and in the production of structure has remained appealing, as will be explored in Chapter 25.

GEORG SIMMEL'S FORMAL APPROACH

It is difficult to trace the direct influence of Georg Simmel's sociology on structuralist sociology.[25] Some structural theorists, such as Peter Blau, acknowledge

---

[24]Marshall D. Sahlins, "On the Delphic Writings of Claude Lévi-Strauss," *Scientific American* 214 (1966), p. 134.

[25]Georg Simmel, *Sociology: Studies in the Forms of Sociation* (1908), has still not been fully translated. For samplings of translated portions, see Kurt Wolf, ed., *The Sociology of Georg Simmel* (New York: Free Press, 1950), as well as Reinhard Bendix, ed., *Conflict and the Web of Group Affiliations* (New York: Free Press, 1955).

their debt to Simmel, but for the most part Simmel's sociology remains an enigma—that is, an approach that is universally known and admired but without obvious contemporary adherents.

Yet Simmel's emphasis on the "forms of interaction" has, I believe, exerted considerable influence on structural thinking. For Simmel, social structure consists of "permanent interactions," and formal sociology seeks to uncover the underlying pattern of these permanent interactions. The content or substantive nature of these interactions is far less significant, sociologically, than their basic form. For although interactions can reveal an enormous variety of contents, the underlying form of relations may be the same. As Simmel emphasized:[26]

> Social groups, which are the most diverse imaginable in purpose and general significance, may nevertheless show identical forms of behavior toward one another on the part of individual members. We find superiority and subordination, competition, division of labor . . . and innumerable similar features in the state, in the religious community, in a band of conspirators, in an economic association, in an art school, in the family. However diverse the interests are that give rise to these associations, the *forms* in which the interests are realized may yet be identical.

Social structure must thus be conceptualized as forms or configurations of interaction that undergird and make possible the wide variety of substantive activities of individuals. This point of emphasis makes Simmel's eclectic approach to sociology more understandable, for although he studied widely diverse substantive matters, he was always seeking to discover the underlying form of interaction. In all of Simmel's most famous essays, this is the message: whatever the surface substance and content of social relations, there is an underlying form or structure. For example, as Simmel examined conflict, he could see that conflicts between nation/states, individual people, and small groups all reveal certain basic elements, or forms; or, to illustrate further, as Simmel examined the effects of increasing numbers of people in an encounter, he could argue that the geometric increase in the number of possible relations (two people can have two ties, three can have six, four can have twelve, and so on) changes the form or structure of the encounter; or, as Simmel examined the effects of money (a neutral and nonspecific medium of exchange), he could conclude that the form of social relations is fundamentally altered.

These kinds of arguments have led many structural theorists to concentrate on underlying forms of relations, rather than on the substantive nature of the units in these relations. Such efforts are not structur*alist* in the modern French tradition of excessive appeals to cognitive structures, but rather the goal is to isolate the observable and generic pattern of actual social relations among different units.[27] As we will see, Peter Blau's macrostructural approach (Chapter

---

[26]Georg Simmel, *Fundamental Problems of Sociology* (1918), portions of which are translated in Wolf, *The Sociology of Georg Simmel*. Quote is from p. 22 of original.

[27]For my review of Simmel's work, see Jonathan H. Turner, Leonard Beeghley, and Charles Powers, *The Emergence of Sociological Theory* (Belmont, CA: Wadsworth, 1989), pp. 248–83.

28) and much network sociology (Chapter 27) are both solidly in this Simmelian tradition. And thus, although we cannot easily find Simmelian sociologists, as we can Marxians, Durkheimians, and Lévi-Straussian structuralists, Simmel's basic approach has had a fundamental, if only implicit, impact on structural thinking in sociology.

INTERACTIONISM AND MICROSTRUCTURALISM

As structural analyses have become more concerned with the interactive processes by which structure is produced and reproduced, they have turned to interactionist theory—primarily ethnomethodology, phenomenology, Chicago School symbolic interactionism, and the dramaturgical work of Erving Goffman.[28] All of these interactionist traditions are built from the conceptual base laid by George Herbert Mead[29] and, to a lesser extent, Alfred Schutz.[30] Let me first examine Mead's ideas that reappear in contemporary structural analysis and then turn to Schutz's approach.

G. H. Mead's Behavioristic Structuralism

As I noted in Chapter 18, Mead was a social behaviorist.[31] By this he meant that those behavioral capacities facilitating adjustment to the social environment will be retained by a maturing individual. They are retained because they provide reinforcement or the rewards that come from adjustment to ongoing patterns of social organization. What, then, are these behavioral capacities that facilitate adjustment to social structures? For Mead they are (1) the capacity to assume the perspective of a variety of others (role taking); (2) the capacity to see oneself as an object in situations and to adjust response in accordance with self-perceptions and evaluations (self); and (3) the capacity to engage in the process of "imaginative rehearsal" of alternative lines of conduct and their potential outcomes (mind). In a very real sense, Mead provided the mechanism for Émile Durkheim's analysis of how the "collective conscience" is connected to individual behavior.

With these basic elements, Mead presents a general image of social organization as constructed and actively sustained through the mutual reading of gestures, the evaluation of oneself from the perspective of generalized others or "communities of attitudes," and the reflexive weighing of alternatives. These ideas are, today, very much a part of most microstructural approaches. Yet most micro approaches tend to underemphasize Mead's view that social structure constrains and circumscribes the options of individuals; in particular, Mead stressed that the ability to role-take with the generalized other and to

---

[28]See Chapters 18, 19, 21, and 22.

[29]George Herbert Mead, *Mind, Self, and Society* (Chicago: University of Chicago Press, 1934).

[30]Alfred Schutz, *The Phenomenology of the Social World* (Evanston, IL: Northwestern University Press, 1967; originally published in 1932).

[31]See early sections of Mead, *Mind, Self, and Society.*

use this community of attitudes as a basis of self-evaluation is crucial to the reproduction of structure. That is, in a very Durkheimian and structuralist fashion, Mead saw structure as being reproduced by people's use of general attitudes (values, beliefs, norms, and other cultural processes in modern terms). Thus structuralists take only part of Mead's legacy, in which emphasis is on situational interpersonal practices. They tend to ignore Mead's more Durkhemian ideas on conformity to generalized others.

Alfred Schutz

Structuralists seem to prefer Schutz's conceptualization of stock knowledge at hand to Mead's notion of generalized others and to modern conceptions of roles, values, beliefs, and norms. I assume that this preference comes from the metaphorical idea of stocks that can be used selectively as the actor wants and desires. Schutz's conception would appear to deemphasize explicit normative constraints and roles. But does it? Let me quote Schutz:

> Now, to the natural man all his past experiences are present as *ordered* [italics in original], as knowledge or as awareness of what to expect, just as the whole external world is present to him as ordered. Ordinarily, and unless he is forced to solve a special kind of problem, he does not ask questions about how this ordered world was constituted. The particular patterns of order we are now considering are synthetic meaning—configurations of already lived experience.[32]

For me Schutz's view emphasizes constraint. Norms, values, beliefs, and roles are indeed highly salient parts of one's implicit and, if need be, explicit interpretation of a situation. They "order" experiences and lines of conduct. But for many the concept of *stocks of knowledge* is meant to connote something much less ordered and structured. For them it is a series of implicit cognitions that are used to construct lines of conduct in strategic conduct.

MODERN STRUCTURAL ANALYSIS

As this brief review of Marx's, Durkheim's, Lévi-Strauss's, Simmel's, Mead's, and Schutz's ideas would indicate, modern structural theory goes in many different directions within structural sociology. It would be impossible to examine all of these directions, and so I have chosen some of the most prominent representatives of different traditions in structural analysis. I will begin with some of the more creative work within the Durkheim-Lévi-Strauss lineage, but I will focus on two theorists—Pierre Bourdieu and Robert Wuthnow—who reject the extremes of French structuralism as it became transformed by Lévi-Strauss. Next I will turn to the structuration theory of Anthony Giddens, who, despite his anti-science and positivistic rhetoric, has creatively synthesized elements of structuralism (in the sense of French lineage) with Marxian ideas

---

[32]Schutz, *The Phenomenology of the Social World*, p. 81.

on dialectics, Meadian interactionism, Schutzian phenomenology, and even psychoanalytic theory. Then I will turn to network analysis, which, in my view, is solidly within the Simmelian tradition of formal sociology. And, finally, I will close with Peter Blau's analysis, which represents a creative extension of Simmel's approach.

CHAPTER 25

◆

Cultural Structuralism

Robert Wuthnow and Pierre Bourdieu*

MODERN STRUCTURALISM

French structuralism has exerted an enormous influence on modern social thought—from anthropology and sociology to literary criticism and many other fields of inquiry in between. These many types of structuralist analyses vary enormously, but they all have a common theme: there is a deep underlying structure to most surface phenomena, and this structure can be conceptualized as a series of generative rules that can create a wide variety of empirical phenomena. That is, empirically observable phenomena—from a literary text to a social structure—are constructed in conformity to an implicit logic. Some see this logic as lodged in the biology of the human brain, whereas others would view this underlying structure as a cultural product.

Purely structuralist analysis has had its greatest impact in linguistics and literary criticism, and for a time it enjoyed considerable popularity in anthropology and sociology.[1] But in recent years rigid and orthodox structuralist approaches emphasizing searches for those deep and universal structures that order all phenomena have declined in sociology; in their place a number of more eclectic perspectives have emerged. These more eclectic theories borrow elements of structuralist analysis and blend them with other conceptual tra-

---

*This chapter was coauthored with Stephan Fuchs.

[1]For some general works reviewing structuralism, see Anthony Giddens, "Structuralism, Post-structuralism and the Production of Culture," in *Social Theory Today*, eds. A. Giddens and J. H. Turner (Stanford, CA: Stanford University Press, 1987); S. Clarke, *The Foundations of Structuralism* (Sussex, England: Harvester, 1981); J. Sturrock, ed., *Structuralism and Science* (Oxford: Oxford University Press, 1979); W. G. Runciman, "What Is Structuralism?" in *Sociology in Its Place* (Cambridge, England: Cambridge University Press, 1970); Ino Rossi, *From the Sociology of Symbols to the Sociology of Signs* (New York: Columbia University Press, 1983) and Ino Rossi, ed., *Structural Sociology* (New York: Columbia University Press, 1982); Jacques Ehrmann, *Structuralism* (New York: Doubleday, 1970); Philip Pettit, *The Concept of Structuralism: A Critical Analysis* (Berkeley: University of California Press, 1977); Charles C. Lemert, "The Uses of French Structuralism in Sociology" and Michelle Lamont and Robert Wuthnow, "Recent Cultural Sociology in Europe and the United States," in *Frontiers of Social Theory*, G. Ritzer, ed. (Columbia University Press, 1990).

ditions, such as conflict theory, interactionism, and phenomenology. There is still an emphasis on symbolic codes and the ways in which these are produced by underlying generative rules, but such codes are causally influenced by material conditions and are subject to interpretation by lay agents. Thus, just as Lévi-Strauss "turned Durkheim on his head" in seeing social structure as a reflection of mental structures, many recent efforts have put Durkheim back on his feet and augmented the emphasis on the structure of symbol systems with Marxian conflict analysis, interactionism and phenomenology, and other traditions in theoretical sociology. The result has been a revival of "cultural sociology" in a less ponderous guise than Parsonian functionalism. In these newer approaches, the structure of cultural codes is causally linked not only to the behavioral and interpersonal activities of individuals but also to the institutional parameters within which such activities are conducted.

Although there are several possible candidates who can be seen as squarely in this more eclectic structuralist approach, two have been selected for review here—the French scholar Pierre Bourdieu and the American Robert Wuthnow. Their work can provide a sample of what will be termed here *cultural structuralism*. Each draws from Durkheim, incorporates some of the insights of Lévi-Strauss without viewing social structure as a mere surface manifestation of cultural codes, and connects the French tradition to theories emphasizing the causal priority of material social conditions.

CULTURAL ANALYSIS: ROBERT WUTHNOW

In recent years there has been a revival of cultural analysis.[2] Unfortunately, the nature of "culture" is vaguely conceptualized in social theory, with the result that just about anything—physical objects, ideas, world views, subjective states, behaviors, rituals, thoughts, emotions, and so on—can be considered "cultural." One recent effort to narrow somewhat the domain of cultural analysis is the work of Robert Wuthnow. Although most of his research has focused on religion, he has sought to develop a more general theoretical approach.[3]

This theoretical effort preaches against positivism, incorrectly portrayed as sheer empiricism, but it nonetheless synthesizes several theoretical traditions and, in so doing, develops some general propositions on cultural processes.[4] It

---

[2]For a review, see Robert Wuthnow and Marsha Witten, "New Directions in the Study of Culture," *Annual Review of Sociology* 14 (1988), pp. 149-67. See also Robert Wuthnow, James Davidson Hunter, Albert Bergesen, and Edith Kurzweil, *Cultural Analysis: The World of Peter L. Berger, Mary Douglas, Michel Foucault, and Jurgen Habermas* (London: Routledge & Kegan Paul, 1984).

[3]For examples of the work on religion, see Robert Wuthnow, *The Consciousness Reformation* (Berkeley: University of California Press, 1976) and *Experimentation in American Religion* (Berkeley: University of California Press, 1978); for a review of a more general theory, see: Robert Wuthnow, *Meaning and Moral Order: Explorations in Cultural Analysis* (Berkeley: University of California Press, 1987). For a review of this work, see: Jonathan H. Turner, "Cultural Analysis and Social Theory," *American Journal of Sociology* 94 (July 1988), pp. 637-44.

[4]Wuthnow would seemingly not view these as laws or principles, but his articulation of such laws (often implicitly) is what makes the work theoretically interesting.

is, therefore, one of the more creative approaches to structuralism, primarily because it blends structuralist concerns about relations among symbolic codes, per se, with other theoretical traditions. Among these other traditions are elements of dramaturgy, institutional analysis, and subjective approaches.

Cultural Structure, Ritual, and Institutional Context

In Wuthnow's view it is wise to avoid "radical subjectivity," for the "problem of meaning may well be more of a curse than a blessing in cultural analysis."[5] It is wise, he argues, to move away from an overemphasis on attitudes, beliefs, and meanings of individuals, since these are difficult to measure. Instead, the structure of culture as revealed through *observable* communications and interactions is a more appropriate line of inquiry. In this way one does "not become embroiled in the ultimate phenomenological quest to probe and describe subjective meanings in all their rich detail."[6] Rather, the structure of cultural codes as produced, reproduced, or changed by interaction and communication is examined. Once emphasis shifts away from meaning, per se, to the structure of culture in social contexts and socially produced texts, other theoretical approaches become useful.

Dramaturgy is one essential supplement because of its emphasis on ritual as a mechanism for expressing and dramatizing symbols. In a sense, individual interpersonal rituals as well as collective rituals express deeply held meanings, but at the same time they affirm particular cultural structures. In so doing, ritual performs such diverse functions as reinforcing collective values, dramatizing certain relations, denoting key positions, embellishing certain messages, and highlighting particular activities.[7]

Another important theoretical supplement is institutional analysis. Culture does not exist as an abstract structure in its own right. Nor is it simply dramatic and ritualized performances; it is also embedded in organized social structures. Culture is produced by actors and organizations that require resources—material, organizational, and political—if they are to develop systems of cultural codes, ritualize them, and transmit them to others. And, once the institutional basis of cultural activity is recognized, then the significance of inequalities in resources, the use of power, and the outbreak of conflict become essential parts of cultural analysis.

In sum, then, Wuthnow seeks to blend a muted subjective approach with structuralism, dramaturgy, and institutional analysis. He tries to view the subjective as manifested in cultural products, dramatic performances, and institutional processes. In attempting this synthesis, Wuthnow defines his topic as *the moral order.*

---

[5]Wuthnow, *Meaning and Moral Order*, p. 64.
[6]Ibid., p. 65.
[7]Ibid., p. 132.

The Moral Order

Wuthnow views the moral order as involving the (1) construction of systems of cultural codes, (2) emission of rituals, and (3) mobilization of resources to produce and sustain these cultural codes and rituals. Let us examine each of these in turn.

The structure of moral codes   A *moral code* is viewed by Wuthnow "as a set of cultural elements that define the nature of commitment to a particular course of behavior." Such sets of cultural elements have an "identifiable structure" involving not so much a "tightly organized or logically consistent system" as some basic "distinctions" that can be used "to make sense of areas in which problems in moral obligations may be likely to arise."[8] Wuthnow sees three such distinctions as crucial to structuring the moral order: (1) moral objects vs. real programs, (2) core self vs. enacted social roles, and (3) inevitable constraints vs. intentional options. Each of these is examined below.

(1) The structure of a moral order distinguishes between (*a*) the *objects* of commitment and (*b*) the activities or *real programs* in which the committed engage. The objects of commitment can be varied—a person, a set of beliefs and values, a text, etc.—and the real programs can be almost any kind of activity. The critical point, Wuthnow argues, is that the objects of moral commitment and the behavior emitted to demonstrate this commitment are "connected" and, yet, "different." For example, one's object of commitment may be "making a better life for one's children," which is to be realized through "hard work" and other activities or real programs. For the structure of a moral order to be effective, it must implicitly distinguish and, at the same time, connect such objects and real programs.

(2) The structure of moral codes must also, in Wuthnow's view, distinguish between (*a*) the person's "real self" or "true self" and (*b*) the various "roles" that he or she plays. Moral structures always link self-worth and behavior but, at the same time, allow them to be distinguished so that there is a "real me" who is morally worthy and who can be separated from the roles that can potentially compromise this sense of self-worth. For example, when someone reveals "role distance," an assertion is being made that a role is beneath one's dignity or self-worth.

(3) Moral codes must also distinguish between (*a*) those forces that are out of people's control and (*b*) those that are within the realm of their will. That is, the *inevitable* must be distinguished from the *intentional*. In this way, cultural codes posit a moral evaluation of those behaviors that can be controlled through intent and will power, while forgiving or suspending evaluation for what is out of a person's control. Without this distinction it would be impossible to know what kinds of behaviors are to be subjected to moral evaluations.

Thus the structure of a moral order revolves around three basic types of codes that denote and distinguish commitments with respect to moral objects/

---

[8]Ibid., p. 66.

real programs, self/roles, and inevitable constraint/intentional options. Such codes indicate what is desirable by separating but also linking objects, behavior, self, roles, constraints, and intentions. Without this denotation of, and a distinction along, these three axes, a moral order and the institutional system in which it is lodged will reveal crises and will begin to break down. If objects and programs are not denoted, distinguished, and yet linked, then cynicism becomes rampant; if self and roles are confused, then loss of self-worth spreads; and if constraints and control are blurred, then apathy or frustration increases. Thus for Wuthnow:

> Morality . . . deals primarily with moral commitment—commitment to an object, ranging from an abstract value to a specific person, that involves behavior, that contributes to self-worth, and that takes place within broad definitions of what is inevitable or intentional. Moral commitment, although in some sense deeply personal and subjective, also involves symbolic constructions—codes—that define these various relations.[9]

The nature of ritual   For Wuthnow, ritual is "a symbolic-expressive aspect of behavior that communicates something about social relations, often in a relatively dramatic or formal manner."[10] A *moral ritual* "dramatizes collective values and demonstrates individuals' moral responsibility for such values."[11] In so doing, rituals operate to maintain the moral order—that is, the system of symbolic codes ordering moral objects/real programs, self/roles, and constraints/options. Such rituals can be embedded in normal interaction as well as in more elaborate collective ceremonies, and they can be privately or publicly performed.[12] But the key point is that ritual is a basic mechanism for sustaining the moral order.

However, as Wuthnow stresses, ritual is also used to cope with uncertainty in the social relations regulated by the codes of the moral order. Whether through increased options, uses of authority, ambiguity in expectations, lack of clarity in values, equivocality in key symbols, or unpredictability in key social relations, rituals are often invoked to deal with these varying bases of uncertainty. Uncertainty is thus one of the sources of escalated ritual activity. However, such uses of ritual are usually tied to efforts at mobilizing resources in institutional contexts in order to create a new moral order—a process that, as we will examine shortly, Wuthnow examines under the rubric of "ideology."

Institutional context   For a moral order to exist, it must be produced and reproduced; and for new moral codes to emerge—that is, ideologies—they too must be actively produced by actors using resources. Thus systems of symbolic codes depend upon material and organizational resources; if a moral order is to persist and if a new ideology is to become a part of the moral order, it

---

[9]Ibid., p. 70.

[10]Ibid., p. 109.

[11]Ibid., p. 140.

[12]Here Wuthnow is drawing from the late Durkheimian tradition.

must have a stable supply of resources for actors to use in sustaining the moral order, or in propagating a new ideology. That is, actors must have the material goods necessary to sustain themselves and the organizations in which they participate; they must have organizational bases that depend not only on material goods, such as money, but also on organizational "know-how," communication networks, and leadership; and at times they must also have power. Thus the moral order is anchored in institutional structures revolving around material goods, money, leadership, communication networks, and organizational capacities.

Ideology   One of the central and yet ambiguous concepts in Wuthnow's analysis is his portrayal of ideology, which he defines as "symbols that express or dramatize something about the moral order."[13] This definition is very close to the one used for ritual, and so it is somewhat unclear as to what Wuthnow has in mind.[14] The basic idea appears to be that an ideology is a subset of symbolic codes emphasizing a particular aspect of the more inclusive moral order. Ideologies are also the vehicles for change in the moral order, because it is through the development and subsequent institutionalization of new ideologies that the moral order is altered. The production and institutionalization of these subsets of symbolic codes depend upon the mobilization of resources (leaders, communication networks, organizations, and material goods) and the creation and emission of rituals. New ideologies must often compete with one another; so, those ideologies with superior resource bases are more likely to survive and become a part of the moral order.

In sum, then, the moral order consists of a structure of codes, a system of rituals, and a configuration of resources that define the manner in which social relations should be constituted.[15] An important feature of the moral order is the production of ideologies, which are subsets of codes, ritual practices, and resource bases. With this conceptual baggage in hand, Wuthnow then turns to the analysis of dynamic processes in the moral order.

The Dynamics of the Moral Order

Wuthnow employs an ecological framework for the analysis of dynamics.[16] When a moral order does not specify the ordering of moral objects/real programs, self/roles, and inevitable constraints/intentional controls, when it cannot specify the appropriate communicative and ritual practices for its affirmation and dramatization, and when, as a result of these conditions, it cannot reduce the risks associated with various activities, the ambiguities of social situations, or the unpredictability of social relations, then the level of uncer-

---

[13]Wuthnow, *Meaning and Moral Order*, p. 145.

[14]See J. H. Turner, "Cultural Analysis and Social Theory," for a more detailed critique.

[15]Wuthnow, *Meaning and Moral Order*, p. 145.

[16]See Chapters 7 and 8 for other theories using such a framework. See Wuthnow, *Meaning and Moral Order*, for the outline of this ecological framework, especially Chapters 5 and 6.

tainty among the members of a population increases. Under conditions of uncertainty, new ideologies are likely to be produced as a way of coping. Such ideological production is facilitated by (1) high degrees of heterogeneity in the types of social units—classes, groups, organizations, etc.—in a social system and by high levels of diversity in resources and their distribution, (2) high rates of change (realignment of power, redistribution of resources, establishment of new structures, creation of new types of social relations), (3) inflexibility in cultural codes (created by tight connections among a few codes), and (4) reduced capacity of political authority to repress new cultural codes, rituals, and mobilizations of resources.

Wuthnow portrays these processes as an increase in "ideological variation" that results in "competition" among ideologies. Some ideologies are "more fit" to survive this competition and, as a consequence, are "selected." Such "fitness" and "selection" are dependent upon an ideology's capacity to (1) define social relations in ways reducing uncertainty (over moral objects, programs, self, roles, constraints, options, risks, ambiguities, and unpredictability), (2) reveal a flexible structure consisting of many elements weakly connected, (3) secure a resource base (particularly money, adherents, organizations, leadership, and communication channels), (4) specify ritual and communicative practices, (5) establish autonomous goals, and (6) achieve legitimacy in the eyes of political authority and in terms of existing values and procedural rules. The more that these conditions can be met, the more likely is an ideology to survive in competition with other ideologies and the more likely is it to be come institutionalized as part of the moral order. In particular, the institutionalization of an ideology depends upon the establishment of rituals and modes of communication affirming the new moral codes within an organizational arrangement that allows for ritual dramatization of new codes reducing uncertainty, that secures a stable resource base, and that eventually receives acceptance by political authority.

Different types of ideological movements will emerge, Wuthnow appears to argue, under varying configurations of these conditions that produce variation, selection, and institutionalization.[17] Although he offers many illustrations of ideological movements, particularly of various kinds of religious movements as well as the emergence of science as an ideology, he does not systematically indicate how varying configurations of these general conditions operate to produce basic types of ideological movements. Yet these variables all appear, in a rather ad hoc and discursive way, in his analysis of ideological movements. And so there is at least an implicit effort to test the theory.

In way of summary, then, Table 25-1 formalizes Wuthnow's theory. Wuthnow would probably reject this formalization as being too "positivistic," but, if his ideas are to be more explanatory and less discursive, formalization along these lines is desirable—hence the provisional effort expressed in Table 25-1.

---

[17]Wuthnow offers examples in *Meaning and Moral Order*, Chapters 5-9, but the variables are woven into discursive text and rearranged in an ad hoc manner.

TABLE 25–1 Wuthnow's Principles of Cultural Dynamics

I. The degree of stability in the moral order of a social system is a positive function of its legitimacy, with the latter being a positive and additive function of:

  A. The extent to which the symbolic codes of the moral order facilitate the ordering of:

    1. moral objects and real programs.
    2. self and roles.
    3. inevitable constraints and intentional control.

  B. The extent to which the symbolic codes of the moral order are dramatized by ritual activities.

  C. The extent to which the symbolic codes of the moral order are affirmed by communicative acts.

II. The rate and degree of change in the moral order of a social system are a positive function of the degree of ideological variation, with the latter being a positive and additive function of:

  A. The degree of uncertainty in the social relations of actors, which in turn is an additive function of:

    1. the inability of cultural codes to order moral objects/real projects, self/roles, and constraints/options.
    2. the inability of rituals to dramatize key cultural codes.
    3. the inability of communicative acts to affirm key cultural codes.
    4. the inability of cultural codes to specify the risks associated with various activities and relations.
    5. the inability of cultural codes to reduce the ambiguity of various activities and relations.
    6. the inability of cultural codes to reduce the unpredictability of various acts and relations.

  B. The level of ideological production and variation, which in turn is a positive and additive function of:

    1. the degree of heterogeneity among social units.
    2. the diversity of resources and their distribution.
    3. the rate and degree of change in institutional structures.
    4. the degree of inflexibility in the sets of cultural codes, which is an inverse function of:

      a. the number of symbolic codes.
      b. the weakness of connections among symbolic codes.

    5. the inability of political authority to repress ideological production.

III. The likelihood of survival and institutionalization of new ideological variants is a positive and multiplicative function of:

  A. The capacity of an ideological variant to secure a resource base, which in turn is a positive and additive function of the capacity to generate:

    1. material resources.
    2. communication networks.
    3. rituals.
    4. organizational footings.
    5. leadership.

  B. The capacity of an ideological variant to establish goals and pursue them.

  C. The capacity of an ideological variant to maintain legitimacy with respect to:

    1. existing values and procedural rules.
    2. existing political authority.

  D. The capacity of an ideological variant to remain flexible, which in turn is a positive function of:

    1. the number of symbolic codes.
    2. the weakness of connections among symbolic codes.

CONSTRUCTIVIST STRUCTURALISM: PIERRE BOURDIEU

Pierre Bourdieu's sociology defies easy classification because it cuts across disciplinary boundaries—sociology, anthropology, education, cultural history, art, science, linguistics, and philosophy—and moves easily between empirical and conceptual inquiry.[18] Yet Bourdieu has characterized his work as *constructivist structuralism* or *structuralist constructivism*; in so doing, he distances himself somewhat from the Lévi-Straussian tradition:

> By structuralism or structuralist, I mean that there exists, within the social world itself and not only within symbolic systems (language, myth, etc.), objective structures independent of the consciousness and will of agents, which are capable of guiding and constraining their practices or their representations.[19]

Such structures constrain and circumscribe volition, but at the same time people use their capacities for thought, reflection, and action to *construct* social and cultural phenomena. They do so within the parameters of existing structures, but these structures are not rigid constraints but materials for a wide variety of social and cultural constructions. Acknowledging his structuralist roots, Bourdieu analogizes to the relation of grammar and language in order to make this point: the grammar of a language only loosely constrains the production of actual speech; in fact, it can be seen as defining the possibilities for new kinds of speech acts. And so it is with social and cultural structures: they exist independently of agents and guide their conduct, and yet they also create options, possibilities, and paths for creative actions and for the construction of new and unique cultural and social phenomena. This perspective is best appreciated by highlighting Bourdieu's criticisms of those theoretical approaches from which he selectively borrows ideas.

Criticisms of Existing Theories

The critique of structuralism   Bourdieu's critique of structuralism is similar to symbolic interactionists' attacks on Parsonian functionalism and its emphasis on norms. According to Bourdieu, structuralists ignore the indeterminacy of situations and the practical ingenuity of agents who are not mechanical rule-following and role-playing robots in standard contexts. Rather, agents use their "practical sense" (*sens pratique*) to adapt to situational contingencies within certain "structural limits" that follow from "objective con-

---

[18]Indeed, Bourdieu has been enormously prolific, having authored some 25 books and hundreds of articles in a variety of fields, including anthropology, education, cultural history, linguistics, philosophy, and sociology. His empirical work covers a wide spectrum of topics—art, academics, unemployment, peasants, classes, religion, sports, kinship, politics, law, and intellectuals. See Loic J. D. Wacquant, "Towards a Reflexive Sociology: A Workshop with Pierre Bourdieu," *Sociological Theory* 7 (1, Spring 1989), pp. 26–63. This article also contains a selected bibliography on Bourdieu's own works as well as secondary analyses and comments on Bourdieu.

[19]Pierre Bourdieu, "Social Space and Symbolic Power," *Sociological Theory* 7 (1, Spring 1989), p. 14.

straints." Social practice is more than the mere execution of an underlying structural "grammar" of action, just as "speech" (*parole*) is more than "language" (*langue*). What is missing, says Bourdieu, are the variable uses and contexts of speech and action.[20] Structuralism dismisses action as mere execution of underlying principles (lodged in the human brain or culture), just as normativism forgets that following rules and playing roles require skillful adjustments and flexible improvisations on the part of creative agents.

Most importantly, Bourdieu argues that structuralism hypostatizes the "objectifying glance" of the outside academic observer. The "Homo academicus" transfers a particular *relation* to the world, the distant and objectifying gaze of the professional academic, onto the very properties of that world.[21] As a result, the outside observer constructs the world as a mere "spectacle," which is subject to neutral observation. The distant and uninvolved observer's relation to the world is not only systematic but also a passive *cognition*, and so the world itself comes to be viewed as consisting of cognition rather than active practices. According to Bourdieu, structuralism and other approaches that "objectify" the world do not simply research the empirical world "out there"; rather, they construct it *as* an "objective fact" through the distancing perspective of the outside observer.

Bourdieu does not, however, reject completely structuralism and other "objectifying" approaches that seek, in Durkheim's words, to discover external and constraining "social facts." As we will come to see, Bourdieu views social classes, and factions within such classes, as "social facts" whose structure can be objectively observed and viewed as external to, and constraining on, the thoughts and activities of individuals.[22] Moreover, Bourdieu at least borrows the metaphor, if not the essence, of structuralism in his efforts to discover the "generative principles" that people use to construct social and cultural phenomena—systems of classification, ideologies, forms of legitimating social practices, and other elements of "constructivist structuralism."[23]

The critique of interactionism and phenomenology   Bourdieu is also critical of interactionism, phenomenology, and other subjective approaches.[24] For Bourdieu, there is more to social life than interaction, and there is more to interaction than the "definitions of situations" in symbolic interactionism or the "accounting practices" in ethnomethodology. The "actor" of symbolic interactionism and the "member" of ethnomethodology are abstractions that fail to realize that members are always incumbents in particular

---

[20]Pierre Bourdieu, *Language and Symbolic Power* (Cambridge, MA: Harvard University Press, 1989).

[21]Pierre Bourdieu, *Homo Academicus* (Stanford, CA: Stanford University Press, 1988).

[22]Pierre Bourdieu, *Distinction: A Social Critique of the Judgement of Taste* (Cambridge, MA: Harvard University Press, 1984).

[23]See Wacquant, "Towards a Reflexive Sociology"; Bourdieu, "Social Space and Symbolic Power," pp. 14–25.

[24]See Wacquant, "Towards a Reflexive Sociology."

groups and classes. Interactions are always interactions-in-contexts, and the most important of these contexts is class location. Even such an elementary feature of interaction as the possibility that it might even occur among individuals varies with class background. Interaction is thus embedded in structure, and the structure constrains what is possible.

Moreover, in addition to this rather widespread critique of interactionism as "astructural," Bourdieu argues that interactionism is too cognitive in its overemphasis on the accounting and sense-making activities of agents. As a result, it forgets that actors have objective class-based interests. And, once again, the biases of "Homo academicus" are evident. For it is in the nature of academics to define, assess, reflect, ponder, and interpret the social world; as a result of this propensity, a purely academic relation to the world is imposed upon real people in social contexts. For interactionists, then, people are merely disinterested lay academics who define, reflect, interpret, and account for actions and situations. But lay interpretations, and academic portrayals of these lay interpretations, cannot accurately describe social reality, for two reasons. First, as noted above, these interpretations are constrained by existing structures, especially class and class factions. Second, these interpretations are themselves part of objective class struggles as individuals seek to construct legitimating definitions for their conduct.[25]

Bourdieu then borrows from Karl Marx the notion that people are located in a class position, that this position gives them certain interests, and that their interpretative actions are often ideologies designed to legitimate these interests. People's "definitions of situations" are neither neutral nor innocent, but are often ideological weapons that are very much a part of the objective class structures and the inherent conflicts of interests generated by such structures.[26]

The critique of utilitarianism   Rational economic theories also portray, and at the same time betray, Homo academicus' relation to the social world. Like academics in general, utilitarian economic theorists[27] see humans as rational, calculating, and maximizing (*sujets ravauts*); and rational exchange theories thus mistake a *model* of the human actor for real individuals, thereby reifying their theoretical abstractions.

Yet Bourdieu does not replace the economic model of rational action with an interpretative model of symbolic action. He does not argue that rational action theory is wrong because it is too rationalistic or because it ignores the interpretative side of action. To the contrary, he holds that rational action theory does not realize that even symbolic action is rational and based upon class interests. Thus, according to Bourdieu, the error of the economic model is not that it presents all action as rational and interested; rather, the big

---

[25]Bourdieu, *Outline of a Theory of Practice* (Cambridge, England: Cambridge University Press, 1977), pp. 21 *ff.*, and *Distinction*.

[26]Ibid.

[27]See Chapter 17.

mistake is to restrict interests and rationality to the immediate material payoffs collected by reflective and profit-seeking individuals.[28]

For Bourdieu all social practices are "interested," even if individual agents are unaware of their interests and even if the stakes of these practices are not material profits. Social practices are attuned to the conditions of particular arenas in which actions may yield profits without deliberate intention. For example, in science it is the most "disinterested" and "pure" research that yields the highest cultural profits—i.e., academic recognition and reputation. In social fields other than economic exchange it is the structural *denial* of any "interests" that often yields the highest gains. It is not that agents cynically deny being interested so as to increase their gains even more; rather, innocence assures that honest disinterestedness nevertheless is the most profitable practice.

For example, gift exchange economies, the subject of Bourdieu's early anthropological research,[29] may illustrate this complex idea. Gift exchange economies are typically embedded in larger social relations and solidarities so that exchange is not purely instrumental and material but has a strong moral quality to it. Economic exchanges are expected to follow the social logic of solidarity and group memberships at least as much as the economic logic of material gain. From the narrowly economic perspective of rational action theory, the logic of solidarity would seem like an intrusion of "nonrational" forces, such as tradition or emotion, into an otherwise purely rational system of exchange; but, in fact, the logic of solidarity points to those processes by which symbolic and social capital is accumulated—a "social fact" that is missed by the narrow economic determinism of rational action theory. But, once the notion of "capital" is extended to include symbolic and social capital, apparently "irrational" practices can now be seen to follow their own interested logic, and, contrary to initial impressions, these practices are not irrational at all. In fact, the structural denial of narrowly economic interests in gift exchange economies conceals that social and economic capital can be increased the more the purely instrumental aspects of exchange move into the background. For instance, birthday and Christmas presents are socially more effective when they appear less material and economic; those who brag about the high costs of their presents do not understand the nature of gift exchanges and, as a consequence, are considered rude, thereby losing symbolic and social capital.

Thus, in broadening economic exchange to include social and symbolic resources, as all sociological exchange theories eventually do,[30] Bourdieu introduces a central concept in his approach: *capital*.[31] Those in different classes

---

[28]See Wacquant, "Towards a Reflexive Sociology," p. 43.

[29]Bourdieu, *Outline of a Theory of Practice.*

[30]See Chapters 12, 17–19.

[31]Pierre Bourdieu, "The Forms of Capital," in *Handbook of Theory and Research in the Sociology of Education,* ed. J. G. Richardson (New York: Greenwood Press, 1986). See also: Michele Lamont and Annette P. Larreau, "Cultural Capital: Allusions, Gaps, and Glissandos in Recent Theoretical Developments," *Sociological Theory* 6 (2, Fall 1988), pp. 153–68.

reveal not only varying levels or amounts of capital but also divergent types and configurations of capital. Bourdieu's view of capital recognizes that the resources individuals possess can be material, symbolic, social, and cultural; moreover, these resources reflect class location and are used to further the interests of those in a particular class position.

Bourdieu's Cultural Conflict Theory

Although Bourdieu has explored many topics, the conceptual core of his sociology is a vision of social classes and the cultural forms associated with these classes.[32] In essence, Bourdieu combines a Marxian theory of objective class position in relation to the means of production with a Weberian analysis of status groups (lifestyles, tastes, prestige) and politics (organized efforts to have one's class culture dominate). The key to this reconciliation of Marx's and Weber's views of stratification is the expanded conceptualization of *capital* as more than economic and material resources, coupled with elements of French structuralism.

      Classes and capital   To understand Bourdieu's view of classes, it is first necessary to recognize a distinction among four types of capital.[33] (1) *economic* capital, or productive property (money and material objects that can be used to produce goods and services); (2) *social* capital, or positions and relations in groupings and social networks; (3) *cultural* capital, or informal interpersonal skills, habits, manners, linguistic styles, educational credentials, tastes, and lifestyles; and (4) *symbolic* capital, or the use of symbols to legitimate the possession of varying levels and configurations of the other three types of capital.

      These forms of capital may be converted into one another, but only to a certain extent. The degree of convertibility of capital on various markets is itself at stake in social struggles. The overproduction of academic qualifications, for example, can decrease the convertibility of educational into economic capital ("credential inflation"). As a result, owners of credentials must struggle to get their cultural capital converted into economic gains, such as high-paying jobs. Likewise, the extent to which economic capital can be converted into social capital is at stake in struggles over control of the political apparatus, and the efforts of those with economic capital to "buy" cultural capital can often be limited by their perceived lack of "taste" (a type of cultural capital).

      The distribution of these four types of capital determines the objective class structure of a social system. The overall class structure reflects the total amount of capital possessed by various groupings. Hence the *dominant class* will possess the most economic, social, cultural, and symbolic capital; the *middle class* will possess less of these forms of capital; and the *lower classes* will have

---

[32]Bourdieu, *Distinction* and *Outline of a Theory of Practice.*
[33]Bourdieu, "The Forms of Capital."

TABLE 25–2 Representation of Classes and Class Factions in Industrial Societies*

Dominant Class:	richest in all forms of capital
Dominant faction:	richest in economic capital, which can be used to buy other types of capital. This faction is composed primarily of those who own the means of production—that is, the classical bourgeoisie.
Intermediate faction:	some economic capital, coupled with moderate levels of social, cultural, and symbolic capital. This faction is composed of high-credential professionals.
Dominated faction:	little economic capital but high levels of cultural and symbolic capital. This faction is composed of intellectuals, artists, writers, and others who possess cultural resources valued in a society.
Middle Class:	moderate levels of all forms of capital
Dominant faction:	highest in this class with respect to economic capital but having considerably less economic capital than the dominant faction of the dominant class. This faction is composed of petite bourgeoisie (small business owners).
Intermediate faction:	some economic, social, cultural, and symbolic capital but considerably less than the intermediate faction of the dominant class. This faction is composed of skilled clerical workers.
Dominated faction:	little or no economic capital and comparatively high social, cultural, and symbolic capital. This class is composed of educational workers, such as schoolteachers, and other low-income and routinized professions that are involved in cultural production.
Lower Class:	low levels of all forms of capital
Dominant faction:	comparatively high economic capital for this general class. Composed of skilled manual workers.
Intermediate faction:	lower amounts of economic and other types of capital. Composed of semi-skilled workers without credentials.
Dominated faction:	very low amounts of economic capital. Some symbolic capital in form of uneducated ideologues and intellectuals for the poor and working person.

*These portrayals are inferences from Bourdieu's more discursive and rambling text. The table captures the imagery of Bourdieu's analysis; however, since he is highly critical of stratification research in America, he would probably be critical of this "layered" portrayal of his argument.

the least amount of these capital resources. The class structure is not, however, a simple lineal hierarchy. Within each class are *factions* that can be distinguished by (1) the composition or configuration of their capital and (2) the social origin and amount of time that individuals in families have possessed a particular profile or configuration of capital resources.

In Table 25-2 an effort is made to represent schematically Bourdieu's portrayal of the factions in three classes. The top faction within a given class controls the greatest proportion of economic or productive capital typical of a class; the bottom faction possesses the greatest amount of cultural and symbolic capital for a class; and the middle faction possesses an intermediate amount of economic, cultural, and symbolic capital. The top faction is the dominant faction within a given class, and the bottom faction is the dominated faction for that class, with the middle faction being both superordinate over the dominated faction and subordinate to the top faction. As factions engage in struggles to control resources and legitimate themselves, they mobilize social capital to form groupings and networks of relations, but their capacity to form such networks is limited by their other forms of capital. Thus the overall distribution of social capital (groups and organizational memberships, network ties, social relations, etc.) for classes and their factions will correspond to the overall distribution of other forms of capital. However, the particular forms of groupings, networks, and social ties will reflect the particular configuration of economic, cultural, and symbolic capital typically possessed by a particular faction within a given class.

Bourdieu borrows Marx's distinction between a class "for itself" (organized to pursue its interests) and one "in itself" (unorganized but having common interests and objective location), and then he argues that classes are not real groups but only "potentialities." As noted earlier, the objective distribution of resources for Bourdieu relates to actual groups as grammar relates to speech: it defines the possibilities for actors but needs actual people and concrete settings to become real. And, it is the transformation of class and class-faction interests into actual groupings that marks the dynamics of a society. Such transformation involves the use of productive material, cultural, and symbolic capital to mobilize social capital (groups and networks); even more importantly, class conflict tends to revolve around the mobilization of symbols into ideologies that seek to legitimate a particular composition of resources.[34] Much conflict in human societies, therefore, revolves around efforts to manipulate symbols in order to make a particular pattern of social, cultural, and productive resources seem the most appropriate. For example, when intellectuals and artists decry the "crass commercialism," "acquisitiveness," and "greed" of big business, this activity involves the mobilization of symbols into an ideology that seeks to mitigate their domination by the owners of the means of production.

But class relations involve more than a simple pecking order. There are also homologies among similarly located factions within different classes. For example, the rich capitalists of the dominant class and the small business owners of the middle class are equivalent in their control of productive resources and their dominant position in their respective classes;[35] similarly, intellectuals, artists, and other cultural elites in the dominant class are equivalent to schoolteachers in the middle class because of their reliance on cultural capital and because of their subordinate position in relation to those who control the material resources of their respective classes. These homologies make class conflict complex, because those in similar objective positions in different classes—say, intellectuals and schoolteachers—will mobilize symbolic resources into somewhat similar ideologies–in this example, emphasizing learning, knowledge for

---

[34]Pierre Bourdieu, "Social Space and the Genesis of Groups," *Theory and Society* 14 (November 1985), pp. 723–44.

[35]Bourdieu makes what in network analysis (see Chapter 27) is termed *regular structural equivalence*. That is, those incumbents in positions that stand in an equivalent (similar) relation to other positions will act in a convergent way and evidence common attributes.

its own sake, and life of the mind and, at the same time, decrying crass materialism. Such ideologies legitimate their own class position and attack those who dominate them (by emphasizing the importance of those cultural resources that they have more of). At the same time their homologous positions are separated by the different *amounts* of cultural capital owned: the intellectuals despise the strained efforts of schoolteachers to appear more sophisticated than they are, while the schoolteachers resent the decadent and irresponsible relativism of snobbish intellectuals. Thus, ideological conflict is complicated by the simultaneous convergence of factions within different classes and by the divergence of these factions by virtue of their position in different social classes.

Moreover, there is an additional complication stemming from the fact that people sharing similar types and amounts of resources can have very different origins and social trajectories. Those who have recently moved to a class faction—say, the dominant productive elite or intermediate faction of the middle class—will have somewhat different styles and tastes than those who have been born into these classes, and these differences in social origin and mobility can create yet another source of ideological conflict. For example, the "old rich" will often comment on the "lack of class" and "ostentatiousness" of the "new rich"; or, the "solid middle class" will be somewhat snobbish toward the "poor boy who made good" but who "still has a lot to learn" or who "still is a bit crude."

All those points of convergence and divergence within and between classes and class factions make the dynamics of stratification complex. Although there is always an "objective class location" as determined by the amount and composition of capital and by the social origins of capital holders, the development of organizations and ideologies is not a simple process. Bourdieu often ventures into a more structuralist mode when trying to sort out how various classes, class factions, and splits of individuals with different social origins within class factions generate categories of thought, systems of speech, signs of distinction, forms of mythology, modes of appreciation, tastes, and lifestyle. The general argument is that objective location—class, faction within class, and social origin—creates interests and structural constraints that, in turn, allow for different social constructions.[36] Such constructions may involve the use of "formal rules" (implicitly known by individuals with varying interests) to construct cultural codes that classify and organize "things," "signs," and "people" in the world. This kind of analysis by Bourdieu has not produced a fine-grained structuralist model of how individuals construct particular cultural codes, but it has provided an interesting analysis of "class cultures." Such "class cultures" are always the dependent variable for Bourdieu (with objective class location being the independent variable and rather poorly conceptualized structuralist proc-

---

[36]Bourdieu is not very clear on the issue of how the structural potentialities of a given objective class location become transformed into actual social groups capable of historical action. Like Lévi-Strauss, Bourdieu pursues the formal analogies between deep structures and actual practices, but he lacks a theory about how and when the transformations are going to be made, and made successfully.

esses of generative rules and cultural codes being the "intervening variables"). Yet the detailed description of these class cultures is perhaps Bourdieu's most unique contribution to sociology and is captured by his concept of *habitus*.

Class cultures and habitus   Those within a given class share certain modes of classification, appreciation, judgment, perception, and behavior. Bourdieu conceptualizes this mediating process between class and individual perceptions, choices, and behavior as *habitus*.[37] In a sense, habitus is the "collective unconscious" of those in similar positions because it provides cognitive and emotional guidelines that enable individuals to represent the world in common ways and to classify, choose, evaluate, and act in a particular manner.

The habitus creates syndromes of taste, speech, dress, manner, and other responses. For example, a preference for particular foods will tend to correspond to tastes in art, ways of dressing, styles of speech, manners of eating, and other cultural actions among those sharing a common class location. There is, then, a correlation between the class hierarchy and the cultural objects, preferences, and behaviors of those located at particular ranks in the hierarchy. For instance, Bourdieu devotes considerable attention to "taste," which is seen as one of the most visible manifestations of the habitus. Bourdieu views "taste" in a holistic and anthropological sense to include appreciation of art, ways of dressing, and preferences for foods.[38] Although taste appears as an innocent, natural, and personal phenomenon, it covaries with objective class location: the upper class is to the working class what an art museum is to TV; the old upper class is to the new upper class what polite and distant elegance is to noisy and conspicuous consumption; and the dominant is to the dominated faction of the upper class what opera is to avant-garde theater. Because tastes are organized in a cultural hierarchy that mirrors the social hierarchy of objective class location, conflicts between tastes are class conflicts.

Bourdieu roughly distinguishes between two types of tastes, which correspond to high versus low overall capital, or high versus low objective class position.[39] The "taste of liberty and luxury" is the taste of the upper class; as such, it is removed from direct economic necessity and material need. With respect to art, the taste of liberty is the philosophy of art for its own sake. Following Kant, Bourdieu calls this aesthetic the "pure gaze." The pure gaze looks at the sheer form of art and places this form over function and content. The upper-class taste of luxury is not concerned with art illustrating or representing some external reality; art is removed from life, just as upper-class life is removed from harsh material necessity. Consequently, the taste of luxury purifies and sublimates the ordinary and profane into the aesthetic and beautiful. The pure gaze confers aesthetic meaning to ordinary and profane objects

---

[37]Bourdieu, *Distinction*.

[38]Ibid.

[39]Ibid. Actually, Bourdieu makes more fine-tuned distinctions, but we focus only on the main oppositions here.

because the taste of liberty is at leisure to relieve objects from their pragmatic functions. Thus, as the distance from basic material necessities increases, the pure gaze or the taste of luxury transforms the ordinary into the aesthetic, the material into the symbolic, the functional into the formal. And, since the taste of liberty is that of the dominant class, it is also the dominant and legitimate taste in society.

In contrast, the working class cultivates a "popular" aesthetic. Their taste is the taste of necessity, for working-class life is constrained by harsh economic imperatives. The popular taste wants art to represent reality and despises formal and self-sufficient art as decadent and degenerate. The popular taste favors the simple and honest rather than the complex and sophisticated. It is downgraded by the "legitimate" taste of luxury as naive and complacent, and these conflicts over tastes are class conflicts over cultural and symbolic capital.

Preferences for certain works and styles of art, however, are only part of "tastes" as ordered by habitus. Aesthetic choices are correlated with choices made in other cultural fields. The taste of liberty and luxury, for example, corresponds to the polite, distant, and disciplined style of upper-class conversation. Just as art is expected to be removed from life, so are the bodies of interlocutors expected to be removed from one another and so is the spirit expected to be removed from matter. Distance from economic necessity in the upper-class lifestyle not only corresponds to an aesthetic of pure form, but it also entails that all natural and physical desires are to be sublimated and dematerialized. Hence, upper-class eating is highly regulated and disciplined, and foods that are less filling are preferred over fatty dishes. Similarly, items of clothing are chosen for fashion and aesthetic harmony, rather than for functional appropriateness. "Distance from necessity" is the motif underlying the upper-class lifestyle as a whole, not just aesthetic tastes as one area of practice.

Conversely, because they are immersed in physical reality and economic necessity, working-class people interact in more physical ways, touching one another's bodies, laughing heartily, and valuing straightforward outspokenness more than distant and "false" politeness. Similarly, the working-class taste favors foods that are more filling and less "refined" but more physically gratifying. The popular taste chooses clothes and furniture that are functional, and this is so not only because of sheer economic constraints but also because of a true and profound dislike of that which is "formal" and "fancy."

In sum, then, Bourdieu has provided a conceptual model of class conflict that combines elements of Marxian, Weberian, and Durkheimian sociology. The structuralist aspects of Bourdieu's conceptualization of habitus as the mediating process between class position and individual behavior have been underemphasized in this review, but it is clear that Bourdieu places Durkheim "back on his feet" by emphasizing that class position determines habitus. But the useful elements of structuralism—systems of symbols as generative structures of codes—are retained and incorporated into a theory of class conflict as revolving around the mobilization of symbols into ideologies legitimating a class position and the associated lifestyle and habitus.

CONCLUSION

"Cultural structuralism" oscillates between the extreme positions of objectivism and subjectivism and is, therefore, able to avoid the conceptual problems of either position. The orthodox structuralism of Lévi-Strauss forgot that underlying generative codes need agents and contexts that transform a system of potentialities into actual historical action. Orthodox structuralism eliminates the "knowledgeable and capable agent" who can do more than simply execute latent schemes of practice. The panstructuralism of Lévi-Strauss opened up "culture" as a systematic area of research about which one can *theorize* but, at the same time, exaggerated the antihumanist notion of "subjectless" structures.

Cultural structuralism also avoids the indeterminate view of culture as a system of symbols that can be defined and negotiated at will by creative and spontaneous agents. Symbolic interactionism in particular exaggerates the capacity of agents to define situations and redefine self-identities in the spontaneous setting of unstructured "situations." Structuralism recovers the materialist notion that definitions of the situation and interpretations of symbols are constrained by external structures that cannot be totally controlled by the will of agents. Cultural practices and symbol use are not self-sufficient systems of unrestrained interpretive creativity; rather, cultural work has its own structures and constraints.

Cultural structuralism, then, does not fall into the traps of subjectless objectivism, and it also avoids the hermeneutic idealism of interpretive approaches to action. It has established the idea that culture has a structure that is itself a reality, *sui generis*, and that can be analyzed like any other reality.

CHAPTER 26

◆

Structuration Theory

Anthony Giddens

Among the recent efforts to redirect sociological theorizing is Anthony Giddens' work on what he terms *structuration theory*.[1] In a critical fashion, Giddens has mounted a consistent line of attack against existing forms of theorizing, particularly functionalism, Marxism, structuralism, phenomenology, portions of traditional symbolic interactionism, and role theory.[2] In their place he has attempted to create a theoretical approach that eliminates the shortcomings of these perspectives and that, at the same time, tries to blend interactionist concepts with those allowing for an understanding of the structural properties of societies.

To appreciate Giddens' strategy, I think that it is wise to begin by outlining his critique of social theory. In this way I can place the strategy and substance of "structuration theory" into its critical context.

THE CRITIQUE OF SOCIAL THEORY

Rejecting Naturalism and Positivism

At the core of Giddens' effort to redirect social theory is the rejection of a "covering law" view of sociological explanation (see Chapter 1). In what I feel is a shortsighted view, he flatly dismisses the idea that sociology can be like

---

[1]The basic theory is presented in numerous places, but the two most comprehensive statements are: Anthony Giddens, *The Constitution of Society: Outline of the Theory of Structuration* (Oxford: Polity Press, 1984) and *Central Problems in Social Theory* (London: Macmillan, 1979). The University of California Press also has editions of these two books. For an excellent overview, both sociologically and philosophically, of Giddens' theoretical project, see Ira Cohen, *Structuration Theory: Anthony Giddens and the Constitution of Social Life* (London: Macmillan, 1989). For a commentary and debate on Giddens' work, see J. Clark, C. Modgil, and S. Modgil, eds., *Anthony Giddens: Consensus and Controversy* (London: Falmer Press, 1990).

[2]See, in particular, Anthony Giddens, *Profiles and Critiques in Social Theory* (London: Macmillan, 1982) and *New Rules of Sociological Method: A Positive Critique of Interpretative Sociologies* (London: Hutchinson Ross, 1976).

the natural sciences. For Giddens, there can be no enduring abstract laws about social processes. In asserting this view, he echoes Herbert Blumer's charge that social organization is changeable by the acts of individuals, and thus there can be no laws about the invariant properties of social organization.[3] And, much like Blumer, he visualizes that "the concepts of theory . . . should for many research purposes be regarded as sensitizing devices, nothing more."[4] I should emphasize, however, that the reasons for this assertion are different from those traditionally advocated by Chicago School symbolic interactionism.

First, Giddens asserts that social theorizing involves a "double hermeneutic" that, stripped of this jargon, asserts that the concepts and generalizations used by social scientists to understand social processes can be employed by agents to alter these processes, thereby potentially obviating the generalizations of "science." We must recognize, Giddens contends, that lay actors are also "social theorists who alter their theories in the light of their experience and are receptive to incoming information."[5] And, thus, social science theories are not often "news" to lay actors; when they are, such theories can be used to transform the very order they describe. For within the capacity of humans to be reflexive—that is, to think about their situation—is the ability to change it.[6]

Second, social theory is by its nature social criticism. Social theory often contradicts "the reasons that people give for doing things" and is, therefore, a critique of these reasons and the social arrangements that people construct in the name of these reasons. Sociology does not, therefore, need to develop a separate body of critical theory, as others have argued (see Chapter 13); it *is* critical theory by its very nature and by virtue of the effects it can have on social processes.

The implications of these facts, Giddens believes, are profound. We need to stop imitating the natural sciences. We must cease evaluating our success as an intellectual activity in terms of whether or not we have discovered "timeless laws." We must recognize that social theory does not exist "outside" our universe. We should accept the fact that what sociologists and lay actors do is, in a fundamental sense, very much the same. And, we must redirect our efforts at developing "sensitizing concepts" that allow us to understand the active processes of interaction among individuals as they produce and reproduce social structures while being guided by these structures.

Obviating Sociological Dualisms

I think that one of the most useful criticisms mounted by Giddens is the rejection of dualisms in social theory—micro versus macro theory, subject (people) versus object (structure), individual versus society, subjectivism versus

---

[3]See Chapter 15.

[4]Giddens, *The Constitution of Society*, p. 326.

[5]Ibid., p. 335.

[6]Giddens lists conditions under which social science will not have "transformational implications" for social organization; see ibid., pp. 341–43.

objectivism, and similar dichotomies around which great debate rages. At the heart of his approach is the assertion that "so-called 'microsociological' study does not deal with a reality that is somehow more substantial than that with which 'macrosociological' analysis is concerned. But neither, on the contrary, is interaction in situations of co-presence simply ephemeral, as contrasted to the solidity of large-scale or long-established institutions."[7] The process of "structur*ation*" is intended to emphasize that the individual/society, subject/object, and micro/macro dichotomies do not constitute a dualism, but a "duality." That is, people in interaction use the rules and resources that constitute social structure in their day-to-day routines in contexts of co-presence, and, in so doing, they reproduce these rules and resources of structure. Thus individual action, interaction, and social structure are all implicated in one another. They do not constitute separate realities, but a duality within the same reality, for "the structural properties of social systems are both the medium and the outcome of the practices that constitute those systems."[8] One cannot understand action and interaction without reference to the rules and resources of social structure, whereas one cannot fully understand large-scale, long-term institutional structures without knowledge of how actors use the rules and resources of these institutional structures in concrete interaction.

At times, Giddens recognizes, it is necessary to "bracket out" consideration of either individuals in interaction or institutional structures. But this is a methodological procedure rather than an ontological assertion. I think that Giddens is correct in his conclusion that too much social theory converts this methodological procedure of bracketing with ontological statements about what is really real in the social universe.

The Critique of Functionalism and Evolutionism

In what I think can be viewed as the merciless flogging of two dying horses, Giddens has consistently attacked (1) functionalism, or the analysis of social phenomena in terms of the needs that structures meet; and (2) evolutionism, or the description of the inexorable stages that societies must pass through on the road to modernity. Functionalist theories are almost all evolutionary (see Chapters 2-8), and thus Giddens rejects them as an appropriate approach to understanding society. Similarly—and here Giddens is more creative—he attacks Marx's work and much contemporary Marxism as both functional and evolutionary and, hence, not very useful.[9]

I think that Giddens' list of criticisms is fairly standard, and we need not review them here. However, I see the general argument as very important: functional analysis tends to ignore the active processes of agents in interaction and to overemphasize social structure as an "external constraint" on actors,

---

[7]Ibid., p. 26.

[8]Giddens, *Central Problems in Social Theory*, p. 69.

[9]See, in particular, Anthony Giddens, *A Contemporary Critique of Historical Materialism: Volume I, Power, Property and the State* (London: Macmillan, 1981).

whereas evolutionary analysis tends to stress the inexorable movement of societies in response to some causal factor, such as size, war, and means of
production. For Giddens, the "duality of structure" is lost in all of this analysis
of functional needs, the constraints of structure, the stages of development,
and the "prime causes" of change. That is, agents in situations of interaction
are not seen as doing very much, and structure is seen to march along disembodied from the actors who are involved in its reproduction or transformation.

The Limits of Interactionism

Giddens draws upon interactionist theory, especially that developed by Erving
Goffman. Yet he is critical of Goffman and other interactionist theories in
several respects. First, they tend to perpetuate the dualism (as opposed to
duality) of individual versus structure. Rarely do interactionists discuss social
structure in a way that allows for an understanding of how it is reproduced
through the processes of interaction. Moreover, some interactionists are openly
chauvinistic, dismissing institutional processes as somehow less "real" than
face-to-face symbolic interaction.

Second, interactionist theories tend to ignore motivation. In other words,
what drives actors to do what they do? There is an implicit view of actors,
especially in Goffman's dramaturgy, as cynical manipulators of gestures in
staging areas. But rarely are the questions asked: Why do actors engage in
certain gesturing practices? Why enter or exit a stage or region? Why use a
"folk method"? Thus, interactionist theory needs a revision in terms of a theory
of motivation that avoids the imprecision and mysticism of psychoanalytic
theory while moving beyond a view of actors as cynical and self-centered or-
chestrators of gestures.

Third, more structurally oriented forms of interactionism, such as structural role theory, begin to visualize roles as the constitutive basis of structure.
In Giddens' view it is actual practices of people or collective units, not roles,
that are the point of articulation between the individual and society. Although
these practices may be influenced by an agent's position, "social systems . . .
are regularized social practices, sustained in encounters." Too often the concept
of role has been used as a substitute for normative constraints on individuals
who passively perform their assigned roles. As a result, the typification of
structure as a system of roles tends to remove the active, reflexive, creative,
and potentially transformative behaviors of agents.

The Critique of Structuralism

Giddens has been very critical of Claude Lévi-Strauss's structuralism (see
Chapter 24), because it simply ignores human agency or the capacity of people
to reflect, monitor, define, and decide. In such structuralist approaches, actors
are pushed, if not compelled, to act in accordance with immanent systems of
codes. Other forms of structural theorizing, such as Peter Blau's macrostructuralism, are chauvinistic and simply define away as relevant to sociology the

reflexive capacities of human agents.[10] In such macrostructuralism, social structure simply requires actors to do its bidding.

In all of these and other structural theories, then, there is a failure to recognize that structure is actively reproduced (or altered) by agents in interaction. For Giddens, structure is not some ex cathedra, external, and constraining force that makes humans into robots and dupes. Rather, structure is implicated in, and reproduced by, the day-to-day routines of people in interaction. It is, in Giddens' words, "both constraining and enabling." One cannot, therefore, define away people in interaction as peripheral to the task of sociological explanation.

Let me now summarize. Giddens criticizes sociological theory for its unwarranted belief that universal laws can be developed, for its unnecessary dualisms, for its functionalism and evolutionism, for its failure to implicate motives and structure in the process of interaction, and for its tendency to view structure and symbols as somehow alien to the actors who produce, reproduce, and transform these structures and symbols. Unlike many critics, Giddens does not dismount his soap box at this point and go home. To his credit, he then tries to develop an alternative mode of theoretical analysis that, he believes, overcomes these deficiencies.

THE "THEORY OF STRUCTURATION"

Since Giddens does not believe that abstract laws of social action, interaction, and organization exist, his "theory of structuration" is not a series of propositions. Instead, I see the theory as a cluster of sensitizing concepts linked together discursively. Giddens' work is thus what I called a *sensitizing conceptual scheme* in Chapter 1. The key concept is "structuration," which is intended to communicate the "duality of structure." That is, social structure is used by active agents; in so using the properties of structure, they transform or reproduce this structure. Thus the process of structuration requires a conceptualization of the nature of structure, of the agents who use structure, and of the ways that these are mutually implicated in each other to produce varying patterns of human organization.

Reconceptualizing Structure and Social System

For Giddens, structure can be conceptualized as the "rules" and "resources" that actors use in "interaction contexts" that extend across "space" and over "time." In so using these rules and resources, actors sustain or reproduce structures in space and time.

Rules are "generalizable procedures" that actors understand and use in various circumstances. For Giddens a rule is a methodology or technique that actors know about, often only implicitly, and that provides a relevant formula

---

[10]See *The Constitution of Society*, pp. 207–13.

for action.[11] From a sociological perspective, the most important rules are those that agents use in the reproduction of social relations over significant lengths of time and across space. These rules reveal certain characteristics: (1) they are frequently used in (a) conversations, (b) interaction rituals, and (c) the daily routines of individuals; (2) they are tacitly grasped and understood and are part of the "stock knowledge" of competent actors; (3) they are informal, remaining unwritten and unarticulated; and (4) they are weakly sanctioned through interpersonal techniques.[12]

With this conceptualization, Giddens subsumes (a) the functionalist's emphasis on institutional norms and cultural values, (b) the ethnomethodologist's emphasis on folk methods, (c) the structuralist's concern with the generative nature of symbols and codes, and (d) just about all other conceptualizations in between. The thrust of Giddens' argument is that rules are part of actors' "knowledgeability." Some may be normative in that actors can articulate and explicitly make reference to them, but many other rules are more implicitly understood and used to guide the flow of interaction in ways that are not easily expressed or verbalized. Moreover, actors can transform rules into new combinations as they confront and deal with one another and the contextual particulars of their interaction.

As the other critical property of structure, resources are facilities that actors use to get things done. For, even if there are well-understood methodologies and formulas—that is, rules—to guide action, there must also be the capacity to perform tasks. Such capacity requires resources, or the material equipment and the organizational ability to act in situations. Giddens visualizes resources as what generates power.[13] Power is not a resource, as much social theory argues. Rather, the mobilization of other resources is what gives actors power to get things done. Thus power is integral to the very existence of structure, for, as actors interact, they use resources; and, as they use resources, they mobilize power to shape the actions of others.

Giddens visualizes rules and resources as "transformational" and as "mediating."[14] What he means by these terms is that rules and resources can be transformed into many different patterns and profiles. Resources can be mobilized in various ways to perform activities and achieve ends through the exercise of different forms and degrees of power; rules can generate many diverse combinations of methodologies and formulas to guide how people communicate, interact, and adjust to one another. Rules and resources are mediating in that they are what tie social relations together. They are what actors use to create, sustain, or transform relations across time and in space. And, because rules and resources are inherently transformational—that is, generative of diverse

---

[11]Ibid., pp. 20–21.

[12]Ibid., p. 22.

[13]Ibid., pp. 14–16.

[14]Here Giddens seems to be taking what is useful from "structuralism" and reworking these ideas into a more sociological approach.

FIGURE 26-1 Social Structure, Social System, and the Modalities of Connection

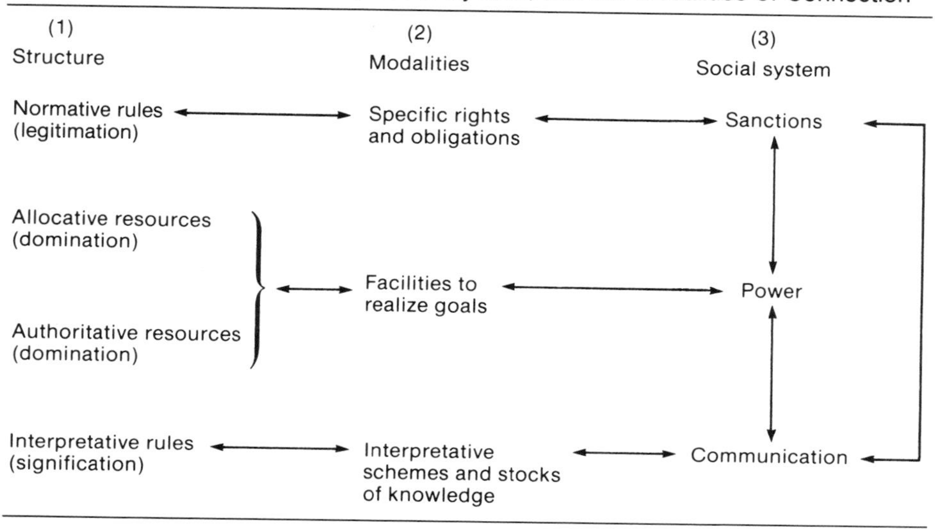

(1) Structure	(2) Modalities	(3) Social system
Normative rules ←→ (legitimation)	Specific rights ←→ and obligations	Sanctions ←
Allocative resources (domination) }←→ Authoritative resources (domination)	Facilities to ←→ realize goals	Power
Interpretative rules ←→ (signification)	Interpretative ←→ schemes and stocks of knowledge	Communication ←

combinations—they can lace together many different patterns of social relations in time and space.

Giddens develops a typology of rules and resources that I think is rather vague and imprecise.[15] He sees the three concepts in this typology—domination, legitimation, and signification—as "theoretical primitives," which, I suspect, is an excuse for defining them imprecisely. The basic idea is that resources are the stuff of domination because they involve the mobilization of material and organizational facilities to do things. Some rules are transformed into instruments of legitimation because they make things seem correct and appropriate. Other rules are used to create signification, or meaningful symbolic systems, because they provide people with ways to see and interpret events. Actually, I think that the scheme makes more sense if the concepts of domination, legitimation, and signification are given less emphasis and the elements of his discussion are selectively extracted to create the typology presented in Figure 26-1.

In the left column of Figure 26-1, structure is viewed by Giddens as composed of rules and resources. Rules are transformed into two basic types of mediating processes: (1) normative, or the creation of rights and obligations in a context; and (2) interpretative, or the generation of schemes and stocks of taken-for-granted knowledge in a context. Resources are transformed into two major types of facilities that can mediate social relations: (1) authoritative resources, or the organizational capacity to control and direct the patterns of interactions in a context; and (2) allocative resources, or the use of material

---

[15]*The Constitution of Society,* p. 29; and *Central Problems in Social Theory,* pp. 97-107.

features, artifacts, and goods to control and direct patterns of interaction in a context.

Giddens sees these types of rules and resources as mediating interaction via three modalities, as is portrayed in column 2 of Figure 26-1: rights and obligations, facilities, and interpretative schemes. I have deviated somewhat from Giddens' discussion, but the idea is the same: that rules and resources are attached to interaction (or "social system" in Giddens' terms) via these three modalities. These modalities are then used to (*a*) generate the power that enables some actors to control others, (*b*) affirm the norms that, in turn, allow actors to be sanctioned for their conformity or nonconformity, and (*c*) create and use the interpretative schemes that make it possible for actors to communicate with one another.

Giddens also stresses that rules and resources are interrelated. In a discussion that I see as very similar to many by Talcott Parsons (see Chapter 3), Giddens emphasizes that the modalities and their use in interaction are separated only analytically. In the actual flow of interaction in the real empirical world, they exist simultaneously, thereby making their separation merely an exercise of analytical decomposition. Thus power, sanctions, and media of communication are interconnected, as are the rules and resources of social structure. In social systems, where people are co-present and interact, power is used to secure a particular set of rights and obligations as well as a system of communication; conversely, power can be exercised only through communication and sanctioning.

I think it reasonable to ask: why does Giddens create this analytic scheme of social structure and social system, especially since it is very vague? I suspect that the answer resides in Giddens' desire to visualize structure in entirely different terms than either functional or structuralist theory. For Giddens, social structure is to be seen as something used by actors, not as some external reality that pushes and shoves actors around. Thus, social structure is defined as the rules and resources that can be transformed as actors use them in concrete settings. But then the question arises: how is structure to be connected to what people actually do in interaction settings, or what Giddens terms "social systems"? The answer is the awkward notion of modalities, whereby rules and resources are transformed into power, sanctions, and communication. Thus structure is not a mysterious system of codes, as Lévi-Strauss and other structural idealists imply; nor is it a set of determinative parameters and external constraints on actors, as Peter Blau and other macrostructuralists contend. In Giddens' conceptualization, social structure is transformative and flexible; it is "part of" actors in concrete situations; and it is used by them to create patterns of social relations across space and through time.

Moreover, I think that this typology allows Giddens to emphasize that, as agents interact in social systems, they can reproduce rules and resources (via the modalities) or they can transform them. Thus social interaction and social structure are reciprocally implicated. Structuration is, therefore, the dual processes in which rules and resources are used to organize interaction across time

TABLE 26–1 The Typology of Institutions

Type of Institution		Rank Order of Emphasis on Rules and Resources
1. Symbolic orders, or modes of discourse, and patterns of communication	are produced and reproduced by	the use of interpretative rules (signification) in conjunction with normative rules (legitimation) and allocative as well as authoritative resources (domination).
2. Political institutions	are produced and reproduced by	the use of authoritative resources (domination) in conjunction with interpretative rules (signification) and normative rules (legitimation).
3. Economic institutions	are produced and reproduced by	the use of allocative resources (domination) in conjunction with interpretative rules (signification) and normative rules (legitimation).
4. Legal institutions	are produced and reproduced by	the use of normative rules (legitimation) in conjunction with authoritative and allocative resources (domination) and interpretative rules (signification).

and in space and, by virtue of this use, to reproduce or transform these rules and resources.

Reconceptualizing Institutions

For Giddens, institutions are systems of interaction in societies that endure over time and that distribute people in space. Giddens uses what I see as rather vague phrases like "deeply sedimented across time and in space in societies" to express the idea that, when rules and resources are reproduced over long periods of time and in explicit regions of space, then institutions can be said to exist in a society. Giddens offers a typology of institutions in terms of the weights and combinations of rules and resources that are implicated in interaction.[16] If signification (interpretative rules) is primary, followed, respectively, by domination (allocative and authoritative resources) and then legitimation (normative rules), a "symbolic order" exists. If authoritative domination, signification, and legitimation are successively combined, political institutionalization occurs. If allocative dominance, signification, and legitimation are ordered, economic institutionalization prevails. And if legitimation, dominance, and signification are rank ordered, institutionalization of law occurs. Table 26–1 summarizes Giddens' argument.

In so many ways I see this typology as imprecise, and, thankfully, it is not used very much in his subsequent analysis of structuralism. But once again I think it reasonable to ask: why generate it in the first place, especially since it smacks of the same kind of concept mongering found in much functional

---

[16]*Central Problems in Social Theory*, p. 107; and *The Constitution of Society*, p. 31.

analysis? One reason appears to be Giddens' desire to reject functional modes of institutional analysis by presenting his alternative. Most functional theories analyze the process of institutional differentiation among, for example, economy, law, polity, education, and kinship (see Chapters 2-5). Such differentiation is typically seen to occur in an evolutionary framework, with each stage of evolution marked by increased separation and autonomy of institutions from one another. Giddens wishes to avoid this mechanical view of institutionalization, in several senses. First, systems of interaction in empirical contexts are a mixture of institutional processes. Economic, political, legal, and symbolic orders are not easily separated; there is usually an element of each in any social system context. Second, institutions are tied to the rules and resources that agents employ and thereby reproduce; they are not external to individuals because they are formed by the use of varying rules and resources in actual social relations. Third, the most basic dimensions of all rules and resources—signification, domination, and legitimation—are all involved in institutionalization; it is only their relative salience for actors that gives the stabilization of relations across time and in space its distinctive institutional character.

To his credit, Giddens is unlike many interactionists because he wants to acknowledge the importance of analyzing stabilized social relations—that is, institutionalization. But, as with all interactionist theory, he wishes to stress that institutionalized social relations are actively reproduced in terms of the creative transformations of rules and resources that are employed by agents in actual interaction. Thus one of the most important structural features of social relations is their institutionalization in space and across time. Such institutionalization moves along four dimensions—legal, economic, political, and symbolic—which are distinguished from one another in terms of the relative use of various rules and resources. I am not sure that he needed the typology in Table 26-1 to make this important point, but it is the vehicle by which he chose to make his argument, and so I have discussed it here.

Structural Principles, Sets, and Properties

The extent and form of institutionalization in societies are related to what Giddens terms *structural principles*.[17] These are the most general principles that guide the organization of societal totalities. These are what "stretch systems across time and space," and they allow for "system integration," or the maintenance of reciprocal relations among units in a society. For Giddens, "structural principles can thus be understood as the principles of organization which allow recognizably consistent forms of time-space distanciation on the basis of definite mechanisms of societal integration."[18] I find such definitions rather vague, to say the least, but the basic idea seems to be that rules and resources are used by active agents in accordance with fundamental principles

---

[17]*The Constitution of Society*, pp. 179-93.
[18]Ibid., p. 181.

of organization. Such principles guide just how rules and resources are transformed and employed to mediate social relations.

On the basis of their underlying structural principles, three basic types of societies have existed: (1) "tribal societies," which are organized in terms of structural principles that emphasize kinship and tradition as the mediating force behind social relations across time and in space; (2) "class-divided societies," which are organized in terms of an urban/rural differentiation, with urban areas revealing distinctive political institutions that can be separated from economic institutions, formal codes of law or legal institutions, and modes of symbolic coordination or ordering through written texts and testaments; and (3) "class societies," which involve structural principles that separate and yet interconnect all four institutional spheres, especially the economic and political.[19]

Structural principles are implicated in the production and reproduction of "structures" or "structural sets." These structural sets are rule/resource bundles, or combinations and configurations of rules and resources, which are used to produce and reproduce certain types and forms of social relations across time and space. Giddens offers the example of how the structural principles of class societies (differentiation and clear separation of economy and polity) guide the use of the following structural set: *private property-money-capital-labor-contract-profit*. The details of his analysis are less important than the general idea that the general structural principles of class societies are transformed into more specific sets of rules and resources that agents use to mediate social relations. The above structural set is used in capitalist societies and, as a consequence, is reproduced. In turn, such reproduction of the structural set reaffirms the more abstract structural principles of class societies.

As these and other structural sets are used by agents and as they are thereby reproduced, societies develop "structural properties," which are "institutionalized features of social systems, stretching across time and space."[20] That is, social relations become patterned in certain typical ways. Thus the structural set of private property-money-capital-labor-contract-profit can mediate only certain patterns of relations; that is, if this is the rule/resource bundle with which agents must work, then only certain forms of relations can be produced and reproduced in the economic sphere. Hence the institutionalization of relations in time and space reveals a particular form, or, in Giddens' terms, structural property.

Structural Contradiction

Giddens always wants to emphasize the inherent "transformative" potential of rules and resources. Structural principles, he argues, "operate in terms of

---

[19]For an extensive discussion of this typology, see Giddens' *A Contemporary Critique of Historical Materialism*.

[20]*The Constitution of Society*, p. 185.

one another but yet also contravene each other."[21] In other words, they reveal contradictions that can be either primary or secondary. A "primary contradiction" is one between structural principles that are formative and constitute a society, whereas a "secondary contradiction" is one that is "brought into being by primary contradictions."[22] For example, there is a contradiction between structural principles that mediate the institutionalization of private profits, on the one hand, and those that mediate socialized production, on the other. If workers pool their labor to produce goods and services, it is contradictory to allow only some to enjoy profits of such socialized labor.

Contradictions are not, Giddens emphasizes, the same as conflicts. Contradiction is a "disjunction of structural principles of system organization," whereas conflict is the actual struggle between actors in "definite social practices."[23] Thus the contradiction between private profits and socialized labor is not, itself, a conflict. It can create situations of conflict, such as struggles between management and labor in a specific time and place, but such conflicts are not the same as contradiction.

I must confess to being somewhat confused by Giddens' discussion of these topics. It is a most imprecise analysis. Yet beneath all the jargon is, I feel, an important insight: the institutional patterns of a society represent the creation and use by agents of very generalized and abstract principles; these principles represent the development of particular rules and the mobilization of certain resources; such principles generate more concrete "bundles" or "sets" of rules and resources that agents actively use to produce and reproduce social relations in concrete settings; and many of these principles and sets contain contradictory elements that can encourage actual conflicts among actors. In this way, structure "constrains" but is not something disembodied from agents. Rather, the "properties" of total societies are not external to individuals and collectivities but are persistently reproduced through the use of structural principles and sets by agents who act. Let me now turn to Giddens' discussion of these active agents.

Agents, Agency, and Action

As is evident, Giddens visualizes structure as a duality, as something that is part of the actions of agents. And so in Giddens' approach it is essential to understand the dynamics of human agency. He proposes a "stratification model," which I see as an effort to synthesize psychoanalytic theory, phenomenology, ethnomethodology, and elements of action theory. This model is depicted in the lower portions of Figure 26-2. For Giddens, "agency" denotes the events that an actor perpetrates rather than "intentions," "purposes," "ends," or other states. Agency is what an actor actually does in a situation that has visible

---

[21]Ibid., p. 193.
[22]Ibid.
[23]Ibid., p. 198.

FIGURE 26-2 The Dynamics of Agency

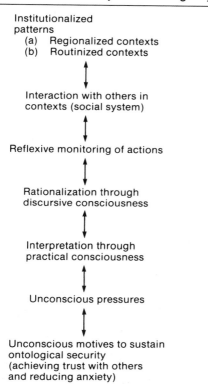

consequences (not necessarily intended consequences). To understand the dynamics of agency requires analysis of each element on the model.

As I have drawn the model in Figure 26-2, it actually combines two overlapping models in Giddens' discussion, but his intent is reasonably clear: humans "reflexively monitor" their own conduct and that of others; in other words, they pay attention to, note, calculate, and assess the consequences of actions.[24] Monitoring is influenced by two levels of consciousness.[25] One is "discursive consciousness," which involves the capacity to give reasons for or rationalize what one does (and presumably to do the same for others' behavior). "Practical consciousness" is the stock of knowledge that one implicitly uses to act in situations and to interpret the actions of others. It is this knowledge-ability that is constantly used, but rarely articulated, to interpret events—one's own and those of others. Almost all acts are indexical, to use Garfinkel's term,

---

[24]Ibid., pp. 5–7; see also *Central Problems in Social Theory*, pp. 56–59.

[25]His debt to Schutz and phenomenology is evident here, but he has liberated it from its subjectivism. See Chapter 18 on the emergence of interactionism.

in that they must be interpreted by their context, and it is this implicit stock of knowledge that provides these contextual interpretations and frameworks.[26]

There are also unconscious dimensions to human agency. There are many pressures to act in certain ways, which an actor does not perceive. Indeed, Giddens argues that much motivation is unconscious. Moreover, motivation is often much more diffuse than action theories portray. That is, there is no one-to-one relation between *an* act and *a* motive. Actors may be able to rationalize through their capacity for discursive consciousness in ways that make this one-to-one relationship seem to be what directs action. But, in fact, much of what propels action lies below consciousness and, at best, provides very general and diffuse pressures to act. Moreover, much action may not be motivated at all; an actor simply monitors and responds to the environment.

In trying to reintroduce the *un*conscious into social theory, Giddens adopts Erik Erikson's psychoanalytic ideas.[27] The basic "force" behind much action is an unconscious set of processes to gain a "sense of trust" in interaction with others. Giddens terms this set of processes the *ontological security system* of an agent. That is, one of the driving but highly diffuse forces behind action is the desire to sustain ontological security or the sense of trust that comes from being able to reduce anxiety in social relations. Actors need to have this sense of trust. How they go about reducing anxiety to secure this sense is often unconscious because the mechanisms involved are developed before linguistic skills emerge in the young and because there may also be psychodynamics, such as repression, that keep these fundamental feelings and their resolution from becoming conscious. In general, Giddens argues that ontological security is maintained through the routinization of encounters with others, through the successful interpretation of acts in terms of practical or stock knowledge, and through the capacity for rationalization that comes with discursive consciousness.

As the top portions of Figure 26–2 emphasize, institutionalized patterns have an effect on, while being a consequence of, the dynamics of agency. As we will see shortly, unconscious motives for ontological security require routinized interactions (predictable, stable over time) that are regionalized (ordered in space). Such regionalization and routinization are the product of past interactions of agents and are sustained or reproduced through the present (and future) actions of agents. To sustain routines and regions, actors must monitor their actions while drawing upon their stock knowledge and discursive capacities. In this way, Giddens visualizes institutionalized patterns implicated in the very nature of agency. Institutions and agents cannot exist without each other, for institutions are reproduced practices by agents, whereas the conscious and unconscious dynamics of agency depend upon the routines and regions provided by institutionalized patterns.

---

[26]Here Giddens is using ideas of phenomenology and ethnomethodology but without all the ontological and metaphorical rhetoric.

[27]*The Constitution of Society*, pp. 45–59.

Routinization and Regionalization of Interaction

Both the ontological security of agents and the institutionalization of structures in time and space depend upon routinized and regionalized interaction among actors. Routinization of interaction patterns is what gives them continuity across time, thereby reproducing structure (rules and resources) and structure*s* (institutions). At the same time, routinization gives predictability to actions and, in so doing, provides for a sense of ontological security. Thus routines become critical for the most basic aspects of structure and human agency. Similarly, regionalization orders action in space by positioning actors in places vis-à-vis one another and by circumscribing how they are to present themselves and act. As with routines, the regionalization of interaction is essential to the sustenance of broader structural patterns and ontological security of actors, because it orders people's interactions in space and time, which in turn reproduces structures and meets an agent's need for ontological security. Let me now elaborate on these general ideas because I think they mark an important contribution to social theory.

Routines   In discussing routines, Giddens sees them as the key link between the episodic character of interactions (they start, proceed, and end), on the one hand, and basic trust and security, on the other.[28] Moreover, "the routinization of encounters is of major significance in binding the fleeting encounter to social reproduction and thus to the seeming 'fixity' of institutions."[29] In a very interesting discussion in which he borrows heavily from Erving Goffman (but with a phenomenological twist), Giddens proposes several procedures, or mechanisms, that humans use to sustain routines: (1) opening and closing rituals, (2) turn taking, (3) tact, (4) positioning, and (5) framing. These are not necessarily exclusive,[30] and I must confess that I am guessing here as to Giddens' actual intent; but the general idea is that people employ these mechanisms to sustain routines and, thereby, reproduce social structures while maintaining ontological security. Let me elaborate briefly on these five routines.

1. Since interaction is serial—that is, it occurs sequentially—there must be symbolic markers of opening and closing. Such markers are essential to the maintenance of routines because they indicate when in the flow of time the elements of routine interaction are to begin and end. There are many such interpersonal markers—words, facial gestures, positions of bodies— and there are physical markers, such as rooms, buildings, roads, and equipment, that also signal when certain routinized interactions are to begin and end (note, for example, the interpersonal and physical markers for a lecture, which is a highly routinized interaction that sustains the ontological security of agents and perpetuates institutional patterns).

---

[28]Ibid., pp. 60–109.

[29]Ibid., p. 72.

[30]I have created this list from what is a much more rambling discussion.

2. Turn taking in a conversation is another process that sustains a routine. All competent actors contain in their practical consciousness, or implicit stock of knowledge, a sense of how conversations are to proceed sequentially. There are "folk methods" that people rely on to construct sequences of talk; in so doing, they sustain a routine and, hence, their psychological sense of security and the larger institutional context (think, for example, about a conversation that did not proceed smoothly in terms of conversational turn taking; recall how disruptive this was for one's sense of order and routine).

3. Tact is, in Giddens' view, "the main mechanism that sustains 'trust' or 'ontological security' over long time-space spans." By tact, Giddens means "a latent conceptual agreement among participants in interaction" about just how each party is to gesture and respond and about what is appropriate and inappropriate.[31] People carry with them implicit stocks of knowledge that signal to them what would be "tactful" and what would be "rude" and "intrusive." And they use this sense of tact to regulate their emission of gestures, their talking, and their relative positioning in situations in order "to remain tactful," thereby sustaining their sense of trust and the larger social order. (Imagine interactions in which tact is not exercised—how they disrupt our routines, our sense of comfort, and our perceptions of an orderly situation.)

4. Giddens rejects the idea of "role" as very useful and substitutes the notion of "position." People bring to situations a position or "social identity that carries with it a certain range of prerogatives and obligations," and they emit gestures in a process of mutual positioning, such as locating their bodies in certain points, asserting their prerogatives, and signaling their obligations. In this way interactions can be routinized, and people can sustain their sense of mutual trust as well as the larger social structures in which their interaction occurs. (For example, examine a student/student or professor/student interaction in terms of positioning, and determine how it sustains a sense of trust and the institutional structure.)

5. Much of the coherence of positioning activities is made possible by "frames," which provide formulas for interpreting a context. Interactions tend to be framed in the sense that there are rules that apply to them, but these are not purely normative in the sense of precise instructions for participants. Equally important, frames are more implicitly held, and they operate as markers that assert when certain behaviors and demeanors should be activated. (For example, compare your sense of how to comport yourself at a funeral, cocktail party, class, and in other contexts that are "framed.")

In sum, social structure is extended across time by these techniques that produce and reproduce routines. In so stretching interaction across time in an orderly and predictable manner, people realize their need for a sense of trust

---

[31]*The Constitution of Society*, p. 75. Surprisingly, Giddens never really defines "tact," assuming that we know what it means. I had to infer this definition.

in others. In this way, then, Giddens connects the most basic properties of structure (rules and resources) to the most fundamental features of human agents (unconscious motives).

Regionalization   Structuration theory is concerned with the reproduction of relations not only across time but also in space. With the concept of regionalization of interaction, Giddens addresses the intersection of space and time.[32] For interaction is not just serial, moving in time; it is also located in space. Again borrowing from Goffman and also from time/space geography, Giddens introduces the concept of "locale" to account for the physical space in which interaction occurs as well as the contextual knowledge about what is to occur in this space. In a locale, actors are not only establishing their presence with respect to one another but they are also using their stocks of practical knowledge to interpret the context of the locale. Such interpretations provide them with the relevant frames, the appropriate procedures for tact, and the salient forms for sequencing gestures and talk.

Giddens classifies locales in terms of their "modes." Locales vary in terms of (1) their physical and symbolic boundaries, (2) their duration across time, (3) their span or extension in physical space, and (4) their character, or the ways they connect to other locales and to broader institutional patterns. Locales also vary in terms of the degree to which they force people to sustain high public presence (what Goffman termed *frontstage*) or allow retreats to back regions where public presence is reduced (Goffman's *backstage*).[33] They also vary in regard to how much disclosure of self (feelings, attitudes, and emotions) they require, some allowing "enclosure" or the withholding of self and other locales requiring "disclosure" of at least some aspects of self.

Regionalization of interaction through the creation of locales facilitates the maintenance of routines. In turn, the maintenance of routines across time and space sustains institutional structures. Thus it is through routinized and regionalized systems of interaction that the reflexive capacities of agents operate to reproduce institutional patterns.

STRUCTURATION THEORY: A SUMMARY AND ASSESSMENT

The most important part of Giddens' theoretical effort is the synthesis of diverse theoretical traditions. Figure 26–3 summarizes the main elements of his theory and indicates the theoretical tradition incorporated into each element. Giddens' conceptualization of structure as rules and resources that are inherently generative of many diverse combinations and his vision that societies reveal general structural principles and sets borrow from French structuralism but, in a very real sense, turn this structuralism on its head. Structure

---

[32]Ibid., pp. 110–44.

[33]See Erving Goffman, *The Presentation of Self in Everyday Life* (Garden City, NY: Doubleday, 1959).

FIGURE 26–3   Key Elements of "Structuration Theory"

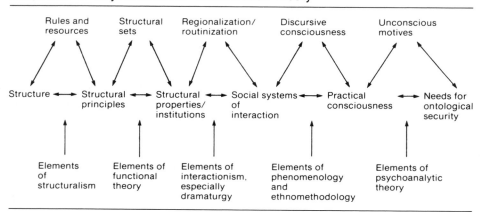

is not something that is lodged in human biology, nor is it a free-floating system of ideas and ideals. Rather, it is actively produced, reproduced, and transformed by the capacities of agents. Structural analysis is thus reattached to the ongoing process of interaction.

The analysis of modalities through which rules and resources influence interaction and the typology of institutions conceptualizes the useful portions of functional analysis: the processes of institutionalization or the creation and maintenance of relatively stable social relations. This emphasis on institutions and institutionalization corrects for deficiencies in most forms of structuralism and interactionism. Some modes of structural analysis, such as the structural idealism of Lévi-Strauss, deny the ontological status of institutions, seeing them as merely derivative of more fundamental codes. Other forms of structural analysis, such as that by Peter Blau (see Chapter 28) or Bruce Mayhew,[34] analyze structure without reference to either the processes behind human agency and interaction or the symbolic rules and resources created and used by actors. At the other extreme, interactionists fail to address the topic of structure, typically viewing it as an "object," "frame," or some other given within which the interpersonal process unfolds. In contrast to the deficiencies of these approaches, Giddens' theory is vitally concerned with the way in which interaction becomes "deeply sedimented in time and space." But, unlike much functional theorizing, institutions do not constitute an "external constraint" but rather, a process of reproduction by actors in situations of co-presence.

The discussion of regionalization and routinization is an important theoretical contribution because it attaches institutional analysis to the interactions of actors in situations of co-presence. By inserting Goffman's dramaturgical concepts, coupled with time/space geography, Giddens anchors the rather vague typology of institutions and their imprecise portrayal as "deeply

---

[34]See, for example, Bruce Mayhew, "Hierarchical Differentiation in Imperatively Coordinated Associations," *Research in the Sociology of Organizations* 2 (1983), pp. 153–229.

sedimented in time and space" to something less slippery: actors orienting their bodies and self in regionalized and routinized situations of co-presence. But, because the staging regions of interaction exist within broader time/space regions as these are shaped by structural principles, sets, and institutions, the processes of interactive dramaturgy are connected to the broader structures of societal totalities. In this way the limitations of most interactionist theory are overcome.

The analysis of discursive and practical consciousness draws from both phenomenology and ethnomethodology, but it connects them conceptually to interactionism, on the one side, and psychoanalytic theory, on the other. Giddens corrects for the extremes of ethnomethodology, which have tended to deny the ontological reality of social structures and phenomenology, which has frequently become an orgy of subjectivism. He does so by emphasizing the reflexive monitoring of action in situations of co-presence, with such monitoring involving layers of conscious and unconscious activity. Actors use rules and resources that are part of both discursive consciousness and practical stocks of knowledge, and this discursive and practical consciousness involves the subtle application of folk or ethno methods in regions of co-presence. And, unlike phenomenology, ethnomethodology, and interactionism, which do not devote much effort to conceptualizing motivation, or unlike functionalism and its intellectual companion, "action theory," which are prone to develop rather mechanical views of motivation (e.g., for each act there is a causative motive), Giddens reworks the psychoanalytic concept of anxiety and unconscious motivation. The result is a perspective that ties diffuse and largely unconscious needs for "ontological security" into actors' reflexive monitoring, use of ethno methods, and participation in routinized interaction.

Such are the strong points of Giddens' theory, as I see them. He has performed, I feel, a remarkable synthesis that demonstrates how very diverse theoretical perspectives can be pulled together into a reasonably coherent theoretical scheme. I do see many problems with his approach, however. Let me now turn to these, cautioning that my perception is guided by my theoretical bias, which Giddens and many others in sociology do not share.

One major problem is that, although Giddens' writing is crisp and eloquent, it is also vague. There is a great deal of jargon, metaphor, and just plain imprecision. It is often difficult to understand in more than a general way what is being said. I sense that "structuration theory" is very much like Parsonian functionalism in at least one sense: one has to "internalize" the perspective with all its imprecision in order to use it. One must become an intellectual convert to Giddens' cause and, in the best tradition of the ethnomethodologists' et cetera principle, accept what he has to say, even though you cannot quite understand some of it. One must "feel," as opposed to "verbalize," elements of structuration theory; this is not, I believe, the way to build social theory.

The above is a remarkable conclusion, I admit, in light of the fact that the theory has been articulated several times in major works, has received considerable and deserved attention,[35] has always offered many definitions of concepts

---

[35]For recent commentaries, see the special volumes of the following journals devoted to Giddens'

and graphic portrayals of their relations, and has even presented us with a glossary of theoretical terms in its latest incarnation.[36] Thus it might be seriously asked: how can I reach the conclusion that the theory is vague? It just is, for several reasons. First, the theory is actually a series of definitions. The linkages among concepts are always rather vaguely stated, despite their suggestiveness. Second, Giddens often argues by example rather than analytically. It is extremely difficult to interpret an analytical discussion that is presented by example. (For example, Giddens offers the same example in several works to illustrate the concepts of structural principles and structural sets; yet the definitions of these ideas and the conceptual discussion of them are decidedly unclear.)

Another problem is that Giddens' theory is, in reality, only a system of concepts. It is very much like Talcott Parsons' strategy of "analytical realism," although I find Giddens' scheme far more interesting and intriguing than Parsons'. Giddens' theory is a series of definitions that are not precisely linked together. Moreover, the scheme tends to become concerned with its own architecture—that is, with the elaboration of additional concepts in a vain effort to "complete" the edifice that, like all conceptual schemes, just keeps growing.

As Giddens admits, this conceptual scheme offers "sensitizing concepts" for the researcher. He does not believe that a natural-science view of explanation is appropriate for the social sciences, and he consistently associates functionalism, naturalism, and objectivism—the unholy alliance or "orthodox consensus" that he and others are breaking down. The irony is that he produces a scheme that in its form looks very much like Parsons'. The substantive thrust is, of course, vastly different from and superior to Parsons'. But it looks like Parsons' scheme in its overconcern with conceptual definitions, in its typological character, and in its need to be intellectually internalized when used to analyze a particular problem. (I should add, however, that both Giddens' and Parsons' interpretation of empirical events is highly stimulating;[37] thus, as others internalize and use Giddens' scheme, I am sure interesting interpretations of empirical events will result.)

These problems, which many would not see as problems at all, stem from what I believe is the major deficiency of Giddens' approach: the rejection of positivism and naturalism. Much of this rejection is made plausible by associating a search for natural laws with functionalism, evolutionism, and other highly deficient forms of theorizing. But guilt by association is not a sufficient reason to reject the search for abstract principles. If one does not believe that there are invariant properties of the social universe that can be articulated in abstract principles, then what is a theorist to do? The answer is: construct conceptual schemes that sensitize us to empirical processes and that allow us

---

work: *Theory, Culture, and Society*, vol. 1 (no. 2, 1982) and *Journal for the Theory of Social Behavior*, vol. 13 (no. 1, 1983). See also John B. Thompson, *Studies in the Theory of Ideology* (Berkeley: University of California Press, 1984). And see especially, Cohen, *Structuration Theory*.

[36]*The Constitution of Society*, pp. 373–77.

[37]In Chapter 6 of ibid., he uses the scheme to reinterpret insightfully several empirical studies.

to describe rather than explain these processes. In fact, descriptions of events are what Giddens and many others mean by explanation.

His view of explanation is misguided, although I must confess that Giddens has produced an important contribution with it (I can only imagine how much better it would be if he dropped this extreme anti-science viewpoint). There are several reasons for this assertion, which, I should emphasize again, many others do not share. First, Giddens' work contradicts his belief that there are no invariant properties of the social world. If there are not basic and fundamental processes, what good is his conceptual scheme? Will it not be outdated as soon as lay actors incorporate it? My answer is no, and so is Giddens', at least implicitly. Giddens has isolated some of the basic properties and processes of the universe; just because lay actors know about them and lock them into their discursive and practical consciousness, these properties will not change. Second—and this is related to the above point—Giddens has a very narrow view of what a law is. For Giddens a law is an empirical generalization—a statement of covariance among empirical events. If this is your vision of law, then it is easy to assert that there are no universal laws, since indeed empirical events change (in accord, I should add, with many of the invariant processes in Giddens' conceptual scheme). Third, I think there are several examples of laws in Giddens' scheme, and it is at just these points where he articulates a law that the scheme takes on more clarity and (for me at least) more interest. Here is one example of a law that Giddens articulates but that he would deny as universal: "The level of anxiety experienced at the level of discursive and practical levels of consciousness is a positive function of the degree of disruption in the daily routines for an actor." There is also a similar proposition about anxiety and unconscious trust and ontological security, but I will for the present ignore this. There are many propositions like this one in Giddens' scheme, and these are universal. If they were not, then his scheme would not make any sense. And, most importantly, the law is not obviated by our knowledge of it, for an actor will not feel less anxiety if day-to-day routines are disrupted. One might even use the law to diagnose the problem and create new, or restore the old, routines, but in the process the law has not been obviated. For, when one's routines are disrupted, the individual will experience anxiety.

I have not extracted these and many other propositions from Giddens, since doing so would violate the essence of his approach. But herein resides the great flaw, and I hope that others working with Giddens' concept are not so antipositivistic as he. For there is too much insight into the basic properties and dynamics of human action, interaction, and organization to use the scheme as a mere "sensitizing device." It has far more potential than Giddens would admit for helping develop a natural science of society—that is, for developing abstract laws of the social universe.

CHAPTER 27

$\blacklozenge$

Network Analysis*

Over the last 20 years, work within anthropology, social psychology, sociology, communications, psychology, geography, and political science has converged on the conceptualization of "structure" in terms of "social networks." During this period, rather metaphorical and intuitive ideas about networks have been reconceptualized in variable types of algebra, graph theory, and probability theory. This convergence has, in some ways, been a mixed blessing. On the one hand, the grounding of concepts in mathematics can give them greater precision and provide a common language for pulling together a common conceptual core of the various overlapping metaphors of different disciplines. On the other hand, the extensive use of mathematics and computer algorithms far exceeds the technical skills of most social scientists; and, more importantly, the use and application of quantitative techniques, per se, have become a preoccupation among many who seem less and less interested in explaining how the actual social world operates.

Nonetheless, despite these drawbacks, the potential for network analysis as a theoretical approach is great because it captures an important property of social structure—patterns of relations among social units, whether people, collectivities, or positions. For, as Georg Simmel emphasized, at the core of any conceptualization of social structure is the notion that structure consists of relations and links among entities. Network analysis forces us to conceptualize carefully the nature of the entities and relations, as well as the properties and dynamics that inhere in these relations.[1]

---

*This chapter is coauthored with Alexandra Maryanski.

[1]For some readable overviews on network analysis, see Barry Wellman, "Network Analysis: Some Basic Principles," *Sociological Theory* (1983), pp. 155–200; Jeremy F. Boisevain and J. Clyde Mitchell, eds., *Network Analysis* (The Hague: Mouton, 1973) and *Social Networks in Urban Situations* (Manchester: Manchester University Press, 1969); J. A. Barnes, "Social Networks" (Addison-Wesley Module, no. 26, 1972); Barry S. Wellman and S. D. Berkowitz, *Social Structures: A Network Approach* (Cambridge, England: Cambridge University Press, 1988). Somewhat more technical summaries of recent network research can be found in Samuel Leinhardt, ed., *Social Networks: A Developing Paradigm* (New York: Academic Press, 1977); Paul Holland and Samuel

THE DIVERSE ORIGINS OF NETWORK SOCIOLOGY

The rationale for network analysis can be found in several of sociology's early masters. For example, Georg Simmel's emphasis on "formal sociology" as an examination of the basic patterns of social relations, irrespective of their content or substance, captures the central thrust of network analysis.[2] For the core task of the network approach is to examine, at least initially, the underlying structure of social relations. Émile Durkheim's[3] analysis of "social morphology," an idea he took from Montesquieu,[4] can also be viewed as an early precursor to the network strategy. For Durkheim, morphological analysis involves "the number, nature, size, arrangement, and interrelations" of parts, and this general idea captures much of the flavor of the network agenda.

Yet it is closer to the present that we must seek the more important sources of inspiration for network concepts. Although a number of diverse scholars can be seen as the early founders of this approach, several figures stand out. Each of their contributions is briefly examined below.

Jacob Moreno and Sociometric Techniques

Jacob Moreno was an eclectic thinker, and we have already encountered his ideas on role and role playing in Chapter 18. But perhaps his more enduring contribution to sociology was the development of *sociograms*.[5] Moreno was interested in the processes of attraction and repulsion in groups, and so he sought a way to conceptualize and measure these processes. What Moreno and subsequent researchers did was to ask group members about their preferences for associating with others in the group. Typically, group members would be asked questions about whom they liked and with whom they would want to spend time or engage in activity. Often subjects were asked to give their first, second, third, etc., choices on these and related issues. The results could then be arrayed in a matrix (this was not always done) in which each person's rating of others in a group is recorded (see Figure 27-1 for a simplified example). The construction of such matrices was to become an important part of network analysis, but equally significant was the development of a sociogram in which

---

Leinhardt, eds., *Perspectives on Social Network Research* (New York: Academic Press, 1979); Ronald S. Burt, "Models of Network Structure," *Annual Review of Sociology* 6 (1980), pp. 79–141; Peter Marsden and Nan Lin, eds., *Social Structure and Network Analysis* (Newbury Park, CA.: Sage, 1982). See also the journal *Social Networks*.

[2]Georg Simmel, *Sociology: Studies in Forms of Sociation* (1908, but incompletely translated; see Jonathan H. Turner, Leonard Beeghley, and Charles Powers, *The Emergence of Sociological Theory* (Belmont, CA: Wadsworth, 1989), for a list of references where portions of this work are translated).

[3]Émile Durkheim, *The Division of Labor in Society* (New York: Free Press, 1947; originally published in 1893), *The Rules of the Sociological Method* (New York: Free Press, 1938; originally published in 1895).

[4]Charles Montesquieu, *The Spirit of the Laws*, 2 vols. (London: Colonial Press, 1900; originally published in 1748).

[5]Jacob L. Moreno, *Who Shall Survive?* (Washington, DC: Nervous and Mental Diseases Publishing Co., 1934; republished in revised form by Beacon House, New York, 1953).

FIGURE 27–1   An Example of an Early Matrix. (*Source:* Constructed from sociogram in J. Moreno, *Who Shall Survive?*, rev. ed., New York: Beacon House, 1953, p. 171.)

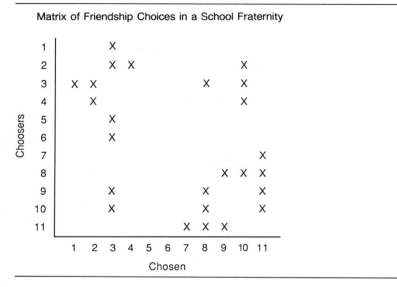

Matrix of Friendship Choices in a School Fraternity

group members were arrayed in a visual space, with their relative juxtaposition and connective lines representing the pattern of choices (those closest and connected being attracted in the direction of the arrows, and those distant and unconnected being less attracted to each other). Figure 27-2 illustrates the nature of Moreno's sociograms.

This visual representation of choices, as pulled from a matrix, captures the "structure" of preferences or, in Moreno's terms, the patterns of attraction and repulsion in groups. The visual array can be viewed as a network, because the "connections" among each individual are what is most significant. Moreover, in looking at the network, structural features emerge that can be observed.

Moreno thus introduced some of the key conceptual ingredients of contemporary network analysis: the mapping of relations among actors in visual space in order to represent the structure of these relations. Yet, alongside Moreno's sociograms, other research and theoretical traditions were developing and pointing toward the same kind of structural analysis.

Studies of Communications in Groups: Alex Bavelas and Harold Leavitt

Alex Bavelas[6] was one of the first to study how the structure of a network influenced the flow of communication in experimental groups. Others such as

---

[6]Alex Bavelas, "A Mathematical Model for Group Structures," *Applied Anthropology* 7 (3) (1948), pp. 16–30.

FIGURE 27–2   An Example of a Sociogram. (*Source:* J. Moreno, *Who Shall Survive?*, rev. ed., p. 171.)

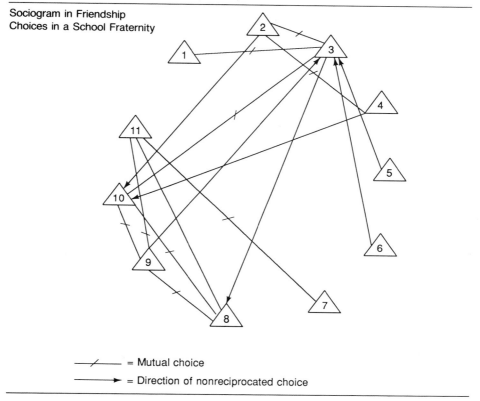

Sociogram in Friendship
Choices in a School Fraternity

—————/——— = Mutual choice

————————▶ = Direction of nonreciprocated choice

Harold Leavitt[7] followed Bavelas' lead and also began to study how communication patterns influence the task performances of people in experimental groups. The network structure in these experiments usually involved artificially partitioning groups in such a way that messages could flow only in certain directions and through particular persons. Emerging from Bavelas' original study was the notion of *centrality*, which was evident when positions lie between other positions in a network. When communications had to flow through this central position, certain styles and levels of task performance prevailed, whereas other patterns of information flow produced different results. Figure 27–3 outlines some of the chains of communication flow that Bavelas originally isolated and that Leavitt later improved upon.

[7]Harold J. Leavitt, "Some Effects of Certain Communication Patterns on Group Performance," *Journal of Abnormal and Social Psychology* 46 (1951), pp. 38–50; Harold J. Leavitt and Kenneth E. Knight, "Most 'Efficient' Solution to Communication Networks: Empirical versus Analytical Search," *Sociometry* 26 (1963), pp. 260–67.

FIGURE 27-3   Types of Communication Structures in Experimental Groups.
(*Source:* Harold J. Leavitt, "Some Effects of Certain
Communication Patterns on Group Performance," *The Journal of
Abnormal and Social Psychology* 56 [1951], p. 40.)

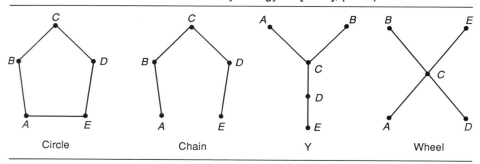

The results of these experiments are perhaps less important than the image of structure that is offered, although we should note in passing that occupying central positions, such as *C* in Figure 27-3, exerted the most influence on the emergence of leadership, task performance, and effective communication. These diagrams in Figure 27-3 resemble the sociograms, but there are some important differences that were to become critical in modern network analysis. First, the network is conceptualized in the communication studies as consisting of positions rather than persons, with the result that the pattern of relations among positions was viewed as a basic or generic type of structure. Indeed, different people could occupy the positions and the experimental results would be the same. Thus there is a real sense that structure constitutes an emergent reality, above and beyond the individuals involved. Second, the idea that the links among positions involve flows of resources—in these studies, information and messages—anticipates the thrust of much network analysis. Of course, we could also see Moreno's sociograms as involving flows of affect and preferences among people, but the idea is less explicit and less embedded in a conception of networks as relations among positions.

Thus these early experimental studies on communication created a new conceptualization of networks as (1) composed of positions, (2) connected together by relations, and (3) involving the flows of resources.

Early Gestalt and Balance Approaches: Heider, Newcomb, Cartwright, and Harary

Fritz Heider,[8] who is often considered the founder of Gestalt psychology, developed some of the initial concepts in various theories of "balance" and "equi-

---

[8]Fritz Heider, "Attitudes and Cognitive Organization," *Journal of Psychology* 2 (1946), pp. 107–12. For the best review of his thought as it accumulated over four decades, see his *The Psychology of Interpersonal Relations* (New York: John Wiley, 1958).

FIGURE 27–4 The Dynamics of Cognitive Balance. (*Source:* Adapted from Fritz Heider with (+) and (−) used instead of Heider's notation.)

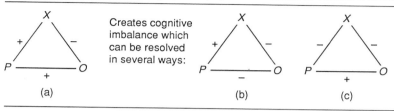

(a)   (b)   (c)

librium" in cognitive perceptions. In Heider's view, individuals seek to balance[9] their cognitive conceptions; in his famous P,O,X model, Heider argued that a person (P) will attempt to balance cognitions toward an object or entity (X) with those of another person (O). If a person (P) has positive sentiments toward an object (X) and another person (O), but O has negative sentiments toward X, then a state of cognitive imbalance exists. A person has two options if the imbalance is to be resolved: (1) to change sentiments toward X or (2) to alter sentiments toward O. By altering sentiments to X toward the negative, cognitive balance is achieved, because P and O now reveal a negative orientation toward X, thereby affirming their positive feelings toward each other. Or, by altering sentiments directed to O toward the negative, cognitive balance is achieved because P has a positive attitude toward X and negative feelings for O, who has a negative orientation to X.

Although Heider did not explicitly do so, this conception of balance can be expressed in algebraic terms, as is done in Figure 27–4 by multiplying the cognitive links in Figure 27–4(a): $(+) \times (-) \times (+) = (-)$ or imbalance. This imbalance can be resolved by changing the sign of the links toward a $(-)$ or a $(+)$, as is done for Figures 27–4(b) and 27–4(c). By multiplying the signs for the lines in 27–4(b) or 27–4(c), a $(+)$ product is achieved, indicating that the relation is now in balance.

Theodore Newcomb[10] extrapolated Heider's logic to the analysis of interpersonal communication. Newcomb argued that this tendency to seek balance applies equally to *inter*personal as well as the *intra*personal situations represented by the P,O,X model, and he constructed an A,B,X model to emphasize this conclusion. A person (A) and another (B) who communicate and develop positive sentiments will, in an effort to maintain balance with each other, develop similar sentiments toward a third entity (X), which can be an object, an idea, or a third person. However, if A's orientation to X is very strong in either a positive or a negative sense and B's orientation is just the opposite,

[9]The process of "attribution" was, along with the notion of "balance," the cornerstones of Heider's Gestalt approach.

[10]Theodore M. Newcomb, "An Approach to the Study of Communicative Acts," *Psychological Review* 60 (1953), pp. 393–404. See his earlier work where these ideas took form: *Personality and Social Change* (New York: Dryden Press, 1943).

FIGURE 27–5 The Dynamics of Interpersonal Balance. (*Source:* Adapted from Theodore Newcomb with alterations to Newcomb's system of notion.)

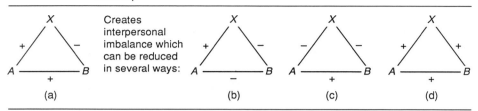

several options are available: (1) A can convince B to change its orientation toward X, and vice versa, or (2) A can change its orientation to B, and vice versa. Figure 27–5 represents this interpersonal situation for A,B,X in the same manner as Heider's P,O,X model in Figure 27–4. Situation 27–5(a) is in interpersonal imbalance, as can be determined by multiplying the signs $(+) \times (+) \times (-) = (-)$ or imbalance. Figures 27–5(b), (c), and (d) represent three options that restore balance to the relations among A, B, and X. (In 27–5(a), (b), and (c), the product of multiplying the signs now equals a $(+)$, or balance.)

Heider's and Newcomb's approach was to stimulate research that would more explicitly employ mathematics as a way to conceptualize the links in interpersonal networks. The key breakthrough had come earlier[11] in the use of the mathematical theory of linear graphs. Somewhat later, in the mid-1950s, Dorin Cartwright and Frank Harary[12] similarly employed the logic of signed-digraph theory to examine balance in larger groups consisting of more than three persons. Figure 27–6 presents a model developed by Cartwright and Harary for a larger set of actors.

The basic idea is much the same as in the P,O,X and A,B,X models, but now the nature of sentiments is specified by dotted (negative) and solid (positive) lines. By multiplying the signs $(+) =$ solid line; $(-) =$ dotted line across all of the lines, points of imbalance and balance can be identified. For Cartwright and Harary, one way to assess balance is to multiply the various *cycles*

[11] For example, D. König, *Theorie der Endlichen und Undlichen Graphen* (Leipzig, 1936 but reissued, New York: Chelsea, 1950), is, as best we can tell, the first work on graph theory. Again, from our reading, it appears that the first important application of this theory to the social sciences came with R. Duncan Luce and A. D. Perry, "A Method of Matrix Analysis of Group Structure," *Psychometrika* 14 (1949), pp. 94–116, followed by R. Duncan Luce, "Connectivity and Generalized Cliques in Sociometric Group Structure," *Psychometrika* 15 (1950), pp. 169–90. Frank Harary's *Graph Theory* (Reading, MA: Addison-Wesley, 1969) later became a standard reference, which had been preceded by Frank Harary and R. Z. Norman, *Graph Theory as a Mathematical Model in Social Science* (Ann Arbor: University of Michigan Institute for Social Research, 1953), and Frank Harary, R. Z. Norman, and Dorin Cartwright, *Structural Models: An Introduction to the Theory of Directed Graphs* (New York: John Wiley, 1965).

[12] Dorin Cartwright and Frank Harary, "Structural Balance: A Generalization of Heider's Theory," *The Psychological Review* 63 (1956), pp. 277–93. For more recent work, see their "Balance and Clusterability: An Overview," in Holland and Leinhardt, eds., *Perspectives on Social Network Research*.

FIGURE 27-6 An *S*-graph of Eight Points. (*Source:* D. Cartwright and F. Harary, "Structural Balance: A Generalization of Heider's Theory," *Psychological Review* 63 (5), 1956, p. 286.)

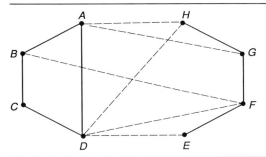

on the graph—for example, *ABCD*, *ABCDEFGH*, *HDFG*, *DFE*, and so on. If multiplying the signs for each connection yields a positive outcome, then this structure is in balance. Another procedure is specified by a theorem:[13] "An *S*-graph is balanced if and only if all paths joining the same pair of points have the same sign."

The significance of introducing graph theory into balance models is that it facilitated the representation of social relations with mathematical conventions—something that Moreno, Heider, and Newcomb failed to do. But the basic thrust of earlier analysis was retained: graph theory could represent directions of links between actors (this is done by simply placing arrows on the lines as they intersect with a point); graph theory could represent two different types of relations between points to be specified by double lines and arrows;[14] it could represent different positive or negative states (the sign being denoted by solid or dotted lines); it offered a better procedure for analyzing more complex social structures; and, unlike the matrices behind Moreno's and others' sociograms, graph theory would make them more amenable to mathematical and statistical manipulation. Thus, although the conventions of graph theory have not remained exactly the same, especially as adopted for network use, the logic of the analysis that graph theory facilitated was essential for the development of the network approach beyond crude matrices and sociograms or simple triadic relations to more complex networks involving the flows of multiple resources in varying directions.

S. F. Nadel and Anthropological Influences on Network Analysis

Several early pioneers[15] in network analysis were anthropologists trying to capture the nature of "structure" in traditional societies. A. R. Radcliffe-

---

[13]Ibid., p. 286.

[14]Early conventional notation would make these lines different colors, but this convention is not always followed since publications are usually in black and white.

[15]For example, Boisevain and Mitchell, eds., *Social Networks in Urban Situations*; Barnes,

Brown's[16] effort to develop a method for analyzing kinship was certainly one line of influence, especially his emphasis on patterns of social relations as the critical element of structure. Yet, in this years between World Wars I and II, Radcliffe-Brown and other anthropologists were still welded to functional analysis, allowing notions of the functions of a structure to distort or short-circuit purely structural analysis (see pp. 42-45 in Chapter 2). Whereas hints of a network approach can be found in a number of anthropological works,[17] S. F. Nadel's *The Theory of Social Structure*[18] was decisive for many anthropologist in separating "structure" and "function"; in so doing, Nadel proposed a mode of analysis compatible with contemporary network analysis. And, since his approach was intended to facilitate the understanding of "structure" in larger populations in natural settings (as opposed to small, contrived experimental groups), Nadel's work encouraged movement out of the psychologist's and sociologist's laboratories into the real world.

Nadel began his argument with the assertion that conceptions of structure in the social sciences are too vague. Indeed, we should begin with a more precise, and yet general, notion of all structure: "structure indicates an ordered arrangement of parts which can be treated as transposable, being relations invariant, while the parts themselves are variable."[19] Thus structure must concentrate on the properties of relations rather than actors, especially on those properties of relations that are invariant and always occur.

From this general conception of all structure, Nadel proposed that "we arrive at the structure of a society through abstracting from the concrete population and its behavior the pattern or network (or system) of relationships obtaining between actors in their capacity of playing roles relative to one another."[20] Within structures exist embedded "subgroups" characterized by certain types of relationships that hold people together. Thus, social structure is to be viewed as layers and clusters of networks—from the total network of a society to varying congeries of subnetworks. The key to discerning structure is to avoid what he termed "the distribution of relations on the grounds of their similarity and dissimilarity" and concentrate, instead, on the "interlocking of relationships whereby interactions implicit in one determine those occurring in others." That is, one should examine specific configurations of link-

---

"Social Networks." See also Jeremy F. Boisevain, "Network Analysis: A Reappraisal," *Current Anthropology* 20 (1979), pp. 392–94; Norman E. Whitten and Alvin W. Wolfe, "Network Analysis," in J. J. Honigmon, ed., *The Handbook of Social and Cultural Anthropology* (Chicago: Rand McNally, 1974); and Alvin Wolfe, "The Rise of Network Thinking in Anthropology," *Social Networks* 1 (1978), pp. 53–64.

[16] A. R. Radcliffe-Brown, "On Social Structure," *Journal of the Royal Anthropological Institute* 70 (1940), pp. 1–2; "Structure and Function in Primitive Society," *American Anthropologist* 37 (1935), pp. 58–72; and *Structure and Function in Primitive Society* (New York: Free Press, 1952).

[17] For examples, see Raymond Firth, *Elements of Social Organization* (London: Watts, 1952); E. E. Evans-Pritchard, *The Nuer* (London: Oxford University Press, 1940); and Meyer Fortes, *The Web of Kinship among the Tallensi* (London: Oxford University Press, 1949).

[18] S. F. Nadel, *The Study of Social Structure* (London: Cohen and West, 1957).

[19] Ibid., p. 8.

[20] Ibid., p. 21.

ages among actors playing roles rather than the statistical distributions of actors in this or that type of role.

From these general ideas, several anthropologists, most notably J. Clyde Mitchell[21] and J. A. Barnes,[22] welded the metaphorical imagery of work like Nadel's to the more specific techniques for conceptualizing the properties of networks. Coupled with path-breaking empirical studies,[23] the anthropological tradition began to merge with work in sociology and social psychology. And in the 1970s, as the use of mathematical approaches and computer algorithms accelerated from the modest beginnings in the late 1940s and early 1950s, network analysis achieved greater distinctiveness as a conceptual orientation and developed a number of basic theoretical concepts.

BASIC THEORETICAL CONCEPTS IN NETWORK ANALYSIS

Emerging from early work is a corpus of concepts that now guide network analysis. Taken together, these concepts provide a new framework for developing theories about social structure. I will now briefly review the most important of these concepts.

Points and Nodes: Persons, Positions, and Actors

Because network analysis is interdisciplinary, the units embedded in the network can be persons, positions, corporate actors, or other entities. In graph theory, as we saw in Figure 27–6, these are conceptualized as points or nodes and symbolized by either letters or numbers—for example, *A, B, C, D,* etc., or 1, 2, 3, 4, etc. The positions and nodes are then arrayed in visual space so as to depict the pattern of relations among them. In a mathematical sense it makes little difference *what* the points or nodes are, and this fact can be a great virtue because it provides a common set of analytical tools for analyzing very diverse phenomena. But, from a theoretical viewpoint, it may make a big difference as to whether the points are individual people, positions (statuses in an organization), or corporate actors (composed of collections of either positions or people). Depending on the nature of the point, very different dynamics may ensue, although it is possible that in some cases the same dynamics operate.[24] If the latter is the case, then network analysis offers a powerful tool for obviating

---

[21]J. Clyde Mitchell, "The Concept and Use of Social Networks" in Boisevain and Mitchell, eds., *Social Networks in Urban Situations.*

[22]J. A. Barnes, "Social Networks." See also his "Network and Political Processes" in Boisevain and Mitchell, eds., *Social Networks in Urban Situations.*

[23]Perhaps the most significant was Elizabeth Bott, *Family and Social Network: Roles, Norms, and External Relationships in Ordinary Urban Families* (London: Tavistock, 1957, 1971). Her basic finding might be expressed as a network "law," which goes something like this: *The flow of resources between actors is reduced to the extent that they are members of dense, but nonoverlapping networks, or to the extent that they occupy distant positions in the same dense network.*

[24]This was, of course, the goal of Simmel's "formal sociology."

micro-vs.-macro debates (since interaction between people and collective actors would reveal the same network dynamics) and for analyzing widely diverse phenomena in the same terms.

Links, Ties, Connections

As Figure 27-6 indicates, points are only one basic element of a network. These points need to be "connected' in some way, as is indicated by lines. These connections were originally viewed as *links*,[25] but more recently, in sociology, they have come to be seen as *ties*. But the question immediately arises as to *what* this tie or link is. Again, in the mathematics of graph theory it does not make much difference, but in the substantive concerns of sociologists it probably does make a difference. If one looks at the large literature, these lines can represent such diverse forces as information, liking, preferences, control, influence, honor/prestige, material things, and ideas. For example, early sociograms by Moreno saw individuals (the nodes, as currently conceptualized) as connected by emotions such as friendship, preference, and liking. Similarly, early work by Gestalt sociologists, such as Heider, Newcomb, and Cartwright, connected actors in terms of sentiments. Other early work by such experimental psychologists as Bavelas and Leavitt examined the flow of messages or information among actors who were assigned by the experimenters to particular positions. More recent work has examined the flows of material resources like money and goods in market networks. For example, "world systems theory" is, in a sense, a network approach to the flow of material resources among nation/states, and terms like "core," "periphery," and "semiperiphery"[26] refer to the position of nations in a world network.

One way to rise above the diversity of resources examined in network analysis is to visualize resource flows in networks in terms of three generic types: materials, symbols, and emotions. That is, what connects persons, positions, and corporate actors in the social world is the flow of (1) symbols (information, ideas, values, norms, messages, etc.), (2) materials (physical things and perhaps symbols, such as money, that give access to physical things), and (3) emotions (approval, respect, liking, pleasure, etc). In nonsociological uses of networks the ties or links may be other types of phenomena, but, when the ties are social, they exist along material, symbolic, and emotional dimensions.

As noted, these ties are represented as lines connecting those positions or nodes represented by letters or numbers, and the lines can constitute a directed graph (or digraph) when the movement of resources is specified by arrows. Moreover, if multiple resources are connecting positions in the graph, multiple lines (and arrows specifying direction) will be used.

The configuration of ties can also be represented as a matrix, as we saw in Figure 27-1. Such matrices are useful for various statistical procedures and

---

[25]For example, see Mitchell, "The Concept and Use of Social Networks."

[26]Immanuel Wallerstein, *The Modern World-System*, Vol. 1 (New York: Academic Press, 1974).

computer algorithms, and they can become very complicated because different ties involving multiple resources are represented in the cells of the matrix. The construction of matrices is generally preliminary for the development of network diagrams like the one in Figure 27–2. And, as we saw, it has been a part of early work, such as sociometry, from the very beginning. Today the matrices are ever more essential for discovering with computer algorithms the patterns and configurations of ties among positions that are of most interest to network analysts. Indeed, network analysis often confines itself exclusively to matrices, avoiding the construction of graphs that would become too complicated and, in essence, unreadable (indeed, many of Moreno's early sociograms were too complicated to be easily discerned).

Patterns and Configurations of Ties

From a network perspective, social structure is conceptualized as the form of ties among positions or nodes. That is, what is the pattern or configuration among what resources flowing among what sets of nodes or points in a graph? To answer questions like this, network sociology addresses a number of properties of networks. The most important of these are number of ties, directedness, reciprocity of ties, transitivity of ties, density of ties, strength of ties, bridges, brokerage, centrality, and equivalence.

Number of ties   An important piece of information in performing network analysis is the total number of ties among all points and nodes. Naturally, the number of potential ties depends upon the number of points in a graph and the number of resources involved in connecting the points. Yet, for any given number of points and resources, it is important to calculate both the actual and potential number of ties that are (and can be) generated. This information can then be used to calculate other dimensions of a network structure.

Directedness   It is important to know the direction in which resources flow through a network; so, as indicated earlier, arrows are often placed on the lines of a graph, making it a digraph. As a consequence, a better sense of the structure of the network emerges. For example, if the lines denote information, we would have a better understanding of how the ties in the network are constructed and maintained, since we could see the direction and sequence of the information flow.

Reciprocity of ties   Another significant feature of networks is the reciprocity of ties among positions. That is, is the flow of resources one way, or is it reciprocated for any two positions? If the flow of resources is reciprocated, then it is conventional to have double lines with arrows pointing in the direction of the resource flow (recall from Moreno's sociogram in Figure 27–2 that he represented reciprocity with a slash across the line). Moreover, if different resources flow back and forth, then this too can be represented. Surprisingly,

FIGURE 27-7 High- and Low-Density Networks

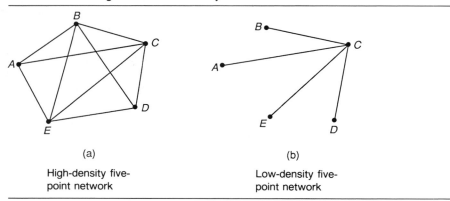

(a)

High-density five-
point network

(b)

Low-density five-
point network

conventions on how to represent this multiplicity of resource flows are not fully developed. One way to denote the flow of different resources is to use varying-colored lines or numbered lines; another is to label the points with the same letter subscripted (i.e., A_1, A_2, A_3, etc.) if similar resources flow and with varying letters (i.e., A, B, C, D) if the resources connecting actors are different. But, whatever the notation, the extent and nature of reciprocity in ties become an important property of a social network.

Transitivity of ties   A critical dimension of networks is the level of transitivity among sets of positions. Transitivity refers to the degree to which there is a "transfer" of a relation among subsets of positions. For example, if nodes A_1 and A_2 are connected with positive affect, and positions A_2 and A_3 are similarly connected, we can ask: will positions A_1 and A_3 also be tied together with positive affect? If the answer to this question is "yes," then the relations among A_1, A_2, and A_3 are transitive (hence Heider and Newcomb were, in essence, examining transitivity). Discovering patterns of transitivity in a network can be important because it helps explain other critical properties of a network, such as density and the formation of cliques.

Density of ties   A significant property of a network is its degree of connectedness, or the extent to which nodes reveal the maximum possible number of ties. The more the actual number of ties among nodes approaches the total possible number among a set of nodes, the greater is the overall density of a network.[27] Figure 27-7 compares the same five-node network under conditions of high and low density of ties.

Of even greater interest are subdensities of ties within a larger network structure. Such subdensities, which are sometimes referred to as *cliques*, reveal

---

[27]There are other ways to measure density; this definition is meant to be illustrative of the general idea.

FIGURE 27–8 A Network with Three Distinct Cliques

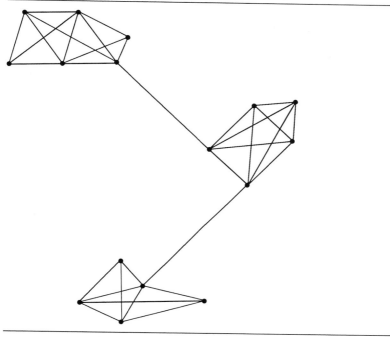

strong, reciprocated, and transitive ties among a particular subset of positions within the overall network.[28] For example, in Figure 27-2, persons 8, 9, and 10 (and perhaps 11) reveal a reciprocity in their friendship choices (high density) and thereby form separate cliques within the overall system of ties that constitute the network. Figure 27-8 also illustrates clique formation.

Strength of ties Yet another crucial aspect of a network is the volume and level of resources that flow among positions. A weak tie is one where few or sporadic amounts of resources flow among positions, whereas a strong tie evidences a high level of resource flow. The overall structure of a network is significantly influenced by clusters and configurations of strong and weak ties. For example, if the ties in the cliques in Figure 27-8 are strong, the network is composed of cohesive subgroupings that have relatively sparse ties to one another. On the other hand, if the ties in these subdensities are weak, then

[28]The terminology on subdensities varies. "Clique" is still the most prominent term, but "alliances" (Linton Freeman, "Alliances: A New Formalism for Primary Groups and Its Relationship to Cliques and to Structural Equivalences," working paper, 1987) has recently been offered as an alternative. Moreover, the old sociological standbys "group" and "subgroup" seem to have made a comeback in network analysis.

the subgroupings will involve less intense linkages,[29] with the result that the structure of the whole network will be very different than would be the case if these ties were strong.

Bridges   When networks reveal subdensities, it is always interesting to know which positions connect the subdensities, or cliques, to one another. For example, in Figure 27-8, those ties connecting subdensities are bridges and are crucial in maintaining the overall connectedness of the network. Indeed, if one removed one of these positions or severed the tie, the structure of the network would be very different; in fact, it would now become three separate networks. These bridging ties are typically weak,[30] since each position in the bridge is more embedded in the flow of resources of a particular subdensity or clique. But, nonetheless, such ties are often crucial to the maintenance of a larger social structure; thus it is not surprising that the number and nature of bridges within a network structure are highlighted in network analysis.

Brokerage   At times a particular position is outside subsets of positions but is crucial to the flow of resources to and from these subsets. This position is often in a brokerage situation because its activities determine the nature and level of resources that flow to and from subsets of positions.[31] In Figure 27-9, position A_6 is potentially a broker for the flow of resources from subsets consisting of positions A_1, A_2, A_3, A_4, and A_5 to B_1, B_2, B_3, B_4, B_5, and B_6. Position A_6 can become a broker if (1) the distinctive resources that pass to, and from, these two subsets are needed or valued by at least one of these subsets and (2) direct ties, or bridges, between the two subsets do not exist. Indeed, a person or actor in a brokerage position often seeks to prevent the development of bridges (as in Figure 27-8) and to manipulate the flow of resources such that at least one, and if possible both, subsets are highly dependent upon its activities.

Centrality   An extremely important property of a network is *centrality*, as was noted for Bavelas' and Leavitt's studies of communication in experimental groups. There are several ways to calculate centrality:[32] (1) the number

---

[29]At one time, "intensity" appears to have been used in preference to "strength." See Mitchell, "The Concept and Use of Social Networks." It appears that Granovetter's classic article shifted usage in favor of "strength" and "weakness." See footnote 30.

[30]See Mark Granovetter, "The Strength of Weak Ties," *American Journal of Sociology* 78 (1973), pp. 1360-80; and "The Strength of Weak Ties: A Network Theory Revisited," *Sociological Theory* (1983), pp. 201-33. The basic network "law" from Granovetter's original study can be expressed as follows: *The degree of integration of a network composed of highly dense subcliques is a positive function of the extensiveness of bridges, involving weak ties, among these subcliques.*

[31]Ronald S. Burt has, perhaps, done the most interesting work here. See, for example, his *Toward a Structural Theory of Action* (New York: Academic Press, 1982) and "A Structural Theory of Interlocking Corporate Directorships," *Social Networks* 1 (1978-79), pp. 415-35.

[32]The definitive work here is Linton C. Freeman, "Centrality in Social Networks: Conceptual Clarification," *Social Networks* 1 (1979), pp. 215-39; and Linton C. Freeman, Douglas Boeder, and Robert R. Mulholland, "Centrality in Social Networks: 11. Experimental Results," *Social Networks* 2 (1979), pp. 119-41. See also Linton C. Freeman, "Centered Graphs and the Structure of Ego Networks," *Mathematical Social Sciences* 3 (1982), pp. 291-304.

FIGURE 27-9 A Network with Brokerage Potential

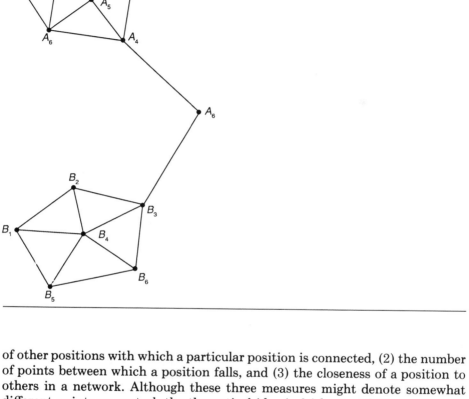

of other positions with which a particular position is connected, (2) the number of points between which a position falls, and (3) the closeness of a position to others in a network. Although these three measures might denote somewhat different points as central, the theoretical idea is fairly straightforward: some positions in a network mediate the flow of resources by virtue of their patterns of ties to other points. For example, in Figure 27–7(b), point C is central in a network consisting of positions A, B, C, D, and E; or, to take another example, points A_5 and B_4 in Figure 27–9 are more central than other positions because they are directly connected to all, or to the most, positions and because a higher proportion of resources will tend to pass through these positions. A network can also reveal several nodes of centrality, as is evident in Figure 27–10. Moreover, as we will see shortly, the patterns of centrality may shift over time. Thus the dynamics of network structure revolve around the nature and pattern of centrality.

Equivalence   When positions stand in the same relation to another position, they are considered *equivalent*. When this idea was first introduced into network analysis, it was termed *structural equivalence* and restricted to

FIGURE 27–10 Equivalence in Social Networks

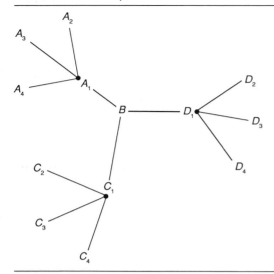

situations in which a set of positions is connected to another position or set of positions in exactly the same way.[33] For example, positions B, A, E, and D in Figure 27-7 are structurally equivalent because they reveal the same relation to position C. Figure 27-10 provides another illustration of structural equivalence. A_2, A_3, and A_4 are structurally equivalent to A_1; similarly, C_2, C_3, and C_4 are structurally equivalent to C_1; D_2, D_3, and D_4 are equivalent to D_1; and A_1, C_1, and D_1 are structurally equivalent to B.

This original formulation of equivalence was limited, however, in that positions could be equivalent only when *actually connected to the same position*. We might also want to consider all positions as equivalent when they are connected to different positions but in the same form, pattern, or manner. For instance, in Figure 27-10, A_2, A_3, A_4, D_2, D_3, D_4, C_2, C_3, and C_4 can all be seen as equivalent because they bear the *same type* of relation to another position— that is $A_{,1}$, D_1, and C_1, respectively. This way of conceptualizing equivalence is termed *regular equivalence*[34] and, in a sense, subsumes the original notion of *structural equivalence*. That is, structural equivalence, wherein the equivalent *positions must actually be connected to the same position in the same way*, is a particular type of a more general equivalence phenomenon. These terms, "structural" and "regular," are awkward, but they have become con-

[33]Francois Lorrain and Harrison C. White, "Structural Equivalence of Individuals in Social Networks," *Journal of Mathematical Sociology* 1 (1971), pp. 49–80; Harrison C. White, Scott A. Boorman, and Ronald L. Breiger, "Social Structure from Multiple Networks: I. Block Models of Roles and Positions," *American Journal of Sociology* 8 (1976), pp. 730–80.

[34]Lee Douglas Sailer, "Structural Equivalence," *Social Networks* 1 (1978), pp. 73–90.

ventional in network analysis and so we are stuck with them. The critical idea is that the number and nature of equivalent positions in a network have important influences on the dynamics of the network.[35] The general hypothesis is that actors in structurally equivalent or regularly equivalent positions will behave or act in similar ways.

The mathematics of network analysis can become quite complicated, as can the computer algorithms used to analyze data sets in terms of the processes outlined above. This listing of concepts is somewhat metaphorical, because it eliminates the formal and quantitative thrust of much network analysis. Indeed, as we mentioned earlier, much network analysis bypasses the conversion of matrices into graphs like those in the various figures presented thus far and, instead, performs mathematical and statistical operations on just the matrices themselves. Yet, if network analysis is to realize its full theoretical (as opposed to methodological) potential, it may be wise to use concepts, at least initially, in a more verbal and intuitive sense. Let us now assess the theoretical potential of network analysis and examine some of the theoretical programs that have used the concepts discussed here.

THE THEORETICAL POTENTIAL OF NETWORK SOCIOLOGY

Few would disagree with the notion that social structure is composed of relations among positions. But is this all that social structure is? Can the concepts denoting nodes, ties, and patterns of ties (number, strength, reciprocity, transitivity, bridges, brokerage, centrality, and equivalence) capture all of the critical properties of social structure?

The answer to these questions is probably "no." Social structure probably involves other crucial processes that are not captured by these concepts. Yet a major property of social structure *is* its network characteristics, as Georg Simmel was perhaps the first to really appreciate. For, whatever other dimensions social structure may be seen to reveal—cultural, behavioral, ecological, temporal, psychological, etc.—its backbone is a system of interconnections among actors who occupy positions vis-à-vis one another and who exchange resources. And, so, network analysis has great potential for theories of social structure. Has this potential been realized? Probably not, for several reasons.

First, as just noted, network analysis is overly methodological and concerned with generating quantitative techniques for arraying data in matrices and then converting the matrices into descriptions of particular networks (whether as graphs or as equations). As long as this is the case, network sociology will remain primarily a tool for empirical description.

---

[35]In many ways Karl Marx's idea that those who stand in a common relationship to the means of production have common interests is an equivalence agreement. Thus the idea of equivalence is not new to sociology—just the formalism used to express it.

Second, there has been little effort to develop principles of network dynamics, per se. Few[36] seem to ask theoretical questions within the network tradition itself. For example, how does the degree of density, centrality, equivalence, bridging, and brokerage influence the nature of the network and the flow of relations among positions in the network? There are many empirical descriptions of events that touch on this question but few actual theoretical laws or principles.[37]

Third, network sociology has yet to translate traditional theoretical concerns and concepts into network terminology in a way that highlights the superiority, or at least the viability, of using network theoretical constructs for mainstream theory in sociology. For example, power, hierarchy, differentiation, integration, stratification, conflict, and many other concerns of sociological theory have not been adequately reconceptualized in network terms, and hence it is unlikely that sociological theory will adopt or incorporate a network approach until this translation of traditional questions occurs.

All of these points, however, need to be qualified by the fact that numerous sociologists have actually sought to develop laws of network processes and to address traditional theoretical concerns with network concepts. Although these efforts are far from constituting a coherent theory of network dynamics, they do illustrate the potential utility of network sociology. Let us examine some of these adaptations of network ideas.

EXCHANGE THEORY AND NETWORK ANALYSIS: THE EMERSON–COOK PROGRAM

Since networks involve the flow of resources among positions that reveal ties, it should not be surprising that exchange theorists have gravitated toward network concepts. Although there are several creative efforts to adopt network concepts to exchange theory,[38] this merger can best be illustrated with the work of the late Richard Emerson[39] and his collaborator, Karen S. Cook.[40]

---

[36]There are, of course, some notable exceptions to this statement. For examples of what we see as the kinds of laws that need to be formulated, see our formal statements on Granovetter's and Bott's work in footnotes 30 and 23, respectively.

[37]Mark Granovetter, "The Theory-Gap in Social Network Analysis" in P. Holland and S. Leinhardt, eds., *Perspectives on Social Network Research.*

[38]See, in particular, David Willer, "The Basic Concepts of the Elementary Theory," in D. Willer and B. Anderson, eds., *Networks, Exchange and Coercion* (New York: Elsevier, 1981); "Property and Social Exchange," *Advances in Group Processes* 2 (1985, pp. 123–42; and *Theory and the Experimental Investigation of Social Structures* (New York: Gordon and Breach, 1986).

[39]Emerson's perspective is best stated in his "Exchange Theory, Part I: A Psychological Basis for Social Exchange" and "Exchange Theory, Part II: Exchange Relations and Network Structures," in *Sociological Theories in Progress*, ed. J. Berger, M. Zelditch, and B. Anderson (New York: Houghton Mifflin, 1972), pp. 38–87. Earlier empirical work that provided the initial impetus to, or the empirical support of, this theoretical perspective includes: "Power-Dependence Relations," *American Sociological Review* 17 (February 1962), pp. 31–41; "Power-Dependence Relations: Two Experiments," *Sociometry* 27 (September 1964), pp. 282–98; John F. Stolte and Richard M. Emerson, "Structural Inequality: Position and Power in Network Structures," in *Behavioral Theory in Sociology*, ed. R. Hamblin (New Brunswick, NJ: Trans-action Books, 1977). Other more

The Overall Strategy

Emerson began by enumerating the basic propositions of operant psychology. Then, through the development of corollaries, he extended these propositions and made them more relevant to human social organization. Finally, he derived from these propositions and their corollaries a series of theorems to account for the operation of different social patterns. At various points in the development of his system of propositions, corollaries, and theorems, new concepts that would be incorporated into the corollaries and theorems were added.

Emerson never did perform the logical operations in deriving corollaries from the basic operant propositions and in developing theorems from these propositions and corollaries. Yet, in contrast to most theory in sociology, Emerson's work is extremely rigorous.[41] Concepts are precisely defined and represented by symbolic notation. Propositions, corollaries, and theorems are stated in terms of covariance among these clearly defined concepts. Thus considerable attention is devoted to concept formation and then to the use of these concepts in a system of propositions, corollaries, and theorems.

Emerson followed the substantive strategy of other exchange theorists by moving from micro processes in simple structures to processes in more complex structures. As the structures under investigation become more complex, additional corollaries and theorems are developed. But the most important difference between Emerson's substantive approach and that of other perspectives is his concern with the *forms* of exchange relations. The theorems delineate the processes inherent in a given form of exchange relationship. The nature of the units in this relationship can be either micro or macro—individual persons or corporate units such as groups, organizations, or nations. Much as Georg Simmel focused on the "forms of sociation" and their underlying exchange basis, so Emerson sought to develop a set of theoretical principles that explains generic social forms. In this way the distinction between micro- and macroanalysis is rendered less obstructive, because it is the *form of the relationship* rather than the properties of the units that is being explained. And

---

conceptual works include "Operant Psychology and Exchange Theory," in *Behavioral Sociology*, eds. R. Burgess and D. Bushell (New York: Columbia University Press, 1969) and "Social Exchange Theory," in *Annual Review of Sociology*, eds. A. Inkeles and N. Smelser, 2 (1976), pp. 335–62.

[40] For example, see Karen S. Cook and Richard Emerson, "Power, Equity and Commitment in Exchange Networks," *American Sociological Review* 43 (1978), pp. 712–39; Karen S. Cook, Richard M. Emerson, Mary R. Gillmore, and Toshio Yamagishi, "The Distribution of Power in Exchange Networks," *American Journal of Sociology* 89 (1983), pp. 275–305; Karen S. Cook and Richard M. Emerson, "Exchange Networks and the Analysis of Complex Organizations," *Research in the Sociology of Organizations* 3 (1984), pp. 1–30. See also Karen S. Cook, "Exchange and Power in Networks of Interorganizational Relations," *Sociological Quarterly* 18 (Winter 1977), pp. 66–82; "Network Structures from an Exchange Perspective," in *Social Structure and Network Analysis*, eds. P. Marsden and N. Lin; and Karen S. Cook and Karen A. Hegtvedt, "Distributive Justice, Equity, and Equality," *American Sociological Review* 9 (1983), pp. 217–41.

[41] There are, of course, notable exceptions to this statement. See, for examples, Alfred Kuhn, *Unified Social Science* (Homewood, IL: Dorsey Press, 1975), and the articles in Berger, Zelditch, and Anderson, eds., *Sociological Theories in Progress* as well as in Willer and Anderson, eds., *Networks, Exchange and Coercion.*

it is this point of emphasis that would make his perspective compatible with network analysis.

Emerson's strategy involves many problems of exposition, however. His concern with rigorous concept formation involves the creation of a new language. Acquiring a familiarity with this language requires considerable time and effort. Moreover, the system of definitions, concepts, propositions, corollaries, and theorems soon becomes exceedingly complex, even though Emerson explores only a few basic types, or forms, of social relations. This analysis of Emerson's work therefore translates terms into more discursive language and omits discussion of certain corollaries and theorems.

The Basic Exchange Concepts

The following is an incomplete list of key concepts in Emerson's exchange perspective:

Actor: An individual or collective unit that is capable of receiving reinforcement from its environment.

Reinforcement: Features of the environment that are capable of bestowing gratification upon an actor.

Behaviors: Actions or movements of actors in their environment.

Exchange: Behaviors by actors that yield environmental reinforcement.

Value: The strength of reinforcers to evoke and reinforce behavioral initiations by an actor, relative to other reinforcers and holding deprivation constant and greater than zero.

Reward: The degree of value attached to a given type of reinforcement.

Alternatives: The number of sources in the environment of an actor that can bestow a given type of reinforcer.

Cost: The magnitude and number of rewards of one type foregone to receive rewards of another type.

Exchange relation: Opportunities across time for an actor to initiate behaviors that lead to relatively enduring exchange transactions with other actors in the environment.

Dependence: A situation in which an actor's reinforcement is contingent upon behaviors on the part of another actor, with the degree of dependence being a dual function of the strength of reinforcement associated with behavior and the number of alternatives for rewards.

Balance: The degree to which the dependency of one actor, A, for rewards from actor B is equal to the dependency of actor B for rewards from actor A.

Power: The degree to which one actor can force another actor to incur costs in an exchange relation.

Resources: Any reward that an actor can use in an exchange relation with other actors.

Several points about this list of concepts should be emphasized. First, Emerson analyzes only the exchange *relation* between actors. This approach bypasses the problem of tautology so evident in much exchange theory by viewing an established relation—not the actors in the relation—as the smallest unit of analysis. In this way questions about each actor's values become less central because attention is focused on the relationship between actors who exchange resources. This line of argument, of course, abandons explanation in terms of an individual actor's values.[42] The emphasis is on the ratio of rewards exchanged among actors and on how this ratio shifts or stabilizes over the course of the exchange relationship. Propositions thus focus on explaining the variables outside the actors in the broader context of the social relationship that might influence the ratio of rewards in a given social relationship. Thus behavior is no longer the dependent variable in propositions; rather, the exchange relationship becomes the variable to be explained. The goal is to discover laws that help account for particular patterns of exchange relations. This approach is contrary to traditional exchange theory, which seeks to explain why a person enters into an exchange relationship in terms of that person's values. But, if the relationship is the unit of analysis, then the question of why the individual enters the relationship is no longer of prime concern. The fact is that the individual has entered a relationship and is willing to exchange rewards with another. When this exchange relationship among actors becomes the unit of analysis, Emerson argued, theory seeks to discover what events could effect variations in the entire relational unit, not in the individual behaviors of actors. For example, in a hypothetical exchange, person A gives esteem and respect to person B in return for advice. With the A, B relationship as the unit of analysis, the question is not what made either A or B enter the relationship— answers to which would take theory into A's and B's cognitive structure and thereby increase the probability of tautologous propositions. Rather, since the A,B unit already exists as an entity, theoretical questions should focus on what events would influence the ratio of esteem and advice exchanged in the A,B unit.

Second, as can be seen from the list of definitions above, the concepts of actor, reinforcement, exchange, value, reward, cost, and resource are all defined in terms of one another; but, since they are not analyzed independently of the exchange relation, the problem of tautology is bypassed. These concepts are the givens of any existing exchange relation. Thus, in contrast to most exchange approaches, social structure is not a theoretical given. Instead, behavioral dispositions are the givens, and it is social structure that is to be explained.

Third, in accordance with the emphasis on the structure of the exchange relation, as opposed to the characteristics of the actors, the concepts of (a) dependence, (b) power, and (c) balance in exchange relations become central. The key questions in Emerson's scheme thus revolve around how dependence,

---

[42]Yet, curiously, Emerson returned to this question in his last article. See Richard M. Emerson, "Toward a Theory of Value in Social Exchange," in Karen S. Cook, ed., *Social Exchange Theory* (Newbury Park, CA: Sage, 1987), pp. 11–46. See, in this same volume, Jonathan H. Turner's critique of this shift in Emerson's thought: "Social Exchange Theory: Future Directions," pp. 223–39.

power, and balance in exchange relations help explain the operation of more complex social patterns.

Fourth, actors are viewed as either individuals or collective units. The same processes in exchange relations are presumed to apply to both individuals and collectivities of individuals, thus obviating many problems in the micro-vs.-macro schism in sociological theorizing. This emphasis is possible when attention is shifted away from the attributes of actors to the *form* of their exchange relationship.

Thus, although this partial list of Emerson's concepts appears to be similar to that developed by George Homans (see Chapter 15) and other behavioristically oriented exchange theorists, there is an important shift in emphasis from concern with the values and other cognitive properties of actors to a concern with the structure of an exchange relation. This concern with structure takes as a given the flow of valued resources among those involved in the exchange. Theoretical attention then focuses on the structural attributes of the exchange relation and on the processes that maintain or change the structural form of an ongoing exchange relationship.

The Basic Exchange Processes

In Emerson's scheme, analysis begins with an existing exchange relation between at least two actors. This relationship has been formed from (1) perceived opportunities by at least one actor, (2) the initiation of behaviors, and (3) the consummation of a transaction between actors mutually reinforcing each other. If initiations go unreinforced, then an exchange relation will not develop. And, unless the exchange transaction between actors endures for at least some period of time, it is theoretically uninteresting.

Emerson's approach thus starts with an established exchange relation and then asks: to what basic processes is this relationship subject? His answer: (1) the use of power and (2) balancing. If exchange relations reveal high dependency of one actor, B, on another actor, A, for reinforcement, then A has what Emerson termed a *power advantage* over B. This conceptualization of power is similar to Peter Blau's formulation (see Chapter 16), although Emerson developed a different set of propositions for explaining its dynamics. To have a power advantage is to use it, with the result that actor A forces increasing costs on actor B within the exchange relationship.

In Emerson's view, a power advantage represents an imbalanced exchange relation. A basic proposition in Emerson's scheme is that, over time, imbalanced exchange relations tend toward balance.[43] He visualized this process as occurring through a number of "balancing operations":

1. A decrease in the value for actor B of reinforcers, or rewards, from actor A.

---

[43]Note here the emphasis on balance, an idea introduced into network analysis by Gestalt psychology.

2. An increase in the number of alternative sources for the reinforcers, or rewards, provided to B by A.
3. An increase in the value of reinforcers provided by B for A.
4. A reduction in the alternative sources for the rewards provided by B for A.

These balancing operations are somewhat similar to the propositions enumerated by Blau on the conditions for differentiation of power (see Table 16–2 on page 334). But in contrast to Blau's emphasis on the inherent and incessant dialectic for change resulting from power imbalances, Emerson stressed that, through at least one of these four balancing operations, the dependency of B and A on each other for rewards will reach an equilibrium. Thus exchange transactions reveal differences in power that, over time, tend toward balance. Naturally, in complex exchange relations involving many actors, A, B, C, D, . . . , n, the basic processes of dependence, power, and balance will ebb and flow as new actors and new reinforcers or resources enter the exchange relations.

The Basic Exchange Propositions

In Table 27–1 the initial propositions that Emerson developed to explain exchange relations are selectively summarized. The general strategy was to begin with behaviorist principles and then to derive theorems from these that explain the basic exchange processes, use of power, and balancing. In turn, corollaries to these theorems can be developed to account for the structural forms of exchange relations. Thus the propositions in Table 27–1 represent only a starting point.

The crucial next step is to derive theorems from these behaviorist principles that pertain to the dynamics of power and balancing. In Table 27–2, Emerson's initial set of theorems is summarized. These theorems describe the dynamics of power as a function of one actor's dependency on another for valued resources (Theorem 4), whereas balance is conceptualized as a process whereby dependency is reduced over time (see definition of balance and Theorem 5). Thus power (P) is a positive function of the dependency (D) of actor B on the resources of actor A, or $P_{AB} = D_{BA}$. Balance is a situation in which B's dependency for resources from A is equal to A's dependency for resources from B, or $D_{BA} = D_{AB}$.

Thus far Emerson has derived some basic theorems on power and balance from a long list of behaviorist principles. This list of behaviorist assumptions now recedes, and the main task is to introduce corollaries and new theorems to account for the structural form of an ongoing exchange relation. It is at this point that Emerson introduced ideas from network analysis.

Structure, Networks, and Exchange

Emerson's portrayal of social networks will be simplified, since for our purposes here the full details of his network terminology need not be addressed. Although

TABLE 27–1   The Operant Propositions and Initial Corollaries

---

Proposition 1: The greater the behavioral repertoire of actor *A* in a situation and the greater the variations in rewards for behaviors, the more likely is *A* to emit those behaviors yielding the greatest rewards.

    Corollary 1.1: The greater the decrease in rewards for *A* in an established exchange relation, the greater the variation in *A*'s behavior.

    Corollary 1.2: The more rewards in an established exchange relation approach a zero level of reinforcement, the fewer the initiations by *A*.

    Corollary 1.3: The greater the power advantage of *A* over *B* in an exchange relation, the more *A* will use its power advantage across continuing transactions.

    Corollary 1.4: The more power is balanced in an exchange relation between *A* and *B*, and the more *A* increases its use of power, the more *B* will increase its use of power.

Proposition 2: The more frequent and valuable the rewards received by actor *A* for a given behavior in a situation, the less likely is actor *A* to emit similar behaviors immediately.

    Corollary 2.1: The more rewards of a given type received by *A*, the less frequent *A*'s initiations for rewards of this type.

Proposition 3: The more actor *A* must emit a given behavior for a given type of reward, and the greater the strength and number of rewards of this type in a situation, the more likely is actor *A* to emit behaviors of a given type in that situation.

    Corollary 3.1: The greater the number of alternatives available to *A* for a given reward, the less dependent is *A* upon that situation.

    Corollary 3.2: The more a situation provides multiple sources of reward for *A*, the more dependent is *A* on that situation.

    Corollary 3.3: The greater the value of rewards received by *A* in a given situation, the greater is the dependency of *A* on that situation.

    Corollary 3.4: The greater the uncertainty of *A*'s ever receiving a given reward in a given situation, and the fewer alternative situations for receiving this reward, the greater is the dependency of *A* on that situation.

    Corollary 3.5: The less the value of a reward for *A* in an *A,B* exchange relation, and the greater the alternative sources of that reward for *A*, the less cohesive is the exchange relation between *A* and *B*; or, conversely, the more the value of a reward for *A*, and the fewer alternative sources of that reward for *A*, the more cohesive is the relationship between *A* and *B*.

    Corollary 3.6: The more an *A,B* exchange relationship at one point in time is transformed to an *A,B,C* relationship, the greater the dependency of *B* upon *A*; also, the fewer the alternatives for *B* in the *A,B* relationship, the more *B*'s dependency upon *A* will be greater than *B*'s dependency upon *C*.

Proposition 4: The more uncertain is an actor *A* of receiving a given type of reward in recent transactions, the more valuable is that reward for actor *A*.

    Corollary 4.1: In a set of potential exchange relations, the more maintenance of one transaction precludes other transactions in this set, the greater the initial costs of this one transaction but the less the costs across continuing transactions.

TABLE 27–2 The Initial Theorems

Theorem 1: The greater the value of rewards to *A* in a situation, the more initiations by *A* reveal a curvilinear pattern, with initiations increasing over early transactions and then decreasing over time. (From Corollaries 1.2 and 2.1.)
Theorem 2: The greater the dependency of *A* on a set of exchange relations, the more likely is *A* to initiate behaviors in this set of relations. (From Propositions 1 and 3.)
Theorem 3: The more the uncertainty of *A* increases in an exchange relation, the more the dependency of *A* on that situation increases, and vice versa. (From Corollary 3.3 and Proposition 4.)
Theorem 4: The greater the dependency of *B* on *A* for rewards in an *A,B* exchange relationship, the greater is the power of *A* over *B* and the more imbalanced is the relationship between *A* and *B*. (From Propositions 1 and 3 and definitions of cost, dependence, and power.)
Theorem 5: The greater the imbalance of an *A,B* exchange relation at one point in time, the more likely it is to be balanced at a subsequent point in time. (From Corollaries 1.1, 1.3, 3.1, 3.3, and Proposition 4.)

Emerson followed the conventions of graph theory and developed a number of definitions, only two definitions are critical:

Actors: Points $A, B, C, \ldots, n$ in a network of relations. Different letters represent actors with different resources to exchange. The same letters— that is, A_1, A_2, A_3, and so forth—represent different actors exchanging similar resources.

Exchange relations: A——B, A——B——C, A_1——A_2, and other patterns of ties that can connect different actors to each other, forming a network of relations.[44]

The next conceptual task is to visualize the forms of networks that can be represented with these two definitions. For each basic form, new corollaries and theorems are added as Emerson documented the way in which the basic processes of dependence, power, and balance operate. His discussion is only preliminary, but it does illustrate the perspective's potential.

Several basic social forms are given special treatment: (*a*) unilateral monopoly, (*b*) division of labor, (*c*) social circles, (*d*) stratification, and, along with Karen Cook, (*e*) centrality in networks. Each of these network forms is discussed below.

Unilateral monopoly In the network illustrated in Figure 27–11, actor *A* is a source of valuable resources for actors B_1, B_2, and B_3. Actors B_1, B_2, and B_3 provide rewards for *A*, but, since *A* has multiple sources for rewards and the *B*s have only *A* as a source for their rewards, the situation is a unilateral monopoly.

Such a structure often typifies interpersonal as well as intercorporate units. For example, *A* could be a female date for three different men, B_1, B_2, and B_3.

---

[44]Emerson usually specified direction in his graphs, but we will keep it simple.

FIGURE 27–11

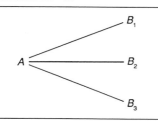

Or A could be a corporation that is the sole supplier of raw resources for three other manufacturing corporations, B_1, B_2, and B_3. Or A could be a governmental body and the Bs dependent agencies. Thus it is immediately evident that, by focusing on the structure of the exchange relationship, many of the micro-vs.-macro problems of exchange analysis, as well as of sociological theory in general, are reduced.

Another important feature of the unilateral monopoly is that, in terms of Emerson's definitions, it is imbalanced and thus its structure is subject to change. Previous propositions and corollaries listed in Table 27–1 provide an initial clue as to what might occur. Corollary 1.3 argues that A will use its power advantage and increase costs for each B. Corollary 1.1 indicates that, with each increment in costs for the Bs, their behaviors will vary and they will seek alternative rewards from A_2, A_3, ... , A_n. If another A can be found, then the structure of the network would change.

Emerson developed additional corollaries and theorems to account for the various ways this unilateral monopoly can become balanced. For instance, if no A_2, A_3, ... , A_n exist and the Bs cannot communicate with each other, the following corollary would apply (termed by Emerson *Exploitation Type I*):

> *Corollary 1.3.1*: The more an exchange relation between A and multiple Bs approximates a unilateral monopoly, the more additional resources each B will introduce into the exchange relation, with A's resource utilization remaining constant or decreasing.

Emerson saw this adaptation as short-lived, since the network will become even more unbalanced. Assuming that the Bs can survive as an entity without resources from A, then Theorem 8 applies (termed by Emerson *Exploitation Type II*):

> *Theorem 8*: The more an exchange relation between A and multiple Bs approximates a unilateral monopoly, the less valuable to Bs the resources provided by A across continuing transactions. (From Corollary 1.3, Theorem 4, and Corollary 4.1.)

This theorem thus predicts that balancing operation 1—a decrease in the value of the reward for those at a power disadvantage—will operate to balance a unilateral monopoly where no alternative sources of rewards exist and where Bs cannot effectively communicate.

FIGURE 27-12

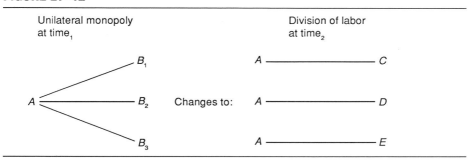

Other balancing operations are possible, if other conditions exist. If Bs can communicate, they might form a coalition (balancing operation 4) and require A to balance exchanges with a united coalition of Bs. If one B can provide a resource not possessed by the other Bs, then a division of labor among Bs (operations 3 and 4) would emerge. Or if another source of resources, A_2, can be found (operation 2), then the power advantage of A_1 is decreased. Each of these possible changes will occur under varying conditions, but Corollary 1.3.1 and Theorem 8 provide a reason for the initiation of changes—a reason derived from basic principles of operant psychology.

Division of labor   The emergence of a division of labor is one of many ways to balance exchange relations in a unilateral monopoly. If each of the Bs can provide different resources for A, then they are likely to use these in the exchange with A and to specialize in providing A with these resources. This decreases the power of A and establishes a new type of network. For example, in Figure 27-12, the unilateral monopoly at the left is transformed to the division of labor form at the right, with B_1 becoming a new type of actor, C, with its own resources; with B_2 also specializing and becoming a new actor, D; and with B_3 doing the same and becoming actor E.

Emerson developed an additional theorem to describe this kind of change, in which each B has its unique resources:

> *Theorem 9*: The more resources are distributed *non*uniformly across Bs in a unilateral monopoly with A, the more likely is each B to specialize and establish a separate exchange relation with A. (From theorems not discussed here and Corollaries 1.1 and 1.3.1.)

Several points should be emphasized. First, the units in this transformation can be individual or collective actors. Second, the change in the structure or form of the network is described in terms of a theorem systematically derived from operant principles, corollaries, and other theorems. The theorem can thus apply to a wide variety of micro and macro contexts. For example, the theorem could apply to workers in an office who specialize and provide A with resources not available from others. It could also apply to a division in a corporation

FIGURE 27-13

FIGURE 27-14

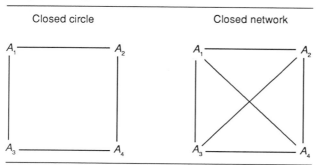

Closed circle

Closed network

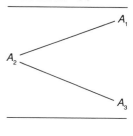

that seeks to balance its relations with the central authority by reorganizing itself in ways that distinguish it, and the services it can provide, from other divisions. Or it could apply to relations between a colonial power (A) and its colonized nations (B_1, B_2, B_3), which specialize (become C, D, and E) in their predominant economic activities in order to establish a less dependent relationship with A.

Social circles Emerson emphasized that some exchanges are intercategory and others intracategory. An intercategory exchange is one in which one type of resource is exchanged for another type—money for goods, advice for esteem, tobacco for steel knives, and so on. The networks discussed thus far have involved *inter*category exchanges between actors with different resources (A, B, C, D, E). An *intra*category is one in which the same resources are being exchanged—affection for affection, advice for advice, goods for goods, and so on. As indicated earlier, such exchanges are symbolized in Emerson's graph approach by using the same letter—A_1, A_2, A_3, and so forth—to represent actors with similar resources. Emerson then developed another theorem to describe what will occur in these intracategory exchanges.

Theorem 10: The more an exchange approximates an intracategory exchange, the more likely are exchange relations to become closed. (From Theorem 5 and Corollaries 1.3 and 1.1.)

Emerson defines "closed" either as a circle of relations (diagramed on the left in Figure 27-13) or as a balanced network in which all actors exchange with one another (diagramed on the right in Figure 27-13). Emerson offered the example of tennis networks to illustrate this balancing process. If two tennis players of equal ability, A_1 and A_2, play together regularly, this is a balanced intracategory exchange—tennis for tennis. However, if A_3 enters and plays with A_2, then A_2 now enjoys a power advantage, as is diagramed in Figure 27-14.

This is a unilateral monopoly, but, unlike those discussed earlier, it is an intracategory monopoly. A_1 and A_3 are dependent upon A_2 for tennis. This relation is unbalanced and sets into motion processes of balance. A_4 may be

FIGURE 27–15

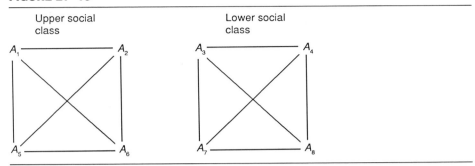

found, creating either the circle or balanced network diagramed in Figure 27–13. Once this kind of closed and balanced network is achieved, it resists entry by others, A_5, A_6, A_7, . . . , A_n, because, as each additional actor enters, the network becomes unbalanced. Such a network, of course, is not confined to individuals; it can apply to nations forming a military alliance or common market, to cartels of corporations, and to other collective units.

Stratified networks   The discussion on how intracategory exchanges often achieve balance through closure can help us understand processes of stratification. If, for example, tennis players A_1, A_2, A_3, and A_4 are unequal in ability, with A_1 and A_2 having more ability than A_3 and A_4, an initial circle may form among A_1, A_2, A_3, and A_4; but, over time, A_1 and A_2 will find more gratification in playing each other, and A_3 and A_4 may have to incur too many costs in initiating invitations to A_1 and A_2. For an A_1 and A_3 tennis match is unbalanced; A_3 will have to provide additional resources—the tennis balls, praise, esteem, self-deprecation. The result will be for two classes to develop:

Upper social class A_1———A_2

Lower social class A_3———A_4

Moreover, A_1 and A_2 may enter into new exchanges with A_5 and A_6 at their ability level, forming a new social circle or network. Similarly, A_3 and A_4 may form new tennis relations with A_7 and A_8, creating social circles and networks with players at their ability level. The result is for stratification to reveal the pattern in Figure 27–15.

Emerson's discussion of stratification processes was tentative, but he developed a theorem to describe these stratifying tendencies:

> *Theorem II*: The more resources are equally valued and the more resources are unequally distributed across a number of actors, the more likely is the network to stratify in terms of resource magnitudes and the more likely are actors with a given level of resources to form closed exchange networks. (From Theorem 5 and Propositions 1 and 4.)

Again, this theorem can apply to corporate units as well as to individuals. Nations become stratified and form social circles, as is the case with the distinctions between the developed and underdeveloped nations and the alliances among countries within these two classes. Or it can apply to traditional sociological definitions of class, since closed networks tend to form among members within, rather than across, social classes.

The dynamics of centrality   As noted earlier, an important concept in network analysis is *centrality*. Indeed, it is considered one of the most critical properties of a network. Centrality is determined by a variety of measures, as mentioned earlier.[45] Most network analyses simply describe centrality, but Karen Cook has tried to use Emerson's theoretical ideas to explain its dynamics.[46] Her argument will be simplified and also rephrased a bit, but the essential logic is the same. We can begin by creating a theorem that summarizes her hypothesis.[47]

Theorem 12: Over time, the distribution of power in complex intracategory networks decentralizes around those actors (points) who possess the highest relative degree of direct access to resources. (From Corollary 1.3.1 and Theorems 4 and 5.)

Using Figure 27-16 as an illustration of this theorem, Cook would predict that power will become increasingly concentrated in A_3, A_4, and A_1. A_2 will become less powerful and, in a more sociological sense, less central. In fact, the entire network will, over time, collapse around A_3, A_4, and A_1. Why should this be so? The basic argument that follows from the dynamics of power and dependence is this: A_3, A_4, and A_1 reveal regular equivalence in that they each have a unilateral monopoly with at least three other As (for example, A_3 has a monopoly exchange relation with A_{11}, A_{12}, and A_{13}). Thus, to get resources, A_2 must bargain with structurally equivalent As—that is, A_1, A_3, and A_4—who can extract more resources from those As over which they possess a monopoly. And so, A_1, A_3, and A_4 will increase exchanges with their monopolized partners and decrease exchanges with A_2, who, in all likelihood, will become like the monopolized As. Hence the network will decentralize around A_1, A_3, and A_4, who possess the highest relative access to resources (by virtue of their respective unilateral monopolies). However, if A_2 provides a unique and highly valued resource—thereby making the exchange intercategory and changing the designation of A_2 to B—then the network may stay centralized around B. For example, if B is a king in a feudal system and provides the organizational know-how to coordinate military defense for all of the As in a hostile environment,

---

[45]Linton C. Freeman, "Centrality in Social Networks: Conceptual Clarification."

[46]Karen S. Cook et al., "The Distribution of Power in Exchange Networks."

[47]This is our wording of a more complex argument presented by Cook, Emerson, and others. See ibid.

FIGURE 27–16

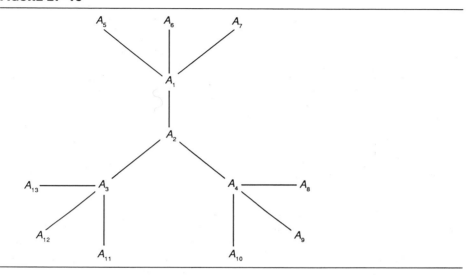

then the structure portrayed in Figure 27–16 would remain in tact. Thus Theorem 12 is most relevant to *intra*category exchanges.[48]

The critical point here is that Cook and her associates are trying to use the basic principles developed by Emerson to address more complex network systems. In so doing, they can explain the process of centrality, and perhaps other network properties, in terms of theoretical deductions. Up to this point, network analysis has typically described centrality using various measures, whereas the Emerson-Cook approach allows these network properties to be explained in terms of an abstract theoretical proposition.

CONCLUSION

There is a curious division between mainstream sociological theory and network analysis. Although the heavily quantitative portions of network sociology are difficult for many social theorists to understand, this fact alone cannot explain the division. For network sociology is doing the very thing that early sociologists and anthropologists saw as crucial—the mapping of the relations that create social structures; and, as we have seen, network analysis can be phrased in nonquantitative terms.

The real reason for the split of network sociology from mainstream theory is that *neither* is very theoretical. Network analysis is a bag of computer algorithms and mathematical formulations whose relevance to the real world or

---

[48]Recall Figure 27–10 on page 556. The prediction here would also be that this network would be decentralized around A_1, C_1, and D_1, *unless* the resources provided by B were highly valued and could not be gotten elsewhere—from a B_1, B_2, B_3, etc.

to traditional theoretical questions in sociology is, at best, tenuous; much mainstream "theory" is now so philosophical and antiscientific that any approach that is too "scientific looking" will be rejected. What is required, then, is for network analysis and much "theory" in sociology to become theoretical in the sense of developing testable laws and models of human organization.

On the network side of this proscription, several tacks are possible. One is to develop formal laws in terms of network concepts—that is, laws on the dynamics of centrality, equivalence, brokerage, bridging, density, clique formation/dissipation, and the like. Another is to translate, much as the Emerson-Cook project has begun to do, traditional sociological concerns—power and stratification, for instance—into network concepts. For either tack it is essential that the tendency to develop "formalisms" for their own sake or their ascetic appeal needs to be tempered by a willingness to communicate the theoretical message in less arcane terms.

Is this likely? It is not clear that it is. Network sociology is part of an interdisciplinary movement that constitutes a world in itself—a big clique, in network terms. Its members talk to one another more than to anyone else, and they appear to be content with this situation. For their part, mainstream theorists have retreated into a variety of camps and perspectives—some soft, others hard and formal—whose members also talk to one another and ignore "outsiders." Thus it is questionable whether mainstream theory can reveal even a central current, or even trickle, in the future, and it is also questionable that network analysis will be widely influential in theory circles in the near future.

Such a scenario would, of course, be unfortunate, since network analysis offers a set of useful tools for examining a critical property of the social universe: social structure. More than any of the "structural perspectives" examined in this section of chapters, network analysis offers the most potential for developing scientific sociology.

CHAPTER 28

◆

Macrostructural Theory

Peter M. Blau

In Chapter 16, I examined Peter Blau's exchange structuralism. In that review of Blau's exchange theory, the tension among processes of differentiation, integration, and conflict figured prominently. In recent years Blau has examined these basic processes with an alternative theory, which he labels *macrostructural.*[1] Although his earlier theory can still prove useful to those working within the exchange-theoretic perspective, Blau's macrostructuralism is clearly intended to replace and supplant his exchange theory. In this recent effort Blau attempts to develop a more rigorous theory that follows from a more clearly articulated view of theory construction than was evident in his earlier exchange theory.[2] In my analysis to follow, I will concentrate, respectively, on the strategy, substance, and theoretical format of Blau's new theory.

BLAU'S THEORETICAL STRATEGY

Blau's new work is explicitly deductive and advocates a "covering law" vision of theory (see Chapter 1). In Blau's view a scientific theory is a "hierarchical system of propositions of increasing levels of generality. All lower order propositions follow in strict logic from higher order ones alone."[3] In developing

---

[1]See Peter M. Blau, *Inequality and Heterogeneity: A Primitive Theory of Social Structure* (New York: Free Press, 1977); "A Macrosociological Theory of Social Structure," *American Journal of Sociology* 83 (July 1977), pp. 26–54; see also "Contrasting Theoretical Perspectives," in *The Micro-Macro Link,* eds. J. C. Alexander, B. Gisen, R. Münch, and N. J. Smelser (Berkeley: University of California Press, 1987); and "Structures of Social Positions and Structures of Social Relations," in J. H. Turner, ed., *Theory Building in Sociology* (Newbury Park, CA: Sage, 1988).

[2]Indeed, he views his work on exchange as a prolegomenon—a kind of pretheory, which, by its nature, would be unsystematic. Indeed, in Chapter 16 I had to extract the theoretical principles.

[3]Peter M. Blau, "Elements of Sociological Theorizing," *Humboldt Journal of Social Relations* 7 (Fall-Winter, 1979–80), p. 105. This article contains Blau's statement of his theoretical strategy that is outlined in this section.

such a theory, a number of steps are critical, and their successful implementation is what typifies good theory.

First, theory must begin with a conception of the basic subject matter—that is, what is to be explained? In deductive theory the phenomenon to be explained by a theory is termed the *explicandum*. For Blau the explicandum is the "pattern of relations between people in society, particularly those of their social relations that integrate society's diverse groups and strata into a distinct coherent social structure."[4]

In addressing this range of phenomena, Blau offers a narrow definition of macrostructure. Discussions of social structure, Blau notes, typically recognize "that there are differences in social positions, that there are social relations among these positions, and that people's positions and corresponding roles influence their social relations."[5] But most theoretical efforts extend the conceptual inventory, adding such notions as psychological needs, cultural values and beliefs, norms, technological forces, class antagonisms, dialectics, and functions. In contrast to these tendencies, Blau prefers to postulate a narrow view of social structure as consisting of the distributions of people among different social positions and their associations.

> To speak of social structure is to speak of differentiation among people. For social structures, as conceptualized, are rooted in the social distinctions people make in their role relations and social associations.[6]

Yet, when one examines an entire society, a community, or any macrostructure, analysis becomes more complicated. The number of people and positions to be analyzed is great. Moreover, people occupy many different positions simultaneously—religious, familial, neighborhood, work, recreational, political, educational, and the like. Thus, for each type of position, there is a separate distribution of the population, making a macrostructure a "multidimensional space of social positions among which people are distributed and which affect their social relations."[7] But how are all these relations among large numbers of people occupying simultaneously many different positions to be analyzed? In Blau's portrayal, network analysis would seek to analyze each and every relation among all people in all positions; but, when the numbers of people, positions, and relations become large, macrostructural analysis must be concerned with the general "patterns of social relations among different social positions occupied by many persons, not with the networks of all relations between individuals."[8] This is done by defining social positions in terms of common attributes of people—age, sex, race, occupation, religion—and then examining their overall rates of association. Note that not every person or association is examined—only general categories of people and overall rates of

---

[4]Ibid., p. 112.
[5]Blau, "A Macrosociological Theory of Social Structure," p. 27.
[6]Ibid., p. 28.
[7]Ibid.
[8]Ibid.

association among people in these categories. In this way, macrostructural theory is distinguished from the microstructural approaches, sociometry, and network analysis. As Blau recently emphasized:[9]

> Whereas network analysis dissects the matrix of possible links between all actors in a group, this is neither feasible nor possible in the study of macrostructures. . . . Macroanalysis is not intended to deal with the multitude of interpersonal links in a large population. . . . It focuses on a population's distribution among differentiated positions in various dimensions and the rates of association and mobility between persons in different positions. These rates indicate how well or poorly the various groups and classes and other subunits are integrated in the larger community or society.

Thus Blau's "concept of the basic subject matter" to be examined is similar to many previous theoretical approaches in that it is concerned with how different groups and social strata come to form integrated or conflictual social wholes. But, in contrast to most efforts on this topic, I see Blau as taking a very narrow view of social structure that excludes many variables, such as the cultural and psychological, that are typically included in traditional orientations. Yet this narrow definition of structure is macro in focus, for, unlike microanalysis of groups in which each and every position and relation is plotted, concern is with the general attributes that distinguish types of people and that influence their overall rates of association.

A second critical element in building theory is "familiarity with relevant empirical knowledge." The often rigid line between deduction and induction is artificial, Blau argues, since deduction rarely occurs in an empirical vacuum. Indeed, the very questions to be answered by a deductive theory are typically the result of interesting empirical findings. Thus, to know what it is that a theory should explain requires familiarity with existing empirical generalizations.[10]

A third element in a theory is the "creative insight" that is typically embodied in a "central theoretical term" or "operator." An operator "supplies the organizing principle for the major theoretical propositions."[11] In Blau's theory the central concept is the *intersection of parameters*. A parameter in Blau's theory is simply the criterion by which a position is distinguished—such as age, sex, race, income, wealth, education, or any basis that people use to distinguish one another and that consequently affects their rates of association. The intersection of parameters denotes the fact that parameters are variously correlated with one another. When parameters intersect, occupancy in one position is not likely to be correlated with, or to predict, occupancy in another.

---

[9]Blau, "Structures of Social Positions and Structures of Social Relations," p. 51. Actually, as network analysis begins to deal with larger populations, they do as Blau advocates. See, for example, Harrison C. White, Scott A. Boorman, and Ronald L. Breiger, "Social Structure from Multiple Networks I," *American Journal of Sociology* 81 (January 1976), pp. 730–80.

[10]Blau's empirical work on organizations and socioeconomic status preceded his theoretical efforts. He has thus followed his own advice.

[11]Blau, "Elements of Sociological Theorizing," p. 107.

For example, race in some societies may not be associated with income; nor is prestige associated with education. Conversely, when parameters consolidate, they reveal high correlations—race and income as well as education and prestige are highly correlated in that being of a particular racial group predicts income or level of education, which in turn assigns a person a corresponding level of prestige. Intersecting parameters reduce divisions in a society because, under this condition, people associate with a diversity of others in at least some of their positions. In contrast, *consolidation of parameters* tends to create divisions in a society since positions cluster together in ways that reduce associations with others outside a particular circle of associations. For instance, the wealthy in most societies live in certain neighborhoods, attend private schools, engage in recreation only with one another, and hold other consolidated positions. The result is that they are separated from others in society who are not wealthy. Thus, for Blau, "the degree to which parameters intersect, or alternatively consolidate differences in social positions through their strong correlations, reflects the most important structural conditions in a society, which have crucial consequences for conflict and for social integration."[12]

A fourth central element of theory building is conceptual clarification of key terms, especially of the operator. For Blau such clarification involves distinguishing patterns of differentiation in terms of two basic types of parameters. One basic type is a "nominal parameter," which divides people into different groups that are not rank ordered. The other type is the "graduated parameter," which differentiates people in terms of hierarchical status rankings. Thus all macrostructures reveal varying degrees of differentiation in terms of nominal and graduated parameters. When people are differentiated into many groups in terms of numerous nominal parameters, then a high degree of "heterogeneity" exists. That is, when two randomly chosen people are unlikely to belong to the same group, heterogeneity is high. When people are differentiated into widely different status positions on the basis of graduated parameters, then "inequality" is high. That is, when the differences in relative standing among people distinguished by a graduated parameter are great, inequality is high. Thus differentiation varies along two axes, heterogeneity and inequality.[13] The specific dynamics of such differentiation are to be understood with additional concepts, which I will discuss shortly when the theory itself is examined, but the notion of intersecting parameters is still the key operator, since it increases heterogeneity and decreases inequality.

A fifth important element in theory construction is the development of a deductive system—that is, the incorporation of concepts into propositions and then the hierarchical ordering of them. Although Blau uses the vocabulary of axiomatic theory, his system of propositions is what I termed *formal theorizing* in Chapter 1, because it does not evidence the rigor of true axiomatic theory (for, as I emphasized, very little theory in the social sciences can meet the

---

[12]Blau, "A Macrosociological Theory of Social Structure," p. 32.
[13]Blau, *Inequality and Heterogeneity*, p. 281.

FIGURE 28-1 Blau's Causal Imagery

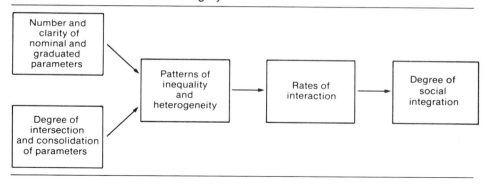

criteria of axiomatic theory). In Blau's view the ordering of propositions involves (*a*) the articulation of a number of simple assumptions as axioms and (*b*) the development of theorems that apply these assumptions to generic types of social relations. For example, Blau's most basic assumption is that social associations are more prevalent among persons in proximate positions than among those in distant social positions. Thus people at similar ranks or in the same group are more likely to interact than those in divergent ranks or different groups. With this and other assumptions to be discussed shortly and with the central operator and additional concepts, Blau can then generate a number of theorems that explain why events in the empirical world should occur. The essence of Blau's theory thus involves the articulation of some simple axioms or assumptions and the incorporation of these into theorems that state relations among carefully defined concepts that have empirical referents.

The final element of theory construction involves "confronting the theory with empirical data." There are two types of data to be used in such confrontations. One is the existing data, which can be interpreted and understood in terms of the theory. Another is the rigorous deduction of hypotheses from the theorems, which serve to organize the collection of new data.[14]

In sum, then, the creative act of theory construction involves (1) a conception of what is to be explained, (2) familiarity with relevant knowledge, (3) a central term or operator, (4) conceptual elaboration and clarification, (5) a deductive system of propositions, and (6) confrontation of the theory with empirical data. This vision has clearly guided the substantive formulation of Blau's theory of macrostructure.

One way to visualize Blau's general theoretical argument is presented in Figure 28-1. The image of social organization revolves around the number and clarity of parameters and their degree of intersection or consolidation, which, together, produce patterns of inequality and heterogeneity. Rates of interaction among people distributed in positions defined by parameters will, in turn, de-

---

[14]Blau has attempted to do so. For example, see his and Joseph E. Schwartz's *Crosscutting Social Circles* (Orlando, FL: Academic Press, 1984).

termine the degree of integration of a macrostructure or, conversely, the degree of isolation, tension, or conflict in a macrostructure.

BLAU'S THEORY OF MACROSTRUCTURE

Blau argues that in some theories the primitive assumptions represent brilliant insights and mark the real contribution of the theory. In contrast, other theories state obvious and simplistic assumptions that are viewed as givens and are taken for granted. In such theories "the contribution of the theory rests on the new knowledge derived from the systematic analysis of the implications of combinations of simple assumptions and definitions."[15] Blau views his theory as belonging to this second type. Simple assumptions, or axioms, serve as premises for theorems, with the result that the power of the theory to explain events resides in its theorems and in subsequent deductions from these theorems to specific empirical cases.

One problem with axiomatic theory in general and with Blau's execution of this strategy is the proliferation of propositions. Blau articulates 21 assumptions, 34 major theorems, and more than 150 subtheorems (corollaries) in many different "sets of theorems." In its present form, then, I find the theory unmanageable. Much of this problem resides, I believe, in the fact that the major theorems are stated at different levels of abstraction and, as a result, overlap a great deal. Indeed, I find some to be little more than empirical generalizations that illustrate a relationship stated in a more abstract theorem. Another problem is Blau's insistence that axioms must represent simple assumptions and that 21 such assumptions are necessary to generate all of the theorems.[16] Moreover, Blau argues, each axiom must represent a primitive assumption that is not derivable from another assumption. Thus any theory that begins with more than 20 assumptions can generate 34 theorems rather easily and many more corollaries besides. But such a theory branches out in so many directions that it loses the elegance and coherence of powerful theories.

Most of these problems can be overcome, I think, if Blau's narrow and restrictive view of axiomatic theory is relaxed. Thus, in presenting the theory, I will remove some of Blau's restrictions. In particular, I will shorten the list of assumptions and include only the most critical ones. Moreover, I will eliminate the redundancy among the theorems. In Table 28–1, I have listed Blau's "axioms" after this necessary surgery.[17] From these axiomatic premises, there are four basic theorems in Blau's theory. I have listed these in Table 28–2. All other theorems in Blau's scheme can, I believe, be derived from these and, hence, constitute corollaries of these four major theorems.

In Table 28–2, Theorem I asserts that, if people occupy different positions in many groups and ranks and that if these positions are not highly correlated,

---

[15]Ibid., p. 245.

[16]See Blau, "Elements of Sociological Theorizing," p. 10.

[17]See my earlier effort along these lines: Jonathan H. Turner, "A Theory of Social Structure: An Assessment of Blau's Strategy," *Contemporary Sociology* 7 (November 1978), pp. 698–705.

TABLE 28–1 A Revised List of Blau's "Axioms"

I. People in a society associate with others not in their group or status rank.

II. Social associations among people in proximate positions are more prevalent than those in distant positions.

    A. In-group associations are more prevalent than out-group associations.
    B. The prevalence of associations declines with increasing status distance.

III. Established relations are resistant to disruption.

IV. Strangers who have common associates become, over time, associates.

V. Associates in other groups or strata often facilitate mobility to these groups or strata.

VI. The distribution of status positions is skewed so that there are always fewer high-ranking positions than middle- and low-ranking positions.

VII. Social associations depend upon opportunities for social contact.

VIII. The influences of various parameters on social associations are partly additive and not always contingent on each other.

TABLE 28–2 A Revised List of Blau's Basic Theorems

I. The more multiple parameters intersect and remain unconsolidated, the greater is the degree of heterogeneity and the less is the level of inequality and, hence, the more likely are social associations among people (from Axioms I, II, and VII).

II. The more multiple parameters intersect or consolidate, the more likely is their effect on social associations to exceed the additive effects of the parameters alone (from Axioms I, III, IV, and VIII).

III. The more parameters intersect, the greater is the rate of mobility and the more likely are social associations (from Axioms I, II, V, and VII).

IV. The more the size differences between two groups or status ranks are distinguished by a parameter, the more likely are social associations among members in the smaller group or rank with members of the larger group to exceed those of members in the larger group with those in the smaller (from Axioms I, II, VI, and VII).

then diverse and varied associations are likely to ensue. Conversely, if occupancy in one group or rank is highly correlated with positions in other groups or ranks, then opportunities for associations with other people are more limited, and associations will be with the same persons. The effects of such intersection and consolidation become amplified, as is stated in Theorem II. Consolidation will become amplified because, as Axioms III and IV stress, people usually come to know and like friends of their friends; once such relations are established, they resist disruption. Intersection tends to become amplified for the same reasons, but this process continually expands. This is because, in contrast to consolidation, which creates closed circles of associations, intersection constantly creates new opportunities for additional social association. Theorem III states that intersection creates opportunities for mobility, since, as Axiom V posits, people tend to help one another be mobile. And, once people are mobile, their opportunities for social association increase. But additionally, as Axiom III argues, they tend to maintain their old social relations that they had before mobility. Theorem IV states a mathematical truism, or the law of proportion. If there are reciprocal relations between groups or ranks of different size, then these reciprocal relations must constitute a larger proportion of the total relations in the smaller group. That is, the smaller group has fewer people who can potentially generate relations, whereas the larger group has more people who can have a greater number of total relations; therefore the same number of reciprocal relations must represent a greater portion of the smaller than the larger group's total relations.

From these four theorems, then, I believe that most other theorems in Blau's many "theorem sets" can be derived.[18] Probably the most interesting of the many examples that I could offer concern those that can be derived from Theorem I on intersecting parameters and Theorem IV on proportions. For example, Blau's Theorem 26 postulates that "the further society's differentiation penetrates into successive subunits of its structural components, the more it promotes the integration of groups by increasing intergroup relations."[19] This theorem is directly derivable from the notion of intersection versus consolidation of parameters (Theorem I above). If a pattern of differentiation—for example, inequality of wealth—repeats itself in every community in a society (this is successive penetration), then the wealthy and not so wealthy people are more likely to interact than if all wealthy people live in one community and all poor people in another (this would be consolidation of wealth with community). Since societies (or any social system) vary in terms of degrees of such successive penetration, their integration will also vary.

Let me take another example. In Table 28-2, Blau's Theorem IV states that members of smaller groups or strata have more relations as a proportion of their total with larger groups or strata than do members of larger groups or strata with those in smaller groups.[20] Thus, since the distribution of people in status ranks will reveal a pyramid form, at least from the middle ranks upward,[21] high-ranking people are more likely to have relations with lower-ranking persons than are lower-ranking with higher-ranking persons. Such a simple artifact of mathematical proportions can help explain, perhaps, the hostility of lower ranks toward higher ranks, with whom they have little contact. Race and ethnic relations can similarly be analyzed in this way, since majority members, who are the larger group, are less likely to have relations with the minority than is the minority with the majority. As a result, stereotypes and prejudice can be more readily maintained when there is little contact by the majority, especially when the processes described by this proportion principle are exacerbated by the processes described in the intersection of parameters principle (since minority status is often highly correlated with other positions that the majority does not occupy).

Blau's basic theorems suggest many ways to analyze concrete empirical cases, as Blau seeks to do throughout his work with various theorems and as

---

[18]I suspect that Blau would disagree with this.

[19]See Blau, *Inequality and Heterogeneity*, p. 183.

[20]See ibid., pp. 42 and 73.

[21]Sometimes the lowest ranks are small, thereby creating a diamond-shaped distribution.

he has done in his research.[22] As he emphasizes, these efforts all address the basic issue of social integration. For he assumes that social contact and association among differentiated positions promote cooperation and accommodation. Integration is thus viewed as the result of high rates of interaction among people in different groups and strata. But differentiation poses barriers to interaction and association. As can be seen by the theorems listed above, or by a review of Blau's numerous propositions, the theorems state structural conditions that promote or retard social associations and, hence, integration. But, unlike many other analyses of this basic topic, Blau's propositions are purely structural and pertain to the number of positions, to their degree of intersection or consolidation or their degree of inequality and heterogeneity, and to the distribution of people in these positions.

The theory is, of course, much more complex than my presentation here, but the general profile and intent of his approach are clear. Such a structural approach, cast into a formal format, offers considerable promise, but it is not without problems. Let me close, therefore, by offering a brief assessment of this theoretical approach.

MACROSTRUCTURALISM: AN ASSESSMENT

Blau has isolated what he defines as two basic properties of the social universe—heterogeneity and inequality—as they are determined by the consolidation or intersection of parameters. He has also developed some abstract principles about the dynamics inherent in these. This is, I believe, the best way to develop theory; and, as I have said elsewhere, I consider this effort one of the most significant of this century.[23] Having said this, however, I will mention what I see as its limitations.

A major criticism of Blau's theory is that it excludes so much. Interaction is conceptualized as a rate rather than as a process; roles are relegated to their effects on parameters; values, beliefs, norms, and other idea systems are eliminated or viewed as relevant only to the extent that they help people define parameters; and the social-psychological properties of the universe, such as self-conceptions and other cognitions of humans, are simply eliminated. Moreover, the theory cannot tell us why a given set of parameters exists in the first place; instead, we are told only about their consequences. I think, therefore, that Blau's theory achieves parsimony by eliminating many critical dynamics of the social universe, including those responsible for variations in his own theoretical variable (nominal and graduated parameters).

Thus, in the end, Blau offers an interesting set of abstract propositions on "rates of interaction" among populations of individuals. This is not, however, a trivial issue, because I believe that many of the dynamics of social structure are affected by such rates of interaction. But, at the same time, much is ex-

---

[22]Blau and Schwartz, *Crosscutting Social Circles.*
[23]Turner, "A Theory of Social Structure," p. 705.

cluded. And so I view Blau's theory as an important contribution to only one dimension of macrostructure, rates of interaction as these are affected by two very critical properties of social structure, inequality and heterogeneity. But this is not all there is to human organization, even when viewed in macrostructural terms; approaches that claim otherwise will, for all their insight, be too limiting.

CHAPTER 29

◆

Approaching Theoretical Synthesis

BELIEVING IN SCIENCE

In this last group of chapters (Part VI), I offer my assessment on the state of sociological theorizing. But, rather than merely commenting on what I like, and do not like, I will attempt to synthesize most of the approaches presented in earlier chapters. When I say "synthesize," however, I do not claim to have a complete vision of how the social universe works; nor do I think it possible for all of the general theories of contemporary sociology to fit together in a nice neat package. My goal is much more modest and involves several tasks: (1) to suggest what properties of the social universe should be the focus of our theories, (2) to indicate what theoretical concepts and propositions in current theory offer the best leads for uncovering the important dynamics of these properties, and (3) to offer a provisional scheme for how these theoretical leads may best fit together into a coherent corpus of theoretical models and principles.

Over the years, as I have written various editions of this book and its companion on the emergence of sociological theory, I have become convinced that sociology can be like any other natural science. This conviction flies in the face of the philosophical skepticism that has cast a long shadow over current social theory. Whether we call this skepticism "postmodernism" or some other such thing, it denies sociology its place at the table of science. It asserts that the social world is "different" from the natural universe; that this world cannot be studied in the same way as the nonhuman universe; that this world constantly changes its fundamental nature and, hence, cannot yield universal laws that transcend time; that this world is only "knowable" through the filter of the rose-colored glasses of subjectivity, culture, ideology, and other distorting prisms; and so it goes. To these skeptics I say: if you will only look, the evidence is before your eyes. For, when humans interact and organize, certain processes always seem to occur, regardless of time, place, and context. The theories discussed in these pages denote these processes and seek to understand them. Thus I am very optimistic that sociology can, in principle, be a natural science.

Whether we become one depends upon our effort to realize the goals of science and, at the same time, to give philosophy back to the philosophers.[1]

Even if what I say here and what I propose to do in the next three chapters leaves you skeptical, if not offended, you might find it interesting—if only as a new "text" for analysis and further philosophical "discourse."

ON THE SHOULDERS OF GIANTS

Now that I have thrown down the gauntlet, let me indicate how I will proceed. I begin by citing Robert K. Merton on the prospects for "grand" theoretical synthesis in sociology. What is needed, Merton argues, is for sociology to avoid the temptation of creating grand theories before it has done the necessary preliminary work. What is needed in sociology are efforts to lay a long research tradition revolving around modest middle-range theories (see Chapter 4), for sociology "has not yet found its Kepler—to say nothing of its Newton, Laplace, Gibbs, Maxwell or Planck."[2] Despite my enormous respect for Merton and his work, I cannot imagine a more incorrect and debilitating statement. Who are Spencer, Marx, Durkheim, Mead, Pareto, Weber, Simmel, and others, if they are not our equivalents of these giants in physics? We have seen our Newton, Laplace, Gibbs, Maxwell, or Planck, but we have not fully recognized them for what they are. We implicitly recognize that they uncovered many of the basic properties of our universe—interaction and social organization—because we constantly read and reread them. Why should we do so? My answer is this: most of the basic properties of our universe were seen by these giants, and each of them gave us a solid lead on how to understand the dynamics of these properties. Indeed, most of contemporary theory has tried to follow up on these leads and to amplify, extend, and fine-tune the ideas of the early masters. What is needed now, then, is not a faint heart and retreat into skepticism or the comfort of narrow middle-range theories but, instead, efforts to synthesize these follow-up efforts of contemporary theorists. We need to keep building upon the early masters, whom I have summarized in the opening chapter of each section in this book—and elsewhere on many occasions.[3] Each of the more contemporary

---

[1]There are organizational reasons for the problems of sociological theorists to believe in science. See Stephen P. Turner and Jonathan H. Turner, *The Impossible Science: An Institutional Analysis of American Sociology* (Newbury Park, CA: Sage, 1990). For a more conceptual statement, see Stephan Fuchs and Jonathan H. Turner, "What Makes A Science Mature?: Organizational Control in Scientific Production," *Sociological Theory* 4 (Fall 1986), pp. 143–50.

[2]Robert K. Merton, *Social Theory and Social Structure* (New York: Free Press, 1968), p. 47.

[3]For example, see Jonathan H. Turner, "Marx and Simmel Revisited: Re-assessing the Foundation of Conflict Theory," *Social Forces* 53 (June 1975), pp. 619–27; "Sociology as a Theory Building Enterprise: Detours from the First Masters," *Pacific Sociological Review* 22 (October 1979), pp. 427–56; "Returning to Social Physics: Illustrations from George Herbert Mead," *Current Perspectives in Social Theory* 2 (1981), pp. 187–208; "Émile Durkheim's Theory of Integration in Differentiated Social Systems," *Pacific Sociological Review* 24 (4, 1981), pp. 379–92; "The Forgotten Theoretical Giant: Herbert Spencer's Models and Principles," *Revue Europeenne Des Sciences Sociales* XIX (59, 1981), pp. 79–98; "A Note on G. H. Mead's Behavioral Theory of Social Structure," *Journal for the Theory of Social Behaviour* 12 (July 1982), pp. 213–22; and "Durkheim's and Spencer's Principles of Social Organization: A Theoretical Note," *Sociological Perspectives* 27 (January 1984), pp. 21–32.

theorists discussed in the preceding pages has "stood on the shoulders of giants," and now it is time to assess and synthesize what these contemporaries have said.

DOMAINS OF THEORIZING

But how is this synthesis to be done? The chapter titles of Part VI give a clue: "micro theorizing," "macro theorizing," and "meso theorizing." Let me review these labels, since they communicate my sense of the basic domains of sociological theory. These domains denote how we should divide up theoretical efforts, for it is unlikely that sociology will produce a "unified" theory of the social universe, at least not for some time. Rather, we will have models and sets of principles that help us understand certain basic facets of social reality. It may, of course, be possible to unify these models and principles, but, since this task has not been completed in even the most advanced hard sciences, we would be better to concentrate on developing sound micro, macro, and meso theories.

Micro Theorizing

Part of our universe is face-to-face interaction among individuals. This portion of social reality represents the micro level of reality, and, as we have seen, it is the topic of many theoretical approaches—from Max Weber's and Talcott Parsons' conceptions of social action through Niklas Luhmann's discussion of "interaction systems" and various theorists' examination of elementary exchange processes to a variety of interactionists' views on the fundamentals of face-to-face behavior. Micro reality is thus critical to our understanding of the social universe, although I question the contention of many that "this is all there is" and that everything else in the social world can be understood in terms of these micro processes (more on this issue shortly).

We must, therefore, develop models and principles about the dynamic properties of this level of social reality. As is evident from the preceding chapters, we have many creative theoretical ideas to choose from—beginning with those of George Herbert Mead, Alfred Schutz, and Émile Durkheim's later work and culminating in the micro theorizing of such scholars as Niklas Luhmann, Randall Collins, Ralph Turner, Erving Goffman, Jurgen Habermas, and others.

But how should we organize their ideas in order to develop a more coherent theory? My position is that we need to develop "sensitizing analytical schemes," which denote the general classes of phenomena about which we should theorize (see Chapter 1 and Figure 1-6). Such schemes inform us about the basic dimensions of micro reality that are to be the subject of theorizing. As I have argued elsewhere,[4] I see the micro social world as best divided into three basic

---

[4]For my best efforts to synthesize the micro universe, see Jonathan H. Turner, *A Theory of Social Interaction* (Stanford, CA: Stanford University Press, 1988); "A Behavioral Theory of Social Structure," *Journal for the Theory of Social Behavior* 18 (4, 1989), pp. 351–72; "A Theory of Microdynamics," *Advances in Group Processes* 6 (1989), pp. 1–34.

dimensions: (1) motivating, or those forces that "energize" individuals to interact with one another; (2) interacting, or those signaling and interpreting processes that "connect" people as they confront one another in face-to-face encounters; and (3) structuring, or those forces that organize interaction in space and over time. In my view, then, micro reality consists of the motivational, interactional, and structuring dimensions of face-to-face encounters among individuals. In Chapter 30 I will outline some ideas on how to use existing theoretical insights to explain this level of social reality.

Macro Theorizing

Another part of the social universe is the organization of larger populations of individuals. This is the macro level of reality; and, as with the micro universe, we have many conceptual leads to follow—beginning with Herbert Spencer, Karl Marx, Max Weber, the early Émile Durkheim, George Simmel, and others, through the modern functional theories of Talcott Parsons, Niklas Luhmann, and Amos Hawley, to those working on exchange, conflict, and structural theories. As with the micro world, we need a sensitizing scheme to guide us in partitioning this macro universe into basic dimensions about which we can theorize.

In my view[5] the macro universe can be divided into three basic dimensions: (1) assembling, or those processes that bring populations of individuals together in space and sustain their presence over time; (2) differentiating, or those forces that divide up populations into various kinds of social units and categories; and (3) integrating, or those procedures by which divided populations are laced together in space and time. Macro theory should thus develop models and abstract principles to explain the dynamics of, and relations among, the assembling, differentiating, and integrating dimensions of the universe. In Chapter 31 some suggestions as to how we might go about realizing this goal will be offered.

Meso Theorizing

One of the great issues in sociological theory is how to "link" the micro and macro realms.[6] For there is a presumption that a "gap" exists between micro face-to-face processes, on the one side, and macro population processes, on the other. Moreover, there is a strongly felt need among many theorists to close this "gap." Efforts to do so will be termed meso theorizing because they seek

---

[5]I have not fully developed a theory of macrodynamics, although I have presented some preliminary ideas. See Jonathan H. Turner, "Analytical Theorizing," in *Social Theory Today*, eds. A. Giddens and J. H. Turner (Stanford, CA: Stanford University Press, 1987), pp. 156–94; "Toward a Theory of Macrodynamics," in *Sociological Theories in Progress*, Vol. III, eds. J. Berger and M. Zelditch (Newbury Park, CA: Sage, 1990), pp. 180–220.

[6]See, for example, Jeffrey C. Alexander et al., eds., *The Micro-Macro Link* (Berkeley: University of California Press, 1987); and Norbert Wiley, "The Micro-Macro Problem in Social Theory," *Sociological Theory* 6 (Fall 1988), pp. 254–61.

to articulate how face-to-face processes can be understood in terms of the organizational features of populations of actors and, conversely, how the processes organizing larger populations can be analyzed in terms of face-to-face encounters. There have been a variety of strategies to deal with this problem, as will be outlined in Chapter 32.[7] There I will also outline my own views on the prospects for meso theorizing. For the present, however, let me mention some of my biases on this issue, because they can help frame my own "resolution" of this theoretical problem.

My first point is that there has been far too much agonizing over the micro/macro "gap" or "link." Physics has worked at the astro and subatomic levels without great worry; microbiologists seem to get along with other types of biologists; micro- and macroeconomists have learned to live with each other; and so on. It is only sociologists who hurl insults at one another, claiming that only face-to-face encounters are "real" and everything else is a reification. And, with the possible exception of the most mature science (physics) and, ironically, one of the most immature (sociology), a certain level of slippage or discontinuity between macro and micro levels of analysis is tolerated. Microanalysis yields one kind of understanding, macroanalysis another; sometimes they can be combined, and at other times they cannot. Sociologists should assume this more tolerant and eclectic attitude.

This conclusion leads to a second point: to reconcile the micro and macro levels of reality presupposes a well-developed set of models and theories about each of these realms. Until we have more mature and precise theories about the micro and macro domains of our universe, it is perhaps premature to worry, at least unduly and excessively, about how to link and reconcile the two.

A third point follows from this observation: without well-developed micro *and* macro theories, meso theory will not be very sophisticated. Hence my proposals in Chapter 32 will be not only premature but also somewhat vague. Yet let me at least hint at what would be involved in mesoanalysis. My view is that micro and macro processes establish parameters for each other; that is, they set limits on what can occur. The assemblage, differentiation, and integration of populations of actors at the macro level are limited by the dynamics of motivating, interacting, and structuring of individuals in face-to-face situations; reciprocally, the existence of macrostructures limits and circumscribes what occurs in any face-to-face encounter. This simple, and perhaps obvious, observation has a number of important implications.

Meso theory will consist of efforts to articulate how macro-level processes circumscribe micro encounters and, conversely, how the dynamics of micro processes limit what can occur at the macro level. Meso theory will not, therefore, involve a decomposition of populations into each and every micro encounter; nor will it consist of an articulation of how the macrostructural context of an encounter determines each person's motives, interaction, and propensity

---

[7]For my review of these, see Jonathan H. Turner, "Theoretical Strategies for Linking Micro and Macro Processes," *Western Sociological Review* 14 (1, 1983), pp. 4–15.

to structure an encounter. Instead, what is possible is this: the development of some principles that specify, on the one hand, how macrostructural processes can influence the values and weights of variables in our micro theories and, on the other, how variables in micro theories load the variables in our macro theories. Maybe this conclusion seems banal, but it is nonetheless fundamental. In fact, much theory already has done just what I am advocating, although the meso-level logic of these efforts is not always recognized. For example, when Marx argues that certain macrostructural processes are "alienating" and, hence, self-transforming for society, he is making a meso-level argument: macrostructures that violate certain human needs are, in the long run, unviable and will cause the disintegration of these macrostructures (the accuracy of Marx's argument is not at issue here—only its logic). Or, when Goffman analyzes how encounters occur within larger social occasions, he is employing a meso argument: more macrostructures set the stage by placing individuals in space, partitioning this space into regions, providing certain general rules of conduct, and defining what is supposed to occur.

Thus mesoanalysis slips into most of our theoretical efforts. What is needed is a more systematic effort to specify the basic and generic ways in which micro and macro processes circumscribe each other.

CONCLUSION

In recent decades, sociologists have become somewhat timid about their earlier pretensions toward science. Many have become convinced that sociology cannot be a science; thus theory is talk about ideas, scholars, and schools, rather than talk about how and why the social universe operates.[8] Twenty years ago, when I first conceived of the first edition of this book, I would not have had to make this distinction so explicit. But times have changed, and the position that has guided my review of theory for almost two decades is now a minority one, at least within American and European "theory circles."[9] One goal of this book in its recent editions has been to try to convince its readers that the original vision of Auguste Comte in the 1830s about a "science of society" is still viable.[10] We can isolate generic properties of our universe (that is, properties that are *always* evident when humans interact and organize), we can develop abstract concepts to denote these, and we can articulate models and propositions about their dynamics. Many do not believe this to be possible, but I remain convinced, and hope to convince others, that we can indeed develop a science of human interaction and organization.

---

[8]For my rather unkind remarks on this tendency, see Jonathan H. Turner, "The Misuse and Use of Metatheory," *Sociological Forum* (March 1990), pp. 346–58.

[9]I suspect that, within the larger sociological community, my position is still in the majority. But social theorists have become decidedly anti-science.

[10]Auguste Comte, *The System of Positive Philosophy*, 3 vols., translated by H. Martineau (London: Bell & Sons, 1896; originally published in translation in 1854 and originally published in French between 1830 and 1842).

To this end, the final chapters represent my best guess about the key properties and dynamics of the micro, macro, and meso domains. We have been given a rich conceptual legacy, and, quite frankly, I do not think that mature theory in sociology is "far off" in the future. We know already what the basic properties of our universe are, and we have come a long way toward developing concepts, models, and principles to explain them. Many choose not to see this, or to retreat from it. But we know a great deal more about the operative dynamics of the social universe than we acknowledge. My goal is to communicate the excitement that I feel in seeing how close we are to developing theories that will make us welcome participants at the table of science.

Finally, let me introduce an important caveat to this last group of chapters. The distinctions among micro, macro, and meso are in theorists' heads. If one looks at the world "out there," one does not see a world composed of levels or layers—that is, micro, meso, and macro. The social world is more seamless and continuous; when we partition theories into micro, meso, and macro, we are doing so for analytical convenience. Thus the very notion of a meso realm is a theoretical construct that denotes certain processes as interesting to focus on and study. It is a construct necessitated by two other constructs, micro and macro, and to view meso processes as linking micro and macro "reality" or as filling in the gap between these realms is also a theoretical construction. We do so because these distinctions allow us to get a "handle" on the flow of reality, but these conceptual handles are not reality itself. I think that the distinctions among micro, macro, and meso are useful for this reason, and yet it may turn out that such is not the case. Maybe it will prove to have been a bad idea of contemporary sociology to divide the social world conceptually into micro and macro, creating the need to bring these together with mesoanalysis. But, for the present, let us see where these distinctions take us.

CHAPTER 30

Micro Theorizing

THE MICRO UNIVERSE

As I indicated in the last chapter, micro theorizing revolves around face-to-face interaction among individuals. There is, however, no precise "line" between micro-level phenomena and macro processes. Rather, face-to-face interaction becomes increasingly difficult to sustain as the number of people, amount of space, and period of time increase. At some point, micro theories become inadequate to explain what is going on, for there are *emergent properties* accompanying increases in number, space, and time. As a result, new kinds of concepts, models, and propositions will become necessary. But, as long as people can remain aware of one another's presence and, if required, initiate face-to-face talk and engagement, micro theorizing will be the most useful.

What, then, are the basic properties of the micro universe, and how should we develop theory about them? My view is that the micro realm can be divided into three constituent processes: (2) motivating, (2) interacting, and (3) structuring. Accordingly, in this chapter I will delineate some of the key ideas that contemporary theorists have about each of these three processes. In so doing, my goal is to suggest the most important ideas for developing a synthetic theory of microdynamics. I have developed my own theory in this regard, but I will not summarize it here.[1] My goal is not to convert others to my theory, but to offer conceptual leads that can be used to create a better theory than my own.

MOTIVATING PROCESSES

Sociology has been somewhat reluctant to address the question of motivation, or those forces that energize behavior. Yet, if we are to understand the micro universe, it is necessary to have concepts, models, and propositions denoting

---

[1]See Jonathan H. Turner, *A Theory of Social Interaction* (Stanford, CA: Stanford University Press, 1988); "A Behavioral Theory of Social Structure," *Journal for the Theory of Social Behavior* 18 (4, 1989), pp. 354–72; and "A Theory of Microdynamics," *Advances in Group Processes* 8 (1989), pp. 1–32.

those processes—however mysterious they may seem—that mobilize individuals to act and interact. I will leave it to psychologists to explain the motivation behind behavior and action, per se; I will concentrate on those motive-states that influence *inter*action. For me, interaction is the most basic unit of sociological analysis, although Max Weber's typology of "action," Talcott Parsons' "unit acts" and typology of action, exchange theory's concern with individual utilities and needs, and a number of other points of view would appear to see individual behavior and action as sociology's basic unit of analysis. I simply disagree, but, obviously, whether the basic micro unit is acts or interactions remains unresolved. But for the present our question is: what forces mobilize and energize people to initiate interaction with others and to sustain interaction over time? That is, what motivates people to interact?

The Concept of "Needs"

Some of the early sociologists were willing to postulate basic "need-states" in humans.[2] Early functional theorists, such as Bronislaw Malinowski (see Chapter 2), went so far as to list basic needs at various system levels—biological, psychological, social, and cultural. Similarly, extreme sociobiology (see Chapter 8) clearly postulates a motivating need-state, "reproductive fitness," or the need to pass on one's genes; and more moderate versions do much the same thing—postulating needs for adaptation and survival. Exchange theories (see Chapters 14–17) also operate with an assumption of need-states, as do the traditions on which such theories are built—behaviorism and utilitarianism. Here humans are seen to have needs that, when met, produce gratification and, when unmet, result in a sense of deprivation. Even traditions that are less explicit about need-states implicitly postulate them. For example, Marx and many conflict traditions, especially those in the critical theory tradition (see Chapters 9 and 13), view humans as having basic needs—freedom from domination and control. Interactionist theorists (see Chapters 18–23) also sneak in a conceptualization of needs—for self-confirmation, for cooperation, and for a sense of reality.

My point here is that, although the notion of "needs" has gotten a bad name in sociological theory, the processes denoted by this term have slipped into sociology through the back door. For almost all theorists see humans as possessing certain basic needs, or psychological and physiological states that, if unsatisfied, produce a sense of deprivation and discomfort.[3] I think that we should revive the notion of "needs"—if only to be conceptually forthright and honest—and develop a profile of what basic needs mobilize individuals to interact.[4] For, ultimately, behavior in interaction is mobilized or motivated; once

---

[2]For example, Georg Simmel discussed "instincts"; Vilfredo Pareto introduced the concepts of "residues" and "deviations" to address the question of instincts; W. I. Thomas postulated "the four wishes"; and so on for many early sociologists.

[3]For more detail along these lines, see J. H. Turner, "A Behavioral Theory of Social Structure."

[4]See, for example, my "Toward a Sociological Theory of Motivation," *American Sociological Review* 52 (February 1987), pp. 15–27.

we seek to understand motivation, our concern shifts to the basic needs of humans.

We could, of course, begin with an inventory of biological needs—sex, sustenance, water, maintenance of body heat, etc. However, from a sociological viewpoint, these are typically met through interaction and are, to a very great extent, mediated by those need-states guiding interaction. Thus, as I indicated earlier, we will concentrate on those need-states that mobilize and energize people during interaction. What, then, do the various theories offer us in the way of an inventory of motives?

An Inventory of Interactive Need-States

Needs for the group   Let us begin with sociobiology (Chapter 8). The basic argument is that organisms seek "inclusive fitness," or the capacity to pass on their genes through reproduction. When examining humans, both the extreme and more muted positions in sociobiology argue much the same thing: humans are "social" or possess "sociality" because they best survive in groups, or, to phrase it more extremely, group membership increases their inclusive fitness to pass on their genes. Thus, in the evolution of humans, there was "selection for" sociality, or needs to be social and part of a group.

This line of argument is not very far removed from various neo-Durkheimian perspectives. For example, Randall Collins' exchange conflict theory (Chapter 12) sees "expectations for group membership" as a primary motivating force in human interaction. That is, people seek to acquire the symbols and material artifacts of group membership, especially higher rank and power, and they get increased emotional energy from interaction in groups. Goffman's dramaturgy (Chapter 22) also posits a similar need: to feel involved in focused and unfocused interactions. If we examine most exchange theories (Chapters 14–17), they too argue that prestige and power in group contexts are what motivate people to interact. Additionally, all interactionist theories (Chapters 18–23) implicitly see adaption to, and cooperation with, others as a motivating force.

The conclusion is thus obvious: one basic need-state of humans is to feel "part of," "included in," or a "member of" ongoing cooperative activities. We do not always need high solidarity or engagement in groups, but we appear to require a "sense of contact" or "involvement" with ongoing affairs. Of course, at times humans do appear to need high degrees of involvement, but not always. How, then, should this need be conceptualized? I propose the term *group inclusion* to denote needs for feeling a part of ongoing social activity, whether this involves high or low degrees of emotional contact and solidarity with others.

Needs for self-confirmation   As George Herbert Mead (Chapter 18) pulled the ideas of William James and Charles Horton Cooley together into a more unified perspective on interaction, he implicitly argued that people seek situations that provide them with confirming and comfortable self-images. Modern interactionism has taken this lead and, in most theories, postulated that a major motivating force in human interaction is the need for self-confirmation in ongoing group activities. There is, however, considerable debate over the nature of the self that people present to others for confirmation. Some argue for a configuration of self-referencing attitudes and feelings that forms a "core self" or "identity" and that is carried from situation to situation; others argue just the opposite: the self that is presented is tied to that situation only. In addition to this debate is one over what aspects—self-esteem, identity, situational images, avoidance of shame and guilt, etc.—of self are most crucial.

For our purposes, it is not necessary to enter into this debate. Rather, we must simply acknowledge that a central motivating force in human affairs is the need to confirm self, however it is conceptualized. Hence we can label a need-state *self-confirmation*, because, when people's images of themselves are not confirmed, they feel deprivation (and many more specific states, such as guilt, anger, shame, anxiety, and frustration, all of which contribute to a person's overall sense of being deprived).

Needs for symbols and materials   All exchange theories (Chapters 14–17) see humans as attempting to secure those symbols and material objects that are valued in a group or larger society. Even if individuals value idiosyncratic symbols and material objects, they nonetheless value symbols and things.

To some extent, simple survival could generate such a need: humans require material objects to maintain their bodies, and they usually need groups to do so, leading them to value the symbols of group membership. But obviously such needs become highly embellished and come to reflect positions (power and rank, for instance) in larger social contexts. Thus, just *what* symbols and objects are valued varies by context and perhaps by activation of other need-states, such as group involvement and self-confirmation. In any case, it is hard to deny that a basic force that drives human interaction is the need to secure symbols and material objects, or what I will call the need for *symbolic and material gratification.*

Needs for facticity   Borrowing from Alfred Schutz (Chapter 18), both Erving Goffman (Chapter 22) and Harold Garfinkel (Chapter 23) see humans as trying to create a sense that they share a common world—both subjectively and externally. Garfinkel used the term *facticity* to describe this sense of a factual, obdurate world "out there." And, in somewhat different ways, both Garfinkel and Goffman see this desire to sustain the appearance of reality as guiding human interaction. For Goffman, self-presentations, frames, and rituals are all directed toward creating the sense—even an illusionary sense—that there is a common world and reality for the individuals involved. For Garfinkel, the use of ethnomethods is directed at producing implicit accounts of a shared universe. And, for both, individuals work very hard, using ethnomethods, tact, etiquette, rituals, and frames, to keep from breaching their shared sense of reality; when breaches do occur, they seek to repair them, often with great energy and dedication.

In a somewhat similar vein, Jurgen Habermas' (Chapter 13) conceptualization of "validity claims" over means/ends efficiency, normative appropriateness, and personal sincerity addresses this same question of facticity. For Habermas, individuals are motivated to create a sense that they are doing the right thing, abiding by similar norms, and being sincere. And so the use of validity claims, and counterclaims, is but an effort to negotiate a sense of a common intersubjective and external world "out there."

Thus people try to create the presumption that they share and participate in a common, agreed-upon world. Whether or not they actually do so is not as important as the fact that they are motivated to create the presumption of a shared world. Such is the power of *needs for facticity* to guide much human interaction.

Needs for ontological security   Related to this need for facticity is what Anthony Giddens (Chapter 26) termed "trust" and "ontological security." People attempt to avoid the anxiety associated with problems in sustaining a sense of trust and predictability as well as a sense that "things are as they appear." Thus people are motivated to feel secure, believing that matters are as they seem and are, thereby, predictable. And, in a vein similar to Garfinkel and Goffman, Giddens argues that this "unconscious motive" mobilizes a considerable amount of energy and explains why people try to avoid breaches in the social order. I will collapse Giddens' discussion of these processes into one need: *ontological security*, or the need to sustain the appearance that matters are as they seem in a situation and that they are predictable.

The Dynamics of Motivation

What emerges from the theories examined in this book, then, is a view of motivation that revolves around needs for (1) group inclusion, (2) self-confirmation, (3) material and symbolic gratification, (4) facticity, and (5) ontological security. As people think about situations and respond to one another, their deliberations and feelings are guided by these need-states.

If sociology is to develop a more adequate theory of micro social processes, it will be necessary to understand in more detail the forces denoted by these five labels. Moreover, we will need to understand how these motives influence thinking and emotions as these affect the course of interaction. I have developed my own views on these matters[5]—translating them into models and propositions. What is essential, I think, are further efforts to explore the interrelations among these motives and, perhaps, to articulate their relations to other motivational forces that investigators might deem crucial. In so doing, sociology will resurrect the concept of "needs" in a way that will lead to a better theory of micro social processes.

Moreover, to the extent that social theorists want to link the micro and macro, the examination of need-states may be useful. For those patterns of

---

[5]See note 1.

social organization that do not allow these needs to be met at some minimal level will, I would hypothesize, be unstable. Conversely, social structures that operate in ways facilitating individuals' efforts to meet these five basic needs will be more viable. Thus a crucial point of articulation between the macro and the micro—that is, where they "loosely couple," in Goffman's terms, or "bump into each other," in my terms—is the extent to which basic need-states can be realized within existing macrostructural arrangements as these constrain micro encounters and face-to-face interaction among individuals.

INTERACTING PROCESSES

As George Herbert Mead (Chapter 18) was the first to emphasize in a true theory of interaction, humans use gestures to signal back and forth their respective lines of conduct, as well as considerable amounts of additional information. I have labeled this mutual reading, interpreting, and acting upon one another's gestures the process of *interacting*. Just how this process occurs is, of course, influenced by motivational forces, both those motives that are brought to an interaction and those that emerge and develop during its course. Yet, no matter what their source, motivational processes "work through" the medium of mutual signaling and interpreting, and these latter processes reveal basic dynamics of their own. What, then, do the theoretical perspectives and theorists examined in earlier chapters have to contribute to a conceptualization of interaction?

The Cognitive Basis of Signaling and Interpreting

What emerges from a review of micro theories is the following conclusion: the process of signaling and interpreting is incredibly complex, and yet individuals interpret the signs of others in a situation and signal their intentions (as well as moods, self, and other states) to these others with relative ease. Certainly part of the answer to the mystery of how people engage in such complex processes so easily resides in their large brains and in their capacity for "mind" or "imaginative rehearsal," in Mead's terms (see Chapter 18), and "rational calculation" or "rational thought," in the vocabulary of utilitarianism and some exchange theories (see Chapter 17). That is, people can covertly think about and deliberate over the consequences of their potential actions and those of others, and they use this anticipation of outcomes to orchestrate their behavior. I prefer the label *deliberative capacity*, but the basic processes denoted by this label are the same for all the various terminologies that have been employed by theorists.

But sheer thinking capacity is obviously not enough. Humans also carry tremendous amounts of information in their "minds," and they reveal complex formulas for assembling information. Alfred Schutz (Chapter 18) coined the term *stocks of knowledge at hand* to capture these additional cognitive processes. Humans have implicit stocks of knowledge that, in a manner not well understood, are warehoused, retrieved, combined and recombined, and used to understand and interpret situations. Anthony Giddens' (Chapter 26) notions of *practical consciousness* and *discursive consciousness* denote similar cognitive processes. What stocks of knowledge allow is for people to understand what gestures and other situational features "mean" in a particular context—what Harold Garfinkel (Chapter 23) termed *indexicality*. The same gestures can "mean" different things in varying contexts, and yet humans' stocks of knowledge enable them to sort out these complex matters and place into context the gestures of others, as well as their own.

Thus, signaling and interpreting depend upon humans' capacity to deliberate, weigh, access, and anticipate, on the one hand, and their ability to invoke (sometimes consciously but more often semi- or unconsciously) the correct information about setting and context, on the other. Without these capacities the incredibly complex and subtle processes of signaling and interpreting could not occur.

The Self-Referencing Basis of Signaling and Interpreting

All interactionist perspectives emphasize (see especially Chapters 19–22) that much of what humans do in situations involves a presentation of self as a certain kind of person deserving of particular responses from others. The way that one presents self is tied to implicit stocks of knowledge about appropriate displays and, at times, to conscious deliberations about what to communicate to others. People store, I suspect, self-references of themselves that must, should, and can be presented in varying types of situations. Thus the flow of signaling and gesturing does not revolve solely around cognitive capacities to deliberate and employ stocks of knowledge; it also revolves around the images of self that individuals construct—sometimes consciously but more often implicitly. These images organize a considerable amount of the actual signaling of gestures, as well as the interpretation of the gestures of others (which are filtered through the prism of a core self-conception or, at a minimum, through the particular image of oneself in a situation).

Types of Signaling and Gesturing

As individuals interact, they delve into their stocks of knowledge, elicit self-references, and think about situations. As they do so, they emit gestures and signals about themselves and what they are going to do, while interpreting those gestures emitted by others. This process of mutual signaling and interpreting occurs along several dimensions.

Role taking and role making   George Herbert Mead (Chapter 18) introduced the concept of *role taking* to explain the basic dynamic of social interaction. By mutually reading each other's gestures, placing oneself mentally in the position of the other, and adjusting conduct so as to accommodate the likely course of action (role) of the other, individuals are able to cooperate with each other. And, when they can also assume the perspective of others who are not present as well as "the attitude" of generalized others, they can cooperate in more extended and complex social structures.[6] Of course, to some extent what one "sees" as a result of role taking is filtered through the prism of one's own self-conception, but, nonetheless, interaction still depends upon humans' facility and accuracy at mutual role taking.

Both Erving Goffman (Chapter 22) and Ralph Turner (Chapter 21) emphasize that people not only "take on" (mentally) the role of others by reading their gestures but also orchestrate their own gestures so as to "make" a role for themselves. This capacity—what Ralph Turner calls *role making*—assumes that individuals possess in their stocks of knowledge inventories of the roles that are relevant, possible, and appropriate in varying types of situations. There are syndromes or strips of behavior that are, in a sense, prepackaged and constitute identifiable roles. Moreover, in their respective efforts at role taking, each individual searches for this pattern or syndrome of gestures, which signals what role the other is trying to assume. Thus one's ability to assume a role in a situation is dependent upon the role-taking abilities of others.

An analysis of interactional processes, then, must begin with a recognition that the dual processes of role taking and role making are fundamental to the signaling and interpreting of individuals. Role taking allows individuals to perceive the self presented by others and the role chosen to meet needs for self-confirmation, ontological security, material and symbolic justification, group inclusion, and facticity. Conversely, role making enables individuals to meet their own needs by carving out a particular role for themselves in a situation.

Framing   In Goffman's (Chapter 22) terms, a significant part of signaling and interpreting revolves around efforts to *frame* a situation. I think that the notion of frame is superior to symbolic interactionists' references to definitions of situations (Chapter 19). The concept of frame emphasizes that actors cognitively enclose situations as containing certain relevant norms, features of the situation, and characteristics of the actors and that, in so doing, they exclude many other materials. The concept of frame also emphasizes that individuals can transform the frame by making it larger or narrower, by keying new implications of the frame, or by fabricating a frame. In fact, it could be argued that individuals must initially frame a situation before their role taking and role making can be effective.

Whether this last statement is true or not is less relevant than the general point that individuals orchestrate and interpret gestures so as to frame situations, thereby delimiting the informational materials in their stocks of knowledge that they will have to draw upon in other signaling and interpreting processes.

---

[6]I should note that Durkheim's notion of "collective conscience" comes close to Mead's notion of "generalized other," and, like Mead, he was searching for the mechanism by which individuals "plug into" the collective conscience. He came to emphasize ritual, whereas Mead stressed role taking. For a more extensive discussion of this point, see Jonathan H. Turner, "A Note on G. H. Mead's Behavioral Theory of Social Structure," *Journal for the Theory of Social Behavior* 12 (July 1982), pp. 213-22.

Staging   As Goffman (Chapter 22) was the first to emphasize and as others, such as Collins (Chapter 12) and Giddens (Chapter 26), have subsequently documented, signaling and interpreting involve the use of physical space, objects, partitions, relative positionings, and movements to inform people about self, roles, frames, and other crucial aspects of a situation. I will term this complex of related processes *staging*, because it concerns the way individuals use the stage (available space, props, partitions) in the situation and how they move about and position themselves on the stage.

Thus a key process of interaction is reading and interpreting staging cues and, at the same time, using staging props, positions, and movements to signal intentions (with respect to such matters as self-presentation, role, and frame). Humans rely upon understandings stored in their stocks of knowledge about what stage cues mean in varying situations in order to interpret the behaviors of others and develop their own line of conduct and self-presentation.

Ritualizing   Goffman (Chapter 22) was the first to adopt Durkheim's (Chapter 24) insights about the significance of ritual for human interaction. More recently, Collins (Chapter 12), Giddens (Chapter 26), and Wuthnow (Chapter 25) have made ritual a central dynamic in their respective theories. Much of what people signal and interpret relies upon stereotyped sequences of gestures that emphasize certain things and focus attention on particular matters. Thus, at the interpersonal level, it is by ritual that people open, close, repair, organize, and sequence their interactions; and each of us is constantly looking for and reading ritualized gestures to tell us what is happening, what will happen next, and what will change in an interaction. I will term this process *ritualizing* to emphasize that other interactional processes—such as role making and taking, framing, self-presenting, and staging—are all dependent upon the effective use of interpersonal rituals.

Claiming   Erving Goffman (Chapter 22) hinted at a crucial interpersonal process when he argued that people try to indicate (and seek confirmation from others about) sincerity and normative appropriateness. Jurgen Habermas (Chapter 13) developed the notion of "validity claims" to extend this idea. His argument is that effective communication and discourse depend upon individuals making validity claims as to their state of sincerity, use of the most efficient means to a given end, and commitment to relevant norms. Others can challenge such claims, forcing individuals to remake or alter their validity claims; and out of this claiming and counterclaiming process, a more open pattern of speech communication emerges.

I doubt if people would seek to negotiate openly their claims as Habermas would have them do (in his ideal critical theory utopia); rather, my sense is that people subtly make claims and others work very hard *not* to challenge them. For to offer a challenge means that a situation will be breached, forcing a renegotiation of sincerity, means/ends efficiency, and normative appropriateness. Thus I would cast Habermas' important insight into a more Schutzian

(Chapter 18) and Goffmanian (Chapter 22) mold, arguing that people's sense of reality and comfort in an interaction depends upon assuming, if at all possible, that others are being sincere, using efficient means, and behaving in terms of appropriate norms. Thus an individual will emit cues—sometimes implicitly, at other times consciously—that make claims. And others in this individual's audience will go along, as long as these cues do not appear grossly out of line.

I will term this process *claiming* and assert that much of what individuals signal (and interpret) is implicit claims about sincerity, means, and norms. This claiming is possible because people possess in their stocks of knowledge information about (1) what is sincere, normatively relevant and appropriate, and instrumentally acceptable in varying types of situations and (2) what gestures and cues are appropriate for making these claims in different situations. Such assurances about sincerity, means, and norms are crucial, I think, not only for ordering interaction but also for meeting need-states, particularly for ontological security and facticity.

Accounting As Alfred Schutz (Chapter 18), Erving Goffman (Chapter 22), and Harold Garfinkel (Chapter 23) recognized, individuals approach situations with a presumption that they share common external and internal worlds, until it is demonstrated otherwise. This presumption is often implicit and unacknowledged, and yet much of what individuals do in interaction is to signal (and search for signals from others) that they share a common world. Thus individuals gesture, usually in very subtle ways, in order to construct an account of "what is real" in the situation. Subtle phrases and gestures are employed to sustain the account, or the presumption and perhaps the illusion, that there is a real world "out there" and that, for the purposes at hand, "we share" similar subjective states. Ethnomethods—et ceteras, conversational turn taking, glosses, searches for normal forms, repairs of breaches—are one way to render this subtle account (see Chapter 23), but there are other techniques as well (see Goffman's discussion of focused and unfocused interaction in Chapter 22). I will term this process *accounting*; when one looks closely, much of the "background filler" of gesturing involves the construction of an account.

This completes my list of basic types of signaling and interpreting. One may wish to add to the list, or perhaps combine the categories that I have emphasized, but the theoretical traditions outlined in this book point to role taking, role making, framing, staging, ritualizing, claiming, and accounting as fundamental to what people do when they signal and interpret gestures. The process of signaling and interpreting thus revolves around the processes outlined here; thus our theoretical objective should be to conceptualize them more precisely, to indicate how they are interrelated, to connect them to motivational processes (or specific need-states), and, as I will examine next, to see how they order an interaction in time and space.

STRUCTURING PROCESSES

Avoiding Micro Chauvinism

The process of *micro*structuring involves the maintenance and organization of face-to-face signaling and interpreting among individuals over time and space. Moreover, the structuring of interaction creates conditions that facilitate the opening of interaction, its closure, and its resumption later on. We should not forget, however, that microstructuring does not involve great numbers of people, large territories of space, or long spans of time. Too often the analysis of those processes sustaining face-to-face interaction in small amounts of space and over relatively short spans of time is seen as also explaining macrostructure. Typically, macrostructure is viewed as "chains" or "aggregations" of micro interactive processes that somehow produce and reproduce macrostructures. Although it is true that macrostructures are made up of micro processes, it does not follow that conceptualization of these micro processes will be useful in explaining the dynamics of populations of actors in larger territories over prolonged periods of time. To phrase the matter in terms of an analogy, knowledge of cell physiology does not explain either the anatomy of an organism or the population-ecology of life forms—despite the fact that all life is composed of cells. And so it is with microstructuring. With this caveat, let me now examine those processes that operate to sustain face-to-face interaction.

Types of Microstructuring Dynamics

How, then, do motivated individuals, who are signaling and interpreting back and forth, organize in space and sustain over time their interaction? Several key processes are involved.

Categorization In his critique of Max Weber (Chapters 2 and 9), Alfred Schutz (Chapter 18) was perhaps the first to conceptualize clearly the significance of categorization of individuals and situations. He employed the term *typification* to denote the tendency of individuals to classify one another as instances of a category and, then, to adjust responses accordingly. When categorization occurs, individuals can more readily emit responses; indeed, they can go on "automatic pilot" and not engage in fine-tuned role taking, framing, accounting, and claiming. Thus, to the degree that a situation is readily categorized, it can be more easily initiated, sustained, left, restarted, and closed.

Many theorists before and after Schutz have come to similar conclusions. For example, although Schutz was highly critical of Max Weber (Chapters 2, 9, and 18), the latter's categorization of types of action hints at Schutz's insight, as does Talcott Parsons' similar typology of action (Chapter 3). Erving Goffman (Chapter 22) and Randall Collins (Chapter 12) also emphasized categorization with their typology of interaction as revolving around a ratio of work-practical, ceremonial, and social considerations.

If we take Schutz's concern with degrees of intimacy (ranging from persons as standard categories to unique intimates) and combine this concern with

Goffman's and Collins' emphasis on the relative amounts of work-practical, ceremonial, and social content, we come up with a typology that is close to what humans do when interacting. That is, they draw upon their respective stocks of knowledge and simultaneously categorize a situation in terms of (1) the level of intimacy appropriate and (2) the relative amounts of work-practical, ceremonial, and social content. When individuals are able to clearly cross-tabulate a situation in these terms, they make it much easier to open, close, pick up, or sustain signaling and gesturing; and they make it much simpler to organize people in space and to regulate their movements. In other words, *categorization* facilitates structuring.

Regionalization   Erving Goffman's (Chapter 22) discussion of staging processes, as modified by theorists such as Giddens (Chapter 26) and Collins (Chapter 12), alerts us to the importance of developing agreements over the ecology and demography of an interaction. People carry in their stocks of knowledge understandings about what physical space, physical objects, spatial partitions, numbers of participants, movements of individuals, and positioning of actors "mean" in varying situations. And, when they can develop stable understandings about how to interpret space, objects, partitions, numbers, po-sitionings, and movements, they will have *regionalized* the situation. Once a situation is regionalized, it is more easily structured because individuals now have clear cues about what they are supposed to do and what the responses of others mean. As a result, they can initiate, close, sustain for a while, leave, and reenter the situation without undue stress and without the necessity of engaging in active role taking, framing, claiming, or accounting.

Normatization   The concept of norms has been central to sociological inquiry since its beginnings. Functional theorizing (especially as portrayed in Chapters 2-6) has been most often associated with the view that social structure is the result of clear and binding norms that guide behavior and interaction. In fact, however, most interactionist theories (Chapters 18-23), structuralist theories (Chapter 25), exchange theories (Chapters 14-17), and even conflict theories (Chapters 9-13) invoke the idea of normative constraints. Why, then, has the concept of norms developed such a bad reputation, as we saw in the virulent criticisms of functionalists like Talcott Parsons (see Chapters 3 and 9)? The problem, I think, results from viewing norms as clear, unambiguous, and attached to each and every social position—a line of argument found not only in functionalism but also in role theory (see Chapter 20, for example).

I think that Goffman's (Chapter 22) analysis helped overcome this problem. There is no doubt, Goffman argued, that norms regulate social conduct, but they are not the only force structuring an interaction. They are, in Herbert Blumer's (Chapter 19) terms, one of many "objects" that individuals take cognizance of as they organize their responses.

Structuralist theorizing (see especially Chapters 24, 25, and 26) can further assist us to resolve the problems in conceptualizing norms. In a sense, norms are constructed and applied to situations, and there are probably "generative rules" (normative "grammars," as it were) for creating norms, applying them to situations, and perhaps adjudicating conflicts among them. Norms are thus assembled from stores or "bits" of information in stocks of knowledge about appropriate rights and duties and perspectives for interpretation for varying types of situations. Giddens' (Chapter 26) analysis of structure as assemblages of "rules and resources" captures much of the sense of this argument, as do the arguments of various structuralisms (Chapter 25).

Once we recognize that norms are assembled, this implies that they can be disassembled, transformed, and reconstituted. But, if an interaction is to be structured, it must be *normatized* in the sense that, for the time being, individuals draw upon their stocks of knowledge and produce for a particular situation agreements about their respective rights and duties, schemes or perspectives for interpretation, and procedures for resolving and repairing disputes. Unless such normatization occurs, interaction will be difficult to structure over time and space.

Ritualization   As sequences of behavior that symbolically denote, mark, and emotionally infuse interaction, rituals have enormous significance for structuring interaction. In effect, they tell us the beginning, sequencing, and ending of the interaction, and they facilitate renewal of interaction at subsequent times. Moreover, they help organize individuals in space, since where people are located, how they move, what props they employ, and what regions they are in are all impregnated with ritual. Émile Durkheim (Chapters 2 and 24) was the first to recognize that rituals are one of the mainstays of social structure. But it was Erving Goffman (Chapter 22) who saw the significance of ordinary, everyday rituals for the structuring of interactions. Goffman's lead has been followed by others, such as Collins (Chapter 12) and Wuthnow (Chapter 25), who have made rituals a central force in their theories. Their basic insight is this: when openings, closings, sequencings, renewals, and repairs of interaction can be *ritualized*—or marked by standardized and stereotyped sequences of gestures that focus attention and emotionally infuse the situation—it is far easier to structure the interaction over time and in space. In a sense, standardized rituals "lead people by the hand" and tell them how and when to act and interact in certain ways; and so, when an interaction can use these standardized rituals for all of its key moments, it is more likely to flow smoothly, persist over time, and remain coherent in space.

Routinization   Anthony Giddens (Chapter 26) has been the most perceptive in recognizing the significance of routines in human affairs. Early social theorists like George Herbert Mead sometimes used the term *habits*, but, however labeled, the basic idea is that people often do the same things, in the same way, and at the same time. Indeed, much social life is routinized and carried out in the same manner without great thought or reflection. In a sense, routines are the opposite of ritual because they do not focus great attention, arouse emotions, or punctuate events; instead, they simply involve "going on automatic pilot" and doing things in pretty much the same way. Most of people's daily activities—work, play, sleep, eating, etc.—are organized as much by routines as by ritual, allowing them to go about their business without undue thought, deliberation, or emotion.

When an interaction can become *routinized*—that is, with individuals signaling and interpreting in similar ways at similar times without undue attention—it is more readily structured. For, with routinization, interactants do not have to engage in much interpersonal work and can pass the time without thinking very much about what is occurring. It is only when routines are disrupted that we fully realize how important they are. When routines are breached, people's sense of order is shaken, forcing them to role-take, role-make, frame, account, stage, and claim in an effort to restructure the interaction. If they can, people avoid such efforts, settling comfortably into routines that not only lessen the interaction burden but also help meet basic needs (primarily, I suspect, needs for ontological security and facticity). Thus interactions that do not involve a significant amount of routinization will, I believe, be difficult to structure over time, organize in space, and pick up again at subsequent points in time.

Stabilization of resource transfers   All interaction involves the giving up of resources by one actor for those of another. This is the crucial point of all exchange theories, whether Collins' conflict perspective (Chapter 12), Homans' behavioristic orientation (Chapter 15), Blau's structural approach (Chapter 16), Hechter's utilitarian emphasis (Chapter 17), or Emerson's and Cook's network analysis (Chapter 27). Much of what an actor does in an interaction—present self, make roles, frame, account, assert claims, and so on—involves an effort to realize material and symbolic resources from others in exchange for resources that are less valuable for that actor. For an interaction to become structured, the nature of the resources exchanged and the rate of exchange must become stabilized in such a way that each participant feels a sense of profit (that they have received more than they have given up). This is what I mean by *stabilization* of resource transfers. When this stabilization does not prevail, interaction will be punctuated with resentment, anger, frustration, tension, and renegotiation. These latter conditions are not conducive to stability and continuity; indeed, they are likely to promote efforts at restructuring the interaction.

In sum, then, the theories of contemporary sociology emphasize a number of microstructuring processes—categorization, regionalization, normatization, ritualization, routinization, and stabilization of resource transfers. Obviously, these are all interconnected—for example, categorization is facilitated by regionalization, and vice versa; normatization (as justice norms or norms of fair exchange) is related to stabilization of resource transfers, and vice versa; routinization is connected to regionalization, and vice versa; and so on for all the possible causal paths that can be drawn among these variables. Moreover, these structuring processes are clearly connected to both the dynamics of signaling and interpreting and the motivational forces that mobilize individuals during interaction. And, of course, there are many feedback loops—for instance, routinization lowers needs for ontological security, stabilization of resource transfers lowers needs for material and symbolic gratification, categorization facilitates framing, and so on for each of these structuring forces as they influence motivation, signaling, and interpreting. Yet, if we simply focus on what structures an interaction in space and across time, ignoring what drives people to do so and just precisely how they do it, the interrelated processes of categorization, regionalization, normatization, ritualization, routinization, and stabilization are crucial. And, to the extent that we seek a theory of microstructuring, we will need to develop a more precise conceptualization of the processes denoted by these concepts and the dynamic relations among them.

CONCLUSION

I have, perhaps, imposed too much of my own thinking on this review of the micro social processes that various theoretical traditions have enumerated. Yet the intent of this exercise is more important than the substance of what I have said. My goal is to demonstrate that a general theory of micro processes can be developed. We have many interesting leads, as I have tried to illustrate. Our task in the future should be to extract these leads, develop precise definitions of concepts, and indicate how these concepts are interrelated. The end result will be a more viable theory of microdynamic processes.

We should not, however, consider this theory an explanation for all of social reality. Rather, we will have developed a theory of "the interaction order"—to use Erving Goffman's label. For, as people congregate, interact, and organize their affairs in larger numbers, time frames, and territories, the concepts of micro theories become inadequate. New levels of reality emerge, requiring their own theories. Such macro theoretical issues are the subject of the next chapter.

CHAPTER 31

◆

Macro Theorizing

THE MACRO UNIVERSE

Compared with micro theories, macro theories focus on (a) greater numbers of individuals, most of whom are not in face-to-face interaction, (b) larger geographical territories, and (c) longer time periods. Macro theories thus involve a shift in perspective to populations of individuals and their organization in space and across time. Although many micro theorists charge that the concepts of macro theory represent reifications and hypostatizations of what is "really real"—that is, individuals in interaction—this allegation could be made by a cell biologist against a physiologist focusing on individual organisms, or by a biochemist to a cell biologist, or by a physicist to a chemist.[1] Thus we must accept that there are emergent properties in the universe and that the concepts useful in one domain of reality need to be supplemented by those appropriate to a new, emergent domain. Once this viewpoint is adopted, it becomes possible to perform macroanalysis, ignoring all of the specific micro interactions that make macrostructures possible. Conversely, when we performed microanalysis in the last chapter, little attention was paid to all of the macrostructures that provided the context and place for micro encounters. Macro- and microanalysis yield different kinds of insights and explanations; the more sociological theory accepts this fact, the more productive theorizing will be.

Thus, in this chapter, I turn to macro processes, shifting the focus to the dynamics involved in organizing populations in space and time. We are returning to where the first sociologists—Auguste Comte, Karl Marx, Herbert Spencer, Émile Durkheim, and Max Weber—began. Our unit of analysis is now "society" as a whole or large-scale portions of it.[2] We have, in essence, put a wide-angled lens on our theoretical camera. What, then, is there to see?

---

[1]Moreover, micro theories often assume, incorrectly, that concepts in their theories are somehow less reified. But, in fact, concepts like "self," "role taking," "framing," etc., are theoretical constructs that imply "what is real."

[2]I think, however, that the macro processes examined in this chapter can be applied to smaller organizational units, such as complex organizations, communities, and perhaps secondary groups.

I think that the early sociologists, especially Herbert Spencer and Émile Durkheim (see Chapter 2), were correct in their view that macro reality can be divided into three constituent processes, which I will label as follows:[3] (1) those forces that *assemble* individual actors in space, (2) those processes that *differentiate* them in various ways, and (3) those mechanisms that *integrate* differentiated actors into a coherent and systemic whole. Our goal in macroanalysis is, therefore, to see what the theories examined in these pages have to contribute to our understanding of assembling, differentiating, and integrating processes. That is, what are the dynamics of assembling individuals in space and over time? What forces cause their differentiation into categories, distinctive types of corporate units, ranks, subcultures, and other axes of differentiation among actors in human societies? And what laces these differentiated units back together to form an identifiable social whole? I am using the vocabulary of functional theories (especially Chapters 2-7), and this is sure to raise objections. But I make no reference to needs or requisites; nor should anyone see my argument as "static" or "conservative." Indeed, in answering the above questions, I will make evident the conditions producing disassembly, dedifferentiation, and disintegration, or some combination of these. We should not, therefore, be too quick to "throw out the baby with the bathwater" when looking at the contributions of functional theorizing. And so, let me now turn to a more detailed review of assembling, differentiating, and integrating processes.

ASSEMBLING PROCESSES

I use the term *assembling* to denote those forces that organize individuals and collective actors (organizations, ethnic populations, communities, etc.) in space and over time. This line of inquiry was most evident in Herbert Spencer's and Émile Durkheim's (Chapter 2) concern with societal growth, and it is a central variable in modern human ecology (Chapter 7). In a sense, the variables involved are obvious but nonetheless fundamental. I will divide them into three classes: (1) those forces influencing *aggregation*, or the organization of actors in space; (2) *size/growth*, or the absolute size of a population and its rate of growth; and (3) *production*, or the gathering, conversion, and distribution of environmental resources into goods and services that are consumed by a population.

Aggregation

Durkheim (Chapter 2) argued that the amount of available space and its organization are crucial for understanding human society. Spencer (Chapter 2)

---

[3] I have developed my own—rather tentative and still formative—theory of macrodynamics. I have taken these labels from that theory. See Jonathan H. Turner, "Analytical Theorizing," in A. Giddens and J. H. Turner, eds., *Social Theory Today* (Stanford, CA: Stanford University Press, 1987) and "Toward a Theory of Macrostructural Dynamics," in M. Zelditch and J. Berger, eds., *Sociological Theories in Progress*, III (Newbury Park, CA: Sage, 1990).

had also noted that ecological confinement (or lack of confinement) of a population is a critical variable influencing social organization. Modern ecological theory (Chapter 7) similarly views size of territory as an important determinant of social organization. And some conflict theories, such as Collins' (Chapter 12) views on the "marchland advantage" of nation/states, also utilize an ecological variable (in Collins' case, whether or not a nation is surrounded by adversaries). In general, the less the amount of space and the more it is contained by natural barriers (mountains, rivers, oceans, etc.), cultural forces (language, values, beliefs, religion), and social forms (contiguous nation/states, military powers, different economic forms), the greater is the degree of aggregation of a population. Of course, if a population is small and is at a steady state of growth, then these barriers will be less critical (unless they delimit a very small territory); but, if a population is large and/or growing, these barriers become determinative of many organizational dynamics. Thus the size and growth rate of a population are important variables and influence the effect of aggregating processes on social organization.

Size and Growth

The absolute size of a population is, as Spencer (Chapter 2) was perhaps the first to recognize, crucial to understanding a society. The greater the size of a population, the greater (and dramatically so) the number of potential social relations—what Durkheim (Chapter 2) was later to term "moral density." Moreover, larger populations create problems of producing sufficient material goods to support themselves as well as problems of coordination and control of the larger social mass.

These same issues—increased numbers of potential social relations, problems of production, and problems of coordination and control—surface as soon as a population begins to grow. There are three basic ways a population can grow: (1) through internal increases or reproduction (increased birth rates relative to death rates), (2) through external influx or migration relative to out migration, and (3) through external incorporation from annexation or conquest of other populations. Durkheim (Chapter 2) emphasized (1) and (2), as does contemporary human ecology (Chapter 7), whereas Spencer (Chapter 2) and, to a lesser extent, Max Weber (Chapter 9) stressed the significance of war, conquest, and annexation, as do present-day theories of geopolitics (Chapter 12). Each of these processes increases production, coordination, and control problems, but (2) and (3) are especially problematic because migrants, conquered subjects, and annexed peoples usually reveal social and cultural differences as well as resentments that create special problems of control; moreover, the indigenous population is often distrustful and hostile toward new members who are "different" (Spencer, in particular, highlighted these problems).

Production

Production is the process of gathering environmental resources, converting them to usable goods and commodities, providing relevant services, and distributing goods and services to the population. Marx (Chapter 9) is, of course, famous for recognizing that the mode and means of production are the central determinants of basic organization and cultural patterns in a society. Other theorists, such as Spencer (Chapter 2), Durkheim (Chapter 2), and Hawley (Chapter 7), also stressed productivity, but they more typically have seen this variable as related to population size and growth, on the one hand, and technological, organizational, and material resources, on the other. Similarly, Weber's (Chapter 9) analysis of capitalism and rationalization stresses these variables. That is, production is related to the size and growth rate of a population and the level of available material resources, the organizational forms extant in a society (markets, money, bureaucracies, financial organizations, mobilization of political power, cultural values and norms), and technologies (knowledge about how to manipulate the environment).

Obviously the variables are interrelated: size is influenced by growth; size and growth increase demands for production; the level of available resources is determined by organizational forms and technologies; organizational forms are determined by social technologies; increased productivity, as stimulated by new organizational forms and technologies, encourages further population growth; and so on for various combinations of these variables. Most macro theorists have recognized these interconnections, although Weber (Chapter 9) and neo-Weberians like Collins (Chapter 12) have been, along with human ecologists (Chapter 7), the most consistently concerned with these kinds of relations. And some modern Marxist-inspired work, which I have not examined in detail,[4] also incorporates some of these causal connections into their theories.

In sum, then, the constituent processes of assembling—aggregation, size/ growth, and production—are perhaps fairly obvious, but they have profound effects on the organization of a population in space and time. One can see these processes as either "prime movers" of society—as Spencer, Durkheim, and human ecologists tend to do—or as background conditions or constraints on what organizational forms are possible for a population. In either case, they are central forces, and sociological theory has been somewhat remiss in conceptualizing them and incorporating them into macro theories.[5]

DIFFERENTIATING PROCESSES

Competition and Conflict

Spencer and Durkheim (Chapter 2) viewed the growing size of a population and its concentration in space as increasing the level of competition among actors over scarce resources. Similarly, human ecology (Chapter 7) sees the

---

[4]For example, modern world systems analysis addresses many of these variables and their interconnections, with a Marxian emphasis on inequality at a world level.

[5]In some geopolitical theories (see Collins' in Chapter 12) and in the analysis of complex organizations, these variables are given some prominence. But in most theories they are ignored or not fully conceptualized in relation to their significance in human affairs.

number of actors and their level of territorial concentration as increasing competition and, at times, conflict. All of these functional approaches borrow an insight from biology: the greater the number of species in an ecological niche, the greater is the competition for resources and the greater is the likelihood of speciation (emergence of new species). Instead of speciation, functional theories stress specialization of individuals and corporate units, or, in more general terms, *social differentiation*.

Expanding production is also seen to cause specialization, as expanded markets increase not only competition but also a greater variety of goods and services. Much of this specialization is the result of competition for jobs, market shares, profits, and other resources, and some is necessary in order to create the division of labor so essential for producing varieties of goods and services. Thus size, ecological concentration, and production are viewed by functional theories as increasing social differentiation by virtue of their effects on competition.

The process of differentiation, once initiated, contains a number of self-reinforcing effects that not only sustain a given level of competition and differentiation but also escalate the degree of social differentiation. The end result of these processes is to differentiate a population along several dimensions: (1) subcategories, (2) subgroupings and subpopulations, (3) subrankings associated with inequality and hierarchy, and (4) subcultures. Let me first examine each of these processes and then return to the processes that produce them.

Subcategories

One pattern of differentiation is the formation of subcategories. Peter Blau's (Chapter 28) analysis of "nominal parameters," Pierre Bourdieu's (Chapter 25) discussion of "habitus," and Amos Hawley's discussion of "categoric units" capture some of this basis of differentiation; for, as populations differentiate, individuals and at times collective units develop distinctive attributes that place them in a designated category. People and groups become distinguished, and responded to in different ways, by virtue of their membership in a category. Because members of a category do not necessarily interact or organize themselves, such categories do not represent actual social structures, although subgroupings can form as a result of being placed in a category.[6] Age and sex are universal categories in even the smallest population, but as a society's size, productivity, and competition among individuals escalate, new categories—race and ethnicity, religious background, income level, styles of taste and consumption, nature of occupation, regional location, level of education, and the like—emerge. *Subcategories* are, therefore, a basic mode of social differentiation in human populations.

---

[6]Especially if discrimination is involved, or other processes that increase rates of intracategory interaction.

Subgroupings and Subpopulations

Subgroupings are subsets of a population that, to varying degrees, are organized. Often such organization corresponds to a basic social category, and equally often the organization of a subgrouping produces a new basis for categorization—feminist, environmentalist, industrialist, etc. Blau's (Chapter 28) notion of nominal parameters would also include subgroupings, but most theories point to the organization of subgroupings—including such diverse forms of organization as communities, states, types of bureaucratic organizations, kinship groups (lineages, clans, moieties), social movements, interest groups, associations of particular categories, and individuals as well as collective units engaged in various types of activities (professional associations, associations of manufacturers, etc.). The elaboration of *subgroupings* is thus the most frequent axis of differentiation in human populations.

Subrankings

All conflict theories (Chapters 9-13) and some structuralist theories (Chapter 25) emphasize the formation of social hierarchies composed of subranks of individuals who possess certain amounts of a particular resource. At times, ranks are categories for labeling individuals in terms of some resource, such as money (for example, rich, poor, middle income). Yet, as all conflict theorists, some functionalists such as Spencer (Chapter 2), and structuralists like Bourdieu (Chapter 25) have stressed, one of the dynamics of hierarchy and ranking is the organization of subranks into subgroupings (mobilized to maintain or redistribute resources). Some theorists, such as Weber (Chapter 9), Collins (Chapter 12) , and Bourdieu (Chapter 25), have emphasized that subranks also develop distinctive cultural characteristics—values, beliefs, orientations, attitudes, linguistic styles, demeanor, and the like—and thereby form distinctive subcultures (to be discussed below). Blau's (Chapter 28) concept of "graduated parameters" encompasses all of these distinctions, since membership in a categorical rank, hierarchically placed subgroup, or differentially valued subculture is a basis for distinguishing people in terms of gradations and responding to them in distinctive ways. Thus the dynamics of differentiation are, to a very great extent, going to revolve around inequalities in resources, hierarchies, and *subranks*. In particular, such variables as the nature of the resources that are unequally distributed, the number of hierarchies, the correlation among hierarchies (that is, does membership in one rank predict membership in another, or, in Dahrendorf's terms, are the ranks "superimposed" on each other?), the distinctiveness and boundedness of a subrank, the relative size of rank, the number of ranks, the degree of mobilization and/or organization of members in a subrank, and the degree of mobility among subranks are all crucial to understanding differentiation and, as we will see, opposition and conflict. As might be expected, then, these variables appear in conflict theories (Chapters 9-13); thus their interrelations and effects on social organization represent the major contribution of this perspective to macro sociology.

Subcultures

As cultural structuralists (Chapter 25), as well as conflict theorists like Weber (Chapter 9) and Collins (Chapter 12), would emphasize, a final basis of differentiation is cultural—values, beliefs, ideologies, religious dogmas, language, deference and demeanor norms, tastes, and the like. Subcultures, or portions of them, can be organized into various subgroupings or subranks, but rarely is an entire subculture so organized. Rather, one usually finds pockets and portions of subcultural organization (in churches, neighborhoods, protest organizations, mutual-aid associations, and the like), but the diverse organizational subgroupings themselves are not tightly interconnected; nor do they reach all people who share the subculture's symbols. Subcultures can also be a basis for categorization, but rarely is a member of a subculture differentiated exclusively on the basis of subcultural categories (other bases of categorization will usually exist for a member). Subcultures vary, of course, in terms of such variables as their deviation from broader values, beliefs, and other population-wide or societal symbols; their level of organization into subgroupings; their membership in distinctive subranks; their physical and social isolation from the organizational mainstream; and their rates of mobility across subgroupings and subrankings. Curiously, with some exceptions (Weber, Collins, Bourdieu and Wuthnow, for example, in Chapters 9, 12, and 25, respectively), conflict theories have tended to focus on the dynamics among subranks, especially those based upon class (as tied to production). But, in fact, much conflict in social systems is over cultural issues (religion and ethnicity, for example), although the conflict is most intense when associated with inequalities and subrankings.

As I indicated earlier, these axes of social differentiation—subcategories, subgroupings, subrankings, and subcultures—are the result of certain self-sustaining processes. The most important of these are (1) exchange of resources and (2) mobilization of power. Each of these will now be examined.

Exchange of Resources

All versions of exchange theory (Chapters 12, 14–17, 27) emphasize that relations among social units, whether individuals or collectivities, involve transfers of valued resources. Competition for resources often forces actors to differentiate in terms of the resources that they can possess and the exchanges that they can undertake. As actors come to be typified by certain resources, they can be further differentiated in terms of: (*a*) a subcategory associated with particular resources, (*b*) a subrank with respect to shares of a given resource and its general value, (*c*) an organizational subgrouping (coalitions, for example) associated with particular types and levels of resources, and (*d*) a subculture displaying symbolic characteristics associated with types and levels of resources. Once established, these differences feed back and influence the nature of exchanges. Frequently they increase competition and, as all conflict theories (Chapters 9–13) and cultural structuralists (Chapter 25) emphasize, they escalate the potential for conflict over shares of resources. Thus competition, varying axes of differentiation, and exchange are mutually interconnected, operating to sustain or increase differentiation and, at times, to produce conflict over resources.

Mobilization of Power

As Spencer (Chapter 2) emphasized, and as ecological theory (Chapter 7) subsequently stressed, competition and differentiation, per se, produce concentrations of power. As problems of coordination and control of conflict escalate with competition and differentiation, one important axis of differentiation is power as a means to regulate the transactions and relations among system units. Moreover, as geopolitical theories, such as those developed by Weber (Chapter 9), Collins (Chapter 12), and Simmel (Chapter 9), emphasize, external conflict with an "enemy" also produces centralization of power.[7] Thus a particularly important hierarchy in a population revolves around the unequal distribution of power.

Exchange theories (see especially Chapters 15, 16, 17, 27) approach the dynamics of power somewhat differently. They see power as inhering in the exchange process: those who have the most valued resources will have power over those who value these resources and who have few viable alternatives for securing them (Chapters 16 and 27). Hechter's rational choice model of exchange shows the way for reconciling Spencer and ecological theory, on the one hand, and modern exchange theory's emphasis on power-dependence relations, on the other. When high levels of competition and conflict prevail in exchanges, those who can mobilize the capacity to control and regulate such conflict will be seen as having a highly valuable and scarce resource; as a result, they will be accorded power. And, if an external enemy is the source of conflict, those who can mobilize and coordinate others to deal with the conflict will be seen as having a particularly valued and scarce resource, leading to their consolidation of power. Of course, once power is mobilized, it can be used to consolidate further power, at least up to the point of resistance or opposition postulated by exchange[8] (Chapters 16 and 27) and conflict theories (Chapters 9–13).

As power is consolidated, it becomes a basis for further differentiation. As Hawley (Chapter 7) emphasized, corporate organizational units (a type of subgroup) will form around the level of power that is controlled (for example, the state will form in large societies) or around the mobilization of opposition (or counterpower) to those subgroupings in which power is concentrated. As Collins (Chapter 12) emphasizes, subcultures will develop around shares of

---

[7]Spencer also saw this as a critical variable. War is, in Spencer's view, the force that created the political power used to organize society.

[8]For most detailed analyses of "resistance" in exchange theory, see Douglas D. Heckathorn, "Extensions of Power-Dependence Theory: The Concept of Resistance," *Social Forces* 61 (1983), pp. 1206–31; David Willer and Barry Markovsky, "The Theory of Elementary Reactions: Its Development and Research Program," presented August 1989 at the *Theory Growth and the Study of Group Processes* meeting, Stanford University.

power because those who give orders think and act very differently than those who take orders. Bourdieu's discussion of the habitus of various class factions makes a similar point (Chapter 25). Subcategories will, to some extent, invoke power as a criterion for differentiation (if only to make a simple distinction between "elites" and "masses"). And, as noted above, a distinctive hierarchy producing subrankings in terms of how much power is held will also emerge. Power is thus a significant force in generating and sustaining all axes of social differentiation; once differentiation revolves around power, it tends to accelerate the concentration of power. For example, as distinctive organizational subgroupings (the "state"), categories ("masses" and "elite"), ranks (the "power elite"), and subcultures (the "aristocracy") emerge, these patterns of differentiation can operate to sustain the mobilization of power. At other times, however, these processes can escalate to the point of mobilizing opposition to power (counterpower), as conflict theory would emphasize, but such counterpower will involve a similar pattern of differentiating organizations, rankings, subcultures, and categories around power.

In sum, then, functional macro theories emphasize that increases in size, production, and aggregation escalate competition. In turn, competition increases the degree of "social speciation" or social differentiation along several axes: (1) subcategories, (2) subgroupings (broadly defined to include all corporate units), (3) subrankings, and (4) subcultures. These four bases of differentiation can overlap or be distinctive, but they are always fueled by two key processes: exchange and power. Exchange theorists stress how power inheres in the respective resources of actors and the value as well as scarcity of these resources. Conflict theories emphasize that power is both the result of past conflicts (to suppress and control it) and the cause of new conflicts (to oppose a given power structure). In a somewhat similar vein, functional theories argue that competition and conflict (both internal and external) create needs for coordination and control that are met by the mobilization and concentration of power. In all these theories, the mobilization of power influences the nature of exchanges and competition that in turn affect the configuration of differentiation with respect to subcategories, subgroupings, subcultures, and subrankings.

The complex interplay among these differentiating processes has yet to be fully worked out. Hence there is a great deal of theoretical work to be done in conceptualizing these processes more precisely and, then, developing models and propositions on their dynamic interconnections. In many ways, macro theory is far less developed than micro-level theories in articulating the interconnections among their constituent processes, and nowhere is this difference more evident than in the analysis of social differentiation—a most talked-about, but poorly conceptualized, property of human organization.

INTEGRATING PROCESSES

The term *integration* has been given a bad name in light of its heavy use by functional theorists (see, especially, Chapters 2 and 3). The term has come to be associated with visions of a functional "need" for integration and with a belief in the inevitability of this need being met. I still like the term, however, because it captures a fundamental property of the social universe: the lacing together of differentiated actors into a coherent system. If integration in this sense does not exist, at some minimal level, then macro sociology has no subject matter. For, in the end, our goal is to understand patterns of social organization; without some degree of integration among social units, social organization could not exist. We should not, therefore, throw out the concept of integration; rather, we should use it in less problematic ways.

As Herbert Spencer[9] was the first to recognize, integration of social units has meaning only when cast against its opposite—disintegration, or the breaking apart of connections, relations, and common symbols holding units of a population together. For, as some conflict theorists emphasized (Chapters 9 and 10), there is a constant tension and dialectic between pressures that hold social units together and those that drive them apart. The former are termed *integrative* processes, the latter *disintegrative*; and we need not be evaluative about those terms, asserting that one or the other is "good" or "bad". Let me turn first to disintegrative processes and, then, to varying types of integrative processes.

Disintegrative Processes

A differentiating population always reveals disintegrative pressures that I label under three headings: (1) problems of *coordinating* subgroupings and, to a lesser extent, subrankings; (2) problems of *symbolically unifying* diverse subcultures and, at times, subcategories; and (3) problems of *politically consolidating* power in the face of opposition from subrankings. I will discuss each of these in turn.

Problems of coordination   As subgroups become differentiated from one another, it becomes increasingly problematic as to how relations among these subgroupings—organizations, communities, associations, and other corporate units—are to be created and maintained. Functional theorists, especially Spencer and Durkheim (Chapter 2), were the first to recognize this problem;[10] since their initial insights, all functional theories, especially that developed by Talcott Parsons (Chapter 3), have viewed the coordination of substructures as a central dilemma of differentiated populations. For, if some degree of coordination does not occur, differentiation leads to disintegration.

---

[9]This aspect of Spencer's thought was not discussed in Chapter 2. See Jonathan H. Turner, *Herbert Spencer: A Renewed Appreciation* (Newbury Park, CA: Sage, 1985) or Jonathan H. Turner, Leonard Beeghley, and Charles Powers, *The Emergence of Sociological Theory* (Belmont, CA: Wadsworth, 1988), Chapter 3.

[10]Actually, the problem was first emphasized by Adam Smith, 100 years earlier, and by Comte, some 60 years previously. But it was with Spencer and Durkheim that the problem became central to sociological work.

Problems of symbolic unification As subcultures and social categories proliferate, questions of how to sustain at least some common symbols—language, values, beliefs, institutional norms, and the like—increase. This problem is aggravated as subgroupings and subranks in hierarchies develop their own unique subcultures and, as a result, become the basis for distinct subcategories. Although the problem was first given forceful advocacy by the founder of utilitarianism (Chapter 2), Adam Smith,[11] it was the French functionalists, especially Émile Durkheim but also Auguste Comte (Chapter 2), who phrased the issue sociologically. For them, a society divided and partitioned into subcultures and subcategories could reveal no unifying force. Durkheim added an additional insight, which was to be adopted by Parsons (Chapter 3), with his famous concept of *anomie*. What Durkheim saw was that, in order for symbols to unify diversely situated and oriented subpopulations, subcultures, subcategories, and subgroupings, it would be necessary for common values and beliefs to "generalize" and become "more abstract." Only in this way could they have relevance to subpopulations that have their own unique experiences and world views. The problem, he noted, is that at times this process—or what Parsons (Chapter 3) termed *value generalization*—occurs too rapidly, *before* more immediate and intermediate subcultures, norms, and other specific constraints and interdependencies among diverse actors and subpopulations are in place. As a consequence, there is a "lack of regulation," or *anomie*. Niklas Luhmann (Chapter 5) was to question whether or not "value consensus" or "symbolic unification" is really all that necessary, but, more typically, functional theorists have emphasized the importance of consensus over symbols. Similarly, most structuralisms also emphasize common symbolic codes as essential to social order (see, especially, Chapters 24 and 25). For, if *symbolic unification* does not occur, at some minimal level, then disintegration of a population is likely.

Problems of political consolidation Beginning with Spencer (Chapter 2), functional theorists have been likely to emphasize the importance of political control and regulation of internal (and external) conflicts and problems of coordination. In contrast, beginning with Marx and, to a lesser extent, Max Weber (Chapter 2), conflict theorists are more likely to argue that the centralization of power will create hierarchies and subranks, with particular subranks exploiting those in lower ranks and, as a result, increasing the probability (or, in some theorists' minds, inevitability) of conflict between those who have and those who do not have power. Similarly, exchange theorists, especially Blau (Chapter 16) and Emerson (Chapter 27), believe that those actors who possess power will use it to their advantage, thereby creating tensions among actors. Thus the very process of *political consolidation* can, despite being necessary to regulate and control differentiated units, become a source of conflict and tension. Without political consolidation, the disintegration of differentiated populations is likely (the functional argument), but at times this very process

---

[11]Adam Smith was particularly concerned with how to provide common "moral sentiments" in a social system differentiated into subunits and subcultures.

of political consolidation, especially the concentration and centralization of power for exploitive purposes, is likely to result in equally significant disintegrative events (the conflict and exchange theory argument).

In sum, then, differentiation creates disintegrative pressures. This has been the fundamental insight of functional theory from its beginnings, and it is still basic to the analysis of the social universe at a macro level. These functional theories (Chapters 2–7, excluding 8) tend to stress structural coordination of subunits and symbolic unification as the most important integrative problems. Conflict and exchange theorists (Chapters 9–13, 14–17, 27) have added to this insight the recognition that the use of power to create and sustain hierarchies and rankings represents not only an axis of differentiation but also a source of disintegration (through conflict). How, then, are these disintegrative problems resolved? In a sense they are never fully resolved, as the dust of historical societies readily documents. At best, integration is a short-term period of "negative entropy"—to use a phrase from general systems theory (Chapter 6)—because, in the long run, entropy or disintegration is inevitable for "living systems." Thus, as we move to the examination of integrative processes, we should keep in mind that these processes are always temporary—at least they have been up to the present moment in history.

But what produces even this short-term integration of a differentiating population? How are the problems of disintegration overcome? How and why do members of a population figure out how to coordinate their activities, symbolically unify themselves, and consolidate power in ways that are seen as appropriate and legitimate by diverse actors? To answer such questions, we need to address an issue that is implicit in all functional theories: social selection.

Selection Pressures and Social Integration

Functional theorizing has always been based upon analogies to biology (see Chapter 2 for details). Only recently have new forms of theorizing, such as those in sociobiology (Chapter 8), gone beyond analogizing and actually viewed human social organization in biological terms. For sociobiologists, human social organization represents an adaptation to environmental forces; hence, human society has been driven by pressures of natural selection. Those "genes," "individuals," or "groups" that have achieved "inclusive fitness" through social organization and culture are more likely to survive than those that have not developed organization and culture.

Although distinctive to present-day sociobiology (Chapter 8), this kind of reasoning has, in actuality, been implicit in all functional arguments from the beginning. Notions of "functional needs" emphasize that certain states—whether integration, adaptation, goal attainment, complexity reduction, inclusive fitness, or whatever—*must* be met *if* a system is to survive. In order to avoid the problems of illegitimate teleology (end states produce their causes) and tautology (circular reasoning), a selection argument is often made, rarely explicitly but often implicitly (see discussion in Chapter 3). For example, when

Talcott Parsons (Chapter 3) indicated that the economy meets "needs for adaptation" in societal systems, he really argues the following: those populations that are able to develop an economy (by luck or chance, imitation, borrowing, or however) are more likely to survive than those that cannot do so. That is, there are *selection pressures* on populations to develop an economy if they are to survive in an environment.

Such arguments are vague in the sense that they do not tell us how, or through what specific historical processes, a population responded to the environmental pressures and developed an economy. Yet they are no more vague than many such selection arguments in biology. Sometimes we cannot know, or can only get hints, about the specific environment of a population or the ways in which it responded to the environment in order to survive. Moreover, many of the cultural and structural features of a population have little to do with survival; rather, they are embellishments and elaborations that have been developed for other reasons having nothing to do with ultimate survival.

Why, then, would we even want to address the issue of selection? My answer is that we need to invoke such arguments to understand, in a very fundamental way, why disintegrative pressures are overcome. When confronted with a disruptive lack of coordination, a distinctive inability to unify symbolically diverse cultures, or a destabilizing political situation, people respond to the "pressure." Sometimes they fail to respond effectively; oftentimes they muddle through to at least a partial solution. What ways a population deals with these pressures and what configuration of these pressures exists are historical questions, tied to the unique and particular circumstances of that population. Responses to disintegration are not theoretical questions when viewed in this way; rather, they are to be answered by examining what a population actually did historically in response to particular exigencies.

But at a more general—albeit vague—level, we can say that the problems of coordination, symbolic unification, and political consolidation create selection pressures to which a differentiating population responds, or perhaps fails to respond. And, if the population is to remain viable—that is, integrated to some minimal degree—it must do certain things in some minimal way. Otherwise, it will not remain integrated or viable. This does not mean, however, that a population will know what to do, or succeed; at some point all populations disintegrate. Thus, functional theories have always tried to say what is essential, at a basic and fundamental level, for a population to remain viable. And this has always been the appeal of functionalism, despite the explanatory problems that notions of functional needs generate.

I have taken some time on this point because we must confront it when doing theoretical analysis at a macro level.[12] We must, in the end, be able to indicate the basic processes that integrate a population and, thereby, hold off the forces of disintegration. In so doing, we invoke a selection argument: dis-

---

[12]This problem also exists at the micro level when micro theorists address the maintenance of an encounter or group structure. Goffman (Chapter 22), for example, often used the terms *function* and *need*.

integrative processes generate selection pressures for a population to respond and develop ways of keeping them in check. If not, a population will be subject to, in Spencer's words, "dissolution." We do not, of course, have to make a very strong functional argument; all we need to say is this: under various configurations of disintegrative pressures, a population will respond, if it can and knows how; if, by luck, rational planning, imitation, trial and error, or however (historically/empirically), it actually does so, these responses will be of certain generic types. It is to isolating these basic types of responses that macro theory is dedicated, and this effort involves the discovery of various integrating processes. Let me review what the theories indicate about these bases of integration.

Integration through Structural Coordination

As the number and diversity of subgroupings and, to a lesser extent, the number of associated subcategories and subcultures increase, problems of coordination escalate. Relations among subgroupings become ever more difficult to articulate, setting into motion "selection pressures" for various patterns of subgroup coordination. What are some of these patterns?

Structural interdependence   Spencer and Durkheim (Chapter 2), contemporary ecological theory (Chapter 7), and exchange theory (Chapters 17 and 27 in particular) all emphasize the creation of interdependencies among subgroupings as fundamental to social organization. With differentiation, subunits specialize and, as a result, must exchange resources with other units in order to sustain themselves. Exchange theories emphasize that built into these transfers of resources are power dynamics, since those units with scarce and valued resources are often in a position to extract compliance from other units. But the more general point is that networks and chains of interdependence (and varying degrees of dependence) can emerge to resolve problems of coordination. Of course, at times these interdependencies cannot be established, or are established in highly exploitive ways, with the result that coordination becomes difficult and releases powerful disintegrative forces.

Structural inclusion   As Niklas Luhmann (Chapter 5) argued, another form of coordination revolves around the process of inclusion, in which one unit is lodged inside a more inclusive unit, with the latter lodged inside an even more inclusive unit, and so on (in Luhmann's case, interaction systems are located inside organizational systems, with the latter being inside societal systems). For example, complex organizations or bureaucracies operate in this fashion by incorporating groups, offices, and divisions inside an organizational structure; or, an institution such as an economy can be seen as incorporating, gathering, producing, distributing, and servicing units. Inclusion not only facilitates the distribution of resources, but, as Simmel (Chapter 14) and later Parsons (Chapter 3), Luhmann (Chapter 5), and Blau (Chapter 16) stressed, it also typically delimits "the media of exchange" used in establishing interdependence. At the same time inclusion can create a common set of cultural symbols, thereby promoting some degree of cultural unification. Thus inclusion provides one potential solution to problems of coordination; conversely, the lack of inclusion escalates disintegrative pressures.

Structural overlap   Another pattern of coordination is structural overlap, in which one unit is partially part of another, either through networks of interdependence or through incomplete inclusion. Simmel (Chapter 24), Blau (Chapter 28), and network theory (Chapter 27) all recognize the significance of cross-cutting social circles or overlapping membership in structural units. Such overlap allows actors to understand one another's perspectives and courses of action, thereby reducing problems of coordination, and perhaps conflict as well. Overlap also creates chains of dual loyalties and obligations, which also make it easier to coordinate activities across subgroupings and subrankings. Without some degree of overlap, then, structural units confront one another as strangers, and, as a result, relations among them become more difficult.

Other coordinating processes   Whether through interdependence, inclusion, or overlap, structural coordination is facilitated by the other integrating processes to be discussed below. Let me just mention them here. First, as Durkheim (Chapter 2), Parsons (Chapter 3), Luhmann (Chapter 5), the early Blau (Chapter 16), various structuralisms (Chapter 25), and structuration theory (Chapter 26) argued in their own unique ways, common symbolic repertoires among individuals and corporate actors greatly facilitate structural interdependence in exchanges (through generalized values, norms of fair exchange, justice, media of exchange), structural inclusion (through the above, plus entrance and exit rules), and structural overlap (through all of the above). Thus, to the degree that symbolic unification becomes possible, structural coordination is facilitated. Second, political consolidation can also encourage structural interdependence, inclusion, and overlap through regulation by laws and agencies of the state or other forms of centralized authority. Such authority must not only consolidate power to do so; it must also be perceived as legitimate. When such is not the case, political processes often disrupt as much as facilitate structural coordination.

Integration through Symbolic Unification

Durkheim (Chapters 2 and 24) and Weber (Chapters 2 and 9) both stressed the importance of common ideas as an integrating force in human populations. More recent functional approaches, inspired not only by Durkheim and Weber but also by Simmel (Chapters 2 and 9), have also recognized that values, codes, norms, information, and symbolic media are essential to the organization of a population (see Chapters 3–8). Even in the negative sense of "false consciousness" or "ideological domination," critically oriented conflict theorists, such as Marx (Chapter 9) and Habermas (Chapter 13), acknowledge the power of ideas to regulate and control. And various structuralisms (Chapter 25) as well as structuration theory (Chapter 26) similarly emphasize the constitution and use of informational and evaluative codes and rules across a population as essential to the social order. Even micro approaches, such as Goffman's (Chapter 22) dramaturgy, recognize the importance of institutional norms and frames for constraining micro encounters. Exchange theories (Chapters 14–17, 27) also acknowledge that norms and perhaps values come to regulate transactions.

There are disagreements among these theorists, however, over what kinds of unifying symbols are most important. For example, Spencer (Chapter 2), Durkheim (Chapters 2 and 24), Parsons (Chapter 3), and the early Blau (Chapter 16) argue for the importance of generalized values that provide common standards of moral evaluation. In contrast, Merton (Chapter 4) stresses the importance of regulatory norms in addition to generalized values—if "anomie" is to be avoided. Similarly, general systems theory, whether Luhmann's neofunctionalism or Miller's (Chapter 6) living systems approach, appears to emphasize codes or norms as regulating conduct. Giddens' structuration theory (Chapter 26) and structuralism (Chapter 25) are also more normative, highlighting rules and codes about rights and duties as well as interpretative schema. Micro interactionist approaches (Chapters 18–23) take a parallel tack, although they use somewhat different terms—definition of situations, accounts, generalized others, reference groups, frames, etc.—to describe the use of interpretation schema and norms to regulate conduct. There are, then, many shades of emphasis on how the process of symbolic unification occurs.

But, however conceptualized, there is the recognition that, as population differentiates, the number and diversity of subgroups, subranks, subcategories, and, in particular, subcultures increase; with this increase, problems of symbolic unification escalate dramatically, setting into motion selection pressures for the production of symbols, whether values, beliefs, norms, or codes. But how are we to conceptualize these processes, given many different points of emphasis? My answer is eclectic: all of these conceptualizations are relevant to understanding the process of symbolic unification. Let me elaborate.

Evaluational symbols I think that Parsons (Chapter 3) is correct in his emphasis that, to some degree, symbolic unification depends upon generalized and abstract values that, because they are general, are relevant across diverse subcategories, subgroups, and subranks. These values provide moral premises for other symbols—indicating in general terms what is "good" and "bad." I am thus arguing against Luhmann's (Chapter 5) position that integration does not require consensus on general values. My view is that, the larger the population to be integrated, the more significant such evaluational symbols are. Norms alone cannot integrate a large population.

Definitional symbols The process of symbolic unification also depends upon sets of symbols that define situations as being of a certain nature and that provide a scheme or perspective for interpreting the situation. Many terms have been used in the theories presented in this book to describe these definitional processes—terms like *frame, cultural codes, perspective, collective conscience, moral order, community of attitudes, common meanings, ideology,* *beliefs, cultural capital, symbolic order, interpretative rules, stocks of knowl-edge,* and so on. Whatever the label, the essential point is that symbolic uni-fication depends, to some extent, on individuals employing a common inter-pretative framework or perspective—what I term here *definitional symbols.*

Regulatory symbols   In addition to evaluative and definitional symbols, symbolic unification requires norms or rules about how actors are to operate in a situation. Functional theories (Chapters 2–8) always emphasize this aspect of symbolic unification, but so do interactionist (Chapters 18–23) and struc-turalist (Chapters 24–26) theories as well. These symbols are typically seen as regulating actions by specifying rights and obligations as well as negative sanc-tions for their violation and procedures for repairing such violations. Once again, many terms—*rules, codes, norms, information,* etc.—have been used to denote this dynamic. I have employed the term *regulatory symbols* to incor-porate these diverse terms and to emphasize that the regulation of actors' responses as well as of interactive relations among actors (both individuals and collectivities) in a population is guided by symbols that specify what is to be done.

Symbolic media   As Simmel (Chapter 14) was the first to articulate clearly for the case of "money," relations among actors are carried out in terms of "generalized media of exchange," to use Parsons' (Chapter 3) terms, or "communication media," to use Luhmann's (Chapter 5) vocabulary. That is, actors often employ a symbolic medium in their activities or in their exchanges. Money is the prototypical symbolic medium (it rarely has intrinsic value but, instead, symbolizes value) for economic exchanges, but other symbolic media have been postulated—power, influence, love, etc. This aspect of symbolic uni-fication has not been well conceptualized—being confined primarily to func-tional theories (especially Chapters 3 and 5), but it seems to capture an im-portant dynamic. For it does appear that relations among actors are conducted in terms of media and that consensus over the use of the appropriate symbolic media is an important force in integrating a large population.

Other symbolically unifying processes   Just as symbolic unification influences the degree of structural coordination, so problems of structural co-ordination stimulate a search for unifying symbols, especially definitional and regulatory codes as well as symbolic media that can be used to coordinate relations among social units. Moreover, the specific nature of coordinating problems and the attempted solutions to these problems influence just what symbols are used to unify a population of actors. Problems of coordination, as well as processes of political consolidation, circumscribe symbolic unification. Efforts to concentrate, centralize, and legitimate power will influence the de-velopment of evaluational and definitional symbols. Additionally, as political processes are employed to regulate action and the coordination of interaction among social units, these political processes will shape what regulatory symbols and what symbolic media are employed in situations defined as crucial by those holding power. And, if there is opposition to consolidated power, then mobilization of this opposition will involve extensive efforts on the part of those initiating conflict to develop new evaluational and definitional symbols that *de*legitimate existing authority (see Chapters 9–13 on conflict theory or the discussion of cultural structuralists in Chapter 25).

In sum, then, problems of cultural unification, coupled with problems of structural coordination and political opposition as well as efforts at political consolidation, create selection pressures for symbolic unification. Despite the recent revival of cultural sociology (Chapter 25), however, these symbolic processes are not well understood. There is, then, much macro-level theoretical work to be done in this area.

Integration through Political Consolidation

Problems of coordination, as Spencer (Chapter 2) was the first to recognize fully, create selection pressures for the differentiation of what he termed "regulatory" structures, or government.[13] Moreover, he stressed that there was a constant dialectic between movement toward *centralized* political authority (decision making concentrated and rank ordered) and movement toward decentralized authority (decision making diffused and less rank ordered); for centralized authority creates resentment and opposition, leading to decentralization, which escalates problems of coordination, causing a movement back to centralized power.[14] Conflict theorists, especially those in the Marxian tradition (Chapter 2), and some forms of exchange theory (Chapters 16 and 27) add the insight that power tends to become *concentrated* in the sense that fewer and fewer social units hold an ever-increasing proportion of the power in a population. Eventually there is resistance to this consolidation, but the important point is that power itself is used to gain ever more power. Max Weber (Chapter 9) and later the functional approaches of Talcott Parsons (Chapter 3), Niklas Luhmann (Chapter 5), and Lewis Coser (Chapter 11), as well as Ralph Dahrendorf's (Chapter 10) dialectical theory and Peter Blau's (Chapter 16) exchange model, all recognized the importance of *legitimation* of power—that is, the perception and belief of those subject to power of the right of those using power to do so.

These three power dynamics—centralization, concentration, and legitimation—are, I believe, fundamental to resolving problems of opposition. Conversely, they can also escalate problems of opposition when ineffectively or exploitively implemented. Let me examine each in more detail.

Centralization of power   Problems of coordination emerging with the differentiation of subgroups and problems of opposition stemming from ine-

---

[13]These were not discussed in Chapter 2; see Turner, Beeghley, and Powers, *The Emergence of Sociological Theory*, Chapter 3.

[14]Spencer's distinction between "militant" (centralized) and "industrial" (decentralized) was meant to communicate these dialectical dynamics.

qualities, hierarchies, and ranking will inevitably create selection pressures for centralization of power. As subunits cannot coordinate their activities, it is likely that the power to make decisions will be vested in a third party; when this occurs, the initial steps toward centralization are taken. Once centralized, those ranking high in power will use their power to usurp ever more decision-making alternatives—at least to a degree. This process is likely to accelerate if problems of coordination are acute. And, when there is considerable tension over the inequalities associated with subrankings, political authority is likely to centralize power even more as a way to "deal with" these tensions.

There is, however, a delicate balance between (1) too much centralization, such that decisions cannot be made because of bottlenecks in the hierarchy of authority, and (2) too little centralization, such that decisions cannot be coordinated. In either case, problems of coordination of subgroups and problems of opposition will increase. Thus centralization of decision-making prerogatives is always a problematic force in integrating a population.

Concentration of power　I am distinguishing consolidation of power from centralization. Concentration of power refers to the proportion of social units holding power, whereas centralization is the extent to which *decision-making prerogatives* are vested in a few units. Consolidation and centralization can be highly correlated, but often they are not. For example, coercive power can be highly concentrated in a few units, but decision-making authority on many matters can be delegated (as in the case when leaders in a declining feudal system delegate the rights to collect taxes and make local decisions in exchange for needed funds). Conversely, many decision-making prerogatives may be highly centralized, but the number of power-holding units can still be great (as is the case, for example, in an oppressive dictatorship during periods of mobilization of a revolutionary opposition about to initiate a civil war). But this "split" between centralization and concentration can go only so far, since too much delegation of decision making bestows an enormous amount of capacity to concentrate (often informally and through "corruption") power; and the concentration of coercive power creates pressures not to delegate too many decision-making prerogatives to others who could use these prerogatives to usurp power.

Concentration of power tends to be self-perpetuating, at least up to the point of resistance, opposition, and mobilization of counterpower (which then deconcentrates power). Concentrated power is thus "self-selecting," but there are other processes encouraging concentration. One is acute problems of coordination, another is external enemies who must be dealt with, and yet another is internal opposition that must be repressed (in the short run this latter process will concentrate power, but in the long run it will encourage renewed efforts at mobilization of counterpower). Thus some degree of concentration is necessary to coordinate, regulate, and control a differentiating population, but this very concentration tends to encourage even more concentration, setting into motion opposition and mobilization of counterpower (which can then deconcentrate power; although, if a successful revolution or civil war is a result, power typically becomes highly concentrated in order to suppress those on the losing side).

Legitimation of power   The volatility of centralized and concentrated power is, to some extent, mitigated by symbolic processes in which evaluational and definitional symbols are codified and disseminated to a population, creating in them a sense that those who hold power and make decisions have the right to do so. When centralized and concentrated power is successfully legitimated (as Weber, in Chapter 9, emphasized), it is a source of stability because it gives certain social units the right to control and regulate other units. And, as a result, problems of coordination and opposition can be mitigated. Moreover, legitimated power can be used to create unifying symbols, which obviate problems of cultural unification. For this reason, most conflict theories (Chapters 9–13) stress the importance of developing counterideologies that call into question the evaluational and definitional symbols legitimating an existing political authority. For only by delegitimating power can individuals become willing to incur the risks of mobilizing counterpower and engaging in conflict.

In sum, then, the concentration, centralization, and legitimation of power constitute a double-edged sword. These processes are "selected for" because of problems in coordination, cultural unification, and opposition; and they can, for a time, mitigate these problems. Yet, sustaining legitimacy in the face of escalating centralization and concentration of power is often problematic, especially in differentiating systems where subgroupings, subcategories, subcultures, and subranks all have somewhat different interests, goals, and orientations.

The Dynamics of Integration

As I have indicated, *integration* is merely a term for denoting the diverse ways in which actors, whether individuals or collective units, are organized. Integrative forces for structural coordination, symbolic unification, and political consolidation are "selected for" by disintegrative pressures caused by differentiating subgroupings, subcategories, subcultures, and subrankings. If some degree of structural coordination, symbolic unification, and political consolidation can emerge and prove effective in mitigating problems of coordination, unification, and opposition, then further differentiation can occur. Of course, increased differentiation can escalate disintegrative pressures beyond the capacity of a population to respond and reintegrate the new level of differentiation.

This line of argument does not need to be made in terms of the evolutionary frameworks of the functionalist perspective, particularly the work of Spencer, Durkheim, Parsons, and Luhmann (Chapters 2, 3, and 5). Nor does this kind of argument need to be illegitimately teleological or tautological, as was the case for many functional arguments couched in terms of system requisites and needs (see discussion in Chapters 3 and 8). And we do not need to view any given configuration of integrative and disintegrative forces as "good" or "bad."

Instead, what is required is an effort to conceptualize more precisely those basic and generic processes that are used to integrate a population—if indeed such integration is possible in particular historical circumstances. I have termed these "integrative processes," and our goal should be to specify the general types of conditions under which they are likely to occur (or be "selected for").

CONCLUSION

In this chapter, I have outlined the cumulated legacy of macro-level theorizing. I have classified this theorizing under the general rubrics of assemblage, differentiation, and integration. Assembling processes revolve around aggregation, population size and growth, and production; differentiating processes are fueled by competition, exchange, and mobilization of power in ways that produce subcategories, subgroups, subcultures, and subranks; and integrative processes represent efforts to deal with disintegrative pressures through structural coordination, symbolic unification, and political consolidation. Obviously, my labels for these processes do not have to be accepted, but the general processes that they denote constitute the subject matter of macro sociology.

The next step is to begin reconceptualizing these processes in more precise terms and, then, developing models and propositions to describe the complex relations among these integrative dynamics. Macro sociology surfaces in many areas—functionalism, historical sociology, ecological theory, world systems theory, geopolitics, and many other subfields—but, unlike micro theorizing, it is not even close to being conceptually unified. Part of the reason for this disarray is that macro theorizing has many more historical roots than micro theorizing; as a result, different terms, assumptions, and strategies are often employed by macro sociologists. The result is that macro theorizing is poorly developed.

Until macro theory is more fully developed and until micro theorizing is fine-tuned, concern over the micro/macro "link" seems, to me at least, premature. Nonetheless, I am probably in the minority on this issue, and so we should close this book with a review of how sociological theorists have approached this linkage problem. For lack of a better term, I will call this *meso*-level theorizing.

CHAPTER 32

Meso Theorizing

THE MICRO/MACRO "GAP"

As I indicated in Chapters 1 and 29, sociological theory has in recent decades become concerned with reconciling micro-level conceptualizations of behavior and interaction among individuals with macro-level analyses of populations of actors. The goal of those working on this problem is to find "the missing link" between these realms and, thereby, to fill in this conceptual "gap." Indeed, there has been a recent burst of activity[1] addressing this micro/macro question, and so it is appropriate that I close this book with a review of various theoretical strategies for resolving this conceptual problem.

I will term these efforts *meso*analysis. That is, how is sociology to conceptualize the middle ground between micro and macro social processes? I will not examine micro chauvinistic positions, which assert that only the micro realm is "really real" and that conceptualizations of social structure, institutions, social systems, societies, and other macro-level phenomena are reifications. Nor will I explore macro chauvinistic arguments that macrostructures are the only relevant topic for sociological analysis and that action and interaction are programmed and constrained by such structures. Rather, my concern is with reviewing in more detail than in Chapter 29 just how sociologists have attempted to link micro- and macroanalysis.

STRATEGIES FOR "MESO" ANALYSIS

Constructing Theories of the "Middle Range"

Robert K. Merton's (Chapter 4) advocacy for "theories of the middle range" did not directly confront the micro-vs.-macro issue, as it is now portrayed. Yet

---

[1] See, for example, Karin Knorr-Cetina and Aaron V. Cicourel, eds., *Advances in Social Theory and Methodology: Towards An Integration of Micro and Macro-sociology* (London: Routledge & Kegan Paul, 1981); Randall Collins, "On the Micro-foundations of Macro-sociology," *American Journal of Sociology* 80 (1981), pp. 984–1014; Jeffrey C. Alexander et al., eds., *The Micro-Macro Link* (Berkeley: University of California Press, 1987). Moreover, there have been special sessions at many professional meetings on the issue: in fact, one of the themes of a recent American Sociological Association annual meeting was the micro/macro question.

it is probably the most prevalent strategy for getting around the problems of reconciling the micro/macro realms. Merton argued that theories should not try to explain everything all at once but, instead, should concentrate on specific topics or classes of phenomena. In Merton's view, sociology should seek to develop "theories of" particular phenomena[2]—deviance, social movements, stratification, demographic processes, institutions (family, economy, religion, etc.), organizations, groups, and so on—and thereby encompass in each theory only a "limited range" of social phenomena. Moreover, these theories should not be so abstract that they cannot be easily operationalized and tested.

Eventually, Merton argued, these "theories of" can be consolidated, but first it is necessary to build up a corpus of useful theoretical constructs and a body of empirical findings. Without this preliminary middle-range theorizing about delimited classes of phenomena studied at midlevels of abstraction, theory will float off into the speculative and philosophical clouds.

Merton's eloquent advocacy has, without doubt, exerted enormous influence on sociological theory and research in the last four decades. Indeed, his middle-range strategy became the legitimating ideology of a "mainstream sociology" that required sociological works to involve the collection and analysis of data on some specific topic. Most of the research following Merton's advocacy has been of the survey variety, although other kinds of empirical work also follow Merton's dictates. The use of surveys enable macro/micro problems to be resolved rather easily: ask questions of people about their attitudes, beliefs, and behavior (the micro part) as well as their race/ethnicity, income, education, and occupation (surrogate measures of the macro part); then, correlate aggregate measures or scales of one with the other. Thus a more macro-level phenomenon like "stratification" (measured by a "scale" incorporating income, occupation, and education) could be linked to a micro-level phenomenon such as reports about behavior and attitudes (also measured with attitude "scales"). At the empirical level this kind of analysis can be highly useful, although I think that mainstream sociology went a bit overboard.[3] But, at a theoretical level, this strategy had some "dysfunctional" and "unanticipated" consequences—if I can turn Merton's words back on him.

One consequence was to create a gap or split between "theorists' theory" and "researchers' theory." Theorists often decline to theorize about some specific processes, and so they have done the very thing that Merton feared: wandered up the ladder of abstraction and into the realm of philosophical discourse. Researchers, on the other hand, have kept their noses to the empirical grindstone, fearing any concept that was not easily operationalized. The result was that much theory in sociology cannot be tested and that much research is

---

[2]I am putting these words in Merton's mouth, but the result of his advocacy was to produce "theories of" specific empirical events. For my criticisms of this point of view, see Jonathan H. Turner, *Societal Stratification* (New York: Columbia University Press, 1984).

[3]See Jonathan H. Turner, "The Disintegration of American Sociology," *Sociological Perspectives*, 32 (4, 1990): 419–33; Stephen Park Turner and Jonathan H. Turner, *The Impossible Science* (Newbury Park, CA: Sage, 1990) and Jonathan H. Turner and Stephen Park Turner, *American Sociology: Its History, Structure, and Substance* (Warsaw: Polish Scientific Publishers, in press).

atheoretical. There is, then, a new "gap"—not so much between micro and macro as between theory and research.

Another consequence of Merton's advocacy was the proliferation of what I have termed "theories of" _____ (fill in the blank with an empirical topic).[4] In seeking to delimit the range of phenomena incorporated into a theory, many thinkers (and researchers as well) developed concepts and propositions about nongeneric processes—mobility in the United States between 1950 and the present, juvenile delinquency in America, family violence, demographic transitions, urbanization in the 20th century, and so on. That is, empirical descriptions of specific events in particular times and places were dressed up to "look theoretical" and to appear more basic and generic. And so, for virtually any area of empirical inquiry in sociology, there is a corresponding set of theories. Few of these are abstract, and few address basic and fundamental processes of human interaction and organization. And, as a result, they are unlikely to be "consolidated" in the future; nor are they likely to create a cumulative theoretical inventory of abstract concepts and propositions.

Yet, having said this, I want to qualify this conclusion. There are several middle-range theories that are not just empirical generalizations. These are highly abstract theories about a delimited phenomenon, or related class of phenomena, presumed to exist in all times and places. For example, to some extent, exchange theory (Chapters 14–17) is of this nature, arguing that an exchange dimension is "always there" in social relations among actors, whether individual people or corporate actors. Similarly, the "balance" theories of Fritz Heider and Theodore Newcomb (see Chapter 27 on networks) might be seen as "middle range" because they address a limited range of phenomena (balance processes) while being highly abstract and relevant to all types of social relations.

There are many such theories in sociology, and I have not examined them in these pages. But they all share a view that there are certain basic processes always operating in the social universe and that it is possible to develop and test theories about these processes. Let me simply list some of the theoretical traditions to communicate what they look like. For example, theories about expectation states,[5] legitimation,[6] and justice[7] processes are highly developed and can perhaps be viewed as the quiescence of middle-range theorizing because they are abstract but delimited, while explaining dynamic processes that are, I sense, always there in social relations. And so, if a meso strategy were to

---

[4]See note 2.

[5]Joseph Berger, Thomas L. Conner, and M. H. Fisek, eds., *Expectation States Theory: A Theoretical Research Program* (Cambridge, MA: Winthrop, 1974); Joseph Berger and Morris Zelditch, *Status, Rewards and Influence* (San Francisco: Jossey-Bass, 1985). For recent research and theory, see also Murray Webster, Jr. and Martha Foschi, eds., *Status Generalization: New Theory and Research* (Stanford, CA: Stanford University Press, 1989).

[6]Morris M. Zelditch and Henry A. Walker, "Legitimacy and the Stability of Authority," *Advances in Group Processes* 1 (1984), pp. 1–25; Henry A. Walker, George Thomas, and Morris M. Zelditch, "Legitimation, Endorsement and Stability," *Social Forces* 64 (1986), pp. 620–43.

[7]Guillermina Jasso, "A New Theory of Distributive Justice," *American Sociological Review* 45 (1980), pp. 3–32; "The Theory of the Distributive-Justice Force in Human Affairs," in *Sociological Theories in Progress* III, J. Berger and M. Zelditch, eds. (Newbury Park, CA: Sage, 1990).

concentrate on (1) isolating these and other basic processes, (2) developing formal theories about their dynamics, (3) testing these theories, and perhaps (4) consolidating them in some fashion, then a middle-range approach could prove to be a viable way to deal with micro and macro questions. For, by isolating basic processes that always occur when individuals and collective actors organize, these types of middle-range theories can bridge, or obviate, the perceived gap.

Building Conceptual Staircases

Next to the "middle range" strategy, the most popular approach for filling in the micro/macro gap is to construct a conceptual "staircase," moving from a conceptualization of behavior or action to interaction and then, from these "elementary states" or "building blocks," to more "structural" and "institutional" levels of analysis. At times theorists begin at the structural end and work back down the staircase to behavior, action, and interaction. Or, they go up and down the staircase, moving up from behavior and interaction to social structure and back down again.

Talcott Parsons (Chapter 3), following Max Weber's (Chapters 2 and 18) lead, best illustrates the action-to-structure strategy. As will be recalled (especially from Figure 3–2), Parsons began with a conceptualization of "modes of orientation," revolving around motivation and values. Various combinations of these produce three types of action—instrumental, expressive, and moral. When oriented actors interact and when this interaction becomes stabilized or "institutionalized," a social system emerges. A social system is composed of status-roles and regulated by norms, while being constrained by personality on the micro side and culture on the macro side. In Parsons' case, he stayed primarily at the social system level; when he did move beyond this level, his preferred topic was culture, although he wrote numerous insightful essays on personality.

In somewhat less elaborate ways, other theorists have employed this staircase approach. Michael Hechter's (Chapter 17) rational choice perspective begins with "rational actors" who, when confronted with organizational problems, create structures. Probably the best illustration of going up and down the micro to macro conceptual staircase is Anthony Giddens' (Chapter 26) structuration theory. Here, much like Parsons, Giddens sees "social systems" consisting of individuals communicating, exercising power, and using sanctions as the point of articulation among structure, structural sets and principles, and institutions, on the macro side, and reflexive monitoring, discursive and practical consciousness, and unconscious needs on the micro side. Giddens then goes back and forth in his discussions between agents and action, on the one hand, and structure and institutions, on the other.

A related staircase argument is Niklas Luhmann's (Chapter 5) typology of the three basic kinds of social systems as "interaction," "organizational," and "societal." Each system is an emergent reality operating in terms of its own unique dynamics. Yet, at the same time, the implication is clear that societal systems are composed of organization systems, which in turn are built from interaction systems. Much like Parsons, Luhmann prefers to stay at the more macro level—organization and societal systems—but there is a sense that these are built from more micro interaction systems. Once created, however, these more macro systems circumscribe the context in which interaction systems operate.

The problem with these kinds of staircase approaches is that they seek to do what I see as impossible—step-by-step linking of ever more macro units from elementary behavior and interaction. Since it is usually acknowledged that structure "emerges," there will inevitably be a "gap" between the emergent system level and the constituent processes that have produced it. And it is at these points, where something emerges, that conceptualization becomes vague. For example, Parsons has an elaborate conceptualization of action and macrostructure, but he says very little about interaction[8] (the way an interactionist would), which is *the* process that is supposed to link action and structure. To illustrate further, Luhmann employs very different vocabularies for each system level, making it unclear how the gaps between them are to be filled. And even Giddens, who is the most insistent that structure must be viewed as rules and resources that individuals use, becomes vague and metaphorical when linking institutions, structural principles, and structural sets to actual interaction in social systems.

Thus, in my view, the conceptual staircase strategy highlights rather than resolves problems of filling in the micro/macro gap. By trying to fill the gap, but in fact exposing it, these theorists emphasize the difficulties of doing meso *theoretical*[9] analysis.

Erecting Deductive Reductions

Another strategy for resolving the micro/macro gap revolves around the logic of deductive theory. By developing higher-order laws or axioms about individual behavior or interaction, and then making deductions from these laws to ever more macro contexts, the micro and macro are linked together. George Homans (Chapter 15) was the most persistent advocate of this strategy, arguing that the laws of behavioral psychology are the axioms from which the laws of sociology are to be deduced. Homans did not deny the relevance of sociological laws[10] about emergent structures, only the contention that these would be the highest-order axioms in sociological explanations. Similarly, sociobiology (Chapter 8) also operates with a reductionist strategy—arguing that the prin-

---

[8]For a more detailed criticism, see Jonathan H. Turner, *A Theory of Social Interaction* (Stanford, CA: Stanford University Press, 1988).

[9]Of course, empirical work can easily be done on the staircase, as was evident for Merton's middle-range strategy. In empirical work, one need only find "measures" of some structure and some behavior or attitude; once these are correlated, the micro/macro gap is obviated. Such studies rarely have much theoretical (as opposed to descriptive) relevance, however.

[10]Nor did he develop many, as his strategy would seem to have dictated.

ciples of sociobiology (fitness, reciprocal altruism, etc.) can explain social structures.

There are problems with this strategy, as can be illustrated with Homans' deductions. First, there are often big gaps in the deductions from the psychological axioms. Terms like *given* ("this or that" condition) followed by *therefore* (it follows that this or that is so) are prominent in Homans' deductions. For as soon as "structure" is introduced into the deductive scheme, the very things that are sociologically relevant—how did the structure emerge, what are its properties, what are its dynamics, etc.?—tend to be sidestepped in favor of showing that the structure is "deducible" from psychological axioms.

This problem is the result of a second issue in axiomatic theory—one that such diverse critics as Herbert Blumer (Chapter 18) and Lee Freese[11] (Chapter 1) have stressed. Axiomatic theory assumes that a formal calculus—such as logic or mathematics—is to be used in making deductions. It is the logic of the calculus that is to dictate the process of making deductions, but, in fact, most axiomatic theories are only "formal theories,"[12] involving verbal statements that are linked together in a discursive way. Once the rules of the calculus in a formal language no longer dictate how the deductions are to be made, a great deal of slippage between "axiomatic" premises and conclusions occurs. The deductions become metaphorical and vague, revealing a gap between the micro-level premises and the macro-level events to be explained in terms of these premises.

It is conceivable, however, that this approach could be fruitful *if* (and this is a big "if") gaps in the deductive logic could be reduced, either by careful wording or, more desirably, by the use of mathematics or symbolic logic. But, thus far, theories expressed in such formal languages tend to be technically elegant, narrow, and substantively less interesting than "sloppy" verbal ones. But the potential is there for connecting the micro and macro, not so much by some substantive linkage (that is, a particular process or social unit) as, more likely, by a logical connection. This latter kind of linkage is, I suspect, going to be less appealing than the metaphorical linkages that currently prevail.[13]

Searching for Formal Isomorphisms

One of the most productive ways to link the micro and the macro levels is to search for processes that occur at both these levels. This was, of course, Georg Simmel's (Chapters 2 and 24) strategy in his "formal sociology" in which the

---

[11]See, for example, Lee Freese, "Formal Theorizing," *Annual Review of Sociology* 6 (1980), pp. 187–212.

[12]Ibid.

[13]Several theorists are working creatively with formal languages. For example, see references to Jasso in footnote 7. See also Thomas J. Fararo and John Skvoretz, "Institutions as Production Systems," *Journal of Mathematical Sociology* 10 (1984), pp. 117–81; and "E-State Structuralism: A Theoretical Method," *American Sociological Review* 51 (1986), pp. 591–602.

unit of analysis (individual, group, organization, society, etc.) was not relevant; instead, the form of the relationship between units (exchange, conflict, competition, subordination, coalition formation, etc.) became the object of theoretical interest. In a sense, this strategy simply sidesteps the micro/macro gap and focuses only on those generic and basic kinds of relations in which the unit of analysis does not make a difference.

Several of the theoretical approaches examined in earlier chapters have employed this approach, as do some of the middle-range strategies discussed earlier. Among the functional perspectives, James G. Miller's (Chapter 6) systems approach clearly exemplifies a formal approach by postulating certain properties common to all "living systems"—reproducer, boundary, ingestor, distributor, encoder, decoder, etc. (see Table 6-2)—and articulating certain basic "laws" for all living systems (see Table 6-3). As I noted back in Chapter 6, this approach duplicates, in many ways, Herbert Spencer's Synthetic Philosophy, which similarly sought isomorphisms among realms of the physical, biological, and social universe.[14]

General systems theory, however, is not the most prominent of these formal approaches. Several exchange theories adopt a formal approach. For example, Peter Blau's early exchange theory (Chapter 16) saw the same basic processes—attraction, exchange, competition, differentiation, integration, and opposition—as occurring at the micro level, where individuals are the units involved in the exchange, as well as at the macro level, where collectivities are the units of analysis. The Emerson-Cook (Chapter 27) exchange network approach similarly employs a strategy seeking isomorphisms, arguing that power, power use, and balancing occur for all exchange actors, whether individual or corporate in nature. Network theory (Chapter 27), in general, and the Emerson-Cook merging of exchange theory with network analysis, in particular, operate with a view that the structure and dynamics of relations among positions in the network are the same, regardless of whether the actors occupying positions are individuals or collectivities. The dynamics of networks inhere in their structure, not in the nature of the actors involved.[15]

Even some interactionist approaches have hinted at a more formal and less reductionistic strategy. For example, at times Herbert Blumer (Chapter 18) appeared to argue that the same processes—symbolically denoting objects, mapping courses of action, self-reflecting and -assessing, defining situations, etc.—occur not only among individuals but among organizations as interacting units.

---

[14]For details on this philosophy, see Jonathan H. Turner, *Herbert Spencer: A Renewed Appreciation* (Newbury Park, CA: Sage, 1985).

[15]For another theory program in this vein, see David Willer, *Theory and Experimental Investigation of Social Structures* (New York: Gordon and Breach, 1987); David Willer and Travis Patton, "The Development of Network Exchange Theory," *Advances in Group Processes* 4 (1987), pp. 84–120; David Willer, Barry Markovsky, and Travis Patton, "The Experimental Investigation of Social Structures," in *Sociological Theories in Progress* III, eds. J. Berger and M. Zelditch (Newbury Park, CA: Sage, 1990); and David Willer and Barry Markovsky, "The Theory of Elementary Relations, Its Development and Research Program," presented at Stanford Conference on Theory Growth, August 1989.

This approach was never pursued by Blumer, but it has become highly prominent in organizational analysis.

The only potential problem in seeking isomorphisms across the micro and macro realms is to assume that their discovery obviates differences between these two realms. For at times the unit of analysis *does* make a difference in how social relations are created, sustained, and changed. Interaction or exchange among individuals, groups, organizations, societies, and other kinds of units may reveal important similarities, but also significant differences. Thus, as long as it is recognized that this search for isomorphisms is not the whole story, there is great potential for this strategy.

Positioning Intermediate Units

Probably the most truly "meso"—that is, "in the middle"—strategy is to visualize certain kinds of structures as being built from face-to-face interaction, on the one hand, and as providing the building blocks for more macro structures, on the other. The best example of this approach is Randall Collins' (Chapter 12) exchange conflict approach.[16] Here interaction rituals are seen to occur in face-to-face encounters, which are linked together in two basic kinds of "meso" units: organizations and stratification systems.[17] Thus interaction rituals are what sustain social classes (and class cultures as well) and organizational systems (bureaucracies, associations, large kinship units). In turn, large structures—societies, world systems, and institutions—are built from these two kinds of meso systems. In this way face-to-face encounters are linked to macrostructures; and, if analysis of macrostructure in terms of interpersonal processes is desired, then one engages in a sampling of micro encounters—what Collins terms "micro translations"—in order to see the kinds of interaction rituals involved in sustaining a macrostructure.

The problem with this approach is that, despite Collins' claims, it does not eliminate the micro/macro gap. Collins' own work best illustrates this fact. For, when he begins to address more macro topics, discussion of interaction rituals virtually vanishes. Even if one samples encounters and makes micro translations, this information does not inform us about how a macrostructure built from mesostructures is actually constructed from face-to-face encounters. Micro translation can, perhaps, give the investigator a better feel for "what is going on" at the interpersonal level, but this kind of understanding is very different from explanations of the properties and dynamics of mesostructures and macrostructures in terms of the content of interaction rituals. There will still remain a gap between the nature of the interaction rituals, on the micro side, and their aggregation and organization over time and space, on the meso and macro side.

---

[16]See also his *Theoretical Sociology* (San Diego: Harcourt Brace Jovanovich, 1988), which is organized around three major sections—macro, micro, and meso.

[17]These are, Collins argues, the basic meso units in Max Weber's effort to link action to macro historical processes.

Nonetheless, to the extent that the gap between micro and macro can be partially bridged, the analysis of the content of micro encounters and, then, an analysis of how they are aggregated and organized into the various meso-structures from which even larger macrostructures are built constitute a reasonable strategy for trying to link the micro and macro realms.

Other perspectives have come, at least implicitly, to the same conclusion. For example, as widely diverse perspectives as Parsonian functionalism (Chapter 3), Mertonian middle-range theory (Chapter 4), Luhmann's neofunctionalism (Chapter 5), role theory (Chapters 20 and 21), and dramaturgy (Chapter 22) all converge on notions of status, roles, and norms in organizational settings as the link to the macro realm. That is, organizations order the status, roles, and norms circumscribing behavior and interaction, while at the same time serving as the building blocks for larger macrostructures. But these analyses do not go much beyond my vague phrasing of the issue in the above sentences. Hence the gap remains.

Conceptualizing Mutual Parameters

As I mentioned in Chapter 29, my favorite approach to resolving micro and macro linkage problems is to view each realm as a parameter or constraint on the other and then conceptualize how micro constrains macro, and vice versa. That is, how do the processes of face-to-face interaction constrain what is possible and viable in organizing populations of actors, and, conversely, how does the existence of macrostructures constrain face-to-face interaction? These kinds of questions have been addressed occasionally, but, surprisingly, it is probably the least visible of the meso approaches.

Among the theorists examined in this book, Erving Goffman (Chapter 22) has been the most explicit in recognizing that both focused and unfocused encounters occur within a larger structural setting—gatherings and occasions lodged in institutions—which organizes physical space, produces norms, establishes appropriate deference and demeanor rituals, dictates frames and relevant keys (and fabrications), and displays other elements of face-to-face interaction. Similarly, Collins' (Chapter 12) view of "cultural capital" and "density" implicitly recognizes the constraints of macrostructure, as does Luhmann's (Chapter 5) recognition that interaction systems are embedded in organization systems and thereby constrained. Moreover, Bourdieu's (Chapter 25) analysis of the effects of classes and class factions on interpersonal behavior also views macrostructure as a parameter imposing limits (but also creating possibilities) for individual behavior.

On the other side of the issue—that is, how does micro constrain macro?—George Homans' (Chapter 15) contention that civilizations exist only to the extent that they fulfill people's needs hints at a source of constraint, although Homans never conceptualizes just what these needs are. Somewhat more specifically, Jurgen Habermas (Chapter 5) views humans' needs for "freedom from domination" as somehow making macrostructures less viable, but it is hard to know if this is an ideology or a factual statement about humans. Functional theories similarly imply micro constraints on the macro. For example, Parsonian theory (Chapter 3) would appear to argue that needs at the personality system level constrain, via the cybernetic hierarchy of control (see Figure 3–7), what can occur at the social system level and, indirectly, at the cultural level. Symbolic interactionist and role theories (especially Chapters 19 and 21) also imply this kind of constraint from the micro level, arguing that needs for the maintenance of self and identity will make social structures unviable to the extent that such structures do not meet these needs.

There are, then, hints at this mutual process of constraint, but none is well developed. Let me try to elaborate this line of argument beyond the current theories by specifying some of the micro constraints on macro, and vice versa.

Turning first to the micro, I would argue that the "need-states" outlined in Chapter 30 represent a fundamental constraint on macrostructures. That is, to the extent that needs for group inclusion, ontological security, self-confirmation, facticity, and symbolic/material gratification are not met by a macrostructure (that is, by a particular pattern of assemblage, differentiation, and integration), the less viable is the macrostructure. Also, the greater the number of individuals whose needs are not met and the greater the number of their needs not satisfied, the less viable is the macrostructure. I am arguing, then, that basic need-states set parameters on macrostructures. They do so through the processes of interacting and microstructuring outlined in Chapter 30; and so, to the degree that macrostructures do not readily permit or facilitate role taking and role making, framing, claim validating, accounting, and ritualizing, such macrostructures are unviable. Further, to the degree that individuals have difficulty in (a) categorizing one another and the situation, (b) regionalizing the props, space, and interpersonal demography, (c) normatizing expectations, (d) developing appropriate rituals, (e) stabilizing exchanges of resources, and (f) routinizing behaviors, macrostructures will be unviable.

Shifting to the macro constraints on the micro, the existence of a particular pattern of assemblage, differentiation, and integration creates (a) a physical environment that imposes itself on spacing, regions, props, and interpersonal demography; (b) a distribution of resources across actors engaged in resource transfers; (c) a set of norms, frames, and interpretive schemes; (d) a system of categories; (e) a pattern of routine; and (f) a set of rituals. Individuals are thus highly constrained by the space, props, resources, norms, routines, rituals, and categories dictated by macrostructures. Furthermore, they seek to discover in each encounter such constraints as these become embodied in traditions and particular groups, networks, and organizations that make up macrostructures. It can be argued, therefore, that people reify macrostructures; they seek to create an external, obdurate reality that not only imposes itself physically but also informs them as to the relevant rituals, categories, props, norms, routines, and resources.

Of course, all this is still rather vague. What is needed, I believe, are more systematic efforts to model just how micro and macro processes set parameters for each other. It would seem possible, for example, to create models and propositions about how the level and scope of deprivation for any need-state (or configuration of need-states) can create problems in sustaining a macro pattern of assemblage, differentiation, and integration. Or, conversely, it would appear possible to develop models and propositions explaining how particular patterns of assemblage, differentiation, and integration constrain microstructural processes.

Sociological theorists have hardly addressed such issues systematically, although one can find many implicit efforts along these lines. One may not want to use my categories of micro and macro processes, but the goal is still the same: how do processes (however conceptualized) at the micro level constrain what is viable at the macro, and vice versa? Vague notions of "cultural capital," "alienation," "anomie," "cybernetic hierarchies of control," "structural embeddedness," "structuration," "interaction ritual chains," "network densities," and the like only hint at these mutually constraining processes. What is needed, I believe, is a more systematic effort to conceptualize what is basic to the micro and macro realms and then to identify the specific ways in which these basic processes impose parameters on each other. Yet, even when this is done, there will still be a gap between the micro and macro realms, but perhaps it will seem less formidable or less important.

CONCLUSION

I have reviewed the various strategies that theorists have employed to close the micro/macro gap—performing middle-range theorizing, constructing conceptual staircases, erecting deductive reductions, searching for formal isomorphisms, positioning intermediate units, and conceptualizing mutual parameters. I have arranged these in order of their importance[18] for mitigating the micro/macro gap. Of course, this ordering is only my point of view, and many would alter or even reverse the order.

As a concluding chapter to a long book, this is far from a "big finish." Indeed, by discussing an unresolved and, in my view, unresolvable issue, I am closing on a problematic note, but in a very real sense this is the status of sociological theory today. There are many different theoretical approaches that I have grouped for convenience under the headings of functional, conflict, exchange, interactionist, and structural theorizing; within and between these general theoretical perspectives are profound disagreements on substantive issues, such as over the primacy of micro or macro and the best way to reconcile the two, over whether or not sociology can be a science, over what kind of science it can be (if it is agreed that sociology can be a science), over the best strategy for developing theory (regardless of sociology's scientific pretensions), and over the relative amounts of critique value-neutrality possible or appropriate. Sociological theory is thus a highly eclectic activity, and this gives it a sense of being problematic when viewed as a whole.

---

[18]I would, however, see the more abstract middle range approach (see footnotes 5, 6, and 7) as more important than the typical middle-range theory.

My biases as to how to make sociological theory less problematic have, I think, been very explicit. Nonetheless, I trust that I have done justice to the theories and to the question of synthesizing them in this last group of chapters. Even if one does not accept my particular biases, I hope that I have communicated *the structure of sociological theory* in the world today.

AUTHOR INDEX

SUBJECT INDEX

◆